THE POPULAR ❧ ❧
NATURAL HISTORY

By The REV. J. G. WOOD, M. A.

Author of "The Illustrated Natural History" and "The Illustrated
Natural History of Man."

WITH 600 ILLUSTRATIONS BY
WOLF, ZWECKER, WEIR, COLEMAN AND OTHERS

A. L. BURT COMPANY ❧ ❧ ❧
❧ ❧ PUBLISHERS, NEW YORK

PREFACE.

IT is now just twenty-five years ago when I was asked to write, for the use of the young, a book on Zoology which should be tolerably comprehensive, intelligible, and free from the conventional errors which had been handed down from one writer to another. Since that time the book has passed through many editions and now takes an entirely new shape, embodying the most recent discoveries in zoology, being much enlarged in size, and illustrated with many additional engravings. If it should be as well received as its predecessors I shall be most satisfied.

J. G. W.

INTRODUCTION.

———◆———

IN order to understand any science rightly, it needs that the student should proceed to its contemplation in an orderly manner, arranging in his mind the various portions of which it is composed, and endeavouring, as far as possible, to follow that classification which best accords with nature. The result of any infringement of this rule is always a confusion of ideas, which is sure to lead to misapprehension. So, in the study of living beings, it is necessary to adhere to some determinate order, or the mind becomes bewildered among the countless myriads of living creatures that fill earth, air, and water.

As a general arranges his army into its greater divisions, and each division into regiments and companies, so does the naturalist separate the host of living beings into greater and smaller groups. The present state of zoological science gives five as the number of divisions of which the animal kingdom is composed. These are called Vertebrates, Molluscs, Articulates, Radiates, and Protozoa. Of each of these divisions a slight description will be given, and each will be considered more at length in its own place.

1st. The VERTEBRATES include Man and all the Mammalia, the Birds, the Reptiles, and the Fish.

The term Vertebrate is applied to them because they are furnished with a succession of bones called "vertebræ," running along the body and forming a support and protection to the nervous cord that connects the body with the brain by means of numerous branches.

2nd. The MOLLUSCA, or soft-bodied animals, include the Cuttle-fish, the Snails, Slugs, Mussels, &c. Some of them possess shells, while others are entirely destitute of such defence. Their nervous system is arranged on a different plan from that of the Vertebrates. They have no definite brain, and no real spinal cord, but their nerves issue from certain masses of nervous substance technically called ganglia.

3rd. The ARTICULATES, or jointed animals, form an enormously large division, comprising the Crustaceans, such as the Crabs and Lobsters, the Insects, Spiders, Worms, and very many creatures so different from each other, that it is scarcely possible to find any common characteristics.

4th. The next division, that of the RADIATED animals, is so named on account of the radiated or star-like form of the body, so well exhibited in the Star-fishes and the Sea-anemones.

5th. The PROTOZOA, or primitive animals, are, as far as we know, devoid of internal organs or external limbs, and in many of them the signs of life are so feeble, that they can scarcely be distinguished from vegetable germs.

B

The Sponges and Infusorial Animalcules are familiar examples of this division.

VERTEBRATES.

The term Vertebrate is derived from the Latin word *vertere*, signifying " to turn " ; and the various bones that are gathered round and defend the spinal cord are named vertebræ, because they are capable of being moved upon each other in order to permit the animal to flex its body.

MAMMALIA.

The vertebrated animals fall naturally into four great classes. These four classes are termed MAMMALS, BIRDS, REPTILES, and FISHES,—their precedence in order being determined by the more or less perfect development of their structure.

QUADRUMANA ;

OR, THE MONKEY TRIBE.

The QUADRUMANOUS, or Four-handed animals, are familiarly known by the titles of Apes, Baboons, and Monkeys.

The APES are at once distinguished from the other Quadrumana by the absence of those cheek-pouches which are so usefully employed as temporary larders by those monkeys which possess them ; by the total want of tails, and of those callosities on the hinder quarters which are so conspicuously characteristic of the baboons.

The first in order, as well as the largest of the Apes, is the enormous ape from Western Africa, the GORILLA. The first modern writer who brought the Gorilla before the notice of the public seems to be Mr. Bowdich, the well-known African traveller ; for it is evidently of the Gorilla that he speaks under the name of Ingheena. The natives of the Gaboon and its vicinity use the name Gina, when mentioning the Gorilla. The many tales, too, that are told of the habits, the gigantic strength, and the general appearance of the Ingheena, are precisely those which are attributed to the Gorilla.

The outline of the Gorilla's face is most brutal in character, and entirely destroys the slight resemblance to the human countenance which the full form exhibits. As in the Chimpansee, an ape which is placed in the same genus with the Gorilla, the colour of the hair is nearly black ; but in some lights, and during the life of the animal, it assumes a lighter tinge of greyish brown, on account of the admixture of variously coloured hairs. On the top of the head and the side of the cheeks it assumes a grizzly hue. The length of the hair is not very great, considering the size of the animal, and is not more than two or three inches in length.

As to the habits of the Gorilla many conflicting tales have been told, and many have been the consequent controversies. In order to settle the disputed questions, Mr. Winwoode Reade undertook a journey to Western Africa, where he remained for a considerable time. After careful investigation, he sums up the history of the animal as follows :—

" The ordinary cry of a Gorilla is of a plaintive character, but in rage it is a sharp, hoarse bark, not unlike the roar of the tiger. Owing to the negro propensity for exaggeration, I at first heard some very remarkable stories about the ferocity of the Gorilla ; but when I questioned the real hunters, I found them, as far as I could judge, like most courageous men, modest, and rather taciturn than garrulous. Their account of the ape's ferocity scarcely bears out those afforded by Drs. Savage and Ford. They deny that the

Gorilla ever attacks man without provocation. 'Leave Njina alone,' they say, 'and Njina leave you alone.' But when the Gorilla, surprised while feeding or asleep, is suddenly brought to bay, he goes round in a kind of half-circle, keeping his eyes fixed on the man, and uttering a complaining, uneasy cry.

GORILLA.—(*Troglodytes Gorilla*).

If the hunter shoots at him, and the gun misses fire, or if the ape is wounded, he will sometimes run away ; sometimes, however, he will charge, with his fierce look, his lowered lip, his hair falling on his brow. He does not, however, appear to be very agile, for the hunters frequently escape from him.

" His charge is made on all-fours : he seizes the offensive object, and, dragging it into his mouth, bites it. The story of his crushing a musket-barrel between his teeth is general, and a French officer told me that a gun was exhibited at the French settlements in the Gaboon, twisted ' comme une papillote.' This, however, is not very wonderful, for the cheap Birmingham guns, with barrels made of ' sham-dam-skelp' iron, which are sold to the natives, might easily be bent and twisted by a strong-jawed animal. I heard a great deal about men being killed by Gorillas, but wherever I went I found that the story retreated to tradition. That a man might be killed by a Gorilla I do not affect to doubt for a moment, but that a man has not been killed by one within the memory of the living, I can most firmly assert.

" I once saw a man who had been wounded by a Gorilla. It was Etia, the Mchaga hunter, who piloted me in the forests of Ngumbi. His left hand was completely crippled, and the marks of teeth were visible on the wrist. I asked him to show me exactly how the Gorilla attacked him. I was to be the hunter, he the Gorilla. I pretended to shoot at him. He rushed towards me on all-fours, and seizing my wrist with one of his hands, dragged it to his mouth, bit it, and then made off. So, he said, the Njina had done to him. It is by these simple tests that one can best arrive at truth among the negroes. That which I can attest from my own personal experience in my unsuccessful attempts to shoot a Gorilla, is as follows :—I have seen the nests of the Gorillas, as I have described them ; I cannot say positively whether they are used as beds, or only as lying-in couches. I have repeatedly seen the tracks of the Gorillas, and could tell by the tracks that the Gorilla goes habitually on all-fours."

CLOSELY connected with the preceding animal is the large black ape which is now well known by the name of CHIMPANSEE.

This creature is found in the same parts of Western Africa as the Gorilla, being very common near the Gaboon. It ranges over a considerable space of country, inhabiting a belt of land some ten or more degrees north and south of the torrid zone.

The title *niger*, or black, sufficiently indicates the colour of the hair which envelops the body and limbs of the Chimpansee. The tint of the hair is almost precisely the same as that of the gorilla, being nearly entirely black ; the exception being a few whiter hairs scattered thinly over the muzzle.

It is a remarkable fact that the Chimpansees are partly groundlings, and are not accustomed to habitual residence among the branches of trees. Although these apes do not avail themselves of the protection which would be afforded by a loftier habitation, yet they are individually so strong, and collectively so formidable, that they dwell in security, unharmed even by the lion, leopard, or other members of the cat tribes, which are so dreaded by the monkey tribes generally.

The food of these creatures appears to be almost entirely of a vegetable nature, and they are very unprofitable neighbours to any one who has the mis-fortune to raise crops of rice, or to plant bananas, plantains, or papaus, within an easy journey of a Chimpansee settlement. As is the case with many of the monkey tribes, the animal will eat food of a mixed character when it is living in a domesticated state.

Many specimens have been brought to Europe, and some to England ; but this insular climate seems to have a more deleterious effect on the constitu-tion of this ape than even on that of the other Quadrumana.

As long as they resist the untoward influence of our climate, the specimens which we have known have always been extremely gentle and docile. Taught by the instinctive dread of cold, they soon appreciate the value of clothing, and learn to wrap themselves up in mats, rugs, or blankets, with perfect gravity and decorum. Dress exercises its fascinations even over the ape, for one of these animals has been known to take such delight in a new and handsome costume, that he repudiated the previous dress, and, in order to

THE CHIMPANSEE.—(*Troglodytes niger.*)

guard against the possibility of reverting to the cast-off garment, tore it to shreds.

The head of the Chimpansee is remarkable for the large development of the ears, which stand prominently from the sides of the head, and gives a curiously peculiar expression to the contour of the head and face.

THE ORANG-OUTAN.

The ORANG-OUTAN is a native of Asia, and only to be found upon a small portion of that part of the globe. Borneo and Sumatra are the lands most favoured by the Orang-outan, which inhabits the woody districts of those islands, and there rules supreme, unless attacked by man.

THE ORANG-OUTAN. —(*Simia Satyrus.*)

There seems to be at least two species of this animal that are found in Borneo, and some zoologists consider the Sumatran ape to be a third species.

The natives distinguish the two Bornean species by the names of Mias-kassar and Mias-pappan, the latter of which animals is the *Simia satyrus*, so well represented in the engraving.

The walk of the Orang-outan is little better than an awkward hobble, and the creature shuffles along uneasily by help of its arms. The hands are placed on the ground, and are used as crutches in aid of the feet, which are often raised entirely from the ground, and the body swung through the arms.

Sometimes it bends considerably backwards, and throwing its long arms over its head, preserves its equilibrium by their means.

Among the trees the Orang-outan is in its element, and traverses the boughs with an ease and freedom that contrasts strongly with its awkward movements when on the ground. It has a curious habit of making for itself a temporary resting-place, by weaving together the branches so as to make a rude platform or scaffold on which it reposes. The powerful limbs of the animal enable it to execute this task in a very short time.

The adult male animal is singularly hideous in aspect, owing much of its repulsiveness to the great projection of the jaws and the callosities that appear on the cheeks. As is the case with all the larger apes, it becomes sullen and ferocious as it approaches its adult state, although in the earlier years of its life it is docile, quiet, and even affectionate. Several young specimens have been brought to Europe, and were quite interesting animals, having many curious tricks, and exhibiting marks of strong affection to any one who treated them kindly. One of these animals learned to take its meals in a civilized manner, using a spoon, or a cup and saucer, with perfect propriety.

When brought to colder climates than that of its native land, the animal covets warmth, and is fond of wrapping itself in any woollen clothes or blankets that it can obtain. On board ship it has been known to rob the sailors or passengers of their bedding, and to resist with much energy any attempt to recover the stolen property.

In its native woods the Orang-outan seems to be an unsocial animal, delighting not in those noisy *conversazione* which rejoice the hearts of the gregarious monkeys and deafen the ears of their neighbours. It does not even unite in little bands of eight or ten as do many species, but leads a comparatively eremitical existence among the trees, sitting in dreamy indolence on the platform which it weaves, and averse to moving unless impelled by hunger, anger, or some motives equally powerful. When it does move, it passes with much rapidity from tree to tree, or from one branch to another, by means of its long limbs, and launches itself through a considerable distance, if the space between the branches be too great for its reach of arm.

The hair of the Orang-outan is of a reddish chestnut hue, deepening here and there into brown. The texture of the hair is coarse, and its length varies according to the part of the body on which it is placed. Over the face, back, breast, shoulders, and arms, it falls in thick profusion, becoming especially long at the elbow-joint, where the hairs of the upper and fore-arm meet. The face is partly covered with a beard, which seems to increase in size as the animal grows older. The hair of the face takes a lighter tinge of red than that of the body, and merges the red or auburn tint in the brown, on the inside of the limbs.

At a little distance the face appears to be black ; but if examined closely, is found to present a bluish tint.

THE GIBBONS.

The GIBBONS possess, although in a small degree, those singular callosities on the hinder quarters which are so conspicuous in the baboon family, and assume such strange tints. The Gorilla, Chimpansee, and the Orangs are entirely destitute of these peculiarities, but the Gibbons are found to possess them, although the callosities are very small, and hidden by the fur from a casual view.

As in the great apes, the arms of the Gibbons are of enormous length, and endowed with exceeding power of muscle, though the strength which resides in these largely-developed limbs is of a different character.

All the Gibbons are gifted with voices as powerful as their limbs, and the creatures seem to lose few opportunities of exercising lungs or limbs. The cry which these animals utter is a singular one, loud and piercing, and has been represented by the syllables " wou-wou," which duplex combination of intonations is often used as a general name common to the whole family. Some writers express the sound by the words " oa-oa," and others as " woo-woo," among which the reader is left to choose.

Of the habits of the Gibbons in a wild state very little is known, as they are shy in their nature, and by means of their wonderful agility escape among the trees in a manner that baffles pursuit or observation. As to the species which is represented in the accompanying engraving, it seems to be the most active of this agile family, and well deserves the name that has been given to it. Rather more has been noticed of this wonderful creature, and a further insight into its habits has been gained, by means of a female specimen, which was captured, and brought safely to London, where it lived for some time.

In their native woods, these animals are most interesting to the observer, if he is only fortunate enough to get near them without being seen by the vigilant creatures. A good telescope affords an excellent mode of watching the customs of animals that are too timid to permit a human being to come near their haunts.

When startled, the Agile Gibbon flits at once to the top of the tree, and then, seizing the branch that seems best adapted to its purpose, it swings itself once or twice to gain an impetus, and launches itself through the air like a stone from a sling, gaining its force very much on the same principle. Seizing another branch, towards which it had aimed itself, and which it reaches with unerring certainty, the creature repeats the process, and flings itself with ease through distances of thirty or forty feet, flying along as if by magic. Those who have seen it urging its flight over the trees, have compared its actions and appearance to those of a bird. Indeed, these creatures seem to pass a life that is more aerial than that of many birds, putting out of question the heavy earth-walking birds which have not the power of raising themselves from the ground even if they had the will.

The colour of this species is extremely variable, and as may be seen by reference to the figure the offspring is not necessarily of the same colour as the parent. This difference of tint is not solely caused by age, for it frequently happens that a cream-coloured mother has a dark infant, and *vice versâ.* Of the specimens in the British Museum, hardly any two are alike in the tint of their soft woolly fur. Some are nearly black, some are brown, and some are of a light cream colour.

A VERY different group of animals now comes before us, separated even by the outer form from the apes,

AGILE GIBBON.—(*Hylobates agilis.*)

The chief distinction which strikes the eye is the presence of a tail, which is of some length, and in several species, among which we may mention the SIMPAI itself, is extremely long and slender in proportion to the body. The arms of these animals are not of that inordinate length which is seen in the limbs of the apes, but are delicate and well proportioned. The hinder paws, or hands, are extremely slender, their thumbs being short, and, as will be seen by reference to the engraving, are twice the length of the fore-paws.

Some of these monkeys are furnished with small cheek-pouches, while others appear to be destitute of these natural pockets. The callosities of the hinder quarters are well shown.

In this group of the Quadrumana, the characteristics of the apes disappear, and the animals betray more clearly their quadrupedal nature. Very seldom do they assume the erect attitude, preferring to run on all fours like a dog, that being their legitimate mode of progression. Even when they do stand on their hind feet, the long tail at once deprives them of that grotesque semblance of the human form, which is so painfully exhibited in the tail-less

THE SIMPAI.—(*Presbytes melalophos.*)

apes. Besides these external distinctions, there are many remarkable peculiarities in the anatomy of the internal organs, which also serve to settle the position of the animal in the order of nature. Among these internal organs, the stomach displays the most remarkable construction, being very large, and divided into compartments that bear some resemblance to those in the stomach of ruminating animals.

These monkeys are distributed through several parts of the world, the Simpai making its residence in Sumatra.

This is a beautiful little animal, and is pleasing both for elegance of shape and the contrasting tints with which its fur is decorated. The prevailing colour of the body is a light-chestnut, with a perceptible golden tinge showing itself when the light falls obliquely on the fur. The inside of the limbs and the abdomen are not so bright as the rest of the body, but take a most sober tint of grey. At the top of the head the hair is straight, and is set on nearly perpendicularly, so as to form a narrow crest. The colour of the crest, together with that of a narrow band running over the eyes and temples, is black. From this conspicuous peculiarity, the Simpai is also called the Black-crested Monkey.

The name Presbytes signifies an old man, and is given to these monkeys on account of the wizened, old-fashioned aspect of their countenances. The term "melalophos" is literally "black-crested," and therefore a very appropriate name for this species.

The length of this animal, measured from the nose to the root of the tail, is about twenty inches, and that of the tail itself is not very far from three feet. Its fur is very soft and glossy.

A well-known example of this group of monkeys is the HOONUMAN or ENTELLUS. This is a consider-

ENTELLUS.—(*Presbytes Entellus.*)

ably larger animal than the Simpai, as the adult Hoonuman measures three or four feet from the nose to the root of the tail, and the tail itself rather exceeds the body in length. The colour of this monkey when young, is a greyish brown, excepting a dark brown line along the back and over the loins. As the animal increases in years, the fur darkens in colour, chiefly by means of black hairs that are inserted at intervals. The face, hands, and feet are black.

It is a native of India, and, fortunately for itself, the mythological religion is so closely connected with it that it lives in perfect security. Monkeys are never short-sighted in spying out an advantage, and the Entellus monkeys are no exception to the rule. Feeling themselves masters of the situation, and knowing full well that they will not be punished for any delinquency, they take up their position in a village with as much complacency as if they had built it themselves. They parade the streets, they mix on equal terms with the inhabitants, they clamber over the houses, they frequent the shops, especially those of the pastrycooks and fruit-sellers, keeping their proprietors constantly on the watch.

The PROBOSCIS MONKEY, or KAHAU, as it is sometimes called, on account

KAHAU.—(*Presbytes carvatus.*)

of its cry bearing some resemblance to that word, is an inhabitant of Borneo, and probably of several neighbouring countries. It is, as may be seen by the engraving, an animal of very unattractive features, principally on account of its enormously lengthened nose. . This feature does not present itself in perfection until the Kahau has reached its maturity.

In size, the Kahau is about equal to the Hoonuman, and seems to be an active animal, leaping from branch to branch, through distances of fifteen feet or more.

For the preternatural ugliness of the countenance the Kahau is partially compensated by the beautiful colouring of its fur, which is thick, but not woolly, nor very long. The principal colour in the body is a bright chestnut red : the sides of the face, part of the shoulders, and under part of the body being of a golden yellow. A rich brown tint is spread over the head and between the shoulders ; the arms and legs taking a whiter tinge than the shoulders.

URSINE COLOBUS.—(*Colobus ursinus*).

THE COLOBUS.

THE scientific name which is given to this genus of monkeys explains—as is the proper office of names—one of the leading peculiarities of the animals. The title " Colobus " is a Greek word, signifying " stunted," or " maimed," and is given to these animals because the thumbs of the two fore-limbs give but little external indication of their presence, so that the hand consists merely of four fingers. They are exclusively African animals. They are rather handsome creatures, and their hair is sufficiently long and silky to be valuable as a fur.

The Ursine, or Bearlike Colobus, is so named because the general colour of its long black fur, and the form of the monkey itself, with the exception of the tail, has something of the bearish aspect. The cheeks and chin of this animal are covered with white hair ; there is a white patch on the hind legs ; and, with the exception of a few inches at its root, which retain the black hue of the body, the tail is of a beautiful white, terminated with a . g and full white tuft.

The little animal, the White-Nose Monkey of Western Africa, is a curious little creature, with an air of quaint conceit, for which it is indebted to the fringe of white hairs that surround; its face, and the conspicuous white spot on the nose, which has earned for it the title of White-Nose. As is so often the case in these animals, the under side of the body and inside of the limbs is of a much lighter tint than the upper portions. This distinction is peculiarly well marked in the long tail, which is nearly black above, and beneath takes a greyish hue.

It is a very graceful little creature, playful, but petulant and coquettish, disliking to be touched, but fond of notice and nuts, and often balanced in curious perplexity between its coy shyness and the charms of an offered dainty. When in perfect health, it is seldom still, but flits with light grace from one spot to another, performing the most difficult muscular efforts with exquisite ease, and profoundly sensible of the admiration which its pretty antics never fail to excite in the spectators.

WHITE-NOSE MONKEY. — (*Cercopithecus Petaurista.*)

It is by no means a large animal, its head and body only measuring fifteen or sixteen inches, the tail being little short of two feet in length.

We now arrive at a group of small monkeys with exceedingly long names. The term "Cercopithecus" is composed from two Greek words, signifying "tailed ape."

It is worth notice that the word "monkey" is derived from the name of one of this group, the Mona. The diminutive of Mona is Monikin, the transition from which word to our "monkey" is sufficiently evident.

The GRIVET, or TOTA, as it is called by some writers, is of a sombre green colour; the green being produced by alternate rings of black and yellow on each hair. The limbs and tail are of a greyer tint than the rest of the body, the yellow portion of the hair being changed to a dull white. The inside of the limbs and the abdomen are slightly tinged with white. In the male animal the canine teeth are rather protuberant, showing themselves beyond the lips. The naked skin of the face, ears, and palms, is black, dashed with that deep violet hue that is found in so many of the monkeys. At each side of the head the white hairs stand out boldly, whisker fashion, and give a very lively character to the head. It is an African animal, and common in Abyssinia.

The left hand figure of the group on the next page represents the GREEN MONKEY, sometimes called the Callithrix, or Beautiful-haired Monkey, on account of the exquisitely delicate marking of each separate hair. The inside of the limbs is nearly white, as is the under surface of the body, and the outer side of the limbs takes a greyish tinge. The hairy fringe that grows over the side of the face is of a delicate golden yellow.

This monkey is a native of Senegal and the neighbouring parts, and is frequently brought to this country.

The VERVET is the last of the figures. This is rather a variable animal in point of colour, some specimens being decidedly pale, while others assume a blackish hue. In general, the colour of the animal is as follows. The

GRIVET.—(*Cercopithecus Engythithia.*)

prevailing tint of the fur is much the same as that of the Grivet, to which animal the Vervet bears a strong resemblance. The head, the throat, and

GREEN MONKEY. VERVET.
(*Cercopithecus sabæus.*) (*Cercopithecus pygerythrus.*)

breast are of a light dun, the paws being very dark. In the male Vervet the canines are rather long, and show their points beyond the lips.

MACAQUES.

The various species of monkeys which are ranged under the common title of MACAQUES are mostly well-known animals ; being plentiful in their native lands, and frequently domesticated both in their own and in foreign countries.

The animal which is shown in the opposite engraving is one of the best known of the monkey tribe ; as it is tolerably hardy, it endures the changeable and chilly European climates better than most of its race.

As its name implies, it is a native of Barbary, where it is found in great numbers, but has also been naturalized upon the Rock of Gibraltar. The

MAGOT, OR BARBARY APE.—(*Macacus Innuus*).

Gibraltar MAGOTS are frequently mentioned in books of travel, and display great ingenuity in avoiding pursuit and discovering food. They keep to the most inaccessible portions of the rock, and scamper away hurriedly on the slightest alarm. But with the aid of a moderately good telescope, their movements may be watched, and are very amusing.

This monkey is not very widely spread, for, with the exception of the Rock of Gibraltar, it seems to be confined to Northern Africa.

It is not a very large animal, as the full-grown males only measure about a yard in length, and the females are rather smaller. The general size of the Magot is about that of an ordinary bull-terrier dog.

The colour of the fur is tolerably uniform, differing chiefly in depth of shade, and is of a clear greyish tint.

Its walk on level ground is rather awkward, the animal making use of feet and hands for that purpose ; but it climbs with ease and agility up trees or rocks, and in a domesticated state is fond of running up and down ropes, and swinging itself about in its cage.

One of the last of the Macaques which we shall notice in this work is the monkey which is well known under the name of WANDEROO, or OUANDEROO, as it is sometimes written.

This very singular animal is a native of the East Indies, and is found commonly enough in Ceylon. The heavy mass of hair that surmounts the head and envelops the entire face gives it a rather dignified aspect, reminding the observer of the huge peruke under whose learned shade the great legal chiefs consider judgment. The hair on the top of the head is black, but

WANDEROO.—(*Silenus veter*).

the great beard that rolls down the face and beneath the chin is of a grey tint, as if blanched by the burden of many years. In some instances this beard is almost entirely white, and then the Wanderoo looks very venerable indeed.

From the form of the tail, which is of a moderate length, and decorated with a hairy tuft at its extremity, the Wanderoo is also known by the name of the Lion-tailed Baboon.

The greater part of the fur of this animal is of a fine black, but the colour assumes a lighter hue on the breast and abdomen. The callosities on the hinder quarters are of a light pink.

It is not a very large animal, being rather less than three feet from the nose to the tip of the tail.

IN the absence of a tail, and in general form, the BLACK MACAQUE bears some resemblance to the Magot, but in colour and arrangement of hair, it is entirely distinct from that animal.

The tint of the fur is as deep a black as that of the Budeng, or Black Colobus. Both these monkeys are possessed of crests which give a peculiar character to the whole aspect. That of the Black Colobus, however, is reverted forward, and curves to a point over the forehead, while that of the animal before us rises from the head and bends backward over the neck in a manner not unlike that of the cockatoo.

BLACK MACAQUE.—(*Macacus niger*).

Like the Magot, the Black Macaque has been called an ape by some writers, and a baboon by others, on account of the apology for a tail with which its hinder quarters are terminated, but not decorated. It is an inhabitant of the Philippines and the neighbouring countries.

BABOONS, OR DOG-HEADED MONKEYS.

A well-marked group of animals now come before us, popularly known by the name of BABOONS.

One distinguishing characteristic of these creatures is that the nostrils are situated at the extremity of the muzzle, instead of lying nearly flat upon its base, and just under the eyes, as in the apes and other quadrumanous animals. The muzzle, too, is peculiar in its form, being, as it were, cut off abruptly, leaving a round and flattened extremity.

Of the Dog-headed Baboons, the species which is most celebrated for its feats of prowess is the well-known animal called the CHACMA, or URSINE BABOON.

This animal, when it has attained its full age, equals in size a large mastiff or an ordinary sized wolf; while, in bodily strength and prowess, it is a match for any two dogs that can be brought to attack it.

The Chacma is a most accomplished robber, executing his burglaries openly whenever he knows that he will meet with no formidable opposition, and having recourse to silent craft when there are dogs to watch for trespassers, and men with guns to shoot them.

With such consummate art do these animals plan, and with such admirable skill do they carry out their raids, that even the watchful band of dogs is comparatively useless; and the cunning robbers actually slip past the vigilant sentries without the stirring of a grass blade or the rustling of a dried twig, to give notice to the open ears of the wakeful but beguiled sentries.

FEW animals present a more grotesque mixture of fantastic embellishments and repulsive ferocity than the baboon which is known under the name of MANDRILL.

The colours of the rainbow are emblazoned on the creature's form, but always in the very spots where one would least expect to see them. A bright azure glows, not in its " eyes of heavenly blue," but on each side of its

THE CHACMA.—(*Cynocephalus porcarius.*)

nose, where the snout is widely expanded, and swollen into two enormous masses. The surfaces of these curious and very unprepossessing projections are deeply grooved, and the ridges are bedizened with the cerulean tint above mentioned. Lines of brilliant scarlet and deep purple alternate with the blue, and the extremity of the muzzle blazes with a fiery red like Bardolph's nose.

That all things should be equally balanced, the opposite end of the body is also radiant with chromatic effect, being plenteously charged with a ruddy violet, that is permitted to give its full effect, by the pert, upright carriage of the tail,

The general colour of the fur is of an olive brown tint, fading into grey on the under side of the limbs, and the chin is decorated with a small yellow pointed beard. The muzzle is remarkable for a kind of rim or border, which is not unlike the corresponding part in a hog, and is well shown in the engraving. The ears are small, devoid of fur, and of a black colour with a tinge of blue.

Only the male Mandrill possesses these strange adornments in their full beauty of size and colour, the females being only gifted with the blue tint upon the muzzle, and even that is of a much less brilliant hue than in the male.

THE MANDRILL.—(*Cynocephalus Mormon.*)

In this country the Mandrill is seldom seen to equal a tolerably large terrier in size, but in its native land a full-grown male measures more than five feet when standing upright, a stature which equals, if not exceeds, that of the Chacma.

In its native land the usual food of the Mandrill is of a vegetable nature, although, in common with the rest of the Baboons, it displays a great liking for ants, centipedes, and similar creatures.

The tail of this animal is a remarkable feature, if it may be so termed, in the general aspect of the baboon. It is short, set high on the back, and curved upwards in a manner that is most singular, not to say ludicrous, in the living animals, and conspicuously noticeable in the skeleton.

It is a very common animal in its own country, but on account of its great strength, cunning, and ferocity, is not so often captured as might be expected. Even when a specimen is made prisoner, it is generally a very young one, which soon loses in captivity the individuality of its being, and learns to accommodate itself to the altered circumstances among which it is placed.

ANOTHER well-known species of the Dog-headed Baboons is the PAPION, an animal of rather a more refined aspect than the Chacma, or, more properly speaking, not quite so brutal.

The face, although unattractive enough, is yet not so repulsive as that of the Chacma, and the colours are rather more bright than those of that animal.

Great reverence was paid to these creatures, and specially to certain selected individuals which were furnished with a safe home in or near their temples, liberally fed while living, and honourably embalmed when dead. Many mummied forms of these baboons have been found in the temple caves of Egypt, swathed, and spiced, and adorned, just as if they had been human beings.

Some authors say that the Thoth Baboon was an

THE PAPION.—(*Cynocephalus Sphinx.*)

object of worship among the Egyptians, but hardly with sufficient reason. Various animal forms were used as visible living emblems of the attributes of deity and the qualities of the human intellect, but were no more objects of idolatrous worship than the lion of England or the eagle of America.

The fur of the Papion is of a chestnut colour ; in some parts fading into a sober fawn, and in others warmed with a wash of ruddy bay. The paws are darker than the rest of the body. When young, it is of a lighter hue, and deepens in colour until it reaches its full age. In the prime of existence its colours are the lightest, but as years begin to lay their burden on the animal, the hairs begin to be flecked with a slight grizzle, and, in process of time, the snows of age descend liberally, and whiten the whole fur with hoary hairs.

AMERICAN MONKEYS.

WE have now taken a rapid survey of the varied forms which the Quadrumana of the Old World assume ; forms so diversified that there hardly seems to be scope for further modifications. Yet the prolific power of nature is so inexhaustible, that the depth of our researches only brings to view objects of such infinite variety of shape that the mind is lost in wonder and admiration. We will now take some of the Quadrumana of the New World.

The COAITA, or QUATA, as the word is frequently written, is one of the best known of this group of animals, which are called by the name of SPIDER MONKEYS on account of their long sprawling limbs, and their peculiar action while walking.

The name "Ateles," which is given to the entire genus to which this animal belongs, signifies "imperfect," and has been applied to the creatures because the fore-paws are devoid of useful thumbs. Sometimes that member is almost entirely absent, and in other instances it only just shows itself.

The Spider Monkeys are also remarkable for the long and prehensile tail. With such singularly delicate sense of touch is it furnished, that it almost seems to be possessed of the power of sight, and moves about among the branches with as much decision as if there were an eye in its tip. Should the monkey discover some prize, such as a nest of eggs, or any little dainty, which lies in a crevice too small for the hand to enter, it is in nowise discon-certed, but inserts the end of its tail into the cranny, and hooks out the desired object.

There is a beautiful formation of the tail of this creature, by means of which the grasp of that member retains its hold even after the death of the owner. If a Spider Monkey be mor-tally wounded and not killed outright, it curls its tail round a branch, and thus suspended yields up its life. The tail does not lose its grasp when the life has departed; and the dead monkey hangs with its head down-wards until decomposition sets in and the rigid muscles are relaxed.

The Coaita is by no means a large animal, measuring very little more than a foot from the nose to the root of the tail, while the tail itself is two feet in length. Its colour is very dark and glossy; so dark, indeed, as to be almost black. The hair varies much in length and density. On the back and the outside of the limbs it hangs in long drooping locks, forming a thick covering through which the skin cannot be seen. But on the abdomen the hair is quite scanty, and is so thinly scattered that the skin is plainly visible. The skin of the face is of a dark copper colour.

COAITA.—(*Ateles Paniscus.*)

ANOTHER example of this wonderful group of monkeys is found in the MARIMONDA; an inhabitant, like the last-named animal, of Central Ame-rica, and found in greatest numbers in Spanish Guiana, where, according to Humboldt, it fills the place of the Coaita.

The general shape, the formation of its limbs, and the long prehensile tail, point it out at once as another of the Spider Monkeys. It is certainly a very appropriate name for these animals. Their heads are so small, their bodies so short, their limbs so slender, and their tail so limb-like, that the mind unconsciously draws a parallel between these monkeys and the long-legged spiders that scuttle so awkwardly over the ground, and are so indifferent respecting their complement of legs.

The resemblance holds good even when the monkey is at rest, or when it only appears before the eye in an illustration. But when the creature begins to walk on level ground, and especially if it be hurried, its clumsy movements are so very spider-like, that the similitude is ten times more striking. Be it remarked, that both creatures are supposed to be placed in uncongenial circumstances. The spider is deft and active enough among the many threads of its air-suspended nets, as is the monkey among the slight twigs of the air-bathed branches. But when both animals are subjected to circumstances which are directly opposed to their natural mode of existence, they become alike awkward, and alike afford subjects of mirth.

MARIMONDA.—(*Ateles Belzebuth.*)

The mode by which a Spider Monkey walks on level ground is rather singular, and difficult to describe, being different from that which is employed by the large apes. They do not set the sole of either paw, or hand, flat upon the ground, but, turning the hinder feet inwards, they walk upon their outer sides. The reverse process takes place with the fore-paws, which are twisted outwards, so that the weight of the animal is thrown upon their inner edges.

It will easily be seen how very awkward an animal must be which is forced to employ so complicated a means for the purpose of locomotion. Although the Spider Monkey has been known to walk in a manner much more steady than that of any other monkey, yet this bipedal progression was only employed for a few paces, and with a haven of rest in view in the shape of a window-sill, on which the creature could rest its hands. The tail is also curled over the head, like the letter S, by way of a balance.

In captivity, the Marimonda is a gentle and affectionate animal, attaching itself strongly to those persons to whom it takes a fancy, and playing many fantastic gambols to attract their attention.   Its angry feelings, although perhaps easily roused, do not partake of the petulant malignity which so often characterises the monkey race, and are quite free from the rancorous vengeance which is found in the baboons.   Very seldom does it attempt to bite, and even when such an event does take place, it is rather the effect of sudden terror than of deliberate malice.

On account of its amiable nature it is often brought into a domesticated state, and, if we may give credence to many a traveller, is trained to become not only an amusing companion, but a useful servant.

The colour of this animal varies much, according to the age of the individual.

When adult, the leading colour is of a uniform dull black, devoid of the glossy lustre which throws back the sunbeams from the coaita's furry mantle. On the back, the top of the head, and along the spine, the hair is of a dense, dead black, which seems to have earned for the animal the very inapposite name with which its nomenclators have thought fit to dedecorate the mild and amiable Marimonda.

The throat, breast, inside of the limbs, and the under side of the tail are much lighter in tint, while in some individuals a large, bright chestnut patch covers the latter half of the sides.

It seems to be of rather a listless character, delighting to bask in the sun's rays, and lying in the strangest attitudes for hours without moving.   One of the postures which is most in vogue is achieved by throwing the head back with the eyes turned up, and then flinging the arms over the head.   The position in which this animal is depicted in the illustration is a very favourite one with most of the Spider Monkeys.

THE animal which is here engraved is an example of the celebrated group of HOWLING MON-KEYS, or ALOUATTES, as they are termed by some naturalists, whose strange customs have been so often noticed by travellers, and whose reverberating cries rend their ears.   Little chance is there that the Howling Monkeys should ever fade from the memory of anyone who has once suffered an unwilling martyrdom from their mournful yells.

URSINE HOWLER.—(*Mycetes ursinus.*)

Several species of Howling Monkeys are known to science, of which the ARAGUATO, as it is called in its own land, or the URSINE HOWLER, as it is popularly named in this country, is, perhaps, the commonest and most conspicuous.   It is larger than any of the New World monkeys which have hitherto been noticed ; its length being very nearly three feet when it is fully grown, and the tail reaching to even a greater length.

The colour of the fur is a rich reddish brown, or rather bay, enlivened by a golden lustre when a brighter ray of light than usual plays over its surface. The beard which so thickly decorates the chin, throat, and neck, is of a deeper colour than that of the body.

Few animals deserve the name which they bear so well as the Howling Monkeys. Their horrid yells are so loud, that they can be heard plainly although the animals which produce them are more than a mile distant ; and the sounds that issue from their curiously-formed throats are strangely simulative of the most discordant outcries of various other animals—the jaguar being one of the most favourite subjects for imitation. Throughout the entire night their dismal ululations resound, persecuting the ears of the involuntarily wakeful traveller with their oppressive pertinacity, and driving far from his wearied senses the slumber which he courts, but courts in vain.

In order that an animal of so limited a size should be enabled to produce sounds of such intensity and volume, a peculiar structure of the vocal organs is necessary.

The instrument by means of which the Howlers make night dismal with their funestral wailings, is found to be the "hyoid bone," a portion of the

THE CAPUCIN.—(*Cebus Apella*)

form which is very slightly developed in man, but very largely in these monkeys. In man, the bone in question gives support to the tongue and is attached to numerous muscles of the neck. In the Howling Monkeys it takes a wider range of duty, and, by a curious modification of structure, forms a bony drum which communicates with the windpipe and gives to the voice that powerful resonance which has made the Alouattes famous.

The CAPUCIN MONKEYS, an example of which is here given, are active little animals, lively and playful. In habits, all the species seem to be very similar, so that the description of one will serve equally for any other. In consequence of their youth and sportive manners they are frequently kept in a domesticated state, both by the native Indians and by European settlers. Like several other small monkeys, the Capucin often strikes up a friendship for other animals that may happen to live in or near its home, the cat being one of the most favoured of their allies. Sometimes it carries its familiarity so far as to turn the cat into a steed for the nonce, and, seated upon her back, to perambulate the premises. More unpromising subjects for equestrian exercise have been pressed into the service by the Capucin. Humboldt mentions one of these creatures which was accustomed to catch a pig every morning, and, mounting upon its back, to retain its seat during the day. Even while the pig was feeding in the savannas its rider remained firm, and bestrode its victim with as much pertinacity as Sinbad's old man of the sea.

There is some difficulty in settling the species of the Capucins, for their fur is rather variable in tint, in some cases differing so greatly as to look like another species. The general tint of the CAPUCIN is a golden olive, a whiter fur bordering the face in some individuals, though not in all.

There are several monkeys known by the name of Sakis, among which are reckoned the Cuxio, a rather odd little animal, and two other species, which are easily distinguished from each other by the colour of their heads,

BLACK YARKE.—(*Pithecia leucocephala.*)

The first of these animals is the BLACK YARKE, or WHITE-HEADED SAKI, and the other the CACAJAO, or BLACK-HEADED SAKI.

The former of these Sakis is a rather elegant creature in form, and of colour more varied than those of the Cuxio. As will be seen from the accompanying engraving, the head is surrounded with a thick and closely-set fringe of white hair, which is rather short in the male, but long and drooping in the female. The top of the head is of a deep black, and the remainder of the body and tail is covered with very long and rather coarse hair of a blackish-brown. Under the chin and throat the hairs are almost entirely absent, and the skin is of an orange hue.

Beside the difference of length in the facial hairs of the female Yarke, there are several distinctions between the sexes, which are so decided as to have caused many naturalists to consider the male and female to belong to different species. The hair of the female Yarke is decorated near the tip with several rings of a rusty brown colour, while the hair of the male is entirely devoid of these marks.

The natural food of these animals is said to consist chiefly of wild bees and their honeycombs. Perhaps the long furry hair with which the Sakis are covered may be useful for the purpose of defending them from the stings of the angry insects. On account of the full and bushy tail with which the members of this group are furnished, they are popularly classed together under the title of Fox-tailed Monkeys.

THE term " Nyctipithecus," or Night-monkey, which is used as the generic title of the DOUROUCOULI, refers to its habits, which are more strictly nocturnal than those of the animals heretofore mentioned. The eyes of this little creature are so sensitive to light, that it cannot endure the glare of day, and only awakes to activity and energy when the shades of night throw their welcome veil over the face of nature.

DOUROUCOULI.—(*Nyctipithecus trivergatus.*)

In its wild state, it seeks the shelter of some hollow tree or other darkened place of refuge, and there abides during the hours of daylight, buried in a slumber so deep that it can with difficulty be aroused, even though the rough hand of its captor drag it from its concealment. During sleep, it gathers all its four feet closely together, and drops its head between its fore-paws. It seems to be one of the owls of the monkey race.

The food of this Douroucouli is mostly of an animal nature, and consists chiefly of insects and small birds, which it hunts and captures in the night season. After dark, the Douroucouli awakes from the torpid lethargy in which it has spent the day, and, shaking off its drowsiness, becomes filled with life and spirit. The large dull eyes, that shrank from the dazzling rays of the sun, light up with eager animation at eventide; the listless limbs are instinct with fiery activity, every sense is aroused to keen perception, and the creature sets off on its nightly quest. Such is then its agile address, that it can capture even the quick-sighted and ready-winged flies as they flit by, striking rapid blows at them with its little paws.

The general colour of the Douroucouli is a greyish-white, over which a silvery lustre plays in certain lights. The spine is marked with a brown line, and the breast, abdomen, and inside of the limbs are marked with a very light chestnut, almost amounting to orange. The face is remarkable for three very distinct black lines, which radiate from each other, and which have earned for the animal the title of "Trivergatus," or "Three-striped." There are but very slight external indications of ears, and in order to expose the organs of hearing, it is necessary to draw aside the fur of the head. On account of this peculiarity, Humboldt separated the Douroucouli from its neighbours, and formed it into a distinct family, which he named "Aötes," or "Earless."

MARIKINA.—(*Jacchus Rosalia.*)

Guiana and Brazil are the countries where this curious little animal is found. Although by no means an uncommon species, it is not taken very plentifully, on account of its monogamous habits. The male and its mate may often be discovered sleeping snugly together in one bed, but never in greater numbers, unless there may be a little family at the time. Its cry is singularly loud, considering the small size of the animal which utters it, and bears some resemblance to the roar of the jaguar. Besides this deep-toned voice, it can hiss or spit like an angry cat. mew with something of a cat-like intonation, and utter a guttural, short, and rapidly-repeated bark. The fur is used for the purpose of covering pouches and similar articles.

AMONG the various members of the monkey tribe there is hardly any species that can compare with the exquisite little MARIKINA, either for grace of form, or soft beauty of colour.

The hair with which this creature is covered is of a bright and lustrous chestnut, with a golden sheen playing over its long glossy locks. To the touch, the fur of the Marikina is peculiarly smooth and silken ; and from this circumstance it is sometimes called the Silky Monkey.

Both for the texture and colour of the hair, the name is happily chosen, for the tint of the Marikina's fur is just that of the orange-coloured silk as it is wound from the cocoon, while in texture it almost vies with the fine fibres of the unwoven silk itself.

Another name for the same animal is the Lion Monkey, because its little face looks out of the mass of hair like a lion from out of his mane.

The colour of the hair is nearly uniform, but not quite so. On the paws it darkens considerably, and it is of a deeper tint on the forehead and the upper surface of the limbs than on the remainder of the body. Some specimens are wholly of a darker hue. In no place is the fur very short ; but on the head, and about the shoulders, it is of very great length in proportion to the size of the animal.

The Marikina is rightly careful of its beautiful clothing, and is fastidious to a degree about preserving its glossy brightness free from stain. Whether when wild, it keeps its own house clean, or whether it has no house at all, is not as yet accurately ascertained ; but in captivity, it requires that all cleansing shall be performed by other hands. This slothfulness is the more peculiar, because the creature is so sensitive on the subject, that if it be in the least neglected, it loses its pretty gaiety, pines away, and dies.

It is fond of company, and can seldom be kept alone for any length of time. The food of the Marikina is chiefly composed of fruits and insects ; but in captivity, it will eat biscuit and drink milk. It is a very timid animal, unable to fight a foe, but quick in escape and adroit in concealment. Its voice is soft and gentle when the animal is pleased, but when it is excited by anger or fear, it utters a rather sharp hiss. The dimensions of the Marikina are much the same as those of the following animal.

THE beautiful little creature which is so well known by the name of the MARMOSET, or OUISTITI, is a native of Guiana. The fur is long and exquisitely soft, diversified with bold stripes of black upon a ground of white and reddish yellow. The tail is long and full ; its colour is white, encircled with numerous rings of a hue so deep that it may almost be called black. A radiating tuft of white hairs springs from each side of the face, and contrasts well with the jetty hue of the head.

On account of the beauty of its fur, and the gentleness of its demeanour when rightly treated, it is frequently brought from its native land and forced to lead a life of compelled civilization in foreign climes. It is peculiarly sensitive to cold, and always likes to have its house well furnished with soft and warm bedding, which it piles up in a corner, and under which it delights to hide itself.

The Marmosets do not seem to be possessed of a very large share of intelligence, but yet are very engaging little creatures if kindly treated. They are very fond of flies and other insects, and will often take a fly from the hand of the visitor. One of these animals, with whom I struck up an acquaintance, took great pleasure in making me catch flies for its use, and taking them daintily out of my hand. When it saw my hand sweep over a doomed fly, the bright eyes sparkled with eager anticipation ; and when I approached the cage, the little creature thrust its paw through the bars as far as the wires would permit, and opened and closed the tiny fingers with restless impatience. It then insinuated its hand among my closed fingers, and never failed to find and capture the imprisoned fly.

Generally, the Marmoset preserves silence ; but if alarmed or irritated, it gives vent to a little sharp whistle, from which it has gained its name of Ouistiti. It is sufficiently active when in the enjoyment of good health, climbing and leaping about from bar to bar with an agile quickness that reminds the observer of a squirrel.

MARMOSET.—(*Jacchus vulgaris.*)

Its food is both animal and vegetable in character ; the animal portion being chiefly composed of various insects, eggs, and, it may be, an occasional young bird, and the vegetable diet ranging through most of the edible fruits. A tame Marmoset has been known to pounce upon a living gold fish and to eat it. In consequence of this achievement, some young eels were given to the animal, and at first terrified it by their strange writhings, but in a short time they were mastered and eaten.

The length of the full-grown Marmoset is from seven to eight inches, exclusive of the tail, which measures about a foot.

LEMURS.

THE form of the monkeys which are known by the name of LEMURS is of itself sufficient to show that we are rapidly approaching the more quadru-pedal mammalia.

The head of all the Lemurs is entirely unlike the usual monkey head, and even in the skull the distinction is as clearly marked as in the living being. Sharp, long, and pointed, the muzzle and jaws are singularly fox-like, while the general form of these animals, and the mode in which they walk, would lead a hasty observer to place them among the true quadrupeds. Yet, on a closer examination, the quadrumanous characteristics are seen so plainly, that the Lemurs can but be referred to their proper position among, or rather at the end of, the monkey tribe.

The word Lemur signifies "a night-wandering ghost," and has been applied to this group of animals on account of their nocturnal habits, and their stealthy noiseless step, which renders their progress almost as inaudible as that of the unearthly beings from whom they derive their name.

The RUFFED LEMUR is one of the handsomest of this family, challenging a rivalship even with the Ring-tailed Lemur in point of appearance.

RUFFLED LEMUR.—(*Lemur Macaco.*)

The texture of the fur is extremely fine, and its colour presents bold contrasts between pure white and jetty blackness. The face of the Ruffed Lemur is black, and a fringe of long white hairs stands out like a ruff round the face.

As is the case with all the Lemurs, it is a native of Madagascar and of the adjacent islands, and seems to take the place of the ordinary monkeys. Of all the Lemurs this species is the largest, its size equalling that of a moderately grown cat. Its voice is a sepulchral, deep roar, peculiarly loud, considering the size of the animal, and can be heard at a great distance in the stilly night.

The SLENDER LORIS is a small animal, measuring only nine inches in length, and possessed of limbs so delicately slender as to have earned for it its popular name. Its colour is grey, with a slight rusty tinge, the under portions of the body fading into white. Round the eyes, the fur takes a darker hue, which is well contrasted by a white streak running along the nose.

SLENDER LORIS.—(*Loris gracilis.*)

Small though it be, and apparently without the power to harm, it is a terrible enemy to the birds and insects on which it feeds, and which it captures "like Fabius, by delay."

Night, when the birds are resting with their heads snugly sheltered by their soft feathers, is the time when the Loris awakes from its daily slumbers, and stealthily sets forth on its search. Its movements are so slow and silent, that not a sound falls on the ear to indicate the presence of a living animal.

Alas for the doomed bird that has attracted the fiery eyes of the Loris! With movements as imperceptible and as silent as the shadow on the dial, paw after paw is lifted from its hold, advanced a step and placed again on the bough, until the destroyer stands by the side of the unconscious victim. Then, the hand is raised with equal silence, until the fingers overhang the bird and nearly touch it. Suddenly the slow caution is exchanged for lightning speed, and with a movement so rapid that the eye can hardly follow it, the bird is torn from its perch, and almost before its eyes are opened from slumber, they are closed for ever in death.

The SLOW-PACED LORIS, or KUKANG, is very similar in its habits to the animal just mentioned, but differs from it in size, colour, and several parts of its form.

The fur is of a texture rather more woolly than that of the Slender Loris, and its colour has something of a chestnut tinge running through it, although some specimens are nearly as grey as the Slender Loris. As may be seen from the engraving, a dark stripe surrounds the eyes, ears, and back of the head, reaching to the corners of the mouth. From thence it runs along the entire length of the spine. The colour of this dark band is a deep chestnut. It is rather larger than the preceding animal, being a little more than a foot in length.

In the formation of these creatures some very curious structures are found, among which is the singular grouping of arteries and veins in the limbs.

KUKANG, OR SLOW-PACED LORIS.—(*Nycticebus Javarincus.*)

Instead of the usual tree-like mode in which the limbs of most animals are supplied with blood—one large trunk-vessel entering the limb, and then branching off into numerous subdivisions—the limbs of the Loris are furnished with blood upon a strangely modified system. The arteries and veins, as they enter and leave the limb, are suddenly divided into a great number of cylindrical vessels, lying close to each other for some distance, and giving off their tubes to the different parts of the limb. It is possible that to this formation may be owing the power of silent movement and slow patience which has been mentioned as the property of these monkeys, for a very similar structure is found to exist in the sloth.

The tongue of the Loris is aided in its task by a plate of cartilage, by which it is supported, and which is, indeed, an enlargement of the tendinous band that is found under the root of the tongue. It is much thicker at its base than at the extremity, which is so deeply notched that it seems to have been slit with a knife. It is so conspicuous an organ that it has been often described as a second tongue. The throat and vocal organs seem to be but little developed, as is consistent with the habits of an animal whose very subsistence depends upon its silence. Excepting when irritated, it seldom or never utters a sound ; and even then its vocal powers seem to be limited to a little monotonous plaintive cry.

In captivity this Loris appears to be tolerably omnivorous, eating both animal and vegetable food, preferring, however, the former. Living animals best please its taste, and the greatest dainty that can be afforded to the creature is a small bird, which it instantly kills, plucks, and eats entirely, the

AVAHI, OR INDRI.—(*Indris laniger.*)

bones included. Eggs are a favourite food with it, as are insects. It will take butcher's meat, if raw, but will not touch it if cooked in any way. Of vegetable substances, sugar appears to take its fancy the most, but it will eat fruits of various kinds, such as oranges and plantains, and has been known to suck gum arabic.

ANOTHER curious inhabitant of Madagascar is the INDRI, or AVAHI, a creature that has sometimes been considered one of the lemurs, and placed among them by systematic naturalists. From the curled and woolly hair with which the body is covered it derives its name of "Laniger," or Wool-bearer. Just over the loins and partly down the flanks, the soft wool-like hair takes a firmer curl than is found to be the case in any other part of the body or limbs. It is but a small animal, the length of its head and body being only a foot, and its tail nine inches. The general colour of the fur is a lightish brown, with a white stripe on the back of the thigh, and a tinge of chestnut in the tail. In some individuals a rusty red, mingled with a yellow hue, takes the place of the brown ; and in all the under parts are lighter than the upper. Its face is black, and the eyes are grey, with a greenish light playing through their large orbs.

The name Indri is a native word, signifying, it is said, "man of the woods." Its voice is not very powerful, but it can be heard at some distance. It is of a melancholy wailing character, and has been likened to the cry of a child.

THERE are two animals which bear a close resemblance to each other, namely, the Galago of Madagascar and the little animal which is here figured. The ears, however, are not so large as those of the Galago, and the tail is less thickly covered with fur, being almost devoid of hair, except at its extremity, where it forms a small tuft. On reference to the figure, it will be seen that

TARSIER.—(*Tarsius spectrum.*)

the hands are of extraordinary length, in proportion to the size of the creature. This peculiarity is caused by a considerable elongation of the bones composing the " Tarsus," or back of the hands and feet, and has earned for the animal the title of TARSIER. This peculiarity is more strongly developed in the hinder than in the fore-paws.

The colour of the Tarsier is a greyish-brown, with slight olive tint washed over the body. A stripe of deeper colour surrounds the back of the head and the face and forehead are of a warmer brown than the body and limbs It is a native of Borneo, Celebes, the Philippine Islands, and Banca. From the latter locality it is sometimes called the Banca Tarsier. Another of the titles by which it is known is the Podji.

It is a tree-inhabiting animal, and skips among the branches with little quick leaps that have been likened to the hoppings of a frog. In order to give the little creature a firmer hold of the boughs about which it is constantly leaping, the palms of the hands are furnished with several cushions. The backs of the hands are covered with soft downy fur, resembling the hair with which the tail is furnished. Excepting on the hands and tail, the fur is very thick and of a woolly character, but at the root of the tail, and at the wrists and ankles, it suddenly changes to the short downy covering.

THE true position of that very rare animal the AYE-AYE seems very doubtful, some naturalists placing it in the position which it occupies in this work, and others, such as Van der Hoeven, considering it to form a link between the monkeys and the rodent animals, the incisor teeth bearing some resemblance to those of the rodents.

AYE-AYE. —(*Cheiromys Madagascariensis.*)

These curious teeth are extremely powerful, and are very deeply set in the jawbones, their sockets extending nearly the entire depth of the bone.

They are used just like the rodent teeth, the animal biting deeply into the trees, and so laying bare the burrows of various wood-boring grubs.

The colour of the animal is a dull black on the upper portions of the body, the under parts, as well as the cheeks and throat, being of a light grey. The paws are nearly black. The fur of the body is thickly set, and is remarkable for an inner coating of downy hair of a golden tint, which sometimes shows itself through the outer coating. On the tail the hair is darker than on the body, greater in length, and in texture much coarser. The tail, which is jetty black, seems to be always trailed at length, and never to be set up over the

D

*b*ody like the well known tail of the squirrel.   The ears are large, and nearly destitute of hair.

The natural food of the Aye-aye, like that of the preceding animals, is of a mixed character, the creature eating fruits and insects indiscriminately. But in its wild state it is said to search the trees for insects as well as fruits, and to drag their larvæ from their concealment by means of its delicate fingers.

The fine specimen in the Zoological Gardens, however, does not touch insects, but feeds on a mixture of honey and hard-boiled eggs beaten into a paste and moistened with milk.   Still she uses her teeth freely on the branches that are placed in her cage, and very soon cuts them to pieces, as if

COLUGO.—(*Galeopithecus volans.*)

in search after grubs.   She is very active, and climbs about the cage or on the branches, in almost any position.   Like the squirrel, she covers herself with her bushy tail when in repose.

It is a nocturnal animal like the Galagos and Lemurs, and seeks its prey by night only, spending the day in sleep, curled up in the dark hollow of a tree, or in some similar spot, where it can retire from view and from light.

As is shown by the scientific name of the Aye-aye, it is a native of Madagascar, and even in that island is extremely scarce, appearing to be limited to the western portions of the country, and to escape even the quick eyes of the natives.

The eyes are of a brownish yellow colour, and very sensitive to light, as may be expected in a creature so entirely nocturnal in its habits. It is not a very small animal, measuring almost a yard in total length, of which the tail occupies one moiety.

THE strange animal which is known by the name of the FLYING LEMUR, or COLUGO, affords an intermediate link of transition between the four-handed and the wing-handed mammals.

By means of the largely-developed membrane which connects the limbs with each other, and the hinder limbs with the tail, the Colugo is enabled to leap through very great distances, and to pass from one bough to another with ease. This membrane is a prolongation of the natural skin, and is covered with hair on the upper side as thickly as any part of the body, but beneath it is almost naked. When the creature desires to make one of its long sweeping leaps, it spreads its limbs as widely as possible, and thus converts itself into a kind of living kite, as is shown in the figure. By thus presenting a large surface to the air, it can be supported in its passage between the branches, and is said to vary its course slightly by the movement of its arms. It is said that the Colugo will thus pass over nearly a hundred yards.

Among other bat-like habits, the Colugo is accustomed to suspend itself by its hinder paws from the branch of a tree, and in this pendent attitude it sleeps. Its slumbers are mostly diurnal, for the Colugo is a night-loving animal, and is seldom seen in motion until the shades of evening draw on. But on the approach of night, the Colugo awakes from its drowsiness, and unhooking its claws from the branch on which it has hung suspended during the hours of daylight, sets off on its travels in search of food.

It is found in many of the islands that belong to the Indian Archipelago, and is tolerably common.

The colour of the fur is very uncertain, even in the same species, some specimens being of a light brown, others of a grey tint, more or less deep ; while many individuals have their fur diversified with irregular marblings or stripes, or spots of different shades and tints.

The Colugo is by no means a small animal, as, when it is full grown, it equals a large cat in size.

CHEIROPTERA;

OR, WING-HANDED ANIMALS, POPULARLY CALLED BATS.

IN general form the Bats are clearly separated from any other group of animals, and by most evident modifications of structure, can be recognised by the most cursory glance.

The first peculiarity in the Bat form which strikes the eye, is the wide ...d delicate membrane which stretches round the body, and which is used in the place of the wings with which birds are furnished.

In order to support this beautiful membrane, ᴜ extend it to its requisite width, and to strike the air with it for the purposes of flight, the bones of the fore-part of the body, and especially those of the arms and hands, undergo a singular modification.

The finger-bones are strangely disproportioned to the remainder of the body, the middle finger being considerably longer than the head and body together. The thumb is very much shorter than any of the fingers, and furnished with a sharp and curved claw. By means of this claw, the Bat is enabled to proceed along a level surface, and to attach itself to any object that may be convenient.

The lower portions of the body and limbs are singularly small in proportion

D 2

to the upper limbs. The legs are short and slender, and so arranged that the feet are rather turned outward, for the purpose of using their sharp claws freely. A kind of slender and spur-like bone is seen to proceed from the heel of each foot.

The VAMPIRE BAT is a native of Southern America, and is spread over a large extent of country. It is not a very large animal, the length of its   dy and tail being only six inches, or perhaps seven in large specimens, and the spread of the wing two feet, or rather more. The colour of the Vampire's fur is a mouse tint, with a shade of brown.

Many tales have been told of the Vampire Bat, and its fearful a   ks upon sleeping men—tales which, although founded on fact, were so sadly exaggerated as to cause a reaction in the opposite direction. It was reported to come silently by night, and to search for the exposed toes of a sound sleeper—its instinct telling it whether the intended victim were thoroughly buried in sleep. Poising itself above the feet of its prey, and fanning them with its extended wings, it produced a cool atmosphere, which, in those hot climates, aided in soothing the slumber into a still deeper repose. The Bat then applied its needle-pointed teeth to the upturned foot, and inserted them into the tip of a toe with such adroit dexterity that no pain was caused by the tiny wound. The lips were them brought into action, and the blood was sucked until the Bat was satiated. It then disgorged the food which it had just taken, and began afresh, continuing its alternated feeding and disgorging, until the victim perished from sheer loss of blood.

THE VAMPIRE BAT.—(*Vampyrus spectrum.*)

For a time this statement gained dominion, but after a while was less and less believed, until at last naturalists repudiated the whole story as a " traveller's tale." However, as usual, the truth seems to have lain between the two extremes ; for it is satisfactorily ascertained, by more recent travellers, that the Vampires really do bite both men and cattle during the night, but that the wound is never known to be fatal, and in most instances causes but little inconvenience to the sufferer.

When they direct their attacks against mankind, the Vampires almost invariably select the foot as their point of operation, and their blood-loving propensities are the dread of both natives and Europeans. With singular audacity, the Bats even creep into human habitations, and seek out the exposed feet of any sleeping inhabitant who has incautiously neglected to draw a coverlet over his limbs.

ONE of the most common, and at the same time the most elegant, of the British Cheiroptera, is the well-known LONG-EARED BAT.

This pretty little creature may be found in all parts of England ; and on account of its singularly beautiful ears and gentle temper has frequently been tamed and domesticated. I have possessed several specimens of this Bat, and in every case have been awarded for the trouble by the curious little traits of temper and disposition which have been exhibited.

One of my Bat favourites was captured under rather peculiar circumstances.

It had entered a grocer's shop, and to the consternation of the grocer and his assistant had got among the sugar loaves which were piled on an upper shelf. So terrible a foe as the Bat (nearly two inches long) put to rout their united forces, and beyond poking at it with a broom as it cowered behind the sugar, no attempts were made to dislodge it. At this juncture my aid was invoked ; and I accordingly drew the Bat from its hiding-place. It did its best to bite, but its tiny teeth could do no damage even to a sensitive skin.

The Bat was then placed in an empty mouse-cage, and soon became sufficiently familiar to eat and drink under observation. It would never eat flies, although many of these insects were offered, and seemed to prefer small bits of raw beef to any other food. It was a troublesome animal to feed, for it

LONG-EARED BAT.—(*Pleistus Communis.*)

would not touch the meat unless it were freshly cut and quite moist ; forcing me to prepare morsels fit for its dainty maw six or seven times daily.

It spent the day at the top or on the side of its cage, being suspended by its hinder claws, and would occasionally descend from its eminence in order to feed or to drink. While eating, it was accustomed to lower itself from the cage roof, and to crawl along the floor until it reached the piece of meat. The wings were then thrown forward so as to envelop the food, and under the shelter of its wings the Bat would drop its head over the meat and then consume it. On account of the sharp surface of its teeth, it could not eat its food quietly, but was forced to make a series of pecking bites, something like the action of a cat in similar circumstances.

It would drink in several ways, sometimes crawling up to the water vessel and putting its head into the water, but usually lowering itself down the side of the cage

HEAD OF LONG-EARED BAT.

until its nose dipped in the liquid. When it had thus satisfied its thirst, it would re-ascend to the roof, fold its wings about itself, and betake itself to slumber once more.

I kept the little animal some time, but it did not appear to thrive, having, in all probability, been hurt by the broom-handle which had been used so freely against it, and at last was found dead in its cage from no apparent cause. Although dead, it still hung suspended, and the only circumstance that appeared strange in its attitude was, that the wings drooped downwards instead of being wrapped tightly round the body.

In the attitude of repose, this Bat presents a most singular figure. The wings are wrapped around and held firmly to the body ; the immense ears are folded back, and the pointed inner ear, or "tragus," stands boldly out giving the creature a totally different aspect.

THE Bats which have heretofore been mentioned feed on animal substances,

FLYING FOX, OR ROUSSETTE.—(*Pteropus rubricollis.*)

insects appearing to afford the principal nutriment, and raw meat or fresh blood being their occasional luxuries.   But the Bats of which the accompanying engraving is an example, are chiefly vegetable feeders, and, in their own land, are most mischievous among the fruit-trees.

They are the largest of the present Bat tribe, some of them measuring nearly five feet in expanse of wing. Their popular name is FLYING FOXES, a term which has been applied to them on account of the red, fox-like colour of the fur, and the very vulpine aspect of the head. Although so superior in size to the Vampires, the Flying Foxes are not to be dreaded as personal enemies, for, unless roughly handled, they are not given to biting animated beings.

But though their attacks are not made directly upon animal life, they are of considerable importance in an indirect point of view, for they are aimed against the fruits and other vegetable substances by which animal life is sustained.

I have often seen the Kalong engaged in eating fruit. It would accept a slice of apple or pear, while suspended by its hind legs. It then bent its head upwards, brought its winged arms forward so as to enclose head and fruit together, and then would devour its meal with a series of snapping bites.

The Kalongs do not seem to care much for dark and retired places of abode ; and pass the day, which is their night, suspended from the trunks of large trees, preferring those which belong to the fig genus. On these boughs they hang in vast numbers, and by an inexperienced observer might readily be taken for bunches of large fruits, so closely and quietly do they hang. If disturbed in their repose, they set up a chorus of sharp screams, and flutter about in a state of sad bewilderment, their night-loving eyes being dazzled by the hateful glare of the sun.

FELIDÆ ;

OR, THE CAT TRIBE.

THE beautiful animals which are known by the general name of the CAT Tribe now engage our attention.

With the exception of one or two of the enigmatical creatures which are found in every group of beings, whether animal, vegetable, or mineral, the Cats, or FELIDÆ as they are more learnedly termed, are as distinct an order as the monkeys or the bats. Pre-eminently carnivorous in their diet, and destructive in their mode of obtaining food, their bodily form is more exquisitely adapted to carry out the instincts which are implanted in their nature.

All the members of the Cat tribe are light, stealthy, and silent of foot, quick of ear and eye, and swift of attack. Most of them are possessed of the power of climbing trees or rocks, but some few species, such as the LION, are devoid of this capability.

Of the magnificent and noble creatures called LIONS, several species are reported to exist, although it is thought by many experienced judges that there is really but one species of Lion, which is modified into permanent varieties according to the country in which it lives.

The best known of these species or varieties is the SOUTH AFRICAN LION, of whom so many anecdotes have been narrated.

The colour of the Lion is a tawny yellow, lighter on the under parts of the body, and darker above. The ears are blackish, and the tip of the tail is decorated with a tuft of black hair. This tuft serves to distinguish the Lion from any other member of the Cat tribe. The male Lion, when fully grown, is furnished with a thick and shaggy mane of very long hair, which falls from the neck, shoulders, and part of the throat and chin, varying in tint according to the age of the animal, and possibly according to the locality which it inhabits. The Lioness possesses no mane, and even in the male Lion it is not properly developed until the animal has completed his third year.

When fully grown, the male Lion measures some four feet in height at the shoulder, and about eleven feet in total length.

The Lioness is a smaller animal than her mate, and the difference of size appears to be much greater than really is the case, because she is devoid of the thick mane which gives such grandeur and dignity to her spouse.

In the attack of large animals, the Lion seldom attempts an unaided assault, but joins in the pursuit of several companions. Thus it is that the stately giraffe is slain by the Lion, five of which have been seen engaged in the chase of one giraffe, two actually pulling down their prey, while the other three were waiting close at hand. The Lions were driven off, and the neck of the giraffe was found to be bitten through by the cruel teeth of the assailants.

Owing to the uniform tawny colour of the Lion's coat, he is hardly distinguishable from surrounding objects even in broad daylight, and by night he walks secure.

THE LION.—(*Leo barbarus.*)

walks secure. Even the practised eyes of an accomplished hunter have been unable to detect the bodies of Lions which were lapping water at some twenty yards' distance, betraying their vicinity by the sound, but so blended in form with the landscape, that they afforded no mark for the rifle even at that short distance.

UPON the African continent the Lion reigns supreme, sole monarch over the feline race. But in Asia his claims to undivided royalty are disputed by the TIGER, an animal which equals the Lion in size, strength, and activity, and certainly excels him in the elegance of its form, the grace of its movements, and the beauty of its fur. The range of the Tiger is not so widely spread as that of the Lion, for it is never found in any portions of the New World, nor in Africa, and, except in certain districts, is but rarely seen even in the countries where it takes up its residence. Some portions of country there are, which are absolutely infested by this fierce animal, whose very appearance is sufficient to throw the natives into a state of abject terror.

In its colour the Tiger presents a most beautiful arrangement of markings and contrast of tints. On a bright tawny yellow ground, sundry dark stripes are placed, arranged, as may be seen by the engraving, nearly at right angles with the body or limbs. Some of these stripes are double, but the greater number are single dark streaks. The under parts of the body, the chest, throat, and the long hair which tufts each side of the face, are almost white, and upon these parts the stripes become very obscure, fading gradually into the light tint of the fur. The tail is of a whiter hue than the upper portions of the body, and is decorated in like manner with dark rings.

So brilliantly adorned an animal would appear to be very conspicuous among even the trees and bushes, and to thrust itself boldly upon the view. But

THE TIGER. — (*Tigris regalis.*)

there is no animal that can hide itself more thoroughly than the Tiger, or which can walk through the underwood with less betrayal of its presence.

The vertical stripes of the body harmonize so well with the dry, dusky jungle grass among which this creature loves to dwell, that the grass and fur are hardly distinguishable from each other except by a quick and experienced eye. A Tiger may thus lie concealed so cleverly, that even when crouching among low and scanty vegetation it may be almost trodden on without being seen.

The Tiger is very clever in selecting spots from whence it can watch the approach of its intended prey, itself being crouched under the shade of foliage or behind the screen of some friendly rock. It is fond of lying in wait by the side of moderately frequented roads, more particularly choosing those spots where the shade is the deepest, and where water may be found at hand wherewith to quench the thirst that it always feels when consuming its prey. From such a point of vantage it will leap with terrible effect, seldom making above a single spring, and, as a rule, always being felt before it is seen or heard.

It is a curious fact that the Tiger generally takes up his post on the side of the road which is opposite his lair, so that he has no need to turn and drag his prey across the road, but proceeds forward with his acquisition to his den. Should the Tiger miss his leap, he generally seems bewildered and ashamed of himself, and instead of returning to the spot to make a second attempt, sneaks off discomfited from the scene of his humiliation. The spots where there is most danger of meeting a Tiger, are the crossings of nullahs, or the deep ravines through which the watercourses run. In these localities the Tiger is sure to find his two essentials, cover and water. So apathetic are the natives, and so audacious are the Tigers, that at some of these crossings a man or a bullock may be carried off daily, and yet no steps will be taken to avert the danger, with the exception of a few amulets suspended about the person. Sometimes the Tigers seem to take a panic, and make a general emigration, leaving, without any apparent reason, the spots which they had long infested, and making a sudden appearance in some locality where they had but seldom before been seen.

Many modes are adopted of killing so fearful a pest as the Tiger, and some of these plans are very ingenious—such as the spring-bow, which is discharged by the movements of the animal itself; the pitfall, from which it cannot escape; the leaves smeared with bird lime, by which the Tiger blinds itself, and so falls an easy prey; the fall-trap, and many others. Among Europeans, however, the Tiger is hunted in due form, the sportsmen being mounted on elephants, and furnished with a perfect battery of loaded rifles. The shell bullet, which explodes as it enters the body, has come much into vogue.

The Tiger is a capital swimmer, and will take to the water with perfect readiness, either in search of prey, or to escape the pursuit of enemies.

It swims rather high in the water, and therefore affords a a mark to those who are quick of aim. The natatory abilities are by no means small, and while swimming it can strike out with its paws most effectively, inflicting deep wounds wherever its outspread talons made good their aim. So cunning is the animal, that if there should be no cause for hurry, it will halt on the river's brink, and deliberately put its paw into the water, so as to ascertain the force of the stream. This point being made clear, it proceeds either up or down the river, as may best suit its purpose, and so makes allowance for the river stream or the ocean tide.

UNLIKE the Tiger, which is confined to the Asiatic portion of the world, the LEOPARD is found in Africa as well as in Asia, and is represented in America by the Jaguar, or, perhaps more rightly, by the Puma.

This animal is one of the most graceful of the graceful tribe of Cats, and, although far less in dimensions than the tiger, challenges competition with that animal in the beautiful markings of its fur and the easy elegance of its movements. It is possessed of an accomplishment which is not within the powers of the lion or tiger, being able to climb trees with singular ability, and even to chase the tree-loving animals among their familiar haunts. On account of this power, it is called by the natives of India " Lakree-baug," or Tree-tiger. Even in Africa it is occasionally called a " Tiger," a confusion of nomenclature which is quite bewildering to a non-zoologist, who may read in one book that there are no tigers in Africa, and in another may peruse a narrative of a tiger-hunt at the Cape. Similar mistakes are made with regard to the American Felidæ, not to mention the numerous examples of miscalled animals that are insulted by false titles in almost every part of the globe. For in America the Puma is popularly known by the name of the Lion, or the Panther, or "Painter" as the American forester prefers to call it, while the Jaguar is termed the "Tiger."

In Africa the Leopard is well known and much dreaded, for it possesses a most crafty brain, as well as an agile body and sharp teeth and claws. It commits sad depredations on flocks and herds, and has sufficient foresight to lay up a little stock of provisions for a future day.

When attacked, it will generally endeavour to slink away, and to escape the observation of its pursuers ; but if it is wounded, and finds no means of eluding its foes, it becomes

LEOPARD.—(*Leopardus varius.*)

furious, and charges at them with such determinate rage, that, unless it falls a victim to a well-aimed shot, it may do fearful damage before it yields up its life. In consequence of the ferocity and courage of the Leopard, the native African races make much of those warriors who have been fortunate enough to kill one of these beasts.

In its own country the Leopard is as crafty an animal as our British fox ; and, being aided by its active limbs and stealthy tread, gains quiet admission into many spots where no less cautious a creature could plant a step without giving the alarm. It is an inveterate chicken-stealer, creeping by night into the hen-roosts, in spite of the watchful dogs that are at their posts as sentinels, and destroying in one fell swoop the entire stock of poultry that happen to be collected under that roof. Even should they roost out of doors they are no less in danger, for the Leopard can clamber a pole or tree with marvellous rapidity, and with his ready paw strike down the poor bird before it is fairly awakened.

There are two titles for this animal, namely the Leopard and Panther, both of which creatures are now acknowledged to be but slight varieties of the same species. The OUNCE, however, which was once thought to be but a longer-haired variety of the

OUNCE.—(*Leopardus uncia.*)

Leopard, is now known to be truly a separate species.

In general appearance it bears a very close resemblance to the leopard, but may be distinguished from that animal by the greater fulness and roughness of its fur, as well as by some variations in the markings with which it is decorated. The spots exhibit a certain tendency to form stripes, and the tail is exceedingly bushy when compared with that of a leopard of equal size. The general colour of the body is rather paler than that of the leopard, being a

JAGUAR.--(*Leopardus Onca.*)

greyish white in which a slight yellow tinge is perceptible. The Ounce is an inhabitant of some parts of Asia, and specimens of this fine animal have been brought from the shores of the Persian Gulf. In size, it equals the ordinary leopard of Asia or Africa.

Passing to the New World, we find the feline races well represented by several most beautiful and graceful creatures, of which the JAGUAR is the largest and most magnificent example.

Closely resembling the leopard in external appearance, and in its arboreal habits, it seems to play the same part in America as the leopard in the transatlantic continents. It is a larger animal than the leopard, and may be distinguished from that animal by several characteristic differences.

In the first place, across the breast of the Jaguar are drawn two or three bold black streaks, which are never seen in the leopard, and which alone serve as an easy guide to the species. But the chief point of distinction is found in a small mark that exists in the centre of the dark spots which cover the body and sides. In many instances this central mark is double, and, in order to give room for it, the rosettes are very large in proportion to those of the leopard. Along the spine runs a line, or chain, of black spots and dashes, extending from the back of the head to the first foot, or eighteen inches of the tail.

In its native land the Jaguar ranges the dense and perfumed forests in search of the various creatures which fall victims to its powerful claws. The list of animals that compose its bill of fare is a large and comprehensive one, including horses, deer, monkeys, capybaras, tapirs, birds of various kinds, turtles, lizards, and fish ; thus comprising examples of all the four orders of vertebrated animals. Nor does the Jaguar confine himself to the vertebrates. Various shell-fish, insects, and other creatures fall victims to the insatiate appetite of this ravenous animal.

FEW animals have been known by such a variety of names as the PUMA of America. Travellers have indifferently entitled it the American Lion, the Panther, the Couguar, the Carcajou (which is an entirely different animal), the Gouazoura, the Cuguacurana, and many other names.

It is rather a large animal, but, on account of its small head, appears to be a less powerful creature than really is the case. The total length of the Puma is about six feet and a half, of which the tail occupies rather more than two feet. The tip of the tail is black, but is destitute of the long tuft of black hair which is so characteristic of the lion.

PUMA.

The colour of the Puma is a uniform light tawny tint, deeper in some individuals than in others, and fading into a greyish white on the under parts. It is remarkable that the young Puma displays a gradual change in its fur, nearly in the same way as has been narrated of the lion cub. While the Puma cubs are yet in their first infancy, their coat is marked with several rows of dark streaks extending along the back and sides, and also bears upon the neck, sides, and shoulders many dark spots resembling those of the ordinary leopard. But, as the animal increases in size, the spots fade away, and, when it has attained its perfect development, are altogether lost in the uniform tawny hue of the fur.

The flesh of this animal is said, by those who have made trial of it, to be a pleasant addition to the diet scale, being white, tender, and of good flavour. When taken young, the Puma is peculiarly susceptible of domestication, and has been known to follow its master just like a dog.   The hunters of the Pampas are expert Puma slayers, and achieve their end either by catching the bewildered animal with a lasso, and then galloping off with the poor creature hanging at the end of the leather cord, or by flinging the celebrated *bolas*—metal balls or stones fastened to a rope—at the Puma, and laying it senseless on the ground with a blow from the heavy weapon.

MANY of the members of the large genus Leopardus are classed together under the title of OCELOTS, or, more popularly of TIGER CATS.   They are all most beautiful animals, their fur being diversified with brilliant contrasts of a dark spot, streak, or dash upon a lighter ground, and their actions filled with easy grace and elegance.

The common OCELOT is a native of the tropical regions of America, where it is found in some profusion.   In length it rather exceeds four feet, of which the tail occupies a considerable portion.   Its height averages eighteen inches.

The ground colour of the fur is a very light greyish fawn, on which are drawn partially broken bands of a very deep fawn colour, edged with black, running along the line of the body.   The band that extends along the spine is unbroken.   On the head, neck, and the inside of the limbs, the bands are broken up into spots and dashes, which are entirely black, the fawn tint in their centre being totally merged in the deeper hue; the ears are black, with the exception of a conspicuous white spot upon the back and near the base of each ear.   Owing to the beauty of the fur, the Ocelot skin is in great request for home use and exporta-

OCELOT.

tion, and is extensively employed in the manufacture of various fancy articles of dress or luxury.

In its habits the Ocelot is quick, active, and powerful, proving itself at all points a true leopard, although but in miniature.

The eye of the Ocelot is a pale yellowish brown, and tolerably full, with the linear pupil smaller than is found in the ordinary Felidæ.

There are several species of these pretty and agile animals, among which the most conspicuous are the Common, the Grey, and Painted Ocelots, and the Margay, or Marjay, as it is sometimes called.   The habits of these animals are very similar.

Although so gentle in its demeanour when domesticated as to have earned for itself the name of "*Mitis*," or "placid," the Chati is, when wild, a sufficiently destructive animal.   It is not quite so large as the ocelots, with which creatures it is a compatriot.

The colour of the Chati resembles that of the leopard, only is paler in general hue.   The dark patches that diversify the body are very irregular— those which run along the back are solid, and of a deep black, while those which are placed along the sides have generally a deep fawn-coloured centre. Towards the extremity of the tail, the spots change into partial rings, which nearly, but not quite, surround the tail.   All specimens, however, are not precisely alike, either in the colour or the arrangement of the markings, but those leading characteristics which have just been mentioned may be found in almost every individual.

When at large in its native woods, it wages incessant and destructive warfare against small quadrupeds and birds, the latter creatures being its favourite prey. The Chati is a vexatious and expensive neighbour to any one who may keep fowls, for it seems to like nothing so well as a plump fowl, and is unceasing in its visits to the henroost. It is so active and lithe an animal that it can climb over any palisade, and insinuate itself through a surprisingly small aperture ; and it is so wary and cautious in its nocturnal raids, that it generally gives no other indication of its movements than that which is left next morning by the vacant perches, and a few scattered feathers flecked with blood-spots.

During the day it keeps itself closely hidden in the dark shades of the forest, sleeping away its time until the sun has set, and darkness reigns over its world. It then awakes from its slumber, and issues forth upon its destructive quest. On moonlight nights, however, it either stays at home, or

THE CHATI.—(*Leopardus mitis.*)

confines its depredations to the limits of its native woods, never venturing near the habitations of man. Stormy and windy nights are the best adapted for its purpose, as it is sheltered from sight by the darkness, and from hearing by the rushing wind, which drowns the slight sounds of its stealthy footsteps. On such nights it behoves the farmer to keep a two-fold watch, and see well to his doors and windows, or he may chance to find an empty henroost in the morning.

In two years, no less than eighteen of these animals were caught by a landowner within a space of five miles round his farm, so that their numbers must be truly great. They do not congregate together, but live in pairs, each pair seeming to appropriate its own hunting-ground.

In captivity it is a singularly gentle, and even affectionate, animal, possessed of most engaging habits, and full of pretty graceful tricks. One of these creatures, which was captured by the above-mentioned landowner, became so entirely domesticated that it was permitted to range at liberty. But, although so gentle and tractable towards its owner that it would sleep on the skirts of its master's gown, its poultry-loving habits were too deeply implanted to be thoroughly eradicated, and it was quietly destructive among his neighbours' fowls. This propensity cost the creature its life, for the irritated farmers caught it in the very deed of robbing their henroosts, and killed it on the spot.

The native name for the Chati is Chibiguazu. It was found by experimenting on the captured Chatis, that the flesh of cats and of various reptiles was harmful to their constitution. Cat's-flesh gave them a kind of mange, which soon killed them, while that of snakes, vipers, and toads caused a continual and violent vomiting, under which they lost flesh and died. Fowls however, and most birds, were ravenously devoured, being caught by the head, and killed by a bite and a shake. The Chatis always stripped the feathers from the birds before beginning to eat them.

FEW of the Felidæ are so widely spread or so generally known as the WILD CAT. It is found not only in this country, but over nearly the whole of Europe, and has been seen in Northern Asia and Nepaul.

THE CAT.—(*Felis domestica.*)

Whether the Wild Cat be the original progenitor of our domestic cat is still a mooted point, and likely to remain so, for there is no small difficulty in bringing proofs to bear on such a subject. There are several points of distinction between the wild and the domestic cat; one of the most decided differences being found in the shape and comparative length of their tails.

CATS' TAILS.

As may be seen from the accompanying figure, the tails of the two animals are easily distinguished from each other. The upper figure represents the tail of the domestic cat, which is long, slender, and tapering, while the lower represents the tail of the Wild Cat, which is much shorter and more bushy.

In the eyes of any one who has really examined and can support the character of the Domestic Cat, she must appear to be a sadly calumniated creature. She is generally contrasted with the dog, much to her disfavour. His docility, affectionate disposition, and forgiveness of injuries ; his trustworthy character, and his wonderful intellectual powers, are spoken of, as truly they deserve, with great enthusiasm and respect. But these amiable traits of character are brought into violent contrast with sundry ill-conditioned qualities which are attributed to the Cat, and wrongly so. The Cat is held up to reprobation as a selfish animal, seeking her own comfort and disregardful of others ; attached only to localities, and bearing no real affection for her owners. She is said to be sly and treacherous, hiding her talons in her velvety paws as long as she is in a good temper, but ready to use them upon her best friends if she is crossed in her humours.

Whatever may have been the experience of those who gave so slanderous a character to the Cat, my own rather wide acquaintance with this animal has led me to very different conclusions. The Cats with which I have been most familiar have been as docile, tractable, and good-tempered as any dog could be, and displayed an amount of intellectual power which would be equalled by very few dogs, and surpassed by none.

RETURNING once more to the savage tribe of animals, we come to a small, but clearly-marked group of Cats, which are distinguishable from their feline relations by the sharply-pointed erect ears, decorated with a tuft of hair of varying dimensions. These animals are popularly known by the title of LYNXES. In all the species the tail is rather short, and in some, such as the Peeshoo, or Canada Lynx, it is extremely abbreviated.

By name, if not by sight, the common LYNX of Europe is familiar to us, and is known as the type of a quick-sighted animal. The eyes of the Lynx and the ears of the " Blind Mole " are generally placed on a par with each other, as examples of especial acuteness of either sense.

The European Lynx is spread over a great portion of the Continent, being found in a range of country which extends from the Pyrenees to Scandinavia. It is also found in the more northern forests of Asia.

The usual colour of the Lynx is a rather dark grey, washed with red, on which are placed sundry dark patches, large and few upon the body, and many and small on the limbs. On the body the spots assume an oblong or oval shape, but upon the limbs they are nearly circular. The tail of the Lynx is short, being at the most only seven or eight inches in length, and sometimes extending only six inches. The length of the body and head is about three feet.

The fur of the Lynx is valuable for the purposes to which the feline skin is usually destined, and commands a fair price in the market. Those who hunt the Lynx for the purpose of obtaining its fur, choose the winter months for the time of their operations, as during the cold season the Lynx possesses a richer and a warmer fur than is found upon it during the warm summer months.

The New World possesses its examples of the Lyncine group as well as the Old World, and even in the cold regions of North America a representative of these animals may be

CANADA LYNX.—(*Lyncus Canadensis.*)

found. This is the CANADA LYNX, commonly termed the "Peeshoo" by the French colonists, or even dignified with the title of *Le Chat.*

The hair of this animal is longer than that of its southern relatives, and is generally of a dark grey, flecked or besprinkled with black.

E

The limbs of this Lynx are very powerful, and the thick heavily made feet are furnished with strong white claws that are not seen unless the fur be put aside. It is not a dangerous animal, and, as far as is known, feeds on the smaller quadrupeds, the American hare being its favourite article of diet.

While running at speed it presents a singular appearance, owing to its peculiar mode of leaping in successive bounds, with its back slightly arched, and all the feet coming to the ground nearly at the same time. It is a good swimmer, being able to cross the water for a distance of two miles or more. Powerful though it be, it is easily killed by a blow on the back, a slight stick being a sufficient weapon wherewith to destroy the animal. The flesh of the Peeshoo is eaten by the natives, and is said, though devoid of flavour, to be agreeably tender. The range of this animal is rather extensive, and, in the wide district where it takes up its residence, is found in sufficient plenty to render its fur an important article of commerce. The length of this animal slightly exceeds three feet.

The CHETAH, YOUZE, or HUNTING CAT, as it is indifferently named, is, like the leopard, an inhabitant of Asia and Africa. It is rather a large animal, exceeding an ordinary leopard in stature.

CHETAH.—(*Gueparda jubata.*)

The title "jubata," or crested, is given to the Chetah on account of a short, mane-like crest of stiff long hairs which passes from the back of the head to the shoulders.

The Chetah is one of those animals which gain their living by mingled craft and agility. Its chief food is obtained from the various deer and antelopes which inhabit the same country, and in seizing and slaying its prey no little art is required. The speed of this animal is not very great, and it has but little endurance; so that an antelope or a stag could set the spotted foe at defiance, and in a quarter of an hour place themselves beyond his reach. But it is the business of the Chetah to hinder the active and swift-footed deer from obtaining those invaluable fifteen minutes, and to strike them down before they are aware of his presence.

In order to obtain this end, the Chetah watches for a herd of deer or antelopes, or is content to address himself to the pursuit of a solitary individual, or a little band of two or three, should they be placed in a position favourable for his purpose. Crouching upon the ground so as to conceal himself as much as possible from the watchful eyes of the intended prey, the Chetah steals rapidly and silently upon them, never venturing to show himself until he is within reach of a single spring. Having singled out one individual from the herd, the Chetah leaps upon the devoted animal and dashes it to the ground. Fastening his strong grip in the throat of the dying animal, the Chetah laps the hot blood, and for the time seems forgetful of time or place.

Of these curious habits the restless and all-adapting mind of man has taken advantage, and has diverted to his own service the wild destructive properties of the Chetah. In fact, man has established a kind of quadrupedal falconry, the Chetah taking the place of the hawk, and the chase being one of earth and not of air. The Asiatics have brought this curious chase to great perfection, and are able to train Chetahs for this purpose in a wonderfully perfect manner.

When a Chetah is taken out for the purpose of hunting game, he is hooded and placed in a light native car, in company with his keepers. When they perceive a herd of deer, or other desirable game, the keepers turn the Chetah's head in the proper direction, and remove the hood from his eyes. The sharp-sighted animal generally perceives the prey at once, but if he fails so to do, the keepers assist him by quiet gestures.

No sooner does the Chetah fairly perceive the deer than his bands are loosened, and he gently slips from the car. Employing all his innate artifices, he approaches the game, and with one powerful leap flings himself upon the animal which he has selected. The keepers now hurry up, and take his attention from the slaughtered animal by offering him a ladleful of its blood, or by placing before him some food of which he is especially fond, such as the head and neck of a fowl. The hood is then slipped over his head, and the blinded animal is conducted patient and unresisting to the car, where he is secured until another victim may be discovered.

The natural disposition of this pretty creature seems to be gentle and placid, and it is peculiarly susceptible of domestication. It has been so completely trained as to be permitted to wander where it chooses like a domestic dog or cat, and is quite as familiar as that animal. Even in a state of semi-domestication it is sufficiently gentle. One sleek and well-conditioned specimen with which I made acquaintance behaved in a very friendly manner, permitting me to pat its soft sides, or stroke its face, and uttering short self-sufficient sounds, like the magnified purr of a gratified cat. Unfortunately, the acquaintance was rudely broken up by an ill-conditioned Frenchman, who came to the front of the cage, and with his stick dealt the poor animal a severe thrust in the side. The Chetah instantly lost its confident expression, and was so irritated by this rough treatment that it would not permit a repetition of the former caresses.

Some time ago, while engaged in examining the larger Felidæ, I wished to investigate the structure of the Chetah's foot, some persons having said that its claws were retractile like those of the cat, while others stated that they were constructed like those of the dog. So I went into the Chetahs' cage at the Zoological Gardens, and rather to the surprise of the animals. Thinking that the Cat tribe were tolerably alike in disposition, and supposing that if I went up to either of them they would be alarmed, I sat down with my back against the wall, and quietly waited, taking no notice whatever of the Chetahs.

In a short time the curiosity of the cat-nature overcame distrust, the two Chetahs came closer and closer, until at last the male, who was larger and stronger than his mate, began to sniff at my hand with outstretched neck. Finding that no harm ensued, he came a little closer, and I began to stroke his nose lightly. This he rather liked, and before long I was able to stroke his head, chin, neck, and back, the animal being as pleased as a cat would have been. Presently he came and sat down by me, and I then got from his neck to his legs, just as Rarey used to " gentle " a horse.

The next move was to lift up his foot and put it down again, and then, taking hold lightly of his wrist, to press the fore-finger on the base of the claws so as to press them from their sockets. This rather startled him, and with a sharp hissing sound he struck smartly forwards. As he struck, I slipped my hand up his leg, so that the blow was ineffectual, and presently

E 2

made another attempt.   He now found out that no harm was intended, and in a very short time I had his paw on my knee, and was allowed to push out the claws as I liked, proving that they were as retractile as those of a cat. The oddest part of the proceeding was, that he appropriated me to himself, and would not allow his mate to come near me, exemplifying the jealousy of all animals when brought into contact with man.

The spots which so profusely stud the body and limbs are nearly round in their form and black in their tint.   Excepting upon the face there seem to be no stripes like those of the tiger, but upon each side of the face there is a bold black streak which runs from the eye to the corner of the mouth.   The hair about the throat, chest, and flanks is rather long, and gives a very determinate look to the animal.

The Chetah is known as an inhabitant of many parts of Asia, including India, Sumatra, and Persia, while in Africa it is found in Senegal and at the Cape of Good Hope.

HYÆNAS.

The group of animals which are so well known by the title of HYÆNAS, are, although most repulsive to the view, and most disgusting in their habits, the very saviours of life and health in the countries where they live, and where there is necessity for their existence.   In this land, and at the present day, there is no need of such large animals as the Hyænas to perform their necessary and useful task of clearing the earth from the decaying carcases which cumber its surface and poison its air, for in our utilitarian age even the very hairs from a cow's hide are turned to account, and the driest bones are made to subserve many uses.

In those countries, as well as in our own, there are carnivorous and flesh-burying insects, which consume the smaller animal substances ; but the rough work is left to those industrious scavengers the Hyænas, which content themselves with the remains of large animals.

In the semi-civilized countries of Africa and Asia, the Hyæna is a public benefactor, swallowing with his accommodating appetite almost every species of animal substance that can be found, and even crushing to splinters between his iron jaws the bones which would resist the attacks of all other carnivorous animals.

Useful as is the Hyæna when it remains within its proper boundaries, and restricts itself to its proper food, it becomes a terrible pest when too numerous to find sufficient nourishment in dead carrion.   Incited by hunger, it hangs on the skirts of villages and encampments, and loses few opportunities of making a meal at the expense of the inhabitants.   It does not openly oppose even a domestic ox, but endeavours to startle its intended prey, and cause it to take to flight before it will venture upon an attack.   In order to alarm the cattle it has a curious habit of creeping as closely as possible to them, and then springing up suddenly just under their eyes.   Should the startled animals turn to flee, the Hyæna will attack and destroy them ; but, if they should turn to bay, will stand still and venture no farther.   It will not even attack a knee-haltered horse.   So it often happens that the Hyæna destroys the healthy cattle which can run away, and is afraid to touch the sickly and maimed beasts which cannot flee, and are forced to stand at bay.

The STRIPED HYÆNA is easily to be distinguished from its relations by the peculiar streaks from which it derives its name.   The general colour of the fur is a greyish brown, diversified with blackish stripes, which run along the ribs and upon the limbs.   A large singular black patch extends over the front of the throat, and single black hairs are profusely scattered among the fur. When young, the stripes are more apparent than in the adult age, and the little animal has something of a tigrine aspect about its face.

In proportion to its size, the Hyæna possesses teeth and jaws of extraordinary strength, and between their tremendous fangs the thigh-bones of an ox fly in splinters with a savage crash that makes the spectator shudder.

The muzzle is but short, and the rough thorn-studded tongue is used, like that of the feline groups, for rasping every vestige of flesh from the bones of the prey.

The SPOTTED HYÆNA, or TIGER WOLF, as it is generally called, is, for a Hyæna, a fierce and dangerous animal, invading the sheepfolds and cattle-pens under the cover of darkness, and doing in one night more mischief than can be remedied in the course of years.

STRIPED OR CRESTED HYÆNA.
(*Hyæna striata.*)

The spots, or rather the blotches with which its fur is marked, are rather scanty upon the back and sides, but upon the legs are much more clearly marked, and are set closer together. The paws are nearly black.

The Tiger Wolf is celebrated for the strange unearthly sounds which it utters when under the influence of strong excitement. The animal is often called the "Laughing Hyæna" on account of the maniacal, mirthless, hysterical laugh which it pours forth, accompanying these horrid sounds with the most absurd gestures of body and limbs. During the time that the creature is engaged in uttering these wild fearful peals of laughter, it dances about in a state of ludicrously frantic excitement, running backwards and forwards, rising on its hind legs, and rapidly gyrating on those members, nodding its head repeatedly to the ground ; and, in fine, performing the most singular antics with wonderful rapidity.

CIVETS.

The CIVET, sometimes, but wrongly, called the Civet Cat, is a native of Northern Africa, and is found plentifully in Abyssinia, where it is eagerly sought on account of the peculiarly scented substance which is secreted in certain glandular pouches. This Civet perfume was formerly considered as a most valuable medicine, and could only be obtained at a very high price ; but in the present day it has nearly gone out of fashion as a drug, and holds its place in commerce more as a simple perfume than as a costly panacea.

The substance which is so prized on account of its odoriferous qualities is secreted in a double pouch,

CIVET.—(*Viverra Civetta.*)

which exists under the abdomen, close to the insertion of the tail. As this curious production is of some value in commerce, the animal which furnishes the precious secretion is too valuable to be killed for the sake of its scent-

pouch, and is kept in a state of captivity, so as to afford a continual supply
of the odoriferous material.

The claws of the Civet are only partially retractile. The eyes are of a dull
brown, very protuberant, and with a curiously changeable pupil, which by
day exhibits a rather broad linear pupil, and glows at night with a brilliant
emerald refulgence. The body is curiously shaped, being considerably flattened on the sides, as if the animal had been pressed between two boards.

Altogether, the Civet is a very handsome animal, the bold dashing of black
and white upon its fur having a very rich effect. The face has a curious
appearance, owing to the white fur which fringes the lips, and the long pure
white whisker hairs of the lips and eyes. When young it is almost wholly
black, with the exception of the white whisker hairs and the white fur of the
lips.

GENETTS.

A small, but rather important, group of the Viverrine animals, is that the
members of which are known by the name of the GENETTS. These crea-

BLOTCHED GENETT.—(*Genetta Tigrina.*)

tures are all nocturnal in their habits, as are the civets, and, like those animals, can live on a mixture of animal and vegetable food, or even on vegetable
food alone. The Genetts possess the musk-secreting apparatus which much
resembles the pouch of the civet, although in size it is not so large, nor does
it secrete so powerfully smelling a substance as that of the civets.

The best known of these animals is the COMMON, or BLOTCHED GENETT,
an inhabitant of Southern Africa and of various other parts of the world,
being found even in the south of France. It is a very beautiful and graceful
animal, and never fails to attract attention from an observer. The general
colour of the fur is grey, with a slight admixture of yellow. Upon this
groundwork dark patches are lavishly scattered, and the full, furry tail is
covered with alternate bands of black and white. The muzzle would be en-

tirely black but for a bold patch of white fur on the upper lip, and a less decidedly white mark by the nose. The feet are supplied with retractile claws, so that the animal can deal a severe blow with its outstretched talons, or climb trees with the same ease and rapidity which is found in the cat tribe.

VERY different from the Genetts in its appearance is the CACOMIXLE, although it is closely allied to them.

It is remarkable as being a Mexican representative of the Genett group of animals, although it can hardly be considered as a true Genett or a true Moongus. The colour of this animal is a light uniform dun, a dark bar being placed like a collar over the back of the neck. In some specimens this bar is double, and in all it is so narrow that when the animal throws its head backwards the dark line is lost in the lighter fur. Along the back runs

CACOMIXLE.—(*Bassaris Astuta.*)

a broad, singular, darkish stripe. The tail is ringed something like that of the Ringed Lemur, and is very full. The term Cacomixle is a Mexican word, and the animal is sometimes called by a still stranger name, "Tepemaxthalon." The scientific title "Bassaris" is from the Greek, and signifies a fox.

ICHNEUMONS.

The ICHNEUMONS appear to be the very reptiles of the mammalian animals, in form, habits, and action, irresistibly reminding the spectator of the serpent. The sharp and pointed snout, narrow body, short legs, and flexible form, permit them to insinuate themselves into marvellously small crevices and to seek and destroy their prey in localities where it might well deem itself secure.

The common Ichneumon, or Pharaoh's Rat, as it is popularly but most improperly termed, is plentifully found in Egypt, where it plays a most useful part in keeping down the numbers of the destructive quadrupeds and dangerous reptiles. Small and insignificant as this animal appears, it is a most dangerous foe to the huge crocodile, feeding largely upon its eggs. Snakes, rats, lizards, mice, and various birds fall a prey to this Ichneumon, which will painfully track its prey to its hiding-place, and wait patiently for hours until it makes its appearance, or will quietly creep up to the unsuspecting animal, and flinging itself boldly upon it destroy it by rapid bites with its long sharp teeth.

Taking advantage of these admirable qualities, the ancient Egyptians were wont to tame the Ichneumon, and admit it to the free range of their houses.

The colour of this animal is a brown, plentifully grizzled with grey, each hair being ringed alternately with grey and brown. The total length of the animal is about three feet three inches, the tail measuring about eighteen

ICHNEUMON.—(*Herpestes Ichneumon.*)

inches. The scent-gland of the Ichneumon is very large in proportion to the size of its bearer, but the substance which it secretes has not as yet been held of any commercial value. The claws are partially retractile.

The word Ichneumon is Greek, and literally signifies " a tracker."

The MOONGUS, sometimes called the INDIAN ICHNEUMON, is, in its Asiatic home, as useful an animal as the Egyptian Ichneumon in Africa. In that country it is an indefatigable destroyer of rats, mice, and the various reptiles, and is on that account highly valued and protected. Being, as are Ichneumons in general, extremely cleanly in manners, and very susceptible of domestication, it is kept tame in many families, and does good service in keeping the houses clear of the various animated pests that render an Indian town a disagreeable and sometimes a dangerous residence.

In its customs it very much resembles the cat, and is gifted with all the inquisitive nature of that animal. When first introduced into a new locality it runs about the place, insinuating itself into every hole and corner, and sniffing curiously at every object with which it comes in contact. Even in its wild state it exhibits the same qualities, and by a careful observer may be seen questing about in search of its food, exploring every little tuft of vegetation that comes in its way, running over every rocky projection, and thrusting its sharp snout into every hollow. Sometimes it buries itself entirely in some little hole, and when it returns to light drags with it a mole, a rat, or some such creature, which had vainly sought security in its narrow domicile.

While eating, the Ichneumon is very tetchy in its temper, and will very seldom endure an interruption of any kind. In order to secure perfect quiet while taking its meals, it generally carries the food into the most secluded hiding-place that it can find, and then commences its meal in solitude and darkness. The colour of the Moongus is a grey, liberally flecked with darker hairs, so as to produce a very pleasing mixture of tints. It is not so large an animal as its Egyptian relative.

MOONGUS.—(*Herpestes Griseus*).

THE last of the great Viverine group of animals is the CRYPTOPROCTA, a creature whose rabbit-like mildness of aspect entirely belies its nature.

It is a native of Madagascar, and has been brought from the southern portions of that wonderful island. It is much to be wished that the zoology of so prolific a country should be thoroughly explored, and that competent naturalists should devote much time and severe labour to the collection of specimens, and the careful investigation of animals while in their wild state.

Gentle and quiet as the animal appears, it is one of the fiercest little creatures known. Its limbs, though small, are very powerful, their muscles being extremely full and well knit together. Its appetite for blood seems to be as insatiable as that of the tiger, and its activity is very great, so that it may well be imagined to be a terrible foe to any animals on whom it may choose to make an attack. For this savage nature it has received the name of " Ferox," or fierce. Its generic name of Cryptoprocta is given to it on account of the manner in which the hinder quarters suddenly taper down and merge themselves in the tail. The word itself is from the **Greek**, the former half of it signifying "hidden," and the latter half, "hind-quarters."

The colour of the Cryptoprocta is a light brown, tinged with red. The

CRYPTOPROCTA.—(*Cryptoprocta ferox*).

ears are very large and rounded, and the feet are furnished with strong claws. The toes are five in number on each foot.

DOGS.

The large and important group of animals which is known by the general name of the Dog Tribe embraces the wild and domesticated Dogs, the Wolves, Foxes, Jackals, and that curious South African animal, the Hunting Dog. Of these creatures, several have been brought under the authority of man, and by continual intermixtures have assumed that exceeding variety of form which is found in the different "breeds" of the domestic Dog.

The original parent of the Dog is very doubtful, some authors considering that it owes its parentage to the Dhole, or the Buansuah of India; others thinking it to be an offspring of the Wolf; and others attributing to the Fox the honour of being the progenitor of our canine friend and ally.

All the various Dogs which have been brought under the subjection of man are evidently members of one single species, *Canis familiaris*, being capable of variation to an almost unlimited extent.

It is hardly possible to conceive an animal which is more entirely formed for speed and endurance than a well-bred GREYHOUND.

The chief use—if use it can be termed—of the Greyhound is in coursing the hare, and exhibiting in this chase its marvellous swiftness and its endurance of fatigue.

The narrow head and sharp nose of the Greyhound, useful as they are for aiding the progress of the animal by removing every impediment to its passage through the atmosphere, yet deprive it of a most valuable faculty, that of chasing by scent. The muzzle is so narrow in proportion to its length, that the nasal nerves have no room for proper development, and hence the animal is very deficient in its powers of scent. The same circumstance may be noted in many other animals.

The large and handsome animal which is called from its native country the NEWFOUNDLAND DOG belongs to the group of spaniels, all of which appear to be possessed of considerable mental powers, and to be capable of instruction to a degree that is rarely seen in animals.

GREYHOUND.—(*Canis familiaris.*)

As is the case with most of the large Dogs, the Newfoundland permits the lesser Dogs to take all kinds of liberties without showing the least resentment ; and if it is worried or pestered by some forward puppy, looks down with calm contempt, and passes on its way.   Sometimes the little conceited · animal presumes upon the dignified composure of the Newfoundland Dog,

NEWFOUNDLAND DOG.—(*Canis familiaris.*)

and, in that case, is sure to receive some quaint punishment for its insolence. The story of the big Dog that dropped the little Dog into the water and then rescued it from drowning, is so well known that it needs but a passing refe-rence.   But I know of a Dog, belonging to one of my friends, which behaved in a very similar manner. Being provoked beyond all endurance by the continued annoyance, it took the little tormentor in its mouth, swam well out to sea, dropped it in the water, and swam back again.

Another of the animals, belonging to a workman, was attacked by a small and pugnacious bull-dog, which sprung upon the unoffending canine giant, and, after the manner of bull-dogs, "pinned" him by the nose, and there hung, in spite of all endeavours to shake it off. However, the big Dog happened to be a clever one, and spying a pailful of boiling tar, he bolted towards it, and deliberately lowered his foe into the pail. The bull-dog had never calculated on such a reception, and made its escape as fast as it could run, bearing with it a scalding memento of the occasion.

POMERANIAN DOG.—(*Canis familiaris.*)

Of late years, a Dog has come into fashion as a house-dog, or as a companion. This is the POMERANIAN FOX DOG, commonly known as the "Loup-loup."

It is a great favourite with those who like a dog for a companion, and not for mere use, as it is very intelligent in its character, and very handsome in aspect. Its long white fur, and bushy tail, give it quite a distinguished appearance, of which the animal seems to be thoroughly aware. Sometimes the coat of this animal is a cream colour, and very rarely is deep black. The pure white, however, seems, to be the favourite. It is a lively little creature, and makes an excellent companion in a country walk.

Of the Spaniel Dogs there are several varieties, which may be classed under two general heads, namely, Sporting and Toy Spaniels; the former being used by the sportsman in finding game for him ; and the latter being simply employed as companions.

The FIELD SPANIEL is remarkable for the intense love which it bears for hunting game, and the energetic manner in which it carries out the wishes of its master. There are two breeds of Field Spaniels, the one termed the " Springer," being used for heavy work among thick and thorny coverts, and the other being principally employed in woodcock shooting, and called in consequence the " Cocker." The Blenheim and King Charles Spaniels derive their origin from the Cocker.

WATER SPANIEL.—(*Canis familiaris.*)

While hunting, the Spaniel sweeps its feathery tail rapidly from side to side, and is a very pretty object to any one who has an eye for beauty of movement. It is a rule, that however spirited a Spaniel may be, it must not raise its tail above the level of its back.

MALTESE DOG.—(*Canis familiaris.*)

A very celebrated but extremely rare " toy " Dog is the MALTESE DOG, the prettiest and most lovable of all the little pet Dogs.

The hair of this tiny creature is very long, extremely silky, and almost unique in its glossy sheen, so beautifully fine as to resemble spun glass. In proportion to the size of the animal, the fur is so long that when it is in rapid movement, the real shape is altogether lost in the streaming mass of flossy hair. One of these animals, which barely exceeded three pounds in weight, measured no less than fifteen inches in length of hair across the shoulders. The tail of the Maltese Dog curls strongly over the back, and adds its wreath of silken fur to the already superfluous torrent of glistening tresses.

As the name implies, it was originally brought from Malta. It is a very scarce animal, and at one time was thought to be extinct; but there are still specimens to be obtained by those who have no objection to pay the price which is demanded for these pretty little creatures.

Of all the domesticated Dogs, the POODLE seems to be, take him all in all, the most obedient and the most intellectual. Accomplishments the most difficult are mastered by this clever animal, which displays an ease and intelligence in its performances that appear to be far beyond the ordinary canine capabilities.

A barbarous custom is prevalent of removing the greater portion of the

THE POODLE.—(*Canis familiaris.*)

Poodle's coat, leaving him but a ruff round the neck and legs, and a puff on the tip of the tail, as the sole relic of his abundant fur.

Such a deprivation is directly in opposition to the natural state of the Dog, which is furnished with a peculiarly luxuriant fur, hanging in long ringlets from every portion of the head, body, and limbs.   The Poodle is not the only Dog that suffers a like tonsorial abridgment of coat ; for under the dry arches of the many bridges that cross the Seine, in Paris, may be daily seen a mournful spectacle.   Numerous dogs of every imaginable and unimaginable breed lie helpless in the shade of the arch, their legs tied together, and their eyes contemplating with woeful looks the struggles of their fellows, who are being shorn of their natural covering, and protesting with mournful cries against the operation.

The very tiniest of the dog family is the MEXICAN LAPDOG, a creature so very minute in its dimensions as to appear almost fabulous to those who have not seen the animal itself.

One of these little canine pets is to be seen in the British Museum, and

MEXICAN LAPDOG.—(*Canis familiaris.*)

always attracts much attention from the visitors. Indeed, if it were not in so dignified a locality, it would be generally classed with the mermaid, the flying serpent and the Tartar lamb, as an admirable example of clever workmanship. It is precisely like those white woollen toy Dogs which sit upon a pair of bellows, and when pressed give forth a nondescript sound, intended to do duty for the legitimate canine bark. To say that it is no larger than these toys would be hardly true, for I have seen in the shop windows many a toy Dog which exceeded in size the veritable Mexican Lapdog.

The magnificent animal which is termed the BLOODHOUND, on account of its peculiar facility for tracking a wounded animal through all the mazes of its devious course, is very scarce in England, as there is now but little need of these Dogs.

BLOODHOUND.—(*Canis familiaris.*)

In the "good old times," this animal was largely used by thief-takers, for the purpose of tracking and securing the robbers who in those days made the country unsafe and laid the roads under a black mail. Sheep-stealers, who were much more common when the offence was visited with capital punishment, were frequently detected by the delicate nose of the Bloodhound, which would, when once laid on the scent, follow it up with unerring precision, unravelling the single trail from among a hundred crossing footsteps, and only to be baffled by water or blood.

FOXHOUND.—(*Canis familiaris.*)

The Bloodhound is generally irascible in temper, and therefore a rather dangerous animal to be meddled with by any one excepting its owner. So fierce is its desire for blood, and so utterly is it excited when it reaches its prey, that it will often keep its master at bay when he approaches, and receive his overtures with such unmistakable indications of anger that he will not venture to approach until his Dog has satisfied its appetite on the carcase of the animal which it has brought to the ground. When fairly on the track of the deer, the Bloodhound utters a peculiar, long, loud, and deep bay, which, if once heard, will never be forgotten.

The colour of a good Bloodhound ought to be nearly uniform, no white being permitted, except on the tip of the tail. The prevailing tints are a blackish tan, or a deep fawn. The tail of this Dog is long and sweeping.

Of all the Dogs which are known by the common title of " hound," the FOX-HOUND is the best known. It is supposed that the modern Foxhound derives its origin from the old English hound, and its various points of perfection from judicious crosses with other breeds. For example, in order to increase its speed the greyhound is made to take part in its pedigree, and the greyhound having already some admixture of the bull-dog blood, there is an infusion of stubbornness as well as of mere speed.

POINTER.—(*Canis familiaris.*)

According to the latest authorities, the best average height for Foxhounds is from twenty-one to twenty-five inches, the female being generally smaller than the male. However the size of the Dog does not matter so much ; but it is expected to match the rest of the pack in height as well as in general appearance.

There are two breeds of the POINTER, namely, the modern English Pointer, and the Spanish Pointer. The latter of these Dogs is now seldom used in the field, as it is too slow and heavily built an animal for the present fast style of sporting.

The modern English Pointer is a very different animal, built on a much lighter model, and altogether with a more bold and dashing air about it. While it possesses a sufficiently wide muzzle to permit the development of the olfactory nerves, its limbs are so light and wiry that it can match almost any dog in speed. Indeed, some of these animals are known nearly to equal a greyhound in point of swiftness.

This quality is specially useful, because it permits the sportsman to walk forward at a moderate pace, while his Dogs are beating over the field to his right and left. The sagacious animals are so obedient to the voice and gesture of their master, and are so well trained to act with each other, that at a wave of the hand they will separate, one going to the right and the other to the left, and so traverse the entire field in a series of " tacks," to speak nautically, crossing each other regularly in front of the sportsman as he walks forward.

When either of them scents a bird, he stops suddenly, arresting even his foot as it is raised in the air, his head thrust forward, his body and limbs fixed, and his tail stretched out straight behind him. This attitude is termed a "point," and on account of this peculiar mode of indicating game, the animal is termed the " Pointer." The Dogs are so trained that when one of them comes to a point he is backed by his companion, so as to avoid the disturbance of more game than is necessary for the purpose of the sportsman.

The most useful variety of the canine species is the sagacious creature on whose talent and energy depends the chief safety of the flock.

As the SHEEP-DOG is constantly exposed to the weather, it needs the protection of very thick and closely-set fur, which in this Dog is rather woolly in its character, and is especially heavy about the neck and breast.

The muzzle of this dog is sharp, its head is of moderate size, its eyes are very bright and intelligent, as might be expected in an animal of so much sagacity and ready resource in time of need. Its feet are strongly made, and sufficiently well protected to endure severe work among the harsh stems of the heather on the hills, or the sharply-cutting stones of the high road. Probably on account

SHEPHERD'S DOG.—(*Canis familiaris.*)

of its constant exercise in the open air, and the hardy manner in which it is brought up, the Sheep-dog is perhaps the most untiring of our domesticated animals.

As a general rule, the Sheep-dog cares very little for any one but his master, and so far from courting the notice or caresses of a stranger, will coldly withdraw from them, and keep his distance. Even with other Dogs he rarely makes companionship, contenting himself with the society of his master alone.

F

The BULL-DOG is said, by all those who have had an opportunity of judging its capabilities, to be, with the exception of the game-cock, the most courageous animal in the world.

Its extraordinary courage is so well known as to have passed into a proverb, and to have so excited the admiration of the British nation that we have been pleased to symbolize our peculiar tenacity of purpose under the emblem of this small but most determined animal. In height the Bull-dog is but insignificant, but in strength and courage there is no Dog that can match him. Indeed, there is hardly any breed of sporting dog which does not owe its high courage to an infusion of the Bull-dog blood; and it is chiefly for this purpose that the pure breed is continued.

It is generally assumed that the Bull-dog must be a very dull and brutish animal, because almost every specimen which has come before the notice of the public has held such a character.

BULL-DOG.—(*Canis familiaris.*)

My own experience does not at all coincide with this notion. I once possessed one of these animals, and a better dog I never had. He was gentle almost to a fault, never taking offence except at an insult by a big dog. He was docile, obedient, and wonderfully intelligent, a good retriever, and one of the most accomplished water-dogs I ever saw. Active and broad-chested as a greyhound, his leaping powers were astonishing; and his brown eyes had a look in them that was almost human.

The shape of this remarkable animal is worthy of notice. The fore-quarters are particularly strong, massive, and muscular; the chest wide and roomy; and the neck singularly powerful. The hind-quarters, on the contrary, are very thin and comparatively feeble; all the vigour of the animal seeming to settle in its fore-legs, chest, and head. Indeed, it gives the spectator an impression as if it were composed of two different Dogs ; the one a large and powerful animal, and the other a weak and puny quadruped, which had been put together by mistake.

The MASTIFF, which is the largest and most powerful of the indigenous English Dogs, is of a singularly mild and placid temper, seeming to delight in employing its great powers in affording protection to the weak, whether they be men or dogs.

Yet, with all this no-
bility of its gentle na-
ture, it is a most deter-
mined and courageous
animal in fight, and,
when defending its
master or his property,
becomes a foe which
few opponents would
like to face. These
qualifications of ming-
led courage and gentle-
ness adapt it especially
for the service of watch-
dog, a task in which
the animal is as likely
to fail by overweening
zeal as by neglect of
its duty. It sometimes
happens that a watch-
dog is too hasty in its
judgment, and attacks
a harmless stranger, on

MASTIFF. —(*Canis familiaris.*)

the supposition that it is resisting the approach of an enemy.

The head of the Mastiff bears a certain similitude to that of the blood-hound and the bull-dog, possessing the pendent lips and squared muzzle of the bloodhound, with the heavy muscular development of the bull-dog. The under-jaw sometimes protrudes a little, but the teeth are not left uncovered by the upper-lip, as is often the case with the latter animal. The fur of the Mastiff is always smooth, and its colour varies between a uniform reddish fawn and different brindlings and patches of dark and white. The voice is peculiarly deep and mellow. The height of this animal is generally from twenty-five to twenty-eight inches, but sometimes exceeds these dimensions. One of these Dogs was no less than thirty-three inches in height at the shoulder, measured fifty inches round his body, and weighed a hundred and seventy-five pounds.

The TERRIER, with all its numerous variations of crossed and mongrel breeds, is more generally known in England than any other kind of Dog. Of the recognized breeds, four are generally acknowledged, namely, the English and Scotch Terriers, the Skye, and the little Toy Terrier.

The ENGLISH TERRIER possesses a smooth coat, a tapering muzzle, a high forehead, a bright intelligent eye, and a strong muscular jaw. As its instinct leads it to dig in the ground, its shoulders and fore-legs are well developed, and it is able to make quite a deep burrow in a marvellously short time, throwing out the loose earth with its feet, and dragging away the stones and other large substances in its mouth. It is not a large Dog, seldom weigh-ing more than ten pounds, and often hardly exceeding the moiety of that weight.

F 2

The colour of the pure English Terrier is generally black and tan, the richness of the two tints determining much of the animal's value. The nose and the palate of the Dog ought to be always black, and over each eye a small patch of tan colour. The tail ought to be rather long and very fine, and the legs as light as is consistent with strength.

ENGLISH TERRIER.—(*Canis familiaris.*)   SCOTCH TERRIER.—(*Canis familiaris*)

The quaint-looking SKYE TERRIER has of late years been much affected by all classes of dog-owners, and for many reasons deserves the popularity which it has obtained.

When of pure breed the legs are very short, and the body extremely long in proportion to the length of limb ; the neck is powerfully made, but of considerable length, and the head is also rather elongated, so that the total length of the animal is three times as great as its height. The "dew-claws" are wanting in this variety of domestic Dog. The hair is long and straight, falling heavily over the body and limbs, and hanging so thickly upon the face that the eyes and nose are hardly perceptible under their luxuriant covering. The quality of the hair is rather harsh and wiry in the pure-bred Skye Terrier.

The size of this animal is rather small, but it ought not to imitate the minute proportions of many "toy" Dogs. Its weight ought to range from ten to seventeen or eighteen pounds. Even amongst these animals there are at least two distinct breeds, while some dog-fanciers establish a third.

It is an amusing and clever Dog, and admirably adapted for the companion-ship of mankind, being faithful and affectionate in disposition, and as brave as any of its congeners, except that epitome of courage, the bull-dog. Some-times, though not frequently, it is employed for sporting purposes, and is said to pursue that vocation with great credit.

THERE are several species of the Jackal, one of which will be noticed and figured in this work.

The common JACKAL, or KHOLAH, as it is termed by the natives, is an inhabitant of India, Ceylon, and neighbouring countries, where it is found in very great numbers, forcing itself upon the notice of the traveller not only by its bodily presence, but by its noisy howling, wherewith it vexes the ears of the wearied and sleepy wayfarer, as he endeavours in vain to find repose. Nocturnal in their habits, the Jackals are accustomed to conceal themselves as much as possible during the daytime, and to issue out on their hunting expeditions together with the advent of night.

Always ready to take advantage of every favourable opportunity, the Jackal is a sad parasite, and hangs on the skirts of the larger carnivora as they roam the country for prey, in the hope of securing some share of the creatures which they destroy or wound. On account of this companionship between the large and small marauders, the Jackal has popularly gained the name of the Lion's Provider. But, in due justice, the title ought to be reversed, for the lion is in truth the Jackal's provider, and is often thereby

JACKAL.—(*Canis aureus.*)

deprived of the chance of making a second meal on an animal which he has slain. Sometimes, it is said, the Jackal does provide the lion with a meal by becoming a victim to the hungry animal in default of better and more savoury prey.

The name of "aureus," or golden, is derived from the yellowish tinge of the Jackal's fur. In size it rather exceeds a large fox, but its tail is not proportionately so long or so bushy as the well-known "brush" of the fox.

WOLVES.

FEW animals have earned so widely popular or so little enviable a fame as the WOLVES. Whether in the annals of history, in fiction, in poetry, or even in the less honoured but hardly less important literature of nursery fables, the Wolf holds a prominent position among animals.

There are several species of Wolf, each of which species is divided into three or four varieties, which seem to be tolerably permanent, and by many observers are thought to be sufficiently marked to be considered as separate species. However, as even the members of the same litter partake of several

minor varieties in form and colour, it is very possible that the so-called
species may be nothing more than very distinctly marked varieties. These
voracious and dangerous animals are found in almost every quarter of the
globe; whether the country which they infest be heated by the beams of the
tropical sun or frozen by the lengthened winter of the northern regions.
Mountain and plain, forest and field, jungle and prairie, are equally infested
with Wolves, which possess the power of finding nourishment for their united
bands in localities where even a single predacious animal might be perplexed
to gain a livelihood.

The colour of the common WOLF is grey, mingled with a slight tinting of
fawn, and diversified with many black hairs that are interspersed among the
lighter coloured fur. In the older animals the grey appears to predominate
over the fawn, while the fur of the younger Wolves is of a warmer fawn tint.

The under parts of the
animal, the lower jaw, and,
the edge of the upper lip
are nearly white, while the
interior facing of the limbs
is of a grey tint. Between
the ears the head is almost
entirely grey, and without
the mixture of black hairs,
which is found in greatest
profusion along the line of
the spine.

When hungry—and the
Wolf is almost always
hungry—it is a bold and
dangerous animal, daring
almost all things to reach
its prey, and venturing to
attack large and powerful
animals,—such as the buf-

WOLF.—(*Canis lupus.*)

falo, the elk, or the wild horse. Sometimes it has been known to oppose
itself to other carnivora, and to attack so unpromising a foe as the bear.

It is by no means nice in its palate, and will eat almost any living animal
—from human beings down to frogs, lizards, and insects. Moreover, it is a
sad cannibal, and is thought by several travellers who have noted its habits
to be especially partial to the flesh of its own kind. A weak, sickly, or
wounded Wolf is sure to fall under the cruel teeth of its companions; who
are said to be so fearfully ravenous that if one of their companions should
chance to besmear himself with the blood of the prey which has just been
hunted down, he is instantly attacked and devoured by the remainder of
the pack.

In their hunting expeditions the Wolves usually unite in bands, larger or
smaller in number, according to circumstances, and acting simultaneously
for a settled purpose. If they are on the trail of a flying animal, the footsteps of their prey are followed up by one or two of the Wolves, while the
remainder of the band take up their positions to the right and left of the
leaders, so as to intercept the quarry if it should attempt to turn from its
course. Woe be to any animal that is unlucky enough to be chased by a pack
of Wolves. No matter how swift it may be, it will most surely be overtaken
at last by the long slouching, tireless gallop of the Wolves; and no matter
what may be its strength, it must at last fail under the repeated and constant
attacks of the sharp teeth.

THE FOX AND HER YOUNG. Page 71.
The Popular Natural History

ACCORDING to some systematic naturalists the FOXES are placed in the genus *Canis*, together with the dogs and the wolves. Those eminent zoologists, however, who have arranged the magnificent collections in the British Museum, have decided upon separating the Foxes from the dogs and wolves, and placing them in the genus *Vulpes*. To this decision they have come for several reasons, among which may be noted the shape of the pupil of the eye, which in the Foxes is elongated, but in the animals which compose the genus *Canis* is circular. The ears of the Foxes are triangular in shape, and pointed, and the tail is always exceedingly bushy.

A very powerful scent is poured forth from the Fox in consequence of some glands which are placed near the root of the tail, and furnish the odorous secretion. Glands of a similar nature, but not so well developed, are found in the wolves.

It is by this scent that the hounds are able to follow the footsteps of a flying Fox, and to run it down by their superior speed and endurance. The Fox, indeed, seems to be aware that its pursuers are guided in their chase by this odour, and puts in practice every expedient that its fertile brain can produce in order to break the continuity of the scent, or to overpower it by the presence of other odours, which are more powerful though not more agreeable.

Even when tamed it preserves its singular cunning. A tame Fox, that was kept in a stable-yard, had managed to strike up a friendship with several of the dogs, and would play with them, but could never induce the cats to approach him. Cats are very sensitive in their nostrils, and could not endure the odour. They would not even walk upon any spot where the

FOX —(*Vulpes vulgaris.*)

Fox had been standing, and kept as far aloof as possible from him.

The crafty animal soon perceived that the cats would not come near him, and made use of his knowledge to cheat them of their breakfast. As soon as the servant poured out the cats' allowance of milk, the Fox would run to the spot and walk about the saucer, well knowing that none of the rightful owners would approach the defiled locality. Day after day the cats lost their milk until the stratagem was discovered, and the milk was placed in a spot where it could not be reached by the Fox.

The Fox resides in burrows, which it scoops out of the earth by the aid of its strong digging paws, taking advantage of every peculiarity of the ground, and contriving, whenever it is possible, to wind its subterranean way among the roots of large trees or between heavy stones. In these "earths," as the as the burrows are called in the sportsman's phraseology, the female Fox produces and nurtures her young, which are odd little snub-nosed creatures, resembling almost any animal rather than a Fox. She watches over her offspring with great care, and teaches them by degrees to subsist on animal food, which she and her mate capture for that purpose.

The colour of the common Fox is a reddish fawn, intermixed with black and white hairs. The hair is long and thick, being doubly thick during the colder months of the year, so that the fur of a Fox which is killed in the winter is more valuable than if the animal had been slain in the hot months. The tail, which is technically termed the "brush," is remarkably bushy, and partakes of the tints which predominate over the body, except at the tip, which is white. The height of this animal is about a foot, and its length about two feet and a half, exclusive of the tail.

One of the most celebrated species of the Foxes is the ARCTIC FOX, called by the Russians PESZI, and by the Greenlanders TERRIENNIAK. This animal is in very great repute in the mercantile world on account of its beautiful silky fur, which in the cold winter months becomes perfectly white. During the summer the fur is generally of a grey or dirty brown, but is frequently found of a leaden grey, or of a brown tint with a wash of blue. Towards the change of the seasons the fur becomes mottled ; and by reason

ARCTIC FOX. —(*Vulpes lagopus.*)

of this extreme variableness has caused the animal to be known by several different titles. Sometimes it is called the White Fox, sometimes the Blue Fox, sometimes the Sooty Fox, sometimes the Pied Fox, and sometimes the Stone Fox.

This animal is found in Lapland, Iceland, Siberia, Kamtschatka, and North America, in all of which places it is eagerly sought by the hunters for the sake of its fur. The pure white coat of the winter season is the most valuable, and the bluish grey fur of the summer months is next to the white the colour that is most in request.

In size, the Arctic Fox is not the equal of the English species, weighing only eight pounds on an average, and its total length being about three feet. The eye is of a hazel tint, and very bright and intelligent. It lives in burrows, which it excavates in the earth during the summer months, and prefers to construct its simple dwellings in small groups of twenty or thirty.

The FENNEC, or ZERDA, is an inhabitant of Africa, being found in Nubia and Egypt. It is a very pretty and lively little creature, running about with much activity, and anon sitting upright and regarding the prospect with marvellous gravity. The colour of the Fennec is a very pale fawn, or "isabel" colour, sometimes being almost of a creamy whiteness. The tail is bushy, and partakes of the general colour of the fur, except at the upper part of the base and the extreme tip, which are boldly marked with black. The size of the adult animal is very inconsiderable, as it measures scarcely more than a foot in length, exclusive of the bushy tail, which is about eight inches long.

FENNEC.—(*Vulpes Zaarensis.*)

It is said that the Fennec, although it is evidently a carnivorous animal, delights to feed upon various fruits, especially preferring the date. Such a predilection is according to vulpine and canine analogies, for the common English Fox is remarkably fond of ripe fruits, such as grapes or strawberries, and the domestic dog is too often a depredator of those very gardens which he was enjoined to keep clear from robbers. But that the animal should enjoy the power of procuring that food in which it so delights is a very extra-ordinary circumstance, and one which would hardly be expected from a creature which partakes so largely of the vulpine form and characteristics. The date-palm is a tree of a very lofty growth, and the rich clusters of the fruit are placed at the very summit of the bare, branchless stem. Yet the Fennec is said to possess the capability of climbing the trunk of the date-palm, and of procuring for itself the coveted luxury.

Like the veritable Foxes, the Fennec is accustomed to dwell in subterranean abodes, which it scoops in the light sandy soil of its native land.

As is the case with the greater number of predacious animals, the Fennec is but seldom seen during the daytime, preferring to issue forth upon its marauding expeditions under the friendly cover of night. Even when it has spent some time in captivity, it retains its restless nocturnal demeanour, and during the hours of daylight passes the greater portion of its time in semi-somnolence or in actual sleep.

THE little animal which is known by the name of the ASSE, or the CAAMA, is an inhabitant of Southern Africa, and is in great request for the sake of its skin, which furnishes a very valuable fur.

It is a terrible enemy to ostriches and other birds which lay their eggs in the ground, and is in consequence detested by the birds whose nests are devastated. The ingenuity of the Caama in procuring the contents of an

ASSE, OR CAAMA.—(*Vulpes caama.*)

ostrich's egg is rather remarkable. The shell of the egg is extremely thick and strong; and as the Caama is but a small animal, its teeth are unable to make any impression on so large, smooth, hard, and rounded an object. In order, therefore, to obviate this difficulty, the cunning animal rolls the egg along by means of its fore-paws, and pushes it so violently against any hard substance that may lie conveniently in its path, or against another egg, that the shell is broken and the contents attainable.

The fur of this animal is highly esteemed by the natives for the purpose of making "karosses," or mantles. As the Asse is one of the smallest of the Foxes, a great number of skins are needed to form a single mantle, and the manufactured article is therefore held in high value by its possessor. Indeed, so valuable is its fur, that it tempts many of the Bechuana tribes to make its chase the business of their lives, and to expend their whole energies in capturing the animal from whose body the much-prized fur is taken.

The continual persecution to which the Caama is subjected, has almost exterminated it in the immediate vicinity of Cape Town, where it was formerly seen in tolerable plenty. Gradually, however, it retreats more and more northward before the tread of civilized man, and at the present day is but very rarely seen within the limits of the colony.

WEASELS.

NEXT in order to the dogs, is placed the large and important family of the WEASELS, representatives of which are found in almost every portion of the earth. There is something marvellously serpentine in the aspect and structure of the members of this family—the Mustelidæ, as they are called, from the Latin word *Mustela*, which signifies a Weasel. Their extremely long bodies and very short legs, together with the astonishing perfection of the muscular powers, give them the capability of winding their little bodies into the smallest possible crevices, and of waging successful battle with animals of twenty times their size and strength.

First on the list of Weasels are placed the agile and lively MARTENS, or MARTEN-CATS, as they are sometimes termed. Two species of British Martens are generally admitted into our catalogues, although the distinction of the species is even as yet a mooted point.

The PINE MARTEN is so called because it is generally found in those localities where the pine-trees abound, and is in the habit of climbing the pines in search of prey. It is a shy and wary animal, withdrawing itself as far as possible from the sight of man ; and although a fierce and dangerous antagonist when brought to bay, is naturally of a timid disposition, and shuns collision with an enemy.

It is a tree-loving animal, being accustomed to traverse the trunks and

PINE MARTEN.—(*Martes Abietum.*)

branches with wonderful address and activity, and being enabled by its rapid and silent movements to steal unnoticed on many an unfortunate bird, and to seize it in its deadly gripe before the startled victim can address itself to flight. It is a sad robber of nests, rifling them of eggs and young, and not unfrequently adding the parent bird to its list of victims.

The damage which a pair of Martens and their young will inflict upon a poultry-yard is almost incredible. If they can only gain an entrance into the fowl-house, they will spare but very few of the inhabitants. They will carry off an entire brood of young chickens, eat the eggs, and destroy the parents.

The magpie's nest is a very favourite resort of the Marten, because its arched covering and small entrance afford additional security. A boy who was engaged in bird-nesting, and had climbed to the top of a lofty tree in order to plunder a magpie's nest, was made painfully sensible of an intruder's presence by a severe bite which was inflicted upon his fingers as soon as he inserted his hand into the narrow entrance. This adventure occurred in Belvoir Park, County Down, in Ireland.

The length of the Pine Marten is about eighteen inches, exclusive of the tail, which measures about ten inches. The tail is covered with long and rather bushy hair, and is slightly darker than the rest of the body, which is covered with brown hair. The tint, however, is variable in different specimens, and even in the same individual undergoes considerable modifications, according to the time of year and the part of the world in which it is found. It has rather a wide range of locality, being a native of the northern parts of Europe and of a very large portion of Northern America.

ONE of the most highly valued of the Weasels is the celebrated SABLE, which produces the richly tinted fur that is in such great request. Several species of this animal are sought for the sake of their fur. They are very closely allied to the Martens that have already been described, and are supposed by some zoologists to belong to the same species. Besides the well known *Martes Zibellina* a North American species is known, together with another which is an inhabitant of Japan. These two creatures, although they are very similar to each other in general aspect, can be distinguished from each other by the different hue of their legs and feet : the American Sable, being tinged with white upon those portions of its person, and the corresponding members of the Japanese Sable being marked with black.

SABLE.—(*Martes Zibellina.*)

The Sable is spread over a large extent of country, being found in Siberia, Kamtschatka, and in Asiatic Russia. Its fur is in the greatest perfection during the coldest months of the year, and offers an inducement to the hunter to brave the fearful inclemency of a northern winter in order to obtain a higher price for his small but valuable commodities. A really perfect Sable skin is but seldom obtained, and will command an exceedingly high price. An ordinary skin is considered to be worth from one to six or seven pounds, but, if it should be of the very best quality, is valued at twelve or fifteen pounds.

In order to obtain these much-prized skins, the Sable hunters are forced to undergo the most terrible privations, and often lose their lives in the snow-covered wastes in which the Sable loves to dwell. A sudden and heavy snowstorm will obliterate in a single half hour every trace by which the hunter had marked out his path, and, if it should be of long continuance, may overwhelm him in the mountain "drifts" which are heaped so strangely by the fierce tempests that sweep over those fearful regions.

The Sables take up their abode chiefly near the banks of rivers and in the thickest parts of the forests that cover so vast an extent of territory in those uncultivated regions. Their holes are usually made in holes which the creatures burrow in the earth, and are generally made more secure by being dug among the roots of trees. Sometimes, however, they prefer to make their nests in the hollows of trees, and there they rear their young. Some authors however, deny that the Sable inhabits subterranean burrows, and assert that its nest is always made in a hollow tree. Their nests are soft and warm, being composed chiefly of moss, dried leaves, and grass.

The Sables are taken in various modes. Sometimes they are captured in traps, which are formed in order to secure the animal without damaging its fur. Sometimes they are fairly hunted down by means of the tracks which their little feet leave in the white snow, and are traced to their domicile. A net is then placed over the orifice, and by means of a certain pungent smoke which is thrown into the cavity, the inhabitant is forced to rush into the open air, and is captured in the net. The hunters are forced to support themselves on the soft and yielding surface of the snow by wearing "snow-shoes," or they would be lost in the deep drifts, which are perfectly capable of supporting so light and active an animal as the Sable, but would engulf a human being before he had made a second step.

It now and then happens that the Sable is forced to take refuge in the branches of a tree, and in that case it is made captive by means of a noose which is dexterously flung over its head.

On examining the fur of the Sable, it will be seen to be fixed to the skin in such a manner that it will turn with equal freedom in all directions, and lies smoothly in whatever direction it may be pressed. The fur is rather long in proportion to the size of the animal, and extends down the limbs to the claws. The colour is a rich brown, slightly mottled with white about the head, and taking a grey tinge on the neck.

POLECAT.—(*Putorius fœtidus.*)

The POLECAT has earned for itself a most unenviable fame, having been long celebrated as one of the most noxious pests to which the farmyard is liable. Slightly smaller than the marten, and not quite so powerful, it is found to be a more deadly enemy to rabbits, game, and poultry, than any other animal of its size.

It is wonderfully bold when engaged upon its marauding expeditions, and maintains an impertinently audacious air even when it is intercepted in the act of destruction. Not only does it make victims of the smaller poultry, such as ducks and chickens, but attacks geese, turkeys, and other larger birds with perfect readiness. This ferocious little creature has a terrible habit of de-stroying the life of every animal that may be in the same chamber with itself, and if it should gain admission into a henhouse will kill every one of the inhabitants, although it may not be able to eat the twentieth part of its victims. It seems to be very fond of sucking the blood of the animals which it destroys, and appears to commence its repast by eating the brains. If several victims should come in its way, it will kill them all, suck their blood, and eat the brains, leaving the remainder of the body untouched.

This animal is not only famous for its bloodthirsty disposition, but for the horrid odour which exhales from its body, and which seems to be partially under the control of the owner. When the Polecat is wounded or annoyed in any way, this disgusting odour becomes almost unbearable, and has the property of adhering for a long time to any substance with which it may come in contact.

The Polecat does not restrict itself to terrestrial game, but also wages war against the inhabitants of rivers and ponds. Frogs, toads, newts, and fish are among the number of the creatures that fall victims to its rapacity. Even the formidably defended nests of the wild bees are said to yield up their honeyed stores to the fearless attack of this rapacious creature.

As to rabbits, hares, and other small animals, the Polecat seems to catch and devour them almost at will. The hares it can capture either by stealing upon them as they lie asleep in their "forms," or by patiently tracking them through their meanderings, and hunting them down fairly by scent. The rabbits flee in vain for safety into their subterranean strongholds, for the Polecat is quite at home in such localities, and can traverse a burrow with greater agility than the rabbits themselves. Even the rats that are found so plentifully about the waterside are occasionally pursued into their holes and there captured. Pheasants, partridges, and all kinds of game are a favourite prey with the Polecat, which secures them by a happy admixture of agility and craft. So very destructive are these animals, that a single family is quite sufficient to depreciate the value of a warren or a covert to no small extent.

The Polecat is a tolerably prolific animal, producing four or five young at a litter. The locality which the mother selects for the nursery of her future family is generally at the bottom of a burrow, which is scooped in light and dry soil, defended if possible by the roots of trees. In this subterranean abode a warm nest is constructed, composed of various dried leaves and of moss, laid with singular smoothness. The young Polecats make their appearance towards the end of May or the beginning of June.

The FERRET is well known as the constant companion of the rat-catcher and the rabbit-hunter, being employed for the purpose of following its prey into their deepest recesses, and of driving them from their strongholds into the open air, when the pursuit is taken up by its master. The mode in which the Ferret is employed is too well known to need a detailed description.

It is a fierce little animal, and is too apt to turn upon its owner and wound him severely before he suspects that the creature is actuated by any ill intentions. I once witnessed a rather curious example of the uncertainty of the Ferret's temper. A lad who possessed a beautiful white Ferret had partially tamed the creature, and thought that it was quite harmless. The Ferret was accustomed to crawl about his person, and would permit itself to be caressed almost as freely as a cat. But on one unfortunate morning, when its owner was vaunting the performances of his *protégée*—for it was a female—the creature made a quiet but rapid snap at his mouth, and drove its teeth through both his lips, making four cuts as sharply defined as if they had been made with a razor.

ON account of its water-loving propensities, the MINK is called by various names that bear relation to water. By some persons it is called the Smaller Otter, or sometimes the Musk Otter, while it is known to others under the title of the Water-Polecat. It also goes by the name of the NUREK VISON. The Mink is spread over a very large extent of country, being found in the most northern parts of Europe, and also in North America. Its fur is usually brown, with some white about the jaws, but seems to be subject to

FERRET.—(*Mustela Furo.*)

considerable variations of tinting. Some specimens are of a much paler brown than others ; in some individuals the fur is nearly black about the

MINK.—(*Vison Lutreola.*)

head, while the white patch that is found on the chin is extremely variable in dimensions. The size, too, is rather variable.

It frequents the banks of ponds, rivers, and marshes, seeming to prefer
the stillest waters in the autumn, and the rapidly flowing currents in spring.
As may be supposed from the nature of its haunts, its food consists almost
wholly of fish, frogs, crawfish, aquatic insects, and other creatures that are
to be found either in the waters or in their close vicinity. The general shape
of its body is not quite the same as that of the marten or ferret; and assumes
something of the otter aspect. The teeth, however, are nearer those of the
polecat than of the otter; and its tail, although not so fully charged with
hair as the corresponding member in the polecat, is devoid of that muscular
power and tapering form which is so strongly characteristic of the otter.
The feet are well adapted for swimming, on account of a slight webbing
between the toes.

The fur of this animal is excellent in quality, and is by many persons
valued very highly. By the furriers it passes under the name of "Mœnk,"
and it is known by two other names, "Tutucuri" and "Nœrs." As it bears a

(WEASEL.—*Mustela Vulgaris.*)

great resemblance to the fur of the sable, it is often fraudulently substituted
for that article,—a deception which is the more to be regretted, as the fur of
the Mink is a really excellent one, handsome in its appearance, and extremely
warm in character. By some authors, the identity of the Mink with the
water-polecat has been doubted, but, as it appears, without sufficient reason.

THERE is hardly any animal which, for its size, is so much to be dreaded
by the creatures on which it preys as the common WEASEL. Although its
diminutive proportions render a single Weasel an insignificant opponent to
man or dog, yet it can wage a sharp battle even with such powerful foes, and
refuses to yield except at the last necessity.

The proportions of the Weasel are extremely small, the male being rather
larger than the opposite sex. In total length, a full-grown male does not
much exceed ten inches, of which the tail occupies more than a fifth, while
the female is rather more than an inch shorter than her mate. The colour
of its fur is a bright reddish brown on the upper parts of the body, and the

under portions are of a pure white, the line of demarcation being tolerably well defined, but not very sharply cut.

It is a terrible foe to many of the smaller rodents, such as rats and mice, and performs a really good service to the farmer by destroying many of these farmyard pests. It follows them wherever they may be, and mercilessly destroys them, whether they have taken up their summer abode in the hedgerows and river-banks, or whether they have retired to winter quarters among the barns and ricks. Many farmers are in the habit of destroying the Weasels, which they look upon as "vermin," but it is now generally thought that, although the Weasel may be guilty of destroying a chicken or duckling now and then, it may yet plead its great services in the destruction of mice as a cause of acquittal. The Weasel is specially dreaded by rats and mice, because there is no hole through which either of these animals can pass which will not quite as readily suffer the passage of the Weasel; and as the Weasel is most determined and pertinacious in pursuit, it seldom happens that rats or mice escape when their little foe has set itself fairly on their track.

The Weasel has been seen to catch and to kill a bunting by creeping quietly towards a thistle on which the bird was perching, and then to leap suddenly upon it before it could use its wings. When it seizes an animal that is likely to make its escape, the Weasel flings its body over that of its victim, as if to prevent it from struggling. In single combat with a large and powerful rat, the Weasel has but little hope of success unless it should be able to attack from behind, as the long chisel-edged teeth of the rat are terrible weapons against so small an animal as the Weasel. The modes of attack employed by the two animals are of a different character, the rat making a succession of single bites, while the Weasel is accustomed to fasten its teeth on the head or neck of its opponent, and there to retain its hold until it has drained the blood of its victim. The fore-legs of the Weasel are of very great service in such a contest, for when it has fixed its teeth, it embraces its opponent firmly in its fore-limbs, and rolling over on its side, holds its antagonist in its unyielding grasp, which is never relaxed as long as a spark of life is left.

Like the polecat and others of the same group of animals, the Weasel is most destructive in its nature, killing many more animals than it can devour, simply for the mere pleasure of killing. It is curious to notice how the savage mind, whether it belongs to man or beast, actually revels in destruction, is maddened to absolute frenzy by the sight of blood, and is urged by a kind of fiery delirium to kill and to pour out the vital fluid. Soldiers in the heat of action have often declared that everything which they saw was charged with a blood-red hue, but that the details of the conflict had entirely passed from their minds. A single Weasel, urged by some such destructive spirit, has been known to make its way into a cage full of freshly-caught song-birds, and to destroy every single bird. The little assassin was ·discovered lying quite at its ease in a corner of the cage, surrounded with the dead bodies of its victims.

To persons who have had but little experience in the habits of wild animals, it is generally a matter of some surprise that the celebrated Ermine fur, which is in such general favour, should be produced by one of those very animals which we are popularly accustomed to rank among "vermin," and to exterminate in every possible way. Yet so it is. The highly-prized Ermine and the much-detested Stoat are, in fact, one and the same animal, the difference in the colour of their coats being solely caused by the larger or smaller proportion of heat to which they have been subjected.

In the summer-time, the fur of the Stoat—by which name the animal will

G

be designated, whether it be wearing its winter or summer **dress—is not** unlike that of the weasel, although the dark parts of the fur are not so ruddy nor the light portions of so pure a white as in that animal.   The **toes and** the edges of the ears are also white.

The change of colour which takes place during the colder months **of the** year is now ascertained, with tolerably accuracy, to be caused by an actual whitening of the fur, and not by the gradual substitution of white for dark hairs, as was for some time supposed to be the case.

The hairs are not entirely white, even in their most completely blanched state, but partake of a very delicate cream-yellow, especially upon the under portions, while the slightly bushy tip of the tail retains its original black tinting, and presents a singular contrast to the remainder of the fur.   In these comparatively temperate latitudes, the Stoat is never sufficiently blanched to render its fur of any commercial value.   As may be supposed from the extreme delicacy of the skin in its wintry whiteness, the capture of the Stoat for the purpose of obtaining its fur is a matter of no small difficulty. The traps which are used for the purpose of destroying the Stoat are **formed**

STOAT OR ERMINE (Winter Dress).

so as to kill the animal by a sudden blow, without wounding the skin ; and many of the beautiful little creatures are taken in ordinary snares.

In this country, where the lowest temperature is considerably above that of the ordinary wintry degrees, the Stoat is very uncertain in its change of fur, and seems to yield to or to resist the effects of the cold weather according to the individuality of the particular animal.

The Stoat is considerably larger than the weasel, measuring rather more than fourteen inches in total length, of which the tail occupies rather more than four inches.   There is, however, considerable difference in the size of various individuals.

It is a most determined hunter, pursuing its game with such pertinacious skill that it very seldom permits its intended prey to escape, and by dint of perseverance can even capture the swift-footed hare.

When the female Stoat is providing for the wants of a young family, **she** forages far and wide for her offspring, and lays up the produce of her chase in certain cunningly-contrived larders. In a wood belonging to Lord Bagot, a Stoat nursery was discovered, having within it no less than six inhabitants, a mother and her five young. Their larder was supplied with five hares and four rabbits, neither of which had been in the least mangled, with the exception of the little wound that had caused their death.

IN the clumsy-looking animal which is called the RATEL, a beautiful adaptation of nature is manifested. Covered from the tip of the nose to the insertion of the claws with thick, coarse, and rough fur, and provided moreover with a skin that lies very loosely on the body, the Ratel is marvellously adapted to the peculiar life which it leads.

Although the Ratel is in all probability indebted for its food to various sources, the diet which it best loves is composed of the combs and young of the honey-bee. So celebrated is the animal for its predilection for this sweet dainty that it has earned for itself the title of Honey Ratel, or Honey Weasel. The reason for its extremely thick coating of fur is now evident. The animal is necessarily exposed to the attacks of the infuriated bees when it lays siege to their fastnesses, and if it were not defended by a coating which is impenetrable to their stings, it would soon fall a victim to the poisoned weapons of its myriad foes.

During the daytime the Ratel remains in its burrow; but as evening begins to draw near, it emerges from its place of repose, and sets off on its bee-hunting expeditions. As the animal is unable to climb trees, a bees' nest that is made in a hollow tree-limb is safe from its attacks. But the greater number of wild bees make their nests in the deserted mansions of the termite, or the forsaken burrows of various animals. It is said that the Ratel finds its way towards the bees' nests by watching the direction in which the bees return towards their homes.

The colour of the Ratel is black upon the muzzle, the limbs, and the whole of the under portions of the body; but upon the upper part of the head, neck, back, ribs, and tail, the animal is furnished with a thick covering of long hairs, which are of an ashy grey colour. A bright grey stripe, about an inch in width, runs along each side and serves as a line of demarcation between the light and the dark portions of the fur. The ears of the Ratel are extremely short. The lighter fur of the back is variously tinted in different individuals, some being of the whitish grey which has been already mentioned, and others remarkable for a decided tinge of red. The length of the Cape Ratel is rather more than three feet, inclusive of the tail, which measures eight or nine inches in length. In its walk it is plantigrade, and has so much of

RATEL.—(*Mellivora Rat l.*)

the ursine character in its movements that it has been called the Indian or Honey Bear. It is sometimes known under the title of " Bharsiah."

The animal which has just been described is an inhabitant of Southern Africa, being found in great profusion at the Cape of Good Hope. There is, however, an Indian species of Ratel, which very closely resembles the African animal, and in the opinion of some writers is identical with it.

The WOLVERENE, more popularly known by the name of the GLUTTON, has earned for itself a world-wide reputation for ferocity, and has given occasion to some of the older writers on natural history to indulge in the most unshackled liberty of description.

It is known that the Glutton feeds largely on the smaller quadrupeds, and that it is a most determined foe to the beaver in the summer months. During the winter it has little chance of catching a beaver, for the animals are quietly ensconced in their home, and their houses are rendered so strong by the intense cold, that the Glutton is unable to break through their ice-hardened walls.

The Wolverene is an inhabitant of Northern America, Siberia, and of a great part of Northern Europe. It was once thought that the Glutton and the Wolverene were distinct animals, but it is now ascertained that they both belong to the same species.

WOLVERENE.—(*Gulo luscus.*)

The general aspect of this animal is not unlike that of a young bear, and probably on that account it was placed by Linnæus among the bears under the title of *Ursus Luscus.* The general colour of the Wolverene is a brownish black ; the muzzle is black as far as the eyebrows, and the space between the eyes of a browner hue. In some specimens a few white spots are scattered upon the under jaw. The sides of the body are washed with a tint of a warmer hue. The paws are quite black, and the contrast between the jetty fur of the feet and the almost ivory whiteness of the claws is extremely curious. These white claws are much esteemed among the natives for the purpose of being manufactured into certain feminine adornments.

The SKUNK, which is so celebrated for the horrible odour which emanates from it, belongs to the Weasel tribe.

SCARCELY less remarkable for its ill-odour than the skunk, the TELEDU is not brought so prominently before the public eye as the animal which has just been described.

It is a native of Java, and seems to be confined to those portions of the country that are not less than seven thousand feet above the level of the sea. On certain portions of these elevated spots, the Teledu, or Stinkard, as it is popularly called, can always be found. The earth is lighter on these spots than in the valleys, and is better suited to the habits of the Teledu, which roots in the earth after the manner of hogs, in search of the worms and insects which constitute its chief food. This habit of turning up the soil renders it very obnoxious to the native agriculturists, as it pursues the worms in their subterraneous meanderings, and makes sad havoc among the freshly-planted seeds. It is also in the habit of doing much damage to the sprouting plants by eating off their roots.

We are indebted to Mr. Horsfield for an elaborate and interesting account of the Teledu, an animal which he contrived to tame and to watch with singular success. The following passages are selected from his memoir :—

"The Mydaus forms its dwelling at a slight depth beneath the surface, in the black mould, with considerably ingenuity. Having selected a spot defended above by the roots of a large tree, it constructs a cell or chamber of a globular form, having a diameter of several feet, the sides of which it makes perfectly smooth and regular ; this it provides with a subterraneous conduit or avenue, about six feet in length, the external entrance to which it conceals with twigs and dry leaves. During the day it remains concealed, like a badger in its hole ; at night it proceeds in search of its food, which consists of insects and other larvæ, and of worms of every kind. It is particularly fond of the common lumbrici, or earth-worms, which abound in the fertile mould. These animals, agreeably to the information of the natives, live in pairs, and the female produces two or three young at a birth.

" The motions of the Mydaus are slow, and it is easily taken by the natives, who by no means fear it. During my abode on the Mountain Prahu, I engaged them to procure me individuals for preparation ; and as they received a desirable reward, they brought them to me daily in greater numbers than I could employ. Whenever the natives surprise them suddenly, they prepare them for food ; the flesh is then scarcely impregnated with the offensive odour, and is described as very delicious. The animals are generally in excellent condition, as their food abounds in fertile mould.

" On the Mountain Prahu, the natives, who were most active in supplying me with specimens of the Mydaus, assured me that it could only propel the fluid to the distance of about two feet. The fetid matter itself is of a viscid

TELEDU. —(*Mydaus meliceps.*)

nature : its effects depend on its great volatility, and they spread through a great extent. The entire neighbourhood of a village is infected by the odour of an irritated Teledu, and in the immediate vicinity of the discharge it is so violent as in some persons to produce syncope. The various species of Mephitis in America differ from the Mydaus in the capacity of projecting the fetid matter to a greater distance.

"The Mydaus is not ferocious in its manners, and, taken young, like the badger, might be easily tamed. An individual which I kept some time in confinement afforded me an opportunity of observing its disposition. It soon became gentle and reconciled to its situation, and did not at any time emit the offensive fluid. I carried it with me from Mountain Prahu to Bladeran, a village on the declivity of that mountain, where the temperature was more moderate. While a drawing was made, the animal was tied to a small stake. It moved about quietly, burrowing the ground with its snout and feet, as if in search of food, without taking notice of the bystanders, or making violent efforts to disengage itself ; on earth-worms (lumbrici) being brought, it ate voraciously ; holding one extremity of a worm with its claws, its teeth were employed in tearing the other. Having consumed about ten or twelve, it became drowsy, and making a small groove in the earth, in which it placed its snout, it composed itself deliberately, and was soon sound asleep."

The colour of the Teledu is a blackish brown, with the exception of the fur upon the top of the head, a stripe along the back, and the tip of the short tail, which is a yellowish-white. The under surface of the body is of a lighter hue. The fur is long and of a silken texture at the base, and closely set together, so as to afford to the animal the warm covering which is needed in the elevated spots where it dwells. The hair is especially long on the sides of the neck, and curls slightly upwards and backwards, and on the top of the head there is a small transverse crest. The feet are large, and the claws of the fore-limbs are nearly twice as long as those of the hinder paws.

In the whole aspect of the Teledu there is a great resemblance to the badger, and, indeed, the animal looks very like a miniature badger, of rather eccentric colours.

ALTHOUGH one of the most quiet and inoffensive of our indigenous animals, the BADGER has been subjected to such cruel persecutions as could not be justified even if the creature were as destructive and noisome as it is harmless or innocuous. For the purposes of so-called "sport," the Badger was captured and put into a cage ready to be tormented at the cruel will of every ruffian who might choose to risk his dog against the sharp teeth of the captive animal.

Being naturally as harmless an animal as can be imagined, it is a terrible antagonist when provoked to use the means of defence with which it is so well provided. Not only are the teeth long and sharp, but the jaws are so formed that when the animal closes its mouth the jaws "lock" together by a peculiar structure of their junction with the skull, and retain their hold without any need of any special effort on the part of the animal.

BADGER.—(*Meles Taxus*).

Unlike the generality of the weasel tribe, the Badger is slow and clumsy in its actions, and rolls along so awkwardly in its gait that it may easily be mistaken for a young pig in the dark of the evening, at which time it first issues from its burrow. The digging capacities of the Badger are very great, the animal being able to sink itself into the ground with marvellous rapidity. For this power the Badger is indebted to the long curved claws with which the fore-feet are armed, and to the great development of the muscles that work the fore-limbs.

In its burrow the female Badger makes her nest and rears her young, which are generally three or four in number.

The food of the Badger is of a mixed character, being partly vegetable and partly animal. Snails and worms are greedily devoured by this creature, and the wild bees, wasps, and other fossorial Hymenoptera find a most destructive foe in the Badger, which scrapes away the protecting earth and devours honey, cells, and grubs together, without being deterred from its meal by the stings of the angry bees.

As is the case with the generality of weasels, the Badger is furnished with an apparatus which secretes a substance of an exceedingly offensive odour, to which circumstance is probably owing much of the popular prejudice against the "stinking brock."

The colours of the Badger are grey, black, and white, which are rather curiously distributed. The head is white, with the exception of a rather broad and very definitely marked black line on each side, commencing near the snout and ending at the neck, including the eye and the ear in its course. The body is of a reddish grey, changing to a white grey on the ribs and tail. The throat, chest, abdomen, legs, and feet are of a deep blackish brown. The average length of the Badger is two feet six inches, and its height at the shoulder eleven inches.

ALTHOUGH by no means a large animal, the OTTER has attained a universal reputation as a terrible and persevering foe to fish. Being possessed of a very discriminating palate, and invariably choosing the finest fish that can be found in the locality, the Otter is the object of the profoundest hate to the proprietors of streams and to all human fishermen.

When the Otter is engaged in eating the fish it has captured, it holds the slippery prey between its fore-paws, and, beginning with the back of the neck, eats away the flesh from the neck towards the tail, rejecting the head, tail, and other portions.

For pursuit of its finny prey the Otter is admirably adapted by nature. The body is lithe and serpentine; the feet are furnished with a broad web that connects the toes, and is of infinite service in propelling the animal through the water; the tail is long, broad, and flat, proving a powerful and effectual rudder by which its movements are directed; and the short, powerful legs are so loosely jointed that the animal can turn them in almost any direction.

The colour of the Otter varies slightly according to the light in which it is viewed, but is generally of a rich brown tint, intermixed with whitish grey. This colour is lighter along the back and the outside of the legs than on the other parts of the body, which are of a paler greyish hue. Its habitation is made on the bank of the river which it frequents, and is rather inartificial in its character, as the creature is fonder of occupying some natural crevice or deserted excavation than of digging a burrow for itself. The nest of the Otter is composed of dry rushes, flags, or other aquatic plants, and is purposely placed as near the water as possible, so that in case of a sudden alarm the mother Otter may plunge into the stream together with her young family, and find a refuge among the vegetation that skirts the

OTTER. — (*Lutra vulgaris.*)

river banks. The number of the young is from three to five, and they make their appearance about March or April.

The fur of the Otter is so warm and handsome that it is in some request for commercial purposes. The entire length of the animal is rather under three feet and a half, of which the tail occupies about fourteen or fifteen inches. On the average it weighs about twenty-three pounds; but there are examples which have far surpassed that weight. Mr. Bell records an instance of a gigantic Otter that was captured in the River Lea, between Hertford and Ware, and which weighed forty pounds.

Although so fierce and savage an animal when attacked, the Otter is singularly susceptible of human influence, and can be taught to catch fish for the service of its masters rather than for the gratification of its own palate. The CHINESE or INDIAN OTTER affords an excellent instance of this capability; for in every part of India the trained Otters are almost as common as trained dogs in England. It seems odd that the proprietors of streams should not press the Otter into their service instead of destroying it, and manage to convert into a faithful friend the animal which at present is considered but as a ruthless enemy.

BEARS.

The BEARS and their allies form a family which is small in point of numbers, but is a very conspicuous one on account of the large size of the greater number of its members.

These animals are found in almost every portion of the earth's surface, and are fitted by nature to inhabit the hottest and the coldest parts of the world. India, Borneo, and other burning lands are the homes of sundry members of this family, such as the Bruang and the Aswail; while the

BROWN BEAR. —(*Ursus Arctos.*)

snowy regions of Northern Europe and the icebound coasts of the Arctic Ocean are inhabited by the Brown Bear and the Nennook or Polar Bear.

The paws of the Bears are armed with long and sharp talons, which are not capable of retraction, but which are most efficient weapons of offence when urged by the powerful muscles which give force to the Bear's limbs. Should the adversary contrive to elude the quick and heavy blows of the paw, the Bear endeavours to seize the foe round the body, and by dint of sheer pressure to overcome its enemy. In guarding itself from the blows which are aimed at it by its adversary the Bear is singularly adroit, warding off the fiercest strokes with a dexterity that might be envied by many a pretender to the pugilistic art.

Several species of Bears are now recognised by systematic naturalists, the principal examples of which will be noticed in the following pages.

The Bear which is most popularly known in this country is the BROWN BEAR; a creature which is found rather plentifully in forests and the mountainous districts of many portions of Europe and Asia. As may be supposed from its title, the colour of its fur is brown, slightly variable in tint in different indviduals, and often in the same individual at various ages.

The size to which a well-fed and undisturbed Brown Bear will grow is really surprising, for although it loses its growing properties after its twentieth year, it seems permanently to retain the capability of enlargement, and when in a favourable situation will live to a very great age. The weight of an adult Brown Bear in good condition is very great, being sometimes from seven to eight hundred pounds when the creature is remarkably fine, and from five to six hundred pounds in ordinary cases. Mr. Falk remarks, that a Bear which he killed was so enormously heavy, that when slung on a pole it was a weighty burden for ten bearers.

Ants form a favourite article of diet with a Bear, which scrapes their nests out of the earth with its powerful talons, and laps up the ants and their so-called "eggs" with its ready tongue. Bees and their sweet produce are greatly to the taste of the Bear, which is said to make occasional raids upon the beehives, and to plunder their contents.

Vegetables of various kinds are also eaten by the Bear, and in the selection of these dainties the animal evinces considerable taste. According to Mr. Lloyd, "the Bear feeds on roots, and the leaves and small limbs of the aspen, mountain-ash, and other trees : he is also fond of succulent plants, such as angelica, mountain-thistle, &c. To berries he is likewise very partial, and during the autumnal months, when they are ripe, he devours vast quantities of cranberries, blueberries, raspberries, strawberries, cloudberries, and other berries common to the Scandinavian forests. Ripe corn he also eats, and sometimes commits no small havoc amongst it ; for seating himself, as it is said, on his haunches in a field of it, he collects with his outstretched arms nearly a sheaf at a time, the ears of which he then devours."

During the autumn the Bear becomes extremely fat, in consequence of the ample feasts which it is able to enjoy, and makes its preparations for passing the cold and inhospitable months of winter. About the end of October the Bear has completed its winter house, and ceases feeding for the year.

A curious phenomenon now takes place in the animal's digestive organs, which gives it the capability of remaining through the entire winter in a state of lethargy, without food, and yet without losing condition.

From the end of October to the middle of April the Bear remains in his den, in a dull lethargic state of existence ; and it is a curious fact that if a hibernating Bear be discovered and killed in its den, it is quite as fat as if it had been slain before it retired to its resting-place. Experienced hunters say that even at the end of its five months' sleep, the Bear is as fat as at its beginning. Sometimes it is said that the Bear partially awakes, and in that case it immediately loses its sleek condition, and becomes extremely thin. During the winter the Bear gains a new skin on the balls of the feet, and Mr. Lloyd suggests that the curious habit of sucking the paws, to which Bears are so prone, is in order to facilitate the growth of the new integument.

The Bear is possessed of several valuable accomplishments, being a wonderful climber of trees and rocks, an excellent swimmer, and a good digger.

The number of cubs which the female Bear produces is from one to four, and they are very small during the first few days of their existence. They make their appearance at the end of January or the beginning of February, and it is a curious fact that, although the mother has at the time been deprived of food for nearly three months, and does not take any more food until the spring, she is able to afford ample nourishment to her young without suffering any apparent diminution in her condition. It is said by those who have had personal experience of the habits of the Bear, that the mother takes the greatest care of her offspring during the summer, but that when winter approaches, she does not suffer them to partake of her residence, but prepares winter quarters for them in her immediate neighbourhood. During the winter another little family is born, and when they issue forth from their home they are joined by the elder cubs, and the two families pass the next winter in the mother's den.

The SYRIAN BEAR, which is otherwise known by the name of DUBB, or RITCK, is doubly interesting to us, not only on account of its peculiarly gentle character, but from the fact that it is the animal which is so often mentioned in the Scriptural writings.

The colour of this animal is rather peculiar, and varies extremely during the different periods of its life. While it is in its earliest years, the colour of its fur is a greyish brown, but as the animal increases in years the fur becomes gradually lighter in tint, and, when the Bear has attained maturity, is nearly white. The hair is long and slightly curled, and beneath the longer hair is a thick and warm covering of closely-set woolly fur, which seems to defend the animal from the extremes of heat or cold. Along the shoulders and front of the neck, the hair is so perpendicularly set, and projects so firmly, that it gives the appearance of a mane, somewhat resembling that of the hyæna.

At the present day the Syrian Bear may be found in the mountainous parts of Palestine, and has been frequently seen upon the higher Lebanon mountains.

The fur of this bear is rather valuable on account of its warmth and beauty; and the fat and the gall are also held in much esteem for various purposes, chiefly medicinal.

AMERICA furnishes several species of the Bear tribe, two of which, the GRIZZLY BEAR and the MUSQUAW, or BLACK BEAR, are the most conspicuous.

THE Black Bear is found in many parts of Northern America, and was formerly seen in great plenty. But as the fur and the fat are articles of great commercial and social value, the hunters have exercised their craft with such determination that the Black Bears are sensibly diminishing in number. The fur of the Black Bear is not so roughly shaggy as that of the European or the Syrian Bear, but is smooth and glossy in its appearance, so that it presents a very handsome aspect to the eye, while its texture is as thick and warm as that of its rougher furred relations.

THERE are few animals which are so widely and deservedly dreaded as the GRIZZLY BEAR. This terrible animal is an inhabitant of many portions of Northern America, and is the acknowledged superior of every animal that ranges over the same country.

The other members of the ursine family are not given to attacking human beings, unless they are alarmed or wounded, but the Grizzly, or " Ephraim," as the creature is familiarly termed by the hunters, displays a most unpleasant readiness to assume the offensive as soon as it perceives a man, be he mounted or on foot, armed or otherwise.

So tenacious of life is the Grizzly Bear, that unless it receives a wound in the head or heart it will continue its furious struggles, even though it be riddled with bullets and its body pierced with many a gaping wound. These warlike capacities render the creature respected by the natives and colonists, and the slaughter of a Grizzly Bear in fair fight is considered an extremely high honour. Among the native tribes that dwell in the northern portions of America, the possession of a necklace formed from the claws of the Grizzly Bear is considered as enviable a mark of distinction as a blue ribbon among ourselves. No one is permitted to wear such an ornament unless the Bear has fallen under his hand ; consequently the value of the decoration is almost incalculable. So largely is this mark of distinction prized, that the Indian who has achieved such a dignity can hardly be induced to part with his valued ornament by any remuneration that can be offered.

The colour of the Grizzly Bear is extremely variable, so much so, indeed, that some zoologists have suggested the existence of two distinct species. Sometimes the colour of the fur is a dullish brown, plentifully flecked with grizzled hairs, and in other specimens the entire fur is of a beautiful steely grey.

GRIZZLY BEAR.—(*Ursus ferox.*)

THERE is a small group of these animals called SUN-BEARS, from their habit of basking in the sun instead of hiding in their dens during the hours of daylight. A very curious example of the Sun-Bears is found in the species which is known by the name of the BRUANG, or MALAYAN SUN-BEAR, and has been rendered famous by the spirited description of its appearance and habits which has been given by Sir Stamford Raffles.

THERE is generally an aquatic member of each group of animals through-out the vertebrate kingdom, and among the Bears this part is filled by the NENNOOK, or POLAR BEAR, sometimes called, on account of its beautiful silvery fur, the WHITE BEAR. As has already been mentioned, the Bears are good swimmers, and are able to cross channels of considerable width, but we have, in the person of the Nennook, an animal that is especially formed for traversing the waters and for passing its existence among the ice-mountains of the northern regions.

So active is this Bear, and so admirable are its powers of aquatic locomotion, that it has been seen to plunge into the water in chase of a salmon, and to return to the surface with the captured fish in its mouth. And when it is engaged in the pursuit of seals, as they are sleeping on a rock or an iceraft, it is said to employ a very ingenious mode of approach. Marking the position in which its intended prey lies, it quietly slips into the water, and diving below the surface, swims in the intended direction, until it is forced to return to the surface in order to breathe. As soon as it has filled its lungs with fresh air, it again submerges itself, and resumes its course, timing its submarine journeys so well, that when it ascends to the surface for the last time, it is in close proximity to the slumbering seal. The fate of the unfortunate victim is now settled, for it cannot take refuge in the water without falling into the clutches of its pursuer, and if it endeavours to escape by land it is speedily overtaken and destroyed by the swifter-footed Bear.

So powerful an animal as the Polar Bear must necessarily be very dangerous when considered in the light of a foe. Sometimes it runs away as soon as it sees or smells a human being, but at others it is extremely malicious, and will attack a man without any apparent reason. As is the case with nearly all the Bears, it is very tenacious of life, and even when pierced with many wounds, will fight in the most desperate manner, employing both teeth and claws in the combat, and only yielding the struggle with its life.

The colour of the Nennook's fur is a silvery white, tinged with a slight yellow hue, rather variable in different individuals. Even in specimens that were confined in the Zoological Gardens there was a perceptible difference in

POLAR BEAR.—(*Thalarctos maritimus.*)

the tint of their fur, the coat of one of them being of a purer white than that of the other. The yellowish tinge which has been just mentioned is very similar to the creamy yellow hue which edges the ermine's fur. The feet are armed with strong claws of no very great length, and but slightly curved. Their colour is black, so that they form a very bold contrast with the white fur that falls over the feet. Even at a considerable distance, and by means of its mere outline, the Polar Bear may be distinguished from every other member of the Bear tribe by its peculiar shape. The neck is, although extremely powerful, very long in proportion to the remainder of the body, and the head is so small and sharp that there is a very snake-like aspect about that portion of the animal's person.

The young of the Nennook are generally two in number, and, when they make their first appearance outside the snow-built nursery in which their few months of existence have been passed, are about the size of shepherds' dogs, and in excellent condition.

PRESERVING somewhat of the ursine aspect and much of the ursine habits, the RACOON, or MAPACH, as it is sometimes named, is an active, spirited, and amusing animal. As it is readily tamed, although rather subject to occasional infirmity of temper, and is inquisitive, quaint, and lively withal, it is a great favourite with such persons as have kept it in captivity.

The colour of this animal is rather peculiar, and not very easy to describe. The general tint of the body and limbs is an undecided blackish grey, the grey and black predominating according to the position of the observer and the arrangement of the fur. The hairs that form the coat of the Racoon are of two kinds, the one of a soft and woolly character, lying next to the skin, and the other composed of long and rather stiff hairs that project through the wool for some distance. The woolly fur is of a uniform grey, while the longer hairs are alternately marked with black and greyish white. Upon the top of the head and across the eyes the fur is of a very dark blackish brown ; and upon the knee-joint of each leg the fur is of a darker tint than on the rest of the body. The tail is rather short and bushy in character, and is marked with five, or sometimes six, blackish rings upon a ground of dark grey.

As is indicated by the peculiar nature of its teeth, the Racoon is capable of feeding on animal or vegetable food, but seems to prefer the latter. Indeed, there seem to be few things which the Racoon will not eat. One of these animals ate a piece of cedar pencil which it snatched out of my hand, and tried very hard to eat the envelope of a letter on which I was making notes. Not succeeding in the attempt, it consoled itself by tearing the paper into minute morsels, employing teeth and paws in the attempt. It did its best to get a ring off my finger, by hitching one of its crooked claws into the ring and pulling with all its strength, which was very considerable in proportion to the size of the animal. Its brown eyes lighted up with animation when engaged in play, and it was very fond of pushing its paw through the bars of its cage, in order to attract attention.

In its native state it is a great devourer of oysters, crabs, and other similar animals, displaying singular ingenuity in opening the stubborn shells of the oysters, or in despatching the crabs without suffering from their ready claws. Sometimes it is said to fall a victim to the oyster, and to be held so firmly by the closing shells that it cannot extricate itself, and perishes miserably by the rising tide. Its oyster-eating propensities have been questioned, but are now clearly proven. The sand and soil that fringe the oyster-beds are frequently seen to be covered with the foot-marks of this animal.

It is always fond of water, drinking largely, and immersing its food, so as to moisten it as much as possible. When engaged in this curious custom it grasps the food in both its

RACOON.—(*Procyon Lotor.*)

fore-paws, and shakes it violently backward and forward in the water. On account of this remarkable habit it has been dignified with the title of *Lotor*, "a washer." The German naturalists term it *Wasch-Bär*, or Washing Bear.

Roving at night through the woods, and being gifted with singular subtlety as well as agility, it is frequently chased by the residents, who think a 'Coon hunt to be one of the most exciting of sports. Certainly, to judge from the animated descriptions of such scenes, the whole affair must be marvellously picturesque to the eye as well as exciting to the mind. The usual plan of hunting the 'Coon is to set an experienced dog on its trail, and to chase it until it takes refuge in a tree. A blazing fire of pine chips is then built under the tree, which illuminates its branches and renders the smallest leaf perceptible. A good climber then ascends the tree, and speedily dislodges the concealed animal.

In size the Racoon equals a small fox, to which animal it bears a slight external resemblance. The number of its young is usually two or three, and they make their appearance in the month of May.

THE animals which compose the curious genus that is known by the name of Narica are easily recognised on account of the singular length of the nose, which is prolonged so as to form a miniature and mobile proboscis. In their general habits and diet they very strongly resemble the racoons, and are as admirable climbers of trees as can be found in the animal kingdom.

The extraordinary snout with which the Coaitis are gifted is very useful to the possessor, being employed for the purpose of rooting in the ground in search of worms and insects, together with other important uses. When they drink, the Coaitis lap the water after the manner of dogs, and when so engaged, turn up their flexible snouts, so as to keep that useful member from being wetted more than is necessary. They are inhabitants of Southern America, and are found in small companies upon the trees among which they reside, and on the thin branches of which they find the greater part of their food. Two examples of the Coaitis will be briefly described.

The COAITI-MONDI, or RED COAITI, derives its name from the reddish chestnut hue which prevails over the greater portion of the fur, and is only broken by the black ears and legs, the maroon-coloured bands upon the tail, and the white hairs which edge the upper jaw and entirely cover the lower. The texture of the fur is rather harsh and wiry, and of no very great importance in commerce. Upon the paws are certain curious tubercles, which alone would serve to identify the animal were it entirely destroyed with the exception of a single foot. It is extremely active in the ascent and descent of trees, and pursues its prey among the limbs with great certainty. Its food consists of sundry vegetable and animal substances, but the creature seems to prefer the latter to the former.

It is a nocturnal animal, and does not show its true liveliness until the shades of evening begin to draw on, but lies curled up in a curious but comfortable attitude, its long and bushy tail serving for blanket and pillow. Towards evening, however, the Coaiti arouses itself from its lethargy, and becomes full of life and vigour, careering about the branches with extraordinary rapidity of movement and certainty of hold, and agitating its mobile nose with unceasing energy, as if for the purpose of discovering by the snout the presence of some welcome food. It is a merci-

COAITI-MONDI.—(*Nasua Rufa.*)

less robber of birds' nests, and will eat parent, eggs, or young with equal appetite.

Although possessed of a very irritable temper, the Coaiti is tamed without difficulty to a certain extent, but is always capricious in its affections, and cannot be trusted without danger. When attacked by men or dogs, the Coaiti fights desperately, and can inflict such dangerous wounds with its double-edged canine teeth, that it is, although so small an animal, no despicable antagonist.

Another species of Coaiti inhabits the same regions as the last-mentioned animal. This is the NARICA, or QUASJE, which is sometimes called the BROWN COAITI, in order to distinguish it from the red species. Sometimes the name is spelled as " Quaschi."

THE singular creature which is known under the title of KINKAJOU, or POTTO, is an inhabitant of Southern America, and is spread over a very large extent of country, so that it is known in different places under different appellations, such as Honey Bear, Manaviri, or Guchumbi. When fully grown, the Kinkajou is equal to a large cat in size, but is very much stronger in proportion to the dimensions of its body. The colour of the animal is a very light dun, obscurely traversed by narrow darker bands, that run over the back towards the ribs, and partly follow their course. Another darker band is observable round the neck, but all these marks are so very indistinct that they can only be seen in a favourable light.

The most remarkable point in this animal is the extreme length and flexibility of the tongue, which it is able to protrude to a marvellous extent, and which it can insinuate into the smallest crevices in search of the insects which have taken shelter therein. It is said that the animal employs its long tongue for the purpose of thrusting that organ into the bee-cells and licking out the sweet contents of the waxen treasury. With its tongue it can perform many offices of an elephant's trunk, and will frequently seize and draw

KINKAJOU, OR POTTO.—(*Cercoleptes caudivolvulus.*)

towards its mouth the articles of food which may be beyond the reach of its lips. It has also been seen to use its tail for the same purpose.

Assisted by its prehensile tail, the Kinkajou is an admirable and fearless climber, possessing the capability of suspending its body by the hinder feet and the tail, and remaining in this inverted attitude for a considerable space of time. It is evidently nocturnal in its habits, being sadly distressed by the effect of daylight upon its eyes.

It is easily tamed, and when domesticated is of a sportful nature, delighting to play with those persons whom it knows and trusts, and making pretence to bite, after the manner of puppies and kittens. It is very susceptible to kindness, and is fond of the caresses which are offered by its friends. In its wild state, however, it is a rather fierce animal, and, when assaulted, offers such a spirited resistance, even to human foes, that it will beat off any but a determined man, supposing him to be unarmed and unassisted.

INSECTIVORA.

THE animals which are comprised in the Insect-eating eating group are well represented in England, in which country we find the Mole, the various Shrews, and the Hedgehog, as examples of the TALPIDÆ, or the family of the MOLES.

Some of these creatures, such as the shrew, present so close an external resemblance to the common mice, that they are popularly supposed to belong to the same class, and are called by the same general name. Many species live beneath the surface of the earth, and seek in that dark hunting-ground the prey which cannot be enticed to the surface in sufficient numbers to supply adequate nourishment for the ever-hungry worm-devourers.

Of all the insect-eating animals, there is none which is better known by name than the common MOLE, and very few which are less known by their true character.

MOLE.—(*Talpa Europæa.*)

On inspecting a living Mole that has been captured on the surface of earth, and comparing it with the multitudinous creatures that find their subsistence on the earth's surface, rejoicing in the full light of day, and free to wander as they please, we cannot but feel some emotions of surprise at the sight of a creature which is naturally debarred from all these sources of gratification, and which passes its life in darkness below the surface of the ground.

Yet this pity, natural though it be, will be entirely thrown away, for there is scarcely any creature that lives which is better fitted for enjoyment, or which is urged by more fiery passions. Dull and harmless as it may appear to be, it is in reality one of the most ferocious animals in existence, and will engage in the fiercest combats upon very slight provocation. While thus employed, its whole faculties are so entirely absorbed in its thirst for revenge, that it will leave the subterraneous shafts which it has been so busily excavating, and join battle with its foe in the full light of day. Should one of the combatants overpower and kill the other, the victorious Mole springs upon the vanquished enemy, tears its body open, and eagerly plunging its nose into the wound, drinks the blood of its slaughtered enemy, and feasts richly on the sanguine banquet.

With the exception of sight, the senses of the Mole seem to be remarkably developed.

The sense of smell is singularly acute, and enables the animal to discover the presence of the earthworms on which it feeds, and to chase them success-fully through their subterranean meanderings.

The hearing of the Mole is proverbially excellent ; and it is probable that the animal is aided in its pursuit of worms by the sense of hearing as well as that of smell. Much of the Mole's safety is probably owing to its exquisite hearing, which gives it timely notice of the approach of any living being, and enables it to secure itself by rapidly sinking below the surface of the earth. To tread so softly that the blind Mole may not hear a footfall, is an expression which has become a household word.

The sense of touch is peculiarly delicate, and seems to be chiefly resident in the long and flexible nose, which is employed by the Mole for other purposes than those of scent. When the creature is placed upon the surface of the ground, and is about to sink one of its far-famed tunnels, it employs its nose for that purpose almost as effectually as its armed fore-paws.

It seldom happens that all the senses of an animal are developed to an equal extent, so that where one or two are singularly acute, it is generally at the expense of the others. Such is the case with the Mole; for although the scent, touch, and hearing are remarkable for their excellence, the sight is so extremely defective that it may almost be considered as a nullity. It is true that the Mole possesses eyes; but those organs of vision are so small, and so deeply hidden in the fur, that they can be but of little use to the owner, except to mark the distinctions between light and darkness. The eyes are so exceedingly small that their very existence has been denied, and it is only by a careful search that they can be seen at all.

The fore paws are extremely large, and furnished with strong and flattened nails. They are turned rather obliquely, as seen in the figure, in order to give free scope to their exertions. The paws are devoid of the soft fur that shields the rest of the body, and are covered with a thick but naked skin. It is chiefly to these paws that any mould is found adherent when the Mole is captured, for the soft and velvet-like fur permits no earthy stain to defile its glossy smoothness.

The Mole is said to be an excellent swimmer, and to be able to cross rivers when led to such an act by any adequately powerful motive. How far true this assertion may be I cannot prove by personal experience; but I think it is likely to be possible, for I have seen a Mole swim across the bend of a brook—a distance of some few yards—and perform its natatory achievement with great ease. I was not near enough to ascertain the mode of its progression, but it seemed to use its fore-paws as the principal instruments of locomotion. This circumstance took place in Wiltshire.

From all accounts the Mole seems to be a thirsty animal, and to stand in constant need of water, drinking every few hours in the course of the day. In order to supply this want it is in the habit of sinking well-like pits in different parts of its "runs," so that it may never be without the means of quenching its thirst. Everything that the Mole does is marked with that air of desperate energy which is so characteristic of the animal. The labourers in different parts of England all unite in the same story, that the Mole works for three hours "like a horse," and then rests for three hours, labouring and resting alternately through the day, and with admirable perception of time.

The well-known "mole-hills," which stud certain lands, and which disfigure them so sadly, however much their unsightliness may be compensated by their real usefulness, are of various kinds, according to the sex and age of the miner. The small hillocks which follow each other in rapid succession are generally made by the female Mole before she has produced her little family, and when she is not able to undergo the great labour of digging in the harder soil. Sometimes the "run" is so shallow as to permit the superincumbent earth to fall in, so that the course which the Mole has followed is little more than a trench. This is said to be produced by the little coquetries that take place between the Mole and its future mate, when the one flies in simulated terror, and the other follows with undisguised determination. Deeper in the soil is often found a very large burrow, sufficiently wide to permit two Moles to pass each other. This is one of the high-roads which lead from one feeding ground to another, and from which the different shafts radiate.

But the finest efforts of talpine architecture are to be found in the central fortress, from which the various roads diverge, and the nest which the maternal Mole forms for the security of her young.

The fortress is of a very peculiar construction, and is calculated to permit the ingress or egress of the Mole from almost any direction, so that when its acute senses give notice of the approach of an enemy, it can make its retreat without difficulty.

The first operation is to build a tolerably large hill of compact and well-trodden earth. Near the summit of this mound the excavator runs a circular gallery, and another near the bottom, connecting the two galleries with five short passages. It then burrows into the centre of the mound, and digs a moderately large spherical hole, which it connects with the lower gallery by three passages. A very large passage, which is a continuation of the high-road, is then driven into the spherical chamber by dipping under the lower gallery, and is connected with the circular chamber from below. Lastly, the Mole drives a great number of runs, which radiate from the rest in all directions, and which all open into the lower circular gallery. It will be seen from this short description, that if a Mole should be surprised in its nest, it can withdraw through its central chamber and so reach the high road at once, or can slip through either of the short connecting galleries and escape into any of the numerous radiatory runs.

In the central or middle chamber of the edifice the Mole places a quantity of dried grass or leaves, upon which it sleeps during its hours of repose. This complicated room is seldom used during the summer months, as at that time the Mole prefers to live in one of the ordinary hillocks.

The nest which the female contrives is not so complicated as the fortress, but is well adapted for its purpose. The hillock in which the nest is made is always a very large one, and is generally placed at some distance from the fortress. Its interior is very large, and is generally filled with dried grass, moss, or other similar substances, and it is said that in some of these nests have been found certain roots on which the young Moles can feed. This statement, however, is scarcely credible. The young are usually born about April, but their appearance in the world is not so determinately settled as that of many animals, as young Moles are found continually from March until August. The average of their number is four or five, although as many as seven young have been found in one nest. There is but one brood in a year.

The colour of the Mole is usually of a blackish grey, but it is extremely variable in the tinting of its fur, and it is not uncommon to find in a single locality specimens of every hue from brown to white. There are specimens in the British Museum of almost every tint, and I have long had in my possession a cream-coloured mole-skin, which was obtained I believe in Wiltshire, as it was furnished by a mole-catcher that resided in that county. The fur is so beautifully smooth and soft that it has sometimes, though rarely, been employed as an article of wearing apparel, or used as a light and delicate coverlet. The fur, or "felt," is best and most glossy if the animal is taken in the winter.

PASSING in a regular gradation from the moles to the shrews and hedgehogs, we pause for a while at the powerfully scented animal that is called, by virtue of its perfumed person, the MUSK RAT of India, and is also known by the titles of MONDJOUROU, and SONDELI.

This animal is a native of various parts of India, and is very well known on account of the extremely powerful scent which exudes from certain glands that are situated in the under parts of the body and on the flanks.

The odoriferous substance which is secreted by the above-mentioned glands is of a musky nature, and possesses the property of penetrating and adhering to every substance over which the Musk Rat has passed. The musky odour clings so pertinaciously to the objects which are impregnated with its tainting contact, that in many cases they become entirely useless. Provisions of all kinds are frequently spoiled by the evil odour with which they are saturated; and of so penetrating a nature is the musky scent, that the combined powers of glass and cork are unable to preserve the contents of bottles from its unpleasant influence. Let a Sondeli but run over a a bottle of wine, and the contained liquid will be so powerfully scented with a musky savour that it will be rendered unfit for civilized palates, and must be removed from the neighbourhood of other wines, lest the contaminating influence should extend to them also.

In colour it is not unlike the common shrew of England, having a slight chestnut or reddish tinge, upon a mouse-coloured ground, fading into grey on

SONDELI.—(*Sorex murinus.*)

the under parts of the body. In size, however, it is much the superior of that animal, being nearly as large as the common brown, or "Hanoverian" rat. The hair is very short, and the peculiar reddish brown hue of the fur is caused by the different tintings of the upper and under fur.

DURING the autumnal months of the year, the country roads and bypaths are frequently rendered remarkable by the presence of little mouse-like animals, with long snouts, that lie dead upon the ground, without mark of external injury to account for the manner of their decease.

These are the bodies of the SHREW MOUSE of England, otherwise known by the name of ERD SHREW.

The head of the Shrew is rather long, and its apparent length is increased by the long and flexible nose which gives so peculiar an aspect to the animal, and serves to distinguish it at a glance from the common mouse, which it so nearly resembles in general shape and colour. The object of this elongated nose is supposed to be for the purpose of enabling the animal to root in the ground after the various creatures on which it feeds, or to thrust its head among the densest and closest herbage. Many insects and their larvæ are found in such localities, and it is upon such food that the Shrew chiefly subsists. Worms are also captured and eaten by the Shrew, which in many of its habits is not unlike the mole.

The bite of the shrew is so insignificant as to make hardly any impress even on the delicate skin of the human hand. Popular prejudice, however, here steps in, and attributes to the bite of the shrew such venomous properties, that in many districts of England the viper is less feared than the little harmless Shrew.

The very touch of the Shrew's foot is considered a certain herald of evil, and animals or men which had been " shrew-struck," were supposed to labour under a malady which was incurable except by a rather singular remedy, which partakes somewhat of a homœopathic principle, that " similia similibus curantur."

The curative power which alone could heal the shrew-stroke lay in the branches of a shrew-ash, or an ash-tree which had been imbued with the

ERD SHREW, OR SHREW MOUSE.—(*Corsira vulgaris.*)

shrewish nature by a very simple process. A living Shrew was captured and carried to the ash-tree which was intended to receive the healing virtues. An auger-hole was made into the trunk, the poor Shrew was introduced into the cavity, and the auger-hole closed by a wooden plug. Fortunately for the wretched little prisoner, the entire want of air would almost immediately cause its death. But were its little life to linger for ever so long a time in the ash-trunk, its incarceration would still have taken place, for where superstition raises its cruel head, humanity is banished.

The nest of the Shrew is not made in the burrow, as might be supposed, but is built in a suitable depression in the ground, or in a hole in a bank. It is made of leaves and other similar substances, and is entered through a hole at the side. In this nest are produced the young Shrews, from five to seven in number, and, as may be imagined, extremely diminutive in size. They are generally born in the spring.

The total length of the adult Shrew is not quite four inches, of which the tail occupies very nearly the moiety.

SIMILAR to the Erd Shrew in general aspect, but easily to be distinguished from that animal by its colour and other peculiarities, the WATER SHREW stands next on our list.

The fur of the Water Shrew is nearly black upon the upper portions of the body, instead of the reddish brown colour which tints the fur of the Erd Shrew. The under parts of the body are beautifully white, and the line of demarcation between the two colours is very distinctly drawn. The fur is very soft and silken in texture, and, when the animal is submerged under the surface of the water, possesses the useful property of repelling moisture, and preserving the body of the animal from the injurious effects of the water.

When the Water Shrew is engaged in swimming, those parts of the fur which are submerged below the surface appear to be studded with an infinite number of tiny silvern beadlets, that give to the whole animal a very singular aspect. This phenomenon is produced by the minute air-bubbles that cling to the fur, and which exude from the space that is left between the hairs. In fact, the Shrew Mouse, when immersed, bears a curious resemblance to the well-known Water Spider.

WATER SHREW.—(*Crossopus Fodiens*).

A further distinction, and one which is more valuable than that which is furnished by the colour of the fur, is the fringe of stiff white hairs which edges the tail and the toes.

In all its movements the Water Shrew is extremely graceful and active, displaying equal agility, whether its movements be terrestrial or aquatic.

I have repeatedly observed the proceedings of a little colony of these animals, and was able to sit within a yard or two of their haunts, without their cognizance of my person. They are most sportive little creatures, and seem to enjoy a game of play with thorough appreciation, chasing each other over the ground and through the water, running up the stems of aquatic plants, and tumbling off the leaves into the water, scrambling hastily over the stones around which the stream ripples, and playing a thousand little pranks with the most evident enjoyment. Then they will suddenly cease their play, and begin to search after insects with the utmost gravity, rooting in the banks and picking up stray flies, as if they never had any other business in view.

From repeated observations, it seems that the Water Shrew is not entirely confined to the neighbourhood of water, neither is it totally dependent for its subsistence on aquatic insects, for it has been frequently seen at some distance from any stream or pond. It must be remarked, however, that a very small rivulet is amply sufficient for the purposes of the Water Shrew, which will take up its residence for several years in succession on the banks of a little artificial channel that is only used for the purpose of carrying water for the irrigation of low-lying fields.

THE largest of the British Shrews is that species which is called the OARED SHREW, on account of the oar-like formation of the feet and tail, which are

edged with even longer and stiffer hairs than those which decorate the same
parts in the Water Shrew.

As may be imagined from this structure, the habits of the animal are
aquatic in their nature, and its manners are so closely similar to those of the
preceding species, that it may easily be mistaken for that animal, when seen
at a little distance, so as to render the difference in size less conspicuous, and
the colour of the under portions of the body less apparent.

It has already been stated that the back of the Water Shrew is of a
velvety black, and the abdomen and under portions of the body of a beautiful
and clearly defined white. In the Oared Shrew, however, the black is profusely sprinkled with white hairs, and the fur of the abdomen and flanks is
blackish grey instead of pure white. The middle of the abdomen, however,
together with that of the throat, is strongly tinged with yellow : the throat
being more of an ashy yellow than the abdomen.

Although not so common as the Erd and the Water Shrew, it is of more

OARED SHREW, OR BLACK WATER SHREW. —(*Crossopus ciliatus.*)

frequent occurrence than is generally supposed, and has been found in many
parts of England where it was formerly supposed to be wanting. The total
length of the Oared Shrew is about five inches and a quarter, the head and
body measuring rather more than three inches, and the tail being about two
inches in length. Its nose is not quite so sharp or narrow as that of the
Water Shrew, and the ears are decorated with a slight fringe of white hair.
The latter third of the tail is flattened, as if for swimming, while the remaining two-thirds are nearly cylindrical, but are slightly squared, as has been
already mentioned of the common Shrew.

The elongation of the nose, which has already been noticed in the Tupaias
of Sumatra, seems to have reached the utmost limit in those curious inhabitants of the Cape that are called, from their elephantine elongation of nose,
the ELEPHANT SHREWS. Several species of Elephant Shrews are known
to exist, all of which, with one exception, are inhabitants of Southern Africa.
The solitary exception, Macroscelides Roretti, is found in Algeria.

The peculiarly long nose of the Elephant Shrew is perforated at its extremity by the nostrils, which are rather obliquely placed, and is supposed to aid the animal in its search after the insects and other creatures on which it feeds. The eyes are rather large in proportion to the size of the animal.

The tail is long and slender, much resembling the same organ in the common mouse, and in some specimens, probably males, is furnished at the base with glandular follicles, or little sacs. The legs are nearly of equal size, but the hinder limbs are much longer than the forelegs, on account of the very great length of the feet, which are capable of affording support to the creature as it sits in an upright position. As might be presumed from the great length of the hinder limbs, the Elephant Shrew is possessed of great locomotive powers, and when alarmed, can skim over the ground with such

ELEPHANT SHREW. —(*Macroscelides Proboscideus.*)

celerity that its form becomes quite obscured by the rapidity of its movement through the air. Its food consists of insects, which it captures in open day.

Although the Elephant Shrew is a diurnal animal, seeking its prey in broad daylight, its habitation is made below the surface of the ground, and consists of a deep and tortuous burrow, the entrance to which is a perpendicularly-sunk shaft of some little depth. To this place of refuge the creature always flies when alarmed, and as it is so exceedingly swift in its movements, it is not readily captured or intercepted.

The colour of the fur is a dark and rather cloudy brown, which is warmed with a reddish tinge upon the side and flanks, and fades on the abdomen and inner portions of the limbs into a greyish-white. The generic name, Macroscelides, is of Greek origin, in allusion to the great length of its hinder limbs, and signifies " long-legged." It is but a small animal, as the length of the head and body is not quite four inches in measurement, and the tail is about three inches and a quarter.

THE extraordinary animal which has been recently brought before the notice of zoologists, under the characteristic name of PEN-TAIL, is a native of Borneo, from which country it was brought by Mr. Hugh Low.

PEN-TAIL.—(*Ptilocercus Lowii.*)

It is about the size of a small rat, but appears to be of greater dimensions on account of its extremely long tail with the remarkable appendage at its extremity. As may be seen from the engraving, the tail is of extraordinary length when compared with the size of the body, and is devoid of hair except at its extremity, where it is furnished with a double row of stiff hairs on each side, which stand boldly out like the barbs of a quill-pen or the feathers of an arrow. The remainder of the tail is covered with scales, which are square in their form, like those of the long-tailed rats, and of considerable size. The colour of the tail is black, and the bristly barbs are white, so that this member presents a peculiarly quaint aspect.

The fur which covers the body of the Pen-tail is extremely soft in texture, and is of a blackish brown tint above, fading into a yellowish grey beneath. As the tips of the hairs are tinged with a yellow hue, the precise tint of the fur is rather indeterminate, and is changeable, according to the position of the hairs which are exposed to view. The specimen which is preserved in the British Museum was captured by Mr. Low in the house of Sir James Brook, the first Rajah of Sarawak.

THE common HEDGEHOG, HEDGE PIG, or URCHIN, is one of the most familiar of our indigenous mammalia, being found in every part of Great Britain which is capable of affording food and shelter.

The hard round spines which cover the upper part of its body are about an inch in length, and of a rather peculiar shape, which is well represented in the accompanying sketch. This form is wonderfully adapted to meet the peculiar objects which the spine is intended to fulfil, as will be seen in the following account.

The spine, which is here given, is supposed to be lying nearly horizontally upon the back of the animal, a position which it assumes whenever the

Hedgehog chooses to relax the peculiar muscle which governs the spines, and which serves to retain the creature in its coiled attitude. The point of the quill or spine is directed towards the tail. It will be seen that the quill is not unlike a large pin, being sharply pointed at one extremity, furnished at the other with a round bead-like head, and rather abruptly bent near the head. If the skin be removed from the Hedgehog, the quills are seen to be pinned as it were, through the skin, being retained by their round heads, which are acted upon by the peculiar muscle which has already been mentioned.

SPINE OF HEDGEHOG.

It is evident, therefore, that whenever the head of the quill is drawn backward by the contraction of the muscle, the point of the quill is erected in proportion to the force which is exerted upon the head, so that when the animal is rolled up, and the greatest tension is employed, the quills stand boldly out from the body, and present the bayonet-like array of points in every direction.

These curiously-formed spines are useful to the Hedgehog for other pur-

HEDGEHOG.—(*Erinaceus Europæus.*)

poses than the very obvious one of protecting the creature from the attacks of its foe. They are extremely elastic, as is found to be the case with hairs and quills of all descriptions, and the natural elasticity is increased by the sharp curve into which they are bent at their insertion into the skin. Protected by this defence, the Hedgehog is enabled to throw itself from considerable heights, to curl itself into a ball as it descends, and to reach the ground without suffering any harm from its fall. A Hedgehog has been seen repeatedly to throw itself from a wall, some twelve or fourteen feet in height, and to fall upon the hard ground without appearing even to be inconvenienced by its tumble. On reaching the ground it would unroll itself and trot off with perfect unconcern.

Marching securely under the guardianship of its thorn-spiked armour, the Hedgehog recks little of any foe, save man. For, with this single exception, there are, in our land at least, no enemies that need be dreaded by so well-protected an animal. Dogs, foxes, and cats are the only creatures which possess the capability of killing and eating the Hedge-hog, and of these foes it is very little afraid. For dogs are but seldom abroad at night while the Hedgehog is engaged in its nocturnal quests after food ; and the fox would not be foolish enough to waste its time and prick its nose in weary endeavours to force its intended prey out of its defences. Cats, too, are even less adapted to such a proceeding than dogs and foxes.

It is indeed said that the native cunning of the Fox enables it to overreach the Hedgehog, and to induce it to unrol itself by an ingenious, but, I fear, an apocryphal process. Reynard is said, whenever he finds a coiled-up Hedgehog, to roll it over and over with his paw towards some runnel, pond, or puddle, and then to souse it unexpectedly into the water. The Hedgehog, fearing that it is going to be drowned, straightway unrols itself, and is immediately pounced on by the cunning fox, which crushes its head with a single bite, and eats it afterwards at leisure. In America the puma is said to eat the Hedgehog in a very curious manner. Seizing the animal by the head, it gradually draws the animal through its teeth, swallowing the body and stripping off the skin.

Man, however, troubles himself very little about the Hedgehog's prickles, and when disposed to such a diet, kills, cooks, and eats it without hesitation.

The legitimate mode of proceeding is to kill the animal by a blow on the head, and then to envelop it, without removing the skin, in a thick layer of well-kneeded clay. The enwrapped Hedgehog is then placed on the fire, being carefully turned by the cook at proper intervals, and there remains until the clay is perfectly dry and begins to crack. When this event has taken place, the cooking is considered to be complete, and the animal is removed from the fire. The clay covering is then broken off, and carries away with it the whole of the skin, which is adherent by means of the prickles. By this mode of cookery the juices are preserved, and the result is pronounced to be supremely excellent.

This primitive but admirable form of cookery is almost entirely confined to gipsies and other wanderers, as in these days there are few civilized persons who would condescend to partake of such a diet. Utilitarians, however, can render the creature subservient to their purposes by using it as a guardian to their kitchens. Its insect-devouring powers are of such a nature that it can be made a most useful inhabitant of the house, and set in charge of the " black beetles."

The rapidity with which it extirpates the cockroaches is most marvellous, for their speed and wariness are so great that the Hedgehog must possess no small amount of both qualities in order to destroy them so easily. A Hedgehog which resided for some years in our house was accustomed to pass a somewhat nomad existence, for as soon as it had eaten all the cockroaches in our kitchen, it used to be lent to a friend, to whom it performed the same valuable service. In a few months those tiresome insects had again multiplied, and the Hedgehog was restored to its former habitation.

The creature was marvellously tame, and would come at any time to a saucer of milk in broad daylight. Sometimes it took a fancy to promenading the garden, when it would trot along in its own quaint style, poking its sharp nose into every crevice, and turning over every fallen leaf that lay in its path. If it heard a strange step, it would immediately curl itself into a ball, and lie in that posture for a few minutes until its alarm had passed away, when it would cautiously unrol itself, and peer about with its bead-like eyes for a moment or two, and then resume its progress.

From all appearances it might have lived for many years had it not come by its death in a rather singular manner. There was a wood-shed in the kitchen garden, where the bean and pea sticks were laid up in ordinary during the greater part of the year, and it seemed, for some unknown reason, to afford a marvellous attraction to the Hedgehog. So partial to this locality was the creature, that whenever it was missing we were nearly sure to find it among the bean-sticks in the wood-shed. One morning, however, on searching for the animal, in consequence of having missed its presence for some days, we found it hanging by its neck in the fork of a stick, and quite dead. The poor creature had probably slipped while climbing among the sticks, and had been caught by the neck in the bifurcation.

The Hedgehog is accused of stealing and breaking eggs, to which indictment it can but plead guilty.

It is very ingenious in its method of opening and eating eggs ; a feat which it performs without losing any of the golden contents. Instead of breaking the shell and running the chance of permitting the contents to roll out, the clever animal lays the egg on the ground, holds it firmly between its fore-feet, bites a hole in the upper portion of the shell, and, inserting its tongue into the orifice, licks out the contents daintily.

Not contenting itself with such comparatively meagre diet as eggs, the Hedgehog is a great destroyer of snakes, frogs, and other animals, crunching them together with their bones as easily as a horse will eat a carrot. Even the thick bone of a mutton chop, or the big bone of the fish, is splintered by the Hedgehog's teeth with marvellous ease. On one account it is rather a valuable animal, for it will attack a viper as readily as a grass-snake, trusting apparently to its prickly armour as a defence against the serpent's fangs.

Whether, in its wild state, it is able to catch the little birds, is not accurately known, but in captivity it eats finches and other little birds with great voracity. One of these animals, that was kept in a state of domestication, ate no less than seven sparrows in the course of a single night, and another of these creatures crushed and ate in the course of twenty-four hours more than as many sparrow heads, eating bones, bill, and neck with equal ease.

Its legitimate prey is found among the insect tribe, of which it consumes vast numbers, being able, not only to chase and capture those which run upon the ground, but even dig in the earth and feed upon the grubs, worms, and various larvæ which pass their lives beneath the surface of the ground. A Hedgehog has been seen to exhume the nest of the humblebee, which had been placed in a sloping bank, as is often the case with the habitation of these insects, and to eat bees, grubs, and honey, unmindful of the anger of the survivors, who, however, appeared to be but little affected by the inroads which the Hedgehog was making upon their offspring and their stores.

The home of the Hedgehog is made in some retired and well-protected spot, such as a crevice in rocky ground, or under the stones of some old ruin. It greatly affects hollow trees, wherever the decayed wood permits it to find an easy entrance, and not unfrequently is found coiled up in a warm nest which it has made under the large gnarled roots of some old tree, where the rains have washed away the earth and left the roots projecting occasionally from the ground. Besides these legitimate habitations, the Hedgehog is frequently found to intrude itself upon the homes of other animals, and has been often captured within rabbit burrows. Perhaps it may be led to these localities by the double motive of obtaining shelter from weather and enemies, and of making prey of an occasional young rabbit.

In its retreat the Hedgehog usually passes the winter in that semi-animate condition which is known by the name of hibernation.

The hibernation of the Hedgehog is more complete than that of the dormouse or any other of our indigenous hibernating quadrupeds, for they always have a stock of food on which they can rely, and of which they sparingly partake during the cold months of the year. The Hedgehog, however, lays up no such stores, nor, indeed, could it do so, for, as has already been mentioned, its food is almost entirely of an animal nature.

The sight of the Hedgehog does not appear to be so excellent as its powers of scent, which are admirably developed, as may be seen by opening the side of a Hedgehog's face.

One of these animals has been seen to chase a partridge across a road, following her through the hedge with perfect precision ; and another was observed to discover the presence of mankind by means of its powers of scent, as it was in a position from whence it could not see its fancied enemies. The Hedgehog had already passed the observers, who remained perfectly quiet in order to watch its proceedings, but after it had run for a few paces, it suddenly stopped, seemed suspicious of some danger, stretched its nose in the air, and stood on its guard. In a few moments it seemed to have set itself at ease, and resumed its course. The spectators then slightly shifted their position, so as to bring the animal again within the range of their " wind," when the creature repeated the same process, and did not appear entirely at its ease for some little time.

MACROPIDÆ.

THE extraordinary animals which are grouped together under the title of Macropidæ are, with the exception of the well-known Opossum of Virginia, inhabitants of Australasia, and the islands of the Indian Archipelago.

The peculiarity which gives the greatest interest to this group of animals is that wonderful modification of the nutrient organs which has gained for them the title of MARSUPIALIA, or pouched animals—a name which is derived from the Latin word *marsupium*, which signifies a purse or pouch. This singular structure is only found in the female Marsupials, and in them is variously developed according to the character of the animal and the mode of life for which it is intended.

The lower part of the abdomen is furnished with a tolerably large pouch, in the interior of which the mammæ, or teats, are placed. When the young, even of so large an animal as the kangaroo, make their appearance in the world, they are exceedingly minute—the young kangaroo being only an inch in length—and entirely unable to endure the rough treatment which they would meet with were they to be nurtured according to the manner in which the young of all other animals are nourished. Accordingly, as soon as they are born they are transferred by the mother into the pouch, when they instinctively attach themselves to the teats, and there hang until they have attained considerable dimensions. By degrees, as they grow older and stronger, they loosen their hold, and put their little heads out of the living cradle, in order to survey the world at leisure. In a few weeks more they gain sufficient strength to leave the pouch entirely, and to frisk about under the guardianship of their mother, who, however, is always ready to receive them again into their cradle if there is any rumour of danger, and if any necessity for flight should present itself, flies from the dangerous locality, carrying her young with her.

At the head of the Macropidæ are placed a small but interesting band of marsupial animals, which are called Phalangistines, on account of the curious manner in which two of the toes belonging to the hinder feet are joined together as far as the " phalanges." The feet are all formed with great powers of grasp, and their structure is intended to fit them for procuring their food among the branches of the trees, on which they pass the greater portion of their existence.

First and least of the Phalangistines, is the beautiful little animal which is called the OPOSSUM MOUSE in some parts of the country, and the FLYING MOUSE in others.

This pretty little creature is about the size of our common mouse, and when it is resting upon a branch, with its parachute, or umbrella of skin, drawn close to the body by its own elasticity, it looks very like the common mouse of Europe, and at a little distance might easily be taken for that animal. In total length it rather exceeds six inches, the length of its head and body being about three inches and a half, and that of the tail not quite

OPOSSUM MOUSE.—(*Acrobates pygmæus.*)

three inches. On account of its minute size this animal is also called the PIGMY PETAURIST.

In the colour of the upper portions of the body the Opossum Mouse is of the well known mouse tint, slightly sprinkled with a reddish hue ; but on the abdomen and under portions of the skin parachute, the fur is beautifully white. The line of demarcation between the hair is very well defined, and there is a narrow stripe of darker brown that marks out the line of juncture. When the animal is at rest, the parachute closes by its own elasticity, and gathers itself into folds, which have a very pretty effect, on account of the delicate white fur which becomes exposed by the action, and which undulates in rich and graceful folds, alternating with the dark fur of the back and the still darker stripe that forms the line of demarcation.

The tail of the Opossum Mouse is nearly as long as the body, very slender, and remarkable for the manner in which the hairs are affixed to it. The hairs that fringe the greater part of the tail are about one-sixth of an inch in length, reddish grey in colour, rather stiff, and are set on the tail in a double row, like the barks of a feather. A similar formation has already been described in the history of the pen-tail of Sarawak. This mode of arrangement is called " distichous."

The food of the Petaurists is generally of a vegetable character, consisting of leaves, fruits, and buds, but the sharply-pointed molars of the Opossum Mouse approach so closely to the insectivorous type that the creature is probably able to vary a vegetable diet by occasional admixture with animal food.

The parachute-like expansion of the skin is of very great service to the animal when it wishes to pass from one branch or from one tree to another without the trouble of descending and the laborious climbing up again. Trusting to the powers of its parachute, the little creature will boldly launch itself into the air, stretching out all its limbs, and expanding the skin to the utmost. Upborne by this membrane, the Opossum Mouse can sweep through very great intervals of space, and possesses no small power of altering its course at will. It cannot, however, support itself in the air by moving its limbs, like the bats, nor can it make any aërial progress when the original impetus of its leap has expired.

ARIEL PETAURUS.—(*Petaurus Ariel.*)

The beautiful little animal which has been called by the expressive name of ARIEL, is about the size of a small rat, and in the hue of the upper portions of the body is not unlike that animal.

The colour of the fur upon the upper portions of the body is a light brown, which darkens considerably upon the parachute membrane. On the under surface it is white, the white fur just turning over the edge of the parachute, and presenting a pretty contrast with the dark brown colour of its upper surface. The tail is nearly of the same colour as the body, with the exception of the tip, which is dark. On account of its graceful movements, and the easy undulating sweep of its passage through the air, it has earned for itself the appropriate name of Ariel, in remembrance of the exquisite and tricksy sprite that animates the world-celebrated drama of the "Tempest."

It is not an uncommon animal, and is frequently seen at Port Essington.

Fox-like in nature as well as in form, the VULPINE PHALANGIST has well earned the name which has been given to it by common consent. It has also been entitled the Vulpine Opossum, and in its native country is popularly called by the latter of these names.

It is an extremely common animal, and is the widest diffused of all the Australian opossum-like animals. Like the preceding animals, it is a nocturnal being, residing during the day in the hollows of decaying trees, and only venturing from its retreat as evening draws on. The nature of its food is of a mixed character, for the creature is capable of feeding on vegetable food, like the Petaurists, and also displays a considerable taste for animal food of all kinds. If a small bird be given to a Vulpine Phalangist, the creature seizes it in its paws, manipulates it adroitly for a while, and then tears it to pieces and eats it. It is rather a remarkable fact, that the animal is peculiarly fond of the brain, and always commences its feast by crushing the head between its teeth, and devouring the brain.

In all probability, therefore, the creature makes no small portion of its meals on various animal substances, such as insects, reptiles, and eggs. As to the birds on which it so loves to feed, it may very probably, although so slow an animal, capture them in the same manner as has been related of the lemurs, viz. by creeping slowly and cautiously upon them as they sleep, and swiftly seizing them before they can awaken to a sense of their danger. It is a tolerably large animal, equalling a large cat in dimensions, and is, therefore, able to make dire havoc among such prey whenever it chooses to issue

VULPINE PHALANGIST.—(*Phalangista vulpina.*)

forth with the intention of making a meal upon some small bird that may chance to be sleeping in fancied security.

The fore-paws of the Vulpine Phalangist are well adapted for such proceedings, as they are possessed of great strength and mobility, so that the animal is able to take up any small object in its paws, and to hold it after the manner of the common squirrel. When feeding, it generally takes its food in its fore-paws, and so conveys it to its mouth. In captivity it does not seem to be a very intelligent animal, even when night brings forth its time of energy, and it but little responds to the advances of its owner, however kind he may be. It will feed on bread and milk, or fruits, or leaves, or buds, or any substance of a similar nature, but always seems best pleased when it is supplied with some small birds or animals, and devours them with evident glee.

The flesh of the Vulpine Phalangist is considered to be very good, and the natives are so fond of it that, notwithstanding the laziness that is engrained in their very beings, except when they are under the influence of some potent excitement, they can seldom refrain from chasing an " opossum," even though they have been well fed by the white settlers. When the fresh body of a Vulpine Phalangist is opened, a kind of camphorated odour is diffused from it, which is probably occasioned by the foliage of the camphor-perfumed trees in which it dwells, and the leaves of which it eats.

The fur of this animal is not valued so highly as that of the Tapoa, probably because it is of more common occurrence, for the colour of the hair is much more elegant, and its quality seems to be really excellent. Some few experiments have been made upon the capabilities of this fur, and, as far as has yet been accomplished, with very great success. Good judges have declared that articles which had been made from this fur presented a great resemblance to those which had been made from Angola wool, but appeared to be of superior quality. The hat-makers have already discovered the value of the fur, and are in the habit of employing it in their trade.

The natives employ the skin of the " opossum " in the manufacture of their scanty mantles, as well as for sundry other purposes, and prepare the skins in a rather ingenious manner. As soon as the skin is stripped from the animal's body, it is laid on the ground, with the hairy side downwards, and secured from shrinking by a number of little pegs which are fixed around its edges. The inner side is then continually scraped with a shell, and by degrees the skin becomes perfectly clean and pliable. When a sufficient number of skins are prepared, they are ingeniously sewn together with a thread that is made from the tendons of the kangaroo, which, when dried, can be separated into innumerable filaments. A sharpened piece of bone stands the sable tailor in place of a needle. From the skin of the same animal is also formed the " kumeel," or badge of manhood, a slight belt, which no one is permitted to wear until he has been solemnly admitted among the assembly of men.

In its colour, the Vulpine Phalangist is rather variable, but the general hue of its fur is a greyish-brown, sometimes tinted with a ruddy hue. The tail is long, thick, and woolly in its character, and in colour it resembles that of the body, with the exception of the tip, which is nearly black. The dimensions of an old male are given by Mr. Bennett as follows : Total length, two feet seven inches ; the head being four inches in length, and the tail nearly a foot.

THE quaint looking animal which is popularly known by the native name of KOALA, or the AUSTRALIAN BEAR, is of some importance in the zoological world, as it serves to fill up the gulf that exists between the phalangistines and the kangaroos.

The Koala is nocturnal in its habits, and is not very unfrequently found, even in the localities which it most affects. It is not nearly so widely spread as most of the preceding animals, as it is never known to exist in a wild state except in the south-eastern regions of Australia.

Although well adapted by nature for climbing among the branches of trees, the Koala is by no means an active animal, proceeding on its way with very great deliberation, and making sure of its hold as it goes along. Its feet are peculiarly adapted for the slow but sure mode in which the animal progresses among the branches by the structure of the toes of the fore-feet or paws, which are divided into two sets, the one composed of the two inner toes, and the other of the three outer, in a manner which reminds the observer of the feet of the scansorial birds and the chameleon. This formation, although well calculated to serve the animal when it is moving among the branches, is but of little use when it is upon the ground, so that the terrestrial progress of the Koala is especially slow, and the creature seems to crawl rather than walk.

It seems to be a very gentle creature, and will often suffer itself to be captured without offering much resistance, or seeming to trouble itself about its captivity. But it is liable, as are many gentle animals, to sudden and unexpected gusts of passion, and when it is excited by rage it puts on a very fierce look, and utters sharp and shrill yells in a very threatening manner. Its usual voice is a peculiar soft bark.

This animal is rather prettily coloured, the body being furnished with fur of a fine grey colour, warmed with a slight reddish tinge in the adult animal, and fading to a whitish grey in the young. The claws are considerably curved, and black ; and the ears are tufted with long white hairs. In size it equals a small bull-terrier dog, being, when adult, rather more than two feet in length, and about ten inches in height, when standing. The circumference of the body is about eighteen inches, including the fur.

KOALA, OR AUSTRALIAN BEAR.—(*Phascolarctos cinereus.*)

On account of the tree-climbing habits of the Koala, it is sometimes called the Australian Monkey as well as the Australian Bear.

THE animals which come next under consideration are truly worthy of the title of Macropidæ, or long-footed, as their hinder feet are most remarkable for their comparative length, and in almost every instance are many times longer than the fore-feet. This structure adapts them admirably for leaping, an exercise in which the Kangaroos, as these creatures are familiarly termed, are pre-eminently excellent.

Among the largest of the Macropidæ is the celebrated KANGAROO, an animal which is found spread tolerably widely over its native land.

This species has also been called by the name of *giganteus*, on account of its very great size, which, however, is sometimes exceeded by the Woolly Kangaroo. The average dimensions of an adult male are generally as follows :—The total length of the animal is about seven feet six inches, counting from the nose to the tip of the tail ; the head and body exceed four feet, and the tail is rather more than three feet in length ; the circumference of the tail at its base is about a foot. When it sits erect after its curious tripedal fashion, supported by its hind-quarters and tail, its height is rather more than fifty inches ; but when it wishes to survey the country, and stands erect upon its toes, it surpasses in height many a well-grown man. The female is very much smaller than her mate, being under six feet in total length, and the difference in size is so great that the two sexes might be well taken for different species.

The weight of a full-grown male, or " Boomer," as it is more familiarly called, is very considerable, one hundred and sixty pounds having often been attained, and even greater weight being on record. The colour of the animal is brown, mingled with grey, the grey predominating on the under portions of the body and the under-faces of the limbs. The fore-feet are black, as is also the tip of the tail.

As the Kangaroo is a valuable animal, not only for the sake of its skin, but on account of its flesh, which is in some estimation among the human inhabitants of the same land, it is eagerly sought after by hunters, both white and black, and affords good sport to both on account of its speed, its vigour, and its wariness. The native hunter, who trusts chiefly to his own cunning and address for stealing unobserved upon the animal and lodging a spear in its body before it is able to elude its subtle enemy, finds the Kangaroo an animal which will test all his powers before he can attain his object, and lay the Kangaroo dead upon he ground.

The male Kangaroo, or " Boomer," is a dangerous antagonist to man and dog, and unless destroyed by missile weapons will often prove more than a mat h for the combined efforts of man and beast.

When the animal finds that it is overpowered in endeavour by the swift and powerful Kangaroo dogs, which are bred for the express purpose of chasing this one kind of prey, it turns suddenly to bay, and placing its back against a tree-trunk, so that it cannot

THE KANGAROO.--(*Macropus major.*)

be attacked from behind, patiently awaits the onset of its adversaries. Should an unwary dog approach within too close a distance of the Kangaroo, the animal launches so terrible a blow with its hinder feet, that the long and pointed claw, with which the hinder foot is armed, cuts like a knife, and has often laid open the entire body of the dog with a single blow. The claw which is thus used is so long, hard, and sharp, that it is sometimes used as a head to a spear.

When running, the creature has a curious habit of looking back every now and then, and has sometimes unconsciously committed suicide by leaping against one of the tree-stumps which are so plentifully found in the districts inhabited by the Kangaroo.

The doe Kangaroo displays very little of these running or fighting capabilities, and has been known, when chased for a very short distance, to lie down and die of fear. Sometimes, when pursued, she contrives to elude the dogs by rushing into some brushwood, and then making a very powerful leap to one side, so as to throw the dogs off the scent. She lies perfectly still as the dogs rush past her place of concealment, and when they have fairly passed her she quietly makes good her escape in another direction. When young, and before she has borne young, the female Kangaroo affords good sport, and is called, from her extraordinary speed, the "Flying Doe."

The Kangaroo is a very hardy animal and thrives well in England, where it might probably be domesticated to a large extent if necessary, and where it would enjoy a more genial climate than it finds in many districts of its native land. One of the favoured localities of this species is the bleak, wet, and snow-capped summit of Mount Wellington.

The eye of the Kangaroo is very beautiful, large, round, and soft, and gives to the animal a gentle, gazelle-like expression that compensates for the savage aspect of the teeth, as they gleam whitely between the cleft lips.

The KANGAROO RAT, called by the natives the POTOROO, is a native of New South Wales, where it is found in very great numbers.

KANGAROO RAT.—(*Hypsiprymnus minor.*)

It is but a diminutive animal, the head and body being only fifteen inches long, and the tail between ten and eleven inches. The colour of the fur is brownish black, pencilled along the back with a grey white. The under parts of the body are white, and the fore-feet are brown. The tail is equal to the body in length, and is covered with scales, through the intervals of which sundry short, stiff, and black hairs protrude.

This little animal frequents the less open districts, and is very quick and lively in its movements, whether it be indulging in its native gamesomeness or engaged in the search for food. Roots of various kinds are the favourite diet of the Kangaroo Rat, and in order to obtain these dainties the animal scratches them from the ground with the powerful claws of the fore-feet.

It is not so exclusively nocturnal as many of the preceding animals, and seems to be equally lively by day as by night. When the animal is sitting upon its hinder portions, the tail receives part of the weight of the body, but is not used in the same manner as the tail of the true Kangaroos, which, when they are moving slowly and leisurely along, are accustomed to support the body on the tail, and to swing the hinder legs forward like a man swinging himself upon crutches.

I 2

The WOMBAT, or AUSTRALIAN BADGER, as it is popularly called by the colonists, is so singularly unlike the preceding and succeeding animals in its aspect and habits, that it might well be supposed to belong to quite a different order.

As might be imagined from its heavy body and short legs, the Wombat is by no means an active animal, but trudges along at its own pace, with a heavy rolling waddle or hobble, like the gait of a very fat bear. It is found in almost all parts of Australia. The fur of the Wombat is warm, long, and very harsh to the touch, and its colour is grey, mottled with black and white. The under parts of its body are greyish white, and the feet are black. The muzzle is very broad and thick. The length of the animal is about three feet, the head measuring seven inches.

WOMBAT.—(*Phascolomys ursinus.*)

It is nocturnal in its habits, living during the day in the depths of a capacious burrow, which it excavates in the earth to such a depth that even the persevering natives will seldom attempt to dig a Wombat out of its tunnel.

The creature seems to be remarkably sensitive to cold, considering the severe weather which often reigns in its native country. It is fond of hay, which it chops into short pieces with its knife-edged teeth. The natives say that if a Wombat is making a journey and happens to come across a river, it is not in the least discomfited, but walks deliberately into the river, across the bed of the stream, and, emerging on the opposite bank, continues its course as calmly as if no impediment had been placed in its way.

The BANDICOOTS form a little group of animals that are easily recognisable by means of their rat-like aspect, and a certain peculiar but indescribable mode of carrying themselves. The gait of the Bandicoot is very singular, being a kind of mixture between jumping and running, which is the result of the formation of the legs and feet.

The food of the LONG-NOSED BANDICOOT is said to be of a purely vegetable nature, and the animal is reported to occasion some havoc among the

LONG-NOSED BANDICOOT. —(*Perameles nasuta.*)

gardens and granaries of the colonists. Its long and powerful claws aid it
in obtaining roots, and it is not at all unlikely that it may, at the same time

that it unearths and eats a
root, seize and devour the
terrestrial larvæ which are
found in almost every
square inch of ground. The
lengthened nose and sharp
teeth which present so
great a resemblance to the
same organs in insectivorous shrews, afford good
reasons for conjecturing
that they may be employed
in much the same manner.

The CHŒROPUS was for-
merly designated by the
specific title of *ecaudatus*, or
"tailless," because the first
specimen that had been
captured was devoid of
caudal appendage, and
therefore its discoverers
naturally concluded that all
its kindred were equally
curtailed of their fair proportions. But as new specimens came before the notice
of the zoological world, it
was found that the Chœropus

CHŒROPUS. —(*Chœropus castanotis.*)

was rightly possessed of a moderately long and somewhat rat-like tail, and
that the taillessness of the original specimen was only the result of accident

to the individual, and not the normal condition of the species. The size of the Chœropus is about equal to that of a small rabbit, and the soft, woolly fur is much of the same colour as that of the common wild rabbit.

It is an inhabitant of New South Wales, and was first discovered by Sir Thomas Mitchell on the banks of the Murray River, equally to the astonishment of white men and natives, the latter declaring that they had never before seen such a creature. The speed of the Chœropus is considerable, and its usual haunts are among the masses of dense scrub foliage that cover so vast an extent of ground in its native country. Its nest is similar to that of the bandicoot, being made of dried grass and leaves rather artistically put together, the grass, however, predominating over the leaves. The locality of the nest is generally at the foot of a dense bush, or of a heavy tuft of grass, and it is so carefully veiled from view by the mode of its construction that it can scarcely be discovered by the eyes of any but an experienced hunter.

The head of the Chœropus is rather peculiar, being considerably lengthened, cylindrically tapering towards the nose, so that its form has been rather happily compared to the neck and shoulders of a champagne bottle. The hinder feet are like those of the bandicoots, and there is a small swelling at the base of the toes of the fore-feet, which is probably the representative of the missing joints, more especially as the outermost toes are always extremely small in the bandicoots, to which the Chœropus is nearly allied.

DASYURE.—(*Dasyurus viverrinus.*)

The ears are very large in proportion to the size of the animal. The pouch opens backwards.

The food of the Chœropus is said to be of a mixed character, and to consist of various vegetable substances and of insects.

THE teeth of the Dasyurines, sharp-edged and pointed, indicate the carnivorous character of those animals to which they belong.

In the common DASYURE the general colour of the fur is brown, of a very dark hue, sometimes deepening into positive black, diversified with many spots of white, scattered apparently at random over the whole of the body, and varying both in their position and dimensions in almost every individual. In some specimens the tail is washed with white spots similar to those of the body, but in many examples the tail is uniformly dark. In all the Dasyures this member is moderately long, but not prehensile, and is thickly covered with hair; a peculiarity which has caused zoologists to give the title of Dasyure, or "hair-tail," to these animals. This species is the animal which is known in Australia by the popular name of Native Cat.

They are all inhabitants of Australasia, the Common Dasyure being found numerously enough in New Holland, Van Diemen's Land, and some parts of Australia. The habits of all the Dasyures are so very similar that there is no need of describing them separately. They are all rather voracious animals, feeding upon the smaller quadrupeds, birds, insects, and other living beings which inhabit the same country. The Dasyure is said to follow the example of several allied animals, and to be fond of roaming along the sea-coasts by night in search of food.

The Dasyures are all nocturnal animals, and very seldom make voluntary excursions from their hiding-places so long as the sun is above the horizon. They do not, like the Tasmanian wolf and the ursine Dasyure, lie hidden in burrows under the earth, or in the depths of rocky ground, but follow the example of the Petaurists, and make their habitations in the hollows of decayed trees.

YELLOW-FOOTED POUCHED MOUSE.—(*Antechinus flavipes.*)

The YELLOW-FOOTED POUCHED MOUSE is a very pretty little creature, its fur being richly tinted with various pleasing hues.

The face, the upper part of the head, and the shoulders, are dark grey, diversified with yellow hairs, and the sides of the body are warmed with a wash of bright chestnut. The under parts of the body, the chin, and the throat are uniform white, and the tail is black. There is often a slight tufting of hair on the extremity of the tail. The total length of the animal is about eight inches, the head and body being rather more than four inches and a half in length, and the tail a little more than three inches.

The MYRMECOBIUS is remarkable for several parts of its structure, and more especially so for the extraordinary number of its teeth, and the manner in which they are placed in the jaw. Altogether, there are no less than fifty-two teeth in the jaws of an adult and perfect specimen of the Myrmecobius, outnumbering the teeth of every other animal, with the exception of one or two cetacea and the armadillo. There is no pouch in this animal, but the tender young are defended from danger by the long hairs which clothe the under portions of the body.

It is a beautiful little animal, the fur being of agreeable tints and diversified by several bold stripes across the back. The general colour of the fur is a bright fawn on the shoulders, which deepens into blackish-brown from the shoulders to the tail, the fur of the hinder portions being nearly black. Across the back are drawn six or seven white bands, broad on the back and tapering off towards their extremities. The under parts of the body are of a yellowish-white. The tail is thickly covered with long bushy hair, and has a grizzled aspect, owing to the manner in which the black and white hairs of which it is composed are mingled together. Some hairs are annulated with white, red-rust, and black, so that the tints are rather variable, and never precisely the same in two individuals.

The length of the body is about ten inches, and the tail measures about seven inches, so that the dimensions of the animal are similar to those of the common water vole of Europe.

It is an active animal, and when running, its movements are very similar to those of the common squirrel. When hurried it proceeds by a series of small jumps, the tail being elevated over its back after the usual custom of squirrels, and at short intervals it pauses, sits upright, and casts an anxious look in all directions before it again takes to flight. Although not a particularly swift animal, it is not an easy one to capture, as it immediately makes for some place of refuge, under a hollow tree or a cleft in rocky ground, and when it has fairly placed itself beyond the reach of its pursuers, it bids defiance to their efforts to drive it from its haven of safety. Not even smoke—the usual resort of a hunter when his prey has gone to "earth" and refuses to come out again—has the least effect on the Myrmecobius, which is either possessed of sufficient smoke-resisting powers to endure the stifling vapour with impunity, or of sufficient courage to yield its life in the recesses of its haven, rather than deliver itself into the hands of its enemies.

MYRMECOBIUS.—(*Myrmecobius fasciatus.*)

The food of the Myrmecobius is supposed chiefly to consist of ants and similar diet, as it is generally found inhabiting localities where ants most abound. For this kind of food it is well fitted by its long tongue, which is nearly as thick as a common black-lead pencil, and is capable of protrusion to some distance. In confinement a specimen of the Myrmecobius was accustomed to feed on bran among other substances. It is known that in the wild state it will eat hay, as well as the "manna" that exudes from the branches of the Eucalypti.

It is a very gentle animal in its disposition, as, when captured, it does not bite or scratch, but only vents its displeasure in a series of little grunts when it finds that it is unable to make its escape. The number of its young is rather various, but averages from five to eight. The usual habitation of the Myrmecobius is placed in the decayed trunk of a fallen tree, or, in default of such lodging, is made in a hollow in the ground. It is a native of the borders of the Swan River.

THERE are very few of the marsupiated animals which are more remarkable for their form, their habits, or their character, than the Opossums of America. They are nearly all admirable climbers, and are assisted in their scansorial efforts by their long prehensile tails, which are covered with scales, through the interstices of which a few short black hairs protrude. The hinder feet are also well adapted for climbing, as the thumb is opposable to the other toes, so that the animal is able to grasp the branch of a tree with considerable force, and to suspend its whole body together with the additional weight of its prey or its young.

The VIRGINIAN or COMMON OPOSSUM is, as its name implies, a native of Virginia as well as of many other portions of the United States of America. In size it equals a tolerably large cat, being rather more than three feet in

OPOSSUM.—(*Didelphys Virginiana.*)

total length, the head and body measuring twenty-two inches, and the tail fifteen. The colour of this animal is a greyish white, slightly tinged with yellow, and diversified by occasional long hairs that are white towards their base, but of a brownish hue towards their points. These brown-tipped hairs

are extremely prevalent upon the limbs, which are almost wholly of the brown hue, which also surrounds the eye to some extent. The under fur is comparatively soft and woolly, but the general character of the fur is harsh and coarse. The scaly portion of the tail is white.

It is a voracious and destructive animal, prowling about during the hours of darkness, and prying into every nook and corner in hope of finding something that may satisfy the cravings of imperious hunger. Young birds, eggs, the smaller quadrupeds, such as young rabbits, which it eats by the brood at a time, cotton rats and mice, reptiles of various kinds, and insects, fall victims to the appetite of the Virginian Opossum, which is often not content with the food which it finds in the open forests, but must needs insinuate itself into the poultry-yard and make a meal on the fowls and their eggs.

Besides the varied animal diet in which the Opossum indulges, it also eats vegetable substances, committing as much havoc among plantations and fruit-trees as among rabbits and poultry. It is very fond of maize, procuring the coveted food by climbing the tall stems, or by biting them across and breaking them down. It also eats acorns, beech-nuts, chestnuts, and wild berries, while its fondness for the fruit of the " persimmon " tree is almost proverbial. While feeding on those fruits it has been seen hanging by its tail, or its hinder paws, gathering the " persimmons " with its fore-paws, and eating them while thus suspended. It also feeds on various roots, which it digs out of the ground with ease.

Its gait is usually slow and awkward, but when pursued it runs with considerable speed, though in a sufficiently clumsy fashion, caused by its habit of using the limbs of the right and left sides simultaneously in a kind of amble. As, moreover, the creature is plantigrade in its walk, it may be imagined to be anything but elegant in its mode of progress upon the ground. Although it is such an adept at "'possuming," or feigning death, it does not put this ruse in practice until it has used every endeavour to elude its pursuers, and finds that it has no possibility of escape. It runs sulkily and sneakingly forward, looking on every side for some convenient shelter, and seizing the first opportunity of slipping under cover.

The nest of the Opossum is always made in some protected situation, such as the hollow of a fallen or a standing tree, or under the shelter of some old projecting roots.

In MERIAN'S OPOSSUM there is no true pouch, and the place of that curious structure is only indicated by a fold of skin, so that during the infancy of its young, the mother is obliged to have recourse to that singular custom which has gained for it the title of *dorsigerus*, or " back-bearing." At a very early age the young Opossums are shifted to the back of their mother, where they cling tightly to their mother's fur with their little hand-like feet, and further secure themselves by twining their own tails round that of the parent. The little group which is here given was sketched from a stuffed specimen in the British Museum, where the peculiar attitude of mother and young is wonderfully preserved, when the very minute dimensions of the young Opossums are taken into consideration.

Many other species of Opossums are in the habit of carrying their young upon their backs, even though they may be furnished with a well-developed pouch ; but in the pouchless Opossums the young are placed on the back at a very early age, and are retained there for a considerable period.

It is a very small animal, measuring when adult only six inches from the nose to the root of the tail, the tail itself being more than seven inches in length, thus exceeding the united measurement of the head and body. Its general appearance is much like that of a very large mouse or a very small rat.

The fur of the Merian's Opossum is very short, and lies closely upon the skin. On the upper portions of the body its colour is a pale greyish brown, fading below into a yellowish white. Round the eyes is a deep brown mark, which extends forward in front of each eye, and forms a small dark patch.

MERIAN'S OPOSSUM.—(*Philander dorsigerus.*)
(Size of Life.)

The forehead, the upper part of the head, and the cheeks, together with the limbs and feet, are of a yellowish white, tending to grey.

Towards the base the tail is clothed with hair of the same texture and colour as that of the upper part of the body, but towards its extremity it becomes white. Its native country is Surinam.

PHOCIDÆ, OR SEALS.

We now arrive at a very wonderful series of animals, which, although they breathe atmospheric air like other mammalia, are yet almost entirely aquatic in their habits, and are never seen except in the water or its immediate vicinity. The first family of these aquatic mammalia is that which is formed of the animals which are popularly known by the name of SEALS.

The COMMON SEAL is spread very widely over many portions of the globe, and is of very frequent occurrence upon our own coasts, where it is found in considerable numbers, much to the annoyance of the fishermen, who look upon it with intense hatred, on account of the havoc which it makes among the fish.

It is rather a handsome animal, with its beautifully mottled skin and large intelligent eyes, and, although not so large as other species which are also found upon the British coasts, yields to none of them in point of beauty. The colour of its fur is generally of a greyish yellow, sprinkled with spots of brown, or brownish black, which are larger and more conspicuous along the back than upon the sides. The under portions of the body are of a much lighter hue. The feet are short, and the claws of the hinder feet are larger than those of the anterior limbs. The total length of the adult Seal is seldom more than five feet, the head being about eight or nine inches long.

This creature is wonderfully active both in water and on land, although its bodily powers are but awkwardly manifested when it is removed from the watery element in which it loves to roam. It is a persevering hunter of fish, chasing and securing them in a manner that greatly excites the wrath of the fishermen, who see their best captives taken away from them without the possibility of resistance. So cunning as well as active is the Common Seal, that one of these animals will coolly hang about the fishing grounds throughout the season, make itself familiar with all the turns and angles of the nets, and avail itself of their help in capturing the fish on which it is desirous to make a meal.

On the British coasts the chase of the Seal is but of local importance, but on the shore of Newfoundland it assumes a different aspect, and becomes an important branch of commercial enterprise, employing

SEAL.—(*Phoca vitulina.*)

many vessels annually. In a successful season the number of Seals which are taken amount to many hundred thousand. A large quantity of oil is obtained from the bodies of the Seals, and is used for various purposes, while their skins are of considerable value either when tanned into leather or when prepared with the fur and used for making various articles of dress and luxury.

The Common Seal is very easily tamed, and speedily becomes one of the most docile of animals, attaching itself with strong affection to its human friends, and developing a beautifully gentle and loving nature, hardly to be expected in such an animal. Many of these creatures have been taken when young, and have been strongly domesticated with their captors, considering themselves to belong of right to the household, and taking their share of the fireside with the other members of the family.

Of late days, performing Seals have come into vogue under various titles, among which the "Talking Fish" is well known. These clever animals have been taught to perform sundry ingenious feats, requiring not only an intelligent mind to comprehend, but an activity of body to execute, apparently incompatible with the conformation of the animal.

AMONG all the strange forms which are found among the members of the Seal family, there is none which presents a more terribly grotesque appearance than that of the WALRUS, MORSE, or SEA HORSE, as this extraordinary animal is indifferently termed.

The most conspicuous part of this animal is the head, with its protuberant muzzle bristling with long wiry hairs, and the enormous canine teeth that project from the upper jaw. These huge teeth measure, in large specimens, from fourteen inches to two feet in length, the girth at the base being nearly seven inches, and their weight upwards of ten pounds each. In ordinary specimens, however, the length is about one foot. In some examples they approach each other towards their points, and in others they diverge considerably, forming in the opinion of some writers two distinct species. As, however, the relative position of these teeth varies slightly in every specimen that has yet been examined, the structure seems to be of hardly sufficient importance for the establishment of a separate species. The ivory which is furnished by these extraordinary weapons is of very fine quality, and commands a high price in the market.

A Walrus is a valuable animal, for even in this country its skin, teeth, and oil are in much request, while among the Esquimaux its body furnishes them with almost every article in common use. Among civilized men, the skin of the Walrus is employed for harness and other similar purposes where a thick and tough hide is required. The tooth furnishes very good ivory, of a beautiful texture, and possessing the advantage of retaining the white hue longer than ivory which is made from the elephant tusk. The oil is delicate, but there is very little to be obtained from each Walrus, the layer of fatty matter being scarcely more than a hand's-breadth in thickness. Among the Esquimaux the Walrus is put to a variety of uses. Fish-hooks are made from its tusks, the intestines are twisted into nets, its

WALRUS, OR MORSE.—(*Trichecus Rosmarus.*)

oil and flesh are eaten, and its bones and skin are also turned to account by these rude but ingenious workmen.

The Walrus is found in vast herds, which frequent the coasts of the arctic and antarctic regions, and which congregate in such numbers that their united roarings have often given timely warning to fog-bewildered sailors, and acquainted them with the near proximity of shore. These herds present a curious sight, as the huge clumsy animals are ever in movement, rolling and tumbling over each other in a strange fashion, and constantly uttering their hoarse bellowings.

The movements of the Walrus when on land are of a very clumsy character, as might be supposed from the huge, unwieldy body of the animal, and the evident insufficiency of the limbs to urge the weighty body forward with any speed. When this creature is hurried or alarmed, it contrives to get over the ground at a pace that, although not very rapid, is yet wonderfully so when the size of the animal is taken into account. The movement is a mixture of jerks and leaps, and the Walrus is further aided in its progress by the tusks. Should it be attacked, and its retreat cut off, the Walrus advances fiercely upon its enemy, striking from side to side with its long tusks, and endeavouring to force a passage into the sea. If it should be successful in its attempt, it hurries to the water's edge, lowers its head, and rolls unceremoniously into the sea, where it is in comparative safety.

WALRUS'S SKULL.

This animal attains to a very great size, so great, indeed, that its dimensions can hardly be appreciated except by ocular demonstration. A full-grown male Walrus is generally from twelve to fifteen feet in length, while there are many specimens that have been known to attain a still greater size. The skin is black and smooth, and is sparingly covered with brown hairs, which become more numerous on the feet.

Another powerful and grotesque Seal now engages our attention. This is the ELEPHANT SEAL, or SEA ELEPHANT, so called not only on account of the strange prolongation of the nose, which bears some analogy to the proboscis of the elephant, but also on account of its elephantine size. Large specimens of this monstrous Seal measure as much as thirty feet in length, and fifteen or eighteen feet in circumference at the largest part of their bodies.

The colour of the Sea Elephant is ather variable, even in individuals of the same sex and age, but is generally as follows. The fur of the male is usually of a blueish grey, which sometimes deepens into dark brown, while that of the female is darker, and variegated with sundry dapplings of a yellow hue. This animal is an inhabitant of the southern hemisphere, and is spread through a considerable range of country. It is extensively hunted for the sake of its skin and its oil, both of which are of very excellent quality, and, from the enormous size of the animal, can be procured in large quantities. It is not exclusively confined to the sea, but is also fond of haunting fresh-water lakes or swampy ground.

It is an emigrating animal, moving southwards as the summer comes on, and northwards when the cold weather of the winter months would make its more southern retreats unendurable. Their first emigration is generally made in the middle of June, when the females become mothers, and remain in charge of their nurseries for nearly two months. During this time the males are said to form a *cordon* between their mates and the sea, in order to prevent them from deserting their young charges. At the expiration of this time, the males relax their supervision, and the whole family luxuriates together in the sea, where the mothers soon regain their lost condition. They then seek the shore afresh, and occupy themselves in settling their matrimonial alliances, which are understood on the principle that the strongest shall make his choice among the opposite sex, and that the weakest may take those that are rejected by his conquerors, or none at all, as the case may be.

SEA ELEPHANT.—(*Morunga proboscidea.*)

During the season of courtship the males fight desperately with each other, inflicting fearful wounds with their tusk-like teeth, while the females remain aloof, as quiet spectators of the combat. They are polygamous animals, each male being lord over a considerable number of females, whom he rules with despotic sway. When the victorious combatants have chosen their mates they are very careful about their safety, and refuse to quit them if they should be in any danger. Knowing this fact, the seal-hunters always direct their attacks upon the females, being sure to capture the male afterwards. If they were to kill the male at first, his harem would immediately disperse and fly in terror, but as long as he lives they will continue to crowd round him.

Although these animals are of so great dimensions and bodily strength, and are furnished with a very formidable set of teeth, they are not nearly such dangerous antagonists as the walrus, and are most apathetic in their habits. When roused from sleep they open their mouths in a threatening manner, but do not seem to think of using their teeth, and if they find that their disturbers do not run away, they take that office upon themselves, and move off deliberately for the water. As they proceed their huge bodies tremble like masses of jelly, in consequence of the fat with which they are so heavily laden. So plentiful is this fat, that a single adult male will furnish about seventy gallons of clear and scentless oil.

WHALES.

The CETACEA, or WHALES, are more thoroughly aquatic than any other animals which have already been described, and are consequently framed in such a very fish-like manner that they have generally been considered as fishes by those who are but little acquainted with the animal kingdom. The entire livelihood of the Whale is obtained in the waters, and their entire structure is only fitted for traversing the waves, so that if they should happen to be cast upon the shore they have no means of regaining their native element, and are sure to perish miserably from hunger.

When the Whales breathe, they are forced to rise to the surface of the sea, and there make a number of huge respirations, which are technically called "spoutings," because a column of mixed vapour and water is ejected from the nostrils, or "blow-holes," and spouts upwards to a great height, some-times as much as twenty feet. In order to enable the animal to respire without exposing itself unnecessarily, the "blow-holes" are placed on the upper part of the head, so that when a Whale is reposing itself on the surface of the sea, there is very little of its huge carcass visible, except the upper portion of the head and a part of the back. The "spoutings" are made with exceeding violence, and can be heard at some distance.

The limbs of the Whales are so modified in their form that they can hardly be recognised by their external appearance alone as the limbs of a veritable mammal. In shape they closely resemble the fins of fish, and it is not until they are stripped of the thick skin which envelopes them that the true limb is developed. They are, in fact, abnormal developments of the legs in order to suit an aquatic life, just as are the wings of the bat for an aërial life. The chief use of these organs seems to be that they assist the animal in preserving its position in the water, for the huge carcass rolls over on its back as soon as it is deprived of the balancing-power of its fins. They are also employed for the purpose of grasping the young whenever the mother Whale is anxious for the safety of her offspring, but they are of little use in urging the animal through the water, that duty being almost entirely performed by the tail.

The skin of the Whales is devoid of hair, and is of a rather peculiar structure, as is need-ful to enable it to resist the enormous pressure to which it is constantly subjected at the vast depths to which the animal descends. The skin is threefold, consisting first of the scarf-skin, or epidermis ; secondly, of the *rete-mucosum*, which gives colour to the animal ; and thirdly, of the true skin, which is modi-fied in order to meet the needs of the creature

FLIPPER OF THE WHALE.

which it defends. The blubber, indeed, is nothing more than the true skin, which is composed of a number of interlacing fibres, capable of containing a very great amount of oily matter. This blubber is never less than several inches in thickness, and in many places is nearly two feet deep, and as elastic as caoutchouc, offering an admirable resistance to the force of the waves and the pressure of the water.   In a large Whale the blubber will weigh thirty tons.

The GREENLAND WHALE, NORTHERN WHALE, or RIGHT WHALE, as it is indifferently termed, is an inhabitant of the Northern Seas, where it is still found in great abundance, although the constant persecutions to which it has been subjected have considerably thinned its numbers.

THE WHALE.—(*Balæna mysticetus.*)

This animal is, when fully grown, about sixty or seventy feet in length, and its girth about thirty or forty feet.   Its colour is velvety black upon the upper part of the body, the fins, and the tail ; grey upon the junction of the tail with the body and the base of the fins, and white upon the abdomen and the fore-part of the lower jaw. The velvety aspect of the body is caused by the oil which exudes from the epidermis, and aids in destroying the friction of the water.   Its head is remarkably large, being one-third of the length of the entire bulk.   The jaw opens very far back, and in a large Whale is about sixteen feet in length, seven feet wide, and ten or twelve feet in height, affording space, as has been quaintly remarked, for a jolly-boat and her crew to float in.

The most curious part of the jaw and its structure is the remarkable substance which is popularly known by the name of whalebone.

The whalebone, or baleen, is found in a series of plates, thick and solid at the insertion into the jaw, and splitting at the extremity into a multitude of hair-like fringes.   On each side of the jaw there are more than three hundred of these plates, which in a fine specimen are about ten or twelve feet long, and eleven inches wide

SKULL OF GREENLAND WHALE.

at their base. The weight of baleen which is furnished by a large Whale is about one ton. This substance does not take its origin directly from the gum, but from a peculiar vascular formation which rests upon it. These masses of baleen are placed along the sides of the mouth for the purpose of aiding the Whale in procuring its food and separating it from the water.

The mode of feeding which is adopted by the Whale is as follows. The animal frequents those parts of the ocean which are the best supplied with the various creatures on which it feeds, and which are all of very small size, as is needful from the size of its gullet, which is not quite two inches in diameter. Small shrimps, crabs, and lobsters, together with various molluscs and medusæ, form the diet on which the vast bulk of the Greenland Whale is sustained. Driving with open mouth through the congregated shoals of these little creatures, the Whale engulphs them by millions in its enormous jaws, and continues its destructive course until it has sufficiently charged its mouth with prey. Closing its jaws and driving out through the interstices of the whalebone the water which it has taken together with its prey, it retains the captured animals which are entangled in the whalebone, and swallows them at its ease.

The Whale is an animal of very great value to civilized and to savage men. The oil which is procured in great quantities from its blubber and other portions of its structure is almost invaluable to us, while the bones and baleen find their use in every civilized land. To the natives of the polar regions, however, the Whale is of still greater value, as they procure many necessaries of life from various parts of its body, eat the flesh, and drink the oil. Repulsive as such a diet may appear to us who live in a comparatively warm region, it is an absolute necessity in those ice-bound lands, such oleaginous diet being needful in order to keep up the heat of the body by a bountiful supply of carbon.

As far as is yet known, the Greenland Whale produces only a single cub at a birth. When first born, the young Whale is without the baleen, depending upon its mother for its subsistence like any other young mammal. The maternal Whale keeps close to her offspring until the baleen is grown, and does not forsake it until it is capable of supporting itself. The young Whales, before the baleen has developed itself, are technically termed "suckers," and when the baleen is six feet in length they are called by the name of "size."

The Cachalot, or Spermaceti Whale, is one of the largest of the Whales, an adult male, or "old bull," as it is called by the whalers, measuring from seventy to eighty feet in length, and thirty feet in circumference. The head is enormously long, being almost equal to one-third of the total length. Upon the back there is rather a large hump, which rises abruptly in front and tapers gradually towards the tail. The colour of the Cachalot is a blackish grey, somewhat tinged with green upon the upper portions of the body. Round the eyes and on the abdomen it is of a greyish white.

This species is chiefly notable on account of the valuable substances which are obtained from its body, including oil and spermaceti. The oil is obtained from the blubber, which is not very thick in this animal, being only fourteen inches in depth on the breast and eleven inches on the other parts of the body, and is therefore not so abundant in proportion to the size of the animal as that which is extracted from the Greenland Whale. Its superior quality, however, compensates fully for its deficiency in quantity. The layer of blubber is by the whalers technically called the "blanket," probably in allusion to its office in preserving the animal heat.

The spermaceti is almost peculiar to a few species of the genus Catodon, and is obtained as follows :—

K

The enormous and curiously formed head is the great receptacle of the spermaceti, which lies in a liquid oily state, in two great cavities that exist in the huge mass of tendinous substance of which the head is chiefly composed. When the Whale is killed and towed to the ship's side, the head is cut off and affixed to tackles for the purpose of supporting it in a convenient position for the extraction of this valuable substance. A large hole is cut in the top of the head, and a number of sailors lower their buckets into the cavity and bale out the liquid matter.

SPERMACETI WHALE.—(*Cátodon Macrocéphalus.*)

When first exposed to the air it has a clear oily appearance, but after it has been subjected to the action of the atmosphere for a few hours, the spermaceti begins to separate itself from the oil, and in a short time is sufficiently firm to be removed and put into a different vessel.

The amount of spermaceti which is produced from the head of a single Whale is very large indeed. From a Cachalot that only measured sixty-four feet in length, and was therefore by no means a large one, twenty-four barrels of spermaceti and nearly one hundred barrels of oil were obtained.

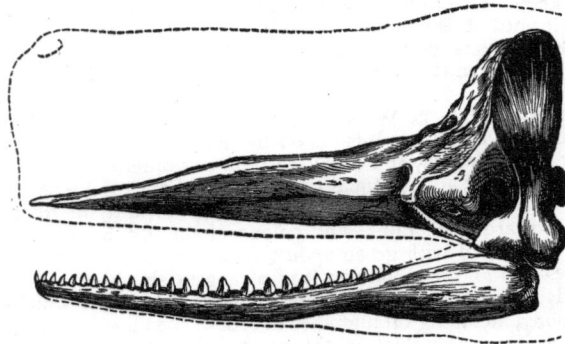

SKULL OF SPERMACETI WHALE.

Ambergris, that curious substance whose origin so long baffled the keenest inquirers, and which was formerly only found at rare intervals floating on the waves or cast upon the shore, is now often discovered within the intestines of the Cachalot, and is supposed to be a morbid secretion peculiar to the animal, and analogous to biliary calculi. Fifty pounds weight of this substance have been found in a single Whale, and on one occasion a single piece of ambergris of the same weight was discovered on the coast of the Bermudas by some sailors, who immediately deserted their ship and escaped to England with their valuable prize. The value of the ambergris is rather variable, but it is always a costly article.

The Spermaceti Whale, when it is in the open seas, lives chiefly on the "squids," or cuttle-fish, which swarm in that ocean, and, when it approaches land, feeds on various fish. It seems, however, to dislike the propinquity of the shore, and is very seldom taken in "soundings." It is a gregarious animal, being seldom seen alone, but in large herds, technically called "schools," and consisting of several hundred in number. The "schools" are generally divided into two bands, the one consisting of young males and the other of females. Each band of females is under the command of several large males, who exercise the strictest discipline over their harems, and will not permit any intruder to join their society. From their office, these leaders are called the schoolmasters.

The teeth of the Spermaceti Whale are conical and slightly curved, sometimes reaching the length of some eight or nine inches. To Europeans these teeth are of great value, but to the Fijians, Tongans, &c., they are almost priceless, a single tooth being thought a present fit for one king to make to another, or to be laid up in a temple as an offering to the idol.

The Spermaceti Whale does not seem to choose any particular portion of the year for the production of its young, but is found at all seasons in charge of its offspring. Moreover, young Whales, or "cubs," are found at all sizes and ages simultaneously roaming the seas, either in company with their parents or turned loose upon the world to shift for themselves. There is but a single cub at a birth. The milk of the animal is exceedingly rich and thick, as indeed is the case with the milk of all Whales.

This animal is very widely spread over the world, as it is found in almost every portion of the aqueous portions of the globe, with the exception of the Polar Seas. Several of these creatures have been discovered off our own coasts, and a few have been stranded on the beach. A Cachalot measuring fifty-four feet in length was driven ashore in the Firth of Forth in 1769, and its appearance off the Orkneys is said to be no very uncommon occurrence.

DOLPHINS.

THE members of this family do not possess the enormous head which characterizes the true whales, and have teeth in both jaws, although they are liable to fall out at an early age. The blow-holes are united together, so as to form a single semilunar opening, which is set transversely on the crown of the head.

The word NARWHAL is derived from the Gothic, signifying "beaked whale," and is a very appropriate term for the SEA UNICORN, as the animal is popularly entitled. The head of the Narwhal is round, and convex in front, the lower jaw being without teeth, and not so wide as the upper jaw. From the upper jaw of the Narwhal springs the curious weapon which has gained for the animal a world-wide reputation.

In the upper jaw of the young or the female Narwhal are found two small or hollow tusks, imbedded in the bone, which, in the female, are generally undeveloped throughout the whole of the animal's existence, but in the male Narwhal are strangely modified. The right tusk remains in its infantine state, excepting that the hollow becomes filled with bony substance ; but the left tusk rapidly increases in length, and is developed into a long, spiral, tapering rod of ivory, sometimes attaining to the length of eight or ten feet. The tusks are supposed to be formed by an excessive growth of the canine teeth, and not of the incisors, as might be supposed from the position which they occupy in the jaw.

K 2

The food of the Narwhal consists chiefly of marine molluscs and of occasional fish, but is found to be generally composed of the same kind of squid, or cuttle-fish, which supplies the gigantic spermaceti whale with subsistence. As the remains of several flat fish have been discovered in the stomach of the Narwhal, it was supposed by some authors that the animal made use of its tusk as a fish-spear, transfixing them as they lay "sluddering" on the mud or sand, after their usual fashion, thus preventing their escape from the toothless mouth into which the wounded fish are then received. However this may be, the force of the tusk is terrific when urged with the impetus of the creature driving through the water at full speed, for the whole combined power of the weight and velocity of the animal is directed along the line of the tusk. A Narwhal has been known to encounter a ship, and to drive its tusk through the sheathing, and deeply into the timbers. The shock was probably fatal to the assailant, for the tooth was snapped by the sudden blow, remaining in the hole which it had made and acting as a plug that effectually prevented the water from gaining admission into the vessel.

The ivory of the Narwhal's tusk is remarkably good in quality, being hard and solid, capable of receiving a high polish, and possessing the property of retaining its beautiful whiteness

NARWHAL.—(*Mónodon Monóceros.*)

for a very long period, so that a large Narwhal horn is of no inconsiderable commercial value.

The native Greenlanders hold the Narwhal in high estimation; for, independently of its value, it is welcomed each succeeding year as the harbinger of the Greenland Whale.

The oil which is extracted from the blubber is very delicate, but is not present in very great amount, as the coating of fatty substance is seldom more than three inches in depth. About half a ton of oil is obtained from a large specimen. The flesh is much prized by the natives, and is not only eaten in its fresh state, but is carefully dried and prepared over the fire.

The colour of this animal is almost entirely black upon the upper surface of the body, but is slightly varied by streaks and patches of a deeper tint. The sides fade into greyish white, diversified with sundry grey marks, and the under portions of the body are white.

MOST familiar of all the Dolphin fraternity is the well-known PORPOISE, or SEA HOG, an animal which may be seen on any of our coasts, tumbling about on the waves and executing various gambols in the exuberance of its sportive feelings.

The Porpoise is a very gregarious animal, herding together in large shoals, and sometimes swimming in " Indian file " as they shoot over the surface of the sea, just showing their black and glossy backs above the water, and keeping such excellent line that they seem to be animated by one spirit and one will.

As might be presumed from the formidable array of sharp teeth with which the jaws are studded, and which are so arranged that the upper and lower sets interlock when the animal closes its mouth, the food of the Porpoise consists entirely of animal substances, and almost wholly of fish, which it consumes in large quantities, much to the disgust of human fishermen. Herrings, pilchards, sprats, and other saleable fish are in great favour with the Porpoise, which pursues its finny prey to the very shores, and, driving among the vast shoals in which these fish congregate, destroys enormous quantities of them. The fish are conscious of the presence of their destroyer, and flee before it in terror, often flinging themselves into the certain death of nets or shallow water in their hope to escape from the devouring jaws of the Porpoise. Even salmon and such large fish fall frequent victims to their pursuer, which twists, turns, and leaps with such continuous agility that it is more than a match for its swift and nimble prey. Not even the marvellous leaping powers of the salmon are sufficient to save them from the voracious Porpoise, which is not to be baffled by any such impotent devices.

PORPOISE, OR PORPESSE. —(*Phocæna communis.*)

The Porpoise seems to keep closely to the coasts, and is seldom seen in mid-ocean. It appears to be a migratory animal, as the season of its disappearance from one locality generally coincides with its arrival on some other coast. It is very widely spread, appearing to inhabit with equal security the warm waters of the Mediterranean, the cool seas of our own coasts, or the icy regions of the high latitudes.

The length of a full-grown Porpoise is extremely variable, the average being from six to eight feet. The colour of the Porpoise is a blue-black on the upper surface of the body, and a bright silvery white below ; so that when the animal executes one of its favourite gyrations the contrasting tints produce a strange effect as they rapidly succeed each other. The iris of the eye is yellowish.

The word Porpoise is corrupted from the French term *Porc-poisson,* i.e. " Hog fish," and bears the same signification as its German name, *Meerschwein.*

THE DOLPHIN is remarkable for the enormous number of teeth which stud its mouth, no less than forty-seven being found on each side of both jaws, the full complement being one hundred and ninety. In the head of one specimen were found fifty teeth on each side of each jaw, making a complement of two hundred in all. Between each tooth there is a space equal to the width of a single tooth, so that when the animal closes its mouth the teeth of both jaws interlock perfectly. All the teeth are sharply pointed and flattened, and slightly curved backwards, so that the entire apparatus is wonderfully adapted for the retention of the slippery marine creatures on which the Dolphin feeds. Fish of various kinds form the usual diet of the Dolphin, which especially delights in the flat fishes of our coasts, and often prowls about the shoals of herrings and pilchards that periodically reach our shores.

The Dolphin is not a very large animal, measuring, when fully grown, from six to ten feet in length, seven feet being the usual average. Its colour is black upon the back, and silvery white upon the abdomen ; while the flanks are greyish white. The beautiful colours which have been said to play about

the body of a dying Dolphin are not entirely mythical, but belong rightly to one of the fishes, the coryphene, or dorado, which is popularly called the Dolphin by sailors.

The eyes of the Dolphin are small, and are supplied with eyelids ; the pupil of the eye is heart-shaped. The ears have but a very minute external aperture, barely admitting an ordinary pin.

DOLPHIN.—(*Delphinus Delphis.*)

It is a lively and playful animal, and being remarkably active in its native element, is fond of gambolling among the waves, and engaging in various sports with its companions. Being of a very gregarious nature, it is seldom seen alone, but prefers to associate in little flocks or herds, and is in the habit of accompanying ships for considerable distances, hovering about the vessel and executing various strange manœuvres.

The Dolphin only produces a single young one at a time, and nurses her offspring with exceeding tenderness and assiduity.

The common Dolphin is found in the European seas, and in the Atlantic and the Mediterranean, and may possibly have a still wider range.

RODENTS.

THE RODENTS, or gnawing animals, derive their name from the peculiar structure of their teeth, which are specially fitted for gnawing their way through hard substances. The jaws of the Rodents are heavily made, and very large in proportion to the head, their size being not only needful for the support of the gnawing teeth, but for their continual development. There are no canines, but a wide gap exists between the incisors and the molars, which are nearly flat on their surfaces, and are well suited for grinding the soft substances on which these animals feed.

The structure of the chisel-edged incisor teeth is very wonderful, and may be easily understood by inspecting the teeth of a rat, mouse, hare, or rabbit.

As their teeth are continually worn out by the severe friction which they undergo continually, there must needs be some provision for renewing their substance, or the creature would soon die of starvation. In order to obviate this calamity, the base of the incisor teeth pass deeply into the jaw-bone, where they are continually nourished by a kind of pulpy substance from which the tooth is formed, and which adds fresh material in proportion to the daily waste.

Something more is needed for the well-being of the animal than the mere growth of its teeth ; for unless their chisel-like edges were continually kept sharp, they would be of little use for cutting their way through the hard substances which the Rodents are in the habit of gnawing. This result is attained as follows :—

The enamel which covers the front face of the incisor teeth is much harder than that which is laid upon the remaining surfaces, while the dentine which makes up the solid mass of each tooth is also harder in front than behind. It is evident that when these teeth are employed in their usual task, the softer enamel and dentine are worn away very much more rapidly than the remainder of the teeth, so that the peculiar chisel-edge of the teeth is continually preserved. Following—perhaps unconsciously—the structure of these teeth, our cutlers have long been accustomed to make their chisels on the same principle, a thin plate of steel being strengthened with a thick backing of iron.

The Rodent animals are widely spread over the entire globe, and are very numerous, comprising nearly one-third of the mammalia.

FEW animals are so well known or so thoroughly detested as the common BROWN RAT, or NORWAY RAT, as it is sometimes erroneously called.

It is an exceedingly voracious animal, eating all kinds of strange food, and not sparing its own species in times of scarcity. The havoc which an army of Rats will make among the corn-ricks is almost incredible, while they carry on their depredation with so much secrecy that an unpractised eye would think the stacks to be sound and unharmed. Fortunately they can easily be dislodged from any rick by taking it down and replacing it on proper "staddles," taking great care that no stray weeds or branches afford a foothold to these persevering marauders. While the rick is being rebuilt, no particular care need be taken to shake the Rats out of the sheaves, for, as they are thirsty animals, they will be forced to leap from the stack in search of water, and then will not be able to return.

Mice can subsist in a stack by means of the rain and dew which moisten the thatch, and may be often seen licking the straws in order to quench their thirst. But the Rats are less tolerant of thirst, and are forced to evacuate their

THE RAT.—(*Mus Décumanus.*)

premises. When mice and Rats are found inhabiting the same stack, the former animals reside in the upper parts, and the Rats in the lower.

Rats are not without their use, especially in large towns, which but for their never-failing appetites would often be in very sad case. Taking for example the metropolis itself, we find that the sewers which underlie its whole extent are inhabited by vast hordes of Rats, which perform the office of scavengers by devouring the mass of vegetable and animal offal which is daily cast into those subterranean passages, and which would speedily breed a pestilence were it not removed by the ready teeth of the Rats. So that, when kept within proper bounds, the Rat is a most useful animal, and will continue to be so until the drainage of towns is constructed in a different manner.

Rats are very cleanly animals, always washing themselves after every meal, and displaying the greatest assiduity in making their toilet. They also exhibit considerable delicacy of palate wherever they find a sufficiency of provisions, although they are in no way nice in their diet when pressed by hunger. If, for example, a party of Rats discover an entrance into a butcher's storehouse, they are sure to attack the best parts of the meat, utterly disdaining the neck, the shin, or other coarse pieces.

There is one peculiarity in the structure of the Rat which is worthy of notice. These animals are able not only to ascend a perpendicular tree or wall by the aid of their sharp hooked claws, but also to descend head foremost with perfect ease. In order to enable them to perform this feat, their hind legs are so made that the feet can be turned outwards, and the claws hitched upon any convenient projections.

However unpromising a subject the Rat may appear, it has often been tamed, and is a very much more educatable animal than could be supposed. It will obey its master's commands with promptitude, and has been known to learn very curious tricks.

For further information on this subject the reader is referred to a work published by Messrs. Routledge and Co., entitled "The Rat," by James Rodwell, in which may be found an elaborate account of the animal and its habits, together with much curious and original information.

" Yᵉ little vulgar MOUSE," as it is quaintly termed by old Topsel, is a truly pretty little creature, with its brown-grey back, grey throat and abdomen, soft velvety fur, its little bright black bead-like eyes, and squirrel-like paws. A detailed description of so familiar an animal would be quite unnecessary, and we will therefore proceed to its habits and manners.

Like the rat, it frequents both town and country, doing an infinity of damage in the former, but comparatively little harm in the latter. In the country it attaches itself mostly to farmyards, where it gains access to the ricks, and, when once firmly established, is not so easily

THE MOUSE.—(*Mus Musculus.*)

dislodged as its larger relative the rat. However, if the rick be kept under cover, the Mice cannot make any lengthened stay, for the cover keeps off the rain, on which they chiefly depend for drink, and they are then obliged to leave the stack in search of water. If the rick be placed on staddles, it will then be safe from these little pests.

They are odd little animals, and full of the quaintest gamesomeness, as may be seen by anyone who will only sit quite still and watch them as they run about a room which they specially affect. They are to the full as inquisitive as cats, and will examine any new piece of furniture with great curiosity.

The Mouse is a marvellously prolific animal, producing its young several times in the course of the year, and at a very early age. The nests are made in any sheltered spot, and formed from any soft substance, such as rags, paper, or wool, that the mother can procure.

SMALLEST, and perhaps the prettiest, of the British mammalia, the elegant little HARVEST MOUSE next claims our attention. The total length of this tiny creature is not quite five inches, its tail being nearly two inches and a half in length. The colour of its fur is a delicate reddish brown, the base of each hair taking the darker tint and the point warming into red, while the under parts of the abdomen are white. The line of demarcation between the brown and white is well defined.

The description which is given of the Harvest Mouse and its wonderful nest, by the Rev. Gilbert White, is so well known that it need only be casually mentioned. I have fortunately had opportunities of verifying his observations by means of a nest which was found in a field in Wiltshire by some mowers.

Independently of its small size, the Harvest Mouse may be distinguished from a young ordinary Mouse by its short ears, narrow head, slender body, and less projecting eyes.

THE short, sturdy, stupid rodent which is so famous under the name of the HAMSTER is widely spread over many parts of Northern Europe, where it is an absolute pest to the agriculturists, who wage unceasing war against so destructive an animal. Before proceeding to the habits and character of the Hamster, a short description of its external appearance will be necessary.

The colour of its fur is a greyish fawn on the back, deepening into black on the under portions of the body, and softening into a yellow hue upon the head and face. The otherwise uniform tinting of the fur is relieved by some patches of whitish yellow upon the cheeks, shoulders, and sides. The creature is furnished with two large cheek-pouches, which are capable of containing a considerable amount of food, and which can be inflated with air at the pleasure of the animal. The length of the adult Hamster is about fifteen inches, the tail being only three inches long.

The Hamster is most destructive to the crops, whether of corn, peas, or beans, and, when the autumn approaches, begins to plunder the fields in a most systematic manner, for the purpose of laying up a winter store of provisions. By dint of dexterous management, the animal fills its cheek-pouches with grain, pressing it firmly with its paws, so as to lose no space, and then carries off its plunder to its subterranean treasury, where it disgorges the contents of the pouches, and returns for another supply. The husbandmen are so well aware of this propensity that they search after the habitation of the Hamster after the harvest is over, and often recover considerable quantities of the stolen grain. The destructive capability of the animal may be gathered from the fact that a single Hamster has been known to hoard no less than sixty pounds of corn in its home, while a hundredweight of beans have been recovered from the storehouses of another specimen.

The skin of the Hamster is of some value in commerce, so that the hunters make a double use of a successful chase, for they not only recover the stolen property of the agriculturist, but gain some profit by selling the skins.

HARVEST MOUSE.—(*Micromys minutus.*)

THERE are many animals which have been saddled with a bad reputation merely on account of an unfortunate resemblance to another animal of really evil character. Among these misused innocents the WATER VOLE, popularly called the WATER RAT, is very conspicuous, as the poor creature has been commonly supposed to be guilty of various poaching exploits which were really achieved by the ordinary brown rat.

It is quite true that rats are often seen on the river-banks in the act of eating captured fish, but these culprits are only the brown rats which have

HAMSTER.—(*Cricêtus frumentarius.*)

migrated from the farmyards for the summer months, and intend to return as soon as autumn sets in. The food of the true Water Rat, or Water Vole, as it is more correctly named, is chiefly of a vegetable nature, and consists almost entirely of various aquatic plants and roots. The common "mare's-

WATER VOLE.—(*Arvicola amphibius.*)

tail," or Equisetum, is a favourite article of diet with the Water Vole, and I have often seen it feeding on the bark of the common rush. Many years ago I shot a Water Vole as it was sitting upon a water-lily leaf and engaged in eating the green seeds; and on noticing the kind of diet on which the animal was feeding, I determined to watch the little creatures with more care. My own testimony coincides precisely with those of other observers, for I never yet saw the true snub-nosed, short-eared, yellow-toothed Vole engaged in eating animal food, although the brown rat may be often detected in such an act.

Many communications have been made to me on the subject, written for the most part by persons who have seen water-side rats engaged in catching and eating fish, and have thought that the delinquents were the true Water Vole. Indeed, the Vole is allied very closely to the beaver, and partakes of the vegetarian character of that animal.

The colour of the Water Vole is a chestnut brown, dashed with grey on the upper parts and fading to grey below. The ears are so short that they are hardly perceptible above the fur. The incisor teeth are of a light yellow, and are very thick and strong. The tail is shorter than that of the common rat, hardly exceeding half the length of the head and body. The average length of a full-grown Water Vole is thirteen inches, the tail being about four inches and three-quarters long.

CAMPAGNOL.—(*Arvicola arvális.*)

THE CAMPAGNOL, or SHORT-TAILED FIELD MOUSE, is even more destructive in the open meadows than the common grey mouse in the barns or ricks; for, not contenting itself with plundering the ripened crops of autumn, it burrows beneath the ground at sowing time, and devours the seed-wheat which has just been laid in the earth. Besides these open-air depredations, it makes inroads into ricks and barns, and by dint of multitudinous numbers does very great harm.

The colour of the Campagnol is ruddy brown on the upper surface of the body, and grey on the abdomen and chest. The ears are rounded and very small, closely resembling those of the water vole. The tail is only one-third the length of the body, and the total length of the animal is rather more than five inches. As it belongs to the same genus as the water vole, and is very closely related to that animal, it sometimes goes by the name of Field Vole.

AT uncertain and distant intervals of time, many of the northern parts of

Europe, such as Lapland, Norway, and Sweden, are subjected to a strange invasion. Hundreds of little, dark, mouse-like animals sweep over the land, like clouds of locusts suddenly changed into quadrupeds, coming from some unknown home, and going no one knows whither. These creatures are the LEMMINGS, and their sudden appearances are so entirely mysterious, that the Norwegians look upon them as having been rained from the clouds upon the earth.

Driven onwards by some overpowering instinct, these vast hordes travel in a straight line, permitting nothing but a smooth perpendicular wall or rock to turn them from their course. If they should happen to meet with any living being, they immediately attack, knowing no fear, but only urged by undiscriminating rage. Any river or lake they swim without hesitation, and rather seem to enjoy the water than to fear it. If a stack or a corn-rick should stand in their way, they settle the matter by eating their way through

LEMMING.—(*Myódes Lemmus.*)

it, and will not be turned from their direct course even by fire. The country over which they pass is utterly devastated by them, and it is said that cattle will not touch the grass on which a Lemming has trodden.

These migrating hosts are accompanied by clouds of predaceous birds, and by many predaceous quadrupeds, who find a continual feast spread for them as long as the Lemmings are on their pilgrimage. While they are crossing the rivers or lakes, the fish come in for their share of the banquet, and make great havoc among their columns. It is a very remarkable fact that the reindeer is often seen in chase of the Lemmings; and the Norwegians say that the deer is in the habit of eating them. This statement, however, seems to be rather of doubtful character. The termination of these extraordinary migrations is generally in the sea, where the survivors of the much-reduced ranks finally perish. Mr. Lloyd mentions that just before his visit to Wermeland, the Lemming had overrun the whole country. The primary cause of these strange migrations is generally thought to be hunger. It is fortunate for the country that these razzias only occur at rare intervals, a space of some ten or fifteen years generally elapsing between them, as if to fill up the places of those which were drowned or otherwise killed in the preceding migration.

The Lemming feeds upon various vegetable substances, such as grass, reeds, and lichens, being often forced to seek the last-named plant beneath the snow, and to make occasional air-shafts to the surface. Even when engaged in their ordinary pursuits, and not excited by the migratorial instinct, they are obstinately savage creatures. Mr. Metcalfe describes them as swarming in the forest, sitting two or three on every stump, and biting the dogs' noses as they came to investigate the character of the irritable little animals. If they happened to be in a pathway, they would not turn aside to permit a passenger to move by them, but boldly disputed the right of way, and uttered defiance in little sharp, squeaking barks.

The colour of the Lemming is dark brownish black, mixed irregularly with a tawny hue upon the back, and fading into yellowish white upon the abdomen. Its length is not quite six inches, the tail being only half an inch long.

THE common BEAVER has earned a world-wide reputation by the wonderful instinct which it displays, independently of its very great value in producing costly fur and perfume.

This animal is found in the northern parts of Europe and Asia, but is found in the greatest profusion in North America. In days long gone by, the Beaver was an inhabitant of our own island.

The Beaver lives in societies, varying considerably in number, and united together in the formation of works which may fairly be considered as belonging to the profession of the engineer. They prefer to make their habitations by small clear rivers and creeks, or close to large springs, although they sometimes take up their abode on the banks of lakes.

Lest they should not have a sufficient depth of water in all weathers and at all seasons, the Beavers are in the habit of building veritable dams, for the purpose of raising the water to the required level. These dams are composed of tree-branches, mud, and stones, and, in order effectually to resist the action of the water, are about ten or twelve feet in thickness at the bottom, although they are only two feet or so wide at the summit.

In forming the dam, the Beaver does not thrust the ends of the stakes into the bed of the river, as is often supposed, but lays them down horizontally, and keeps them in their place by heaping stones and mud upon them. The logs of which the dam is composed are about three feet in length, and vary extremely in thickness. Generally, they are about six or seven inches in diameter, but they have been known to measure no less than eighteen inches in diameter. An almost incredible number of these logs are required for the completion of one dam, as may be supposed from the fact that a single dam will sometimes be three hundred yards in length, ten or twelve feet thick at the bottom, and of a height varying according to the depth of water.

Before employing the logs in this structure, the Beavers take care to separate the bark, which they carry away, and lay up for a winter store of food.

Near the dams are built the beaver-houses or "lodges," as they are termed; edifices as remarkable in their way as that which has just been mentioned. They are chiefly composed of branches, moss, and mud, and will accommodate five or six Beavers together. The form of an ordinarily sized Beaver's lodge is circular, and its cavity is about seven feet in diameter by three feet in height. The walls of this structure are extremely thick, so that the external measurement of the same lodges will be fifteen or twenty feet in diameter, and seven or eight feet in height. The roofs are all finished off with a thick layer of mud, laid on with marvellous smoothness, and carefully renewed every year. As this compost of mud, moss, and branches is congealed into a solid mass by the severe frosts of a North American winter, it forms a very sufficient defence against the attacks of the Beaver's great enemy, the wolverene, and cannot readily be broken through, even with the help of iron tools. The precise manner in which the Beavers perform their various tasks is not easy to discern, as the animals work only in the dark.

Around the lodges the Beavers excavate a rather large ditch, too deep to be entirely frozen, and into this ditch the various lodges open, so that the inhabitants can pass in or out without hindrance. This precaution is the more necessary, as they are poor pedestrians, and never travel by land as long as they can swim by water. Each lodge is inhabited by a small number of Beavers, whose beds are arranged against the wall, each bed being separate, and the centre of the chamber being left unoccupied.

BEAVER.—(*Castor Fiber.*)

In order to secure a store of winter food, the Beavers take a vast number of small logs, and carefully fasten them under water in the close vicinity of their lodges. When a Beaver feels hungry he dives to the store-heap, drags out a suitable log, carries it to a sheltered and dry spot, nibbles the bark away, and then either permits the stripped log to float down the stream, or applies it to the dam.

We must now bestow a little time on the curious odoriferous substance which is called "castoreum" by the learned, and "bark-stone" by the trappers. This substance is secreted in two glandular sacs which are placed near the root of the tail, and gives out an extremely powerful odour.

To the castoreum the trapper is mostly indebted for his success, for the Beavers are strangely attracted by this substance, and if their nostrils perceive its distant scent, the animals will sit upright, sniff about in every direction, and absolutely squeal with excitement. Taking advantage of this curious propensity, the hunter always carries a supply of castoreum, in a closed vessel, and when he comes to a convenient spot for placing his trap, he sets the trap and then proceeds to manufacture his bait. This process is simple enough, consisting merely of taking a little twig of wood about nine inches long, chewing one end of it, and dipping it in the castoreum. The trap is now laid so as to be covered by about six inches of water, and the stick arranged so that its perfumed tip projects from the water. Any Beaver which scents this bait will most certainly come to it, and will probably be captured in the trap.

THE ONDATRA, MUSQUASH, or MUSK RAT, is a native of Northern America, where it is found in various places above the twentieth degree of north latitude.

The colour of this animal is a dark brown on the upper portions of its body, tinged with a reddish hue upon its neck, ribs, and legs, the abdomen being ashy grey; the tail is of the same dark hue as the body. In total length it rather exceeds two feet, of which measurement the tail occupies about ten inches. The incisor teeth are bright yellow, and the nails are white. The whole colouring of the animal is so wonderfully like the hue of the muddy banks on which it resides, that a practised naturalist has often mistaken the Ondatras for mere lumps of mud until they began to move, and so dispelled the illusion. The hinder feet of the Ondatra are well webbed, and their imprint on the soft mud is very like that of a common duck.

The food of the Ondatra in a wild state appears to be almost wholly of a vegetable nature ; although, when confined in a cage, one of these animals has been seen to eat mussels and oysters, cutting open the softest shells, and extracting the inmates, and waiting for the hard-shelled specimens until they either opened of their own accord or died. Although the Ondatra is a clumsy walker, it will sometimes travel to some distance from the water-side, and has been noticed on a spot nearly three-quarters of a mile from any water. These animals have also been detected in ravaging a garden, which they had

MUSQUASH, OR MUSK RAT, OR ONDATRA.—(*Fiber Zibéthicus.*)

plundered of turnips, parsnips, carrots, maize, and other vegetables. The mischievous creatures had burrowed beneath them, bitten through their roots and carried them away to their subterranean storehouses. The maize they had procured by cutting the stalks near the level of the ground.

The Ondatra lives mostly in burrows, which it digs in the banks of the river in which it finds its food, but sometimes takes up its abode in a different kind of habitation, according to the locality and the soil. In the stiff clay banks of rivers the Ondatra digs a rather complicated series of tunnels, some of them extending to a distance of fifteen or twenty yards, and sloping upwards. There are generally three or four entrances, all of which open under water, and unite in a single chamber, where the Ondatra makes its bed. The couch of the luxurious animal is composed of sedges, water-lily leaves, and similar plants, and is so large as to fill a bushel basket. On marshy ground, and especially if it be supplied by springs, the Ondatra builds little houses that rise about three or four feet above the water, and look something like small haycocks.

THE PORCUPINE has long been rendered famous among men by the extraordinary armoury of pointed spears which it bears upon its back, and which it was formerly fabled to launch at its foes with fatal precision.

This animal inhabits many parts of the world, being found in Africa, Southern Europe, and India. The spines or quills, with which it is furnished, vary considerably in length, the longest quills being flexible and not capable of doing much harm to an opponent. Beneath these is a plentiful supply of shorter spines, from five to ten inches in length, which are the really effective weapons of this imposing array. Their hold on the skin is very slight, so that when they have been struck into a foe, they remain fixed in the wound, and, unless immediately removed, work sad woe to the sufferer. For the quill is so constructed that it gradually bores its way into the flesh, burrowing deeper at every movement, and sometimes even causing the death of the wounded creature. In Africa and India, leopards and tigers have frequently been killed, in whose flesh were pieces of Porcupine quills that had penetrated deeply into the body, and had even caused suppuration to take place. In one instance a tiger was found to have his paws, ears, and head filled with the spines of a Porcupine, which he had vainly been endeavouring to kill.

The Porcupine is a nocturnal animal, seldom venturing out of its retreat as long as the sun is above the horizon, and is therefore not often seen even in the localities which it most prefers. It is said not to require the presence of water, but to quench its thirst by eating the succulent roots and plants which it digs out of the ground. Its food is entirely of a vegetable nature, and consists of various kinds of herbage, as well as of bark, fruit, and roots. This animal takes up its abode in deep burrows which it excavates, and in which it is supposed to undergo a partial hibernation.

As the spines of the Porcupines are of some commercial value, and are used for many purposes, the chase of the animal is rather popular in the countries which it inhabits, and derives a further interest from the fact that the Porcupine, although a timid creature, can make a very powerful resistance when it is driven to despair.

The upper parts of the body are covered with hair instead of quills, and upon the head and neck there is a kind of crest, composed of very long stiff hairs, which can be erected or depressed at pleasure. Like the hedgehog, it can coil itself into a ball when it is surprised at a distance from its haven of refuge, and can present such an array of threatening spikes that it is quite safe from any enemy excepting man. When, however, the animal is at peace, it is capable of depressing the bristling spears, and can squeeze itself through

PORCUPINE.—(*Hystrix Cristáta.*)

an opening which would appear at first sight to be hardly large enough to permit the passage of an animal of only half its size.

The total length of the common Porcupine is about three feet six inches, the tail being about six inches long. Its gait is plantigrade, slow, and clumsy, and as it walks, its long quills shake and rattle in a very curious manner. Its muzzle is thick and heavy, and its eyes small and pig-like.

THE URSON, CAWQUAW, or CANADIAN PORCUPINE, is a native of North America, where it is most destructive to the trees among which it lives.

Its chief food consists of living bark, which it strips from the branches as cleanly as if it had been furnished with a sharp knife. When it begins to feed, it ascends the tree, commences at the highest branches, and eats its way regularly downward. Having finished one tree, it takes to another, and then to a third, always choosing those that run in the same line ; so that its path through the woods may easily be traced by the line of barked and dying trees which it leaves in its track. A single Urson has been known to destroy a hundred trees in a single winter, and another is recorded as having killed some two or three acres of timber.

The Urson is not so fully defended with spines as the preceding animal, but is covered with long, coarse, blackish brown hair, among which the short pointed quills are so deeply set, that, except in the head, tail, and hinder quarters, they are scarcely perceptible. These spines are dyed of various colours by the American Indians, and are then used in the decoration of their hunting-pouches, mocassins, and other articles, and after the quills are extracted, the remainder of the fur is sufficiently soft to be used for clothing. The flesh of the Urson is considered eatable, and is said to bear some resemblance to flabby pork.

The length of the Urson is not quite four feet, the head and body measuring rather more than three feet, and the tail about nine inches. The teeth are of a bright orange.

CANADIAN PORCUPINE OR URSON.
Erethizon dorsátum.

THE AGOUTI is a native of Brazil, Paraguay, Guiana, and other neighbouring countries, but its numbers have been considerably thinned in many spots where cultivation has been industriously carried on. In some of the

L

Antilles, where it formerly swarmed, it is now nearly extirpated, and in St. Domingo is but rarely seen.

All its movements are sharp, quick, and active, and even while sitting upright and engaged in feeding itself by the assistance of its fore-paws, its head is continually being turned from side to side, and its bright eyes glance in every direction in order to guard against a surprise. As it is a nocturnal animal, and spends the whole of the day in its dark hiding-place, its ravages take place under cover of night, and are the more difficult to be repelled. Its usual resting place is in the cleft of a rock, or in the hollow of some decaying tree, where twenty or thirty of these animals may be found living amicably together.

AGOUTI.—(*Dasyprocta Agouti.*)

In these dark recesses the young Agoutis are born, and are laid upon a soft bed of leaves, where they remain for a few weeks, and then sally out with their parents on their nocturnal expedition. There are generally two broods in each year, and the number of young at a birth is from three to six.

The name *Dasyprocta*, which has been given to the genus, refers to the thick hair which falls over the hind quarters, and nearly conceals the little pointed stump of a tail. The hair of this part of the body is a bright golden brown, but on the back and sides the fur has a curious speckled aspect, on account of the black, brown, and yellow tints with which each hair is marked. On the greater part of the body the fur is only about one inch in length, but the golden brown hair of the hinder parts is more than four inches long. In character it is coarse, though glossy.

THE CAPYBARA is a native of tropical America, and is by far the largest of all the living rodent animals, rather exceeding three feet in total length, and being so bulkily made that when it walks its abdomen nearly touches the ground. The muzzle of this animal is heavy and blunt, the eyes are set high in the head and are moderate in size, the tail is wanting, and the toes are partially connected together by a development of the skin. The colour of the Capybara is rather indeterminate, owing to the manner in which the hairs are marked with black and yellow, so that the general idea which its coat presents is a dingy, blackish grey, with a tinge of yellow.

CAPYBARA.—(*Hydrochœrus Capybara.*)

It is a water-loving animal, using its webbed feet with great power, and fleeing instinctively to the stream when terrified by real or imaginary danger. It not only swims well, but is a good diver ; and when endeavouring to escape from a foe, always tries to evade its pursuer by diving as long as its breath will hold out, and only permitting the top of its head to appear above the surface when it rises for the purpose of respiration. As, however, it can remain under water for a space of eight or ten minutes, it finds no difficulty in escaping from any ordinary foe, if it can only gain the shelter of the welcome stream. The food of this animal is exclusively vegetable, and its curious teeth are needed in order to bruise the herbage on which it feeds into a mass sufficiently pulpy to enable it to pass through the very narrow throat.

FEW animals have received less appropriate names than the GUINEA PIG ; for it is not a pig, but a rodent, and does not come from Guinea, but from Southern America. Being a very pretty little creature, it is in some favour as a domestic pet : and as it is remarkably prolific, it very rapidly increases in numbers, if it is well defended from cold and preserved from damp.

The food of the Guinea Pig is exclusively of a vegetable nature, and while feeding it generally sits on its hinder feet, and carries its food to its mouth with its fore-paws.

An idea of the extreme fecundity of this animal may be formed from the fact that it begins to breed at ten months of age, that each brood consists of an average of six or eight, and that in less than six weeks after the birth of

GUINEA PIG.—(*Cávia Aperéa.*)

the young family they are driven to shift for themselves, and the mother is then ready for another brood. The young Guinea Pigs are born with their eyes open, and covered with hair, and do not attain their full dimensions until they have reached the age of eight or nine months.

The colour of the Guinea Pig is very variable ; but is generally composed of white, red, and black, in patches of different size and shape in each individual. The bare portions of the skin are flesh-coloured, and the eye is brown. The animal is of little direct use to mankind, as its flesh is held in very low estimation, and its hair is so slightly attached to the skin that its coat is useless to the furrier.

THE common HARE is known from the rabbit by the redder hue of its fur, the great proportionate length of its black-tipped ears, which are nearly an inch longer than the head ; by its very long hind legs, and its large and prominent eyes. When fully grown it is of considerable size, weighing on the average about eight or nine pounds, and sometimes attaining the weight of twelve or even thirteen pounds. In total length it rather exceeds two feet, the tail being about three inches long. The colour of the common Hare is greyish brown on the upper portions of the body, mixed with a dash of yellow; the abdomen is white, and the neck and breast are yellowish white. The tail is black on the upper surface and white underneath, so that when the creature runs it exhibits the white tail at every leap. Sometimes the colour of the Hare deepens into black, and there are many examples of albino specimens of this animal.

L 2

It is a wonderfully cunning animal, and is said by many who have closely studied its habits to surpass the fox in ready ingenuity. Appearing to understand the method by which the hounds are enabled to track its footsteps, it employs the most crafty manœuvres for the purpose of throwing them off the scent. Sometimes it will run forward for a considerable distance, and then, after returning for a few hundred yards on the same track, will make a great leap at right angles to its former course, and lie quietly hidden while the hounds run past its spot of concealment. It then jumps back to its track, and steals quietly out of sight in one direction, while the hounds are going in the other.

The Hare does not live in burrows, like the rabbit, but only makes a slight depression in the ground, in which she lies so flatly pressed to the earth that

HARE.—(*Lepus timidus.*)

she can hardly be distinguished from the soil and dried herbage among which she has taken up her temporary abode.

It is a tolerably prolific animal, beginning to breed when only a year old, and producing four or five young at a litter. The young Hares, or "leverets," as they are technically termed, are born with their eyes open, and covered with hair. For the space of four or five weeks they remain under the care of their mother, but after that time they separate, and depend upon themselves for subsistence.

RESEMBLING the hare in general appearance and in many of its habits, the RABBIT is readily distinguished from that animal by its smaller dimensions, its different colour, its shorter and uniformly brown ears, and its shorter limbs.

The Rabbit is one of the most familiar of British quadrupeds, having taken firm possession of the soil into which it has been imported, and multiplied to so great an extent that its numbers can hardly be kept within proper bounds without annual and wholesale massacres. As it is more tameable than the hare, it has long been ranked amongst the chief of domestic pets, and has been so modified by careful management that it has developed itself into many permanent varieties, which would be considered as different species by one who saw them for the first time.

The burrows in which the Rabbit lives are extremely irregular in their construction, and often communicate with each other to a remarkable extent.

From many of its foes the Rabbit escapes by diving suddenly into its burrow ; but there are some animals, such as the stoat, weasel, and ferret, which follow it into its subterranean abode and slay it within the precincts of its own home.

When the female Rabbit is about to become a mother, she quits the ordinary burrows, and digs a special tunnel for the purpose of sheltering her young family during their first few weeks of life. At the extremity of the burrow she places a large quantity of dried herbage intermixed with down which she plucks from her own body, so as to make a soft and warm bed for the expected occupants. The young Rabbits are about seven or eight in

RABBIT.—(*Lepus cuniculus.*)

number, and are born without hair and with their eyes closed. Not until they have attained the age of ten or twelve days are they able to open their eyelids and to see the world into which they have been brought.

Rabbits are terribly destructive animals, as is too well known to all residents near a warren, and are sad depredators in field, garden, and plantation, destroying in very wantonness hundreds of plants which they do not care to eat. They do very great damage to young trees, delighting in stripping them of the tender bark as far as they can reach while standing on their hind feet. Sometimes they eat the bark, but in many cases they leave it in heaps upon the ground, having chiselled it from the tree on which it grew, and to which it afforded nourishment, merely for the sake of exercising their teeth and keeping them in proper order, just as a cat delights in clawing the legs of chairs and tables.

In its native state the fur of the Rabbit is nearly uniform brown, but when the animal is domesticated its coat assumes a variety of hues, such as pure white, jetty black, pied dun, slated grey, and many other tints.

THE CHINCHILLA, so well known for its exquisitely soft and delicate fur, belongs to the group of animals which are known to zoologists under the title of Jerboidæ, and which are remarkable for the great comparative length of their hinder limbs, and their long hair-clothed tails.

The Chinchilla is an inhabitant of Southern America, living chiefly among the higher mountainous districts, where its thick silken fur is of infinite

service in protecting it from the cold.   It is a burrowing animal, digging its
subterranean homes in the valleys which intersect the hilly country in which
it lives, and banding together in great numbers in certain favoured localities.
The food of the Chinchilla is exclusively of a vegetable nature, and consists
chiefly of various bulbous roots.   While feeding it sits upon its hinder feet,
and conveys the food to its mouth with its fore-feet, which it uses with singular
adroitness.   It is a most exquisitely cleanly animal, as might be supposed
from the beautiful delicacy of its fur, for we may always remark, that when-

CHINCHILLA.—(*Chinchilla lániger.*)

ever an animal is remarkable for the colouring or the texture of its natural
robes, it is always most assiduous in preserving them from any substance
that might stain their purity or clog their fibres.

The fur of the Chinchilla is of a delicate clear grey upon the back,
softening into a greyish white on the under portions, and its texture is
marvellously soft and fine.

THE GERBOAS bear a curious resemblance to the kangaroos, not only in
their general appearance, but in many of their habits.   Like those animals,
they leap over distances which are absolutely enormous when the size of their
bodies is taken into consideration ; they constantly sit upright in order to
observe surrounding objects, their food is of the same nature, and they carry
it to their mouths in a similar manner.   Their fore-limbs are extremely short,
while their hinder legs are developed to a very great extent, and they are all
furnished with a long, hair-clad tail, which serves to aid them in preserving
their balance while shooting through the air.

The most familiar of these singular creatures is the common GERBOA of
Northern Africa.   This beautiful and active little animal is hardly larger than
an ordinary English rat, although its peculiar attitudes and its extremely long
tail give it an appearance of greater dimensions than it really possesses.
The general colour of its fur is a light dun, washed with yellow, the abdomen
being nearly white.   The tail is of very great proportionate length, is

cylindrical in shape, and tufted at its extremity with stiff black hairs, the extreme tip being white.

The Gerboa is a burrowing animal, and lives in society, so that it forms large natural "warrens" in those parts of the country where it takes up its residence. It is much hunted by the natives, who set some store by its rather unpalatable flesh, and is captured by stopping up as many burrows as can conveniently be reached, and killing the Gerboas as they rush affrighted from the open entrances. This is, indeed, almost the only successful mode of capturing these fleet and agile creatures ; for if they can once leap away from the immediate vicinity of their pursuers, they scour over the ground with such wonderful speed that they can hardly be overtaken even by a trained grey hound.

GERBOA.—(*Dipus Ægyptius.*)

The food of these animals consists chiefly of roots and similar substances, which it digs out of the earth, but it also feeds on various kinds of grain.

NEXT in order to the Gerboas is placed the small group of animals which are sufficiently familiar by the name of DORMICE. This term signifies "Sleepy Mouse," and is most appropriate to the lethargic little creatures, which spend the greater part of their time in somnolency.

The common DORMOUSE is abundantly found in many districts of England, as well as on the Continent, and is in great favour as a domestic pet. The total length of this pretty little animal is rather more than five inches, the tail being two inches and a half long. The colour of its fur is a light reddish brown upon the back, yellowish white upon the abdomen, and white on the throat. These tints belong to the adult animal only, as in the juvenile Dormouse the fur is nearly of the same colour as that of the common mouse, the ruddy tinge only appearing on the head and sides. It is not until the little creatures have nearly completed a year of existence that they assume the beautiful hues of adult age. The tail is thickly covered with hair, which is arranged in a double row throughout its length, and forms a slight tuft at the extremity. The head is rather large in proportion to the body, the ears are large and broad, and the eye full, black, and slightly prominent.

The Dormouse is a nocturnal animal, passing the whole of the day in its warm and neatly constructed nest, which is generally built in the most retired spot of some thick bush or small tree. It is a very active little creature, leaping from branch to branch, and traversing the intricate mazes of the brushwood with such ready featness that it can scarcely be taken by a human hand. The food of the Dormouse consists of various fruits and seeds, such as acorns, nuts, haws, and corn.

As the animal is one of the hibernators, it is in the habit of gathering together a supply of dried food, to afford occasional nourishment during the long wintry months when it lies in its bed, imprisoned in the bands of irresistible sleep. Like many other hibernating animals, the Dormouse becomes exceedingly fat towards the end of autumn, and is therefore enabled to withstand the severity of the winter season better than if it retired into its home in only its ordinary condition. As soon as the weather becomes cold, the Dormouse retires into its nest, and there slumbers throughout the entire winter, waking up for a short period whenever a milder temperature breaks the severity of the frost, and, after taking a little nourishment, sinking again into its former lethargy.

The Dormouse is rather gregarious in its habits, so that whenever one nest is discovered several others may generally be found at no great distance. These nests are of considerable dimensions, being about six inches in diameter, and are composed of grass, leaves, and similar substances. The entrance to the nest is at the side.

The young animals are generally three or four in number at a birth, and make their appearance about the end of spring or the beginning of summer. It is probable that there may be a second brood towards the end of autumn, as Mr. Bell received from one locality in the month of September one half-grown Dormouse, which had evidently been born in the spring, and three very little specimens, which were apparently not more than a week or two old. They are born blind, but are able to see in a very few days, and in a remarkably short space of time become independent of their parents.

THE beautiful and active group of animals of which our English SQUIRREL is so familiar an example, are found in almost every portion of the globe, and, with one or two exceptions, live almost exclusively among the branches of trees. In order to enable them to maintain a firm clasp upon the branches and bark, they are furnished with long finger-like toes upon the fore-feet, which are armed with sharp curved claws.

DORMOUSE. — (*Muscardinus avellanárius.*)

IN the FLYING SQUIRRELS, of which the TAGUAN is a good example, the skin of the flanks is modified in a method similar to that which has already been noticed in the Petaurists. The skin is so largely developed, that when the animal is sitting at its ease, its paws but just appear from under the soft folds of the delicate and fur-clad membrane. When the creature intends to make one of its marvellous leaps, it stretches all its four limbs to their fullest extent, and is upborne through the air on the parachute-like expansion which extends along its sides.

This animal is a native of India, where it is tolerably common.

It is rather a large species, as its total length is nearly three feet, the tail occupying about one foot eight inches, measured to the extremity of the long hairs with which it is so thickly clothed. The general colour of this animal is a clear chestnut, deepening into brown on the back, and becoming more ruddy on the sides. The little pointed ears are covered with short and soft fur of a delicate brown, and the tail is heavily clad with bushy hairs, greyish black on the basal portions of that member, and sooty black towards the extremity. The parachute membrane is delicately thin, scarcely thicker than ordinary writing-paper, when it is stretched to its utmost, and is covered with hair on both its surfaces, the fur of the upper side being chestnut, and that

TAGUAN FLYING SQUIRREL.—(*Ptéromys Petaurista.*)

of the lower surface nearly white. A stripe of greyish black hairs marks the edge of the membrane, and the entire abdomen of the animal, together with the throat and the breast, is covered with beautiful silver greyish white fur.

The true Squirrels possess no parachute flying membrane, as do the Flying Squirrels, nor are they furnished with cheek-pouches, as is the case with the Ground Squirrels in America.

ONE of the most handsome of the Squirrels is the JELERANG, or JAVAN SQUIRREL, a native of Java, part of India, and Cochin China. Its total length is about two feet, the tail and body being equal to each other in measurement. In colour it is one of the most variable of animals, so that it has been more than once described under different names. In the British Museum are several specimens of this animal, and all of them present many varieties in point of colour, while some are so very unlike each other that most persons would consider them to be separate species. Some specimens of this animal are pale yellow, while others are deep brown ; in some the colour is tolerably uniform, while in others it is variously pied, but in all there seems to be a tolerably decided contrast between a darker and lighter tint. From this circumstance it has sometimes been termed *Sciúrus bicolor*, or the Two-coloured Squirrel.

In general, the darker hue prevails on the back and upper portions of the body, and the lighter tint is abruptly separated from it by a decided line of demarcation. The usual colour of the Jelerang is a dark brownish black on the back, the top of the head yellowish, and the sides and abdomen golden yellow.

The Jelerang is rather common in the countries which it inhabits, and as it is very retiring in its habits, and dreads the proximity of mankind, it is not so mischievous a neighbour as is the case with the greater number of the Squirrels. It lives chiefly in the depths of the forests, and feeds upon the wild fruits that grow without any aid from the hand of mankind. It is easily tamed, and being an active, amusing animal, as well as possessed of a beautifully marked coat, is often domesticated among the inhabitants of the same country. The flesh of the Jelerang is thought to be very good, and is eaten by the natives.

EVERYONE is familiar with the lively English SQUIRREL, which makes the woods joyous with its active gambols, and is so often repaid for its gaiety by being captured and compelled to make sport for its owner within the narrow precincts of a wire cage.

This little animal is plentiful in many parts of England, and indeed, is generally found wherever there is a tolerably large copse, or a wood of moderate dimensions. In private grounds and parks it luxuriates, knowing instinctively that

JELERANG, OR JAVAN SQUIRREL.—(*Sciúrus Javensis.*)

it may wander at its own will, unchecked and unharmed. Among the tree branches its powers of activity are absolutely surprising, for it will fling itself through such distances, and at such a height, that it seems likely to be dashed to pieces every instant. Yet it very seldom makes a false step, and even if it should lose its foothold, it is not at all disconcerted, but spreads out its legs and bushy tail to their utmost expansion, so that it presents a large surface to the air, and comes quite lightly to the earth, even though it may have leaped from a considerable height.

On the ground it is not so much at its ease as when it is careering amid the branches of some large tree, and, as soon as it feels alarmed, always makes the best of its way towards the nearest tree-trunk. Its gait is a kind of semi-gallop, and even when ascending a perpendicular tree-stem it maintains the same galloping movements, and ascends to a considerable height in a very small space of time.

During the hotter hours of the day the Squirrel is never seen, being quietly asleep in its lofty nest ; but in the early morning, or in the cooler hours of the afternoon, it comes from its retreat, and may be seen leaping about the branches in search of the various fruits on which it feeds.

The nest of the Squirrel is an admirable specimen of natural architecture and is almost invariably placed in the fork of some lofty branch, where it is concealed from the view of anyone passing under the tree, and is out of the reach of any ordinary foe, even if its situation were discovered. Sometimes it is built in the hollow of a decayed bough, but it is always admirably con-

SQUIRREL.—(*Sciurus Europæus.*)

cealed from sight. In form it is nearly spherical, and is made of leaves, moss, grass, and other substances, woven together in so artistic a manner that it is impermeable to rain, and cannot be dislodged from its resting-place by the most violent wind. A single pair of Squirrels inhabit the same nest, and seem to consider some particular tree as their home, remaining in it year after year.

The female Squirrel produces about three or four young at a litter, the little ones being born in the middle of summer, and remaining under the care of their parents until the spring of the succeeding year, when they separate and shift for themselves.

The food of the Squirrel is usually of a vegetable nature, and consists of nuts, acorns, wheat, and other fruits and seeds. Being a hibernating animal, the Squirrel is in the habit of laying up a winter store of provisions, and towards the end of autumn, while acorns and nuts are in their prime, becomes very busy in gathering certain little treasures, which it hides in all kinds of nooks, crevices, and holes, near the tree in which it lodges. The creature must be endowed with a very accurate memory, for it always remembers the spots where it has deposited its store of food, and even when the snow lies thickly upon the earth, and has covered the ground with a uniform white mantle, the Squirrel betrays no perplexity, but whenever it requires nourishment, goes straight to the hidden storehouse, scratches away the snow, and disinters its hidden treasures.

Sometimes the food of the Squirrel is not limited to vegetable substances, as the animal possesses something of the carnivorous nature, and has been often found guilty of killing and eating sundry animated things. Young birds, eggs, and various insects are eaten by the Squirrel, who has been detected in the very act of plundering a nest, and carrying off one of the young birds.

The usual colour of the Squirrel's fur is a ruddy brown upon the back, and a greyish white on the under portions of the body. It is, however, a most variable animal in point of colour, the tint of its fur changing according to the country which it inhabits. Even in England the ruddy fur is sometimes changed to grey during a severe winter, and in Siberia it is generally of a bluish grey. The feathery tufts of hair which fringe the ears are liable to great modification, being very long and full in winter and in cold climates, and almost entirely lost during the hotter summer months of our own country.

THERE are so many species of the Squirrel tribe, that even a cursory notice of each animal would be wholly impracticable in a work of the present dimensions, and we must content ourselves with a brief description of those species which stand out more boldly from the rest, by reason of form, colour, or peculiar habits.

GROUND SQUIRREL, OR HACKEE.—(*Támias Lysteri.*)

THE HACKEE, or CHIPPING SQUIRREL, as it is sometimes termed, is one of the most familiar of North American quadrupeds, and is found in great numbers in almost every locality. It is a truly beautiful little creature, and deserving of notice both on account of the dainty elegance of its form and the pleasing tints with which its coat is decked. The general colour of the Hackee is a brownish grey on the back, warming into orange-brown on the forehead and the hinder quarters. Upon the back and sides are drawn five longitudinal black stripes and two streaks of yellowish white, so that it is a most conspicuous little creature, and by these peculiar stripes may easily be distinguished from any other animal. The abdomen and throat are white. It is slightly variable in colour according to the locality in which it exists, and has been known to be so capricious of hue as to furnish specimens of pure white and jet black. As a fur it is extremely elegant, and if it were not quite so common would long since have taken nearly as high a rank as the sable or ermine.

The length of the Hackee is about eleven inches, the tail being about four inches and a half in length. It is, however, slightly variable in dimensions as well as in colour.

The Hackee is one of the liveliest and briskest of quadrupeds, and by reason of its quick and rapid movements, has not inaptly been compared with the wren. It is chiefly seen among brushwood and small timber; and as it whisks about the branches, or shoots through their interstices with its peculiar quick, jerking movements, and its odd, quaint, little clucking cry, like the chip-chipping of newly-hatched chickens, the analogy between itself and the bird is very apparent. As it is found in such plenty, and is a bold little creature, it is much persecuted by small boys, who, although they are not big or wise enough to be entrusted with guns, wherewith to work the destruction of larger game, arm themselves with long sticks, and by dexterous management knock down many a Hackee as it tries to escape from its pursuers by running along the rail fences. Among boys the popular name of the Hackee is the "Chipmuck."

It is a burrowing animal, making its little tunnels in various retired spots, but generally preferring an old tree, or the earth which is sheltered by a wall, a fence, or a bank. The burrows are rather complicated, and as they run to some length, the task of digging the animal out of its retreat is no easy one.

THE well-known PRAIRIE DOG, as it is called, is not a dog at all, but belongs to the Marmots. It is found in very great plenty along the coast of the Missouri and its tributaries, and also near the River Platte. It congregates together in vast numbers in certain spots where the soil is favourable to its subterranean habits of life, and the vegetation is sufficiently luxuriant to afford it nourishment. The colour of this animal is reddish brown upon the back, mixed with grey and black in a rather vague manner. The abdomen and throat are greyish white, and the short tail is clothed for the first half of its length with hair of the same tint as that of the body, and for the remaining half is covered with deep blackish brown hair, forming a kind of brush. The cheek-pouches are rather small, and the incisor teeth are large and protruding from the mouth. The length of the animal rather exceeds sixteen inches, the tail being a little more than three inches long. The cheek-pouches are about three-quarters of an inch in depth, and are half that measurement in diameter.

The Prairie Dog is a burrowing animal, and as it is very gregarious in its habits, the spot on which it congregates is literally honeycombed with its tunnels. There is, however, a kind of order observed in the "dog-towns," as these warrens are popularly called, for the animals always leave certain roads or streets in which no burrow is made. The affairs of the community seem to be regulated by a single leader, called the Big Dog, who sits before the entrance of his burrow, and issues his orders from thence to the community. In front of every burrow a small heap of mould is raised, which is made from the excavated soil, and which is generally employed as a seat for the occupant of the burrow.

As long as no danger is apprehended the little animals are all in lively motion, sitting upon their mounds, or hurrying from one tunnel to another as eagerly as if they were transacting the most important business. Suddenly a sharp yelp is heard, and the peaceful scene is in a moment transformed into a whirl of indistinguishable confusion. Quick barks resound on every side, the air is filled with a dust-cloud, in the midst of which is indistinctly seen an intermingled mass of flourishing legs and whisking tails, and in a moment the populous "town" is deserted. Not a "dog" is visible, and the whole spot is apparently untenanted. But in a few minutes a pair of dark eyes are seen gleaming at the entrance of some burrow, a set of glistening teeth next shine through the dusky recess, and in a few minutes first one and then another Prairie Dog issues from his retreat, until the whole community is again in lively action.

PRAIRIE DOG, OR WISH-TON-WISH.—(*Spermóphilus Ludoviciánus.*)

The title of Prairie Dog has been given to this animal on account of the sharp yelping sounds which it is in the habit of uttering, and which have some resemblance to the barking of a very small and very peevish lap-dog. Every time that it yelps it gives its tail a smart jerk. This peculiar sound is evidently employed as a cry of alarm ; for as soon as it is uttered all the Prairie Dogs dive into their burrows, and do not emerge again until they hear the shrill whistle which tells them that the danger is past.

The burrows of the Prairie Dogs are generally made at an angle of forty degrees, and after being sunk for some little distance, run horizontally, or even rise towards the surface of the earth. It is well known that these burrows are not only inhabited by the legitimate owners and excavators, but are shared by the burrowing owl and the rattlesnake. According to popular belief, the three creatures live very harmoniously together ; but careful observations have shown that the snake and the owl are interlopers, living in the burrows because the poor owners cannot turn them out, and finding an easy subsistence on the young Prairie Dogs. A rattlesnake has been killed near a burrow, and when the reptile was dissected, a Prairie Dog was found in its stomach.

THE common MARMOT is about the size of an ordinary rabbit, and not very unlike that animal in colour. The general tint of the fur is greyish yellow upon the back and flanks, deepening into black-grey on the top of the head, and into black on the extremity of the tail. It is very common in all the mountainous districts of Northern Europe, where it associates in small societies. The Marmot is an expert excavator, and digs very large and rather complicated burrows, always appearing to reserve one chamber as a store-house for the heap of dried grasses and other similar substances which it amasses for the purpose of sustaining life during the winter. The chamber in which the animal lives and sleeps is considerably larger than the store-house, measuring, in some cases, as much as seven feet in diameter. The tunnel which leads to these chambers is only just large enough to admit the body of the animal, and is about six feet in length.

MARMOT.—(*Ar'ctomys Marmotta.*)

To these burrows the Marmot retires about the middle of September, and after closing the entrance with grass and earth, enters into the lethargic hibernating state, and does not emerge until the beginning of April. Like other hibernating animals, they are very fat just before they take up their winter quarters, and as their fur is then in the best condition, they are eagerly sought after by the human inhabitants of the same country. The burrow of the Marmot is always dug in dry soil, and is seldom known to be at all above, or very much below, the line of perpetual snow.

AT the end of the rodents are placed the singular animals which are grouped together under the title of Aspalácidæ, or Mole Rats, the word *Aspalax*, or *Spalax*, being the Greek term for a mole. The incisor teeth of these animals are extremely large, and project beyond the lips. The external ears are either wholly wanting or are of very small dimensions. The eyes are small, and in some species are concealed by the skin. The body is heavily and clumsily made, the tail is either very short or entirely absent, and the head is large and rounded.

THE common MOLE RAT, which is also known by its Russian name of

SLEPEZ, is a native of Southern Russia, Asia Minor, Mesopotamia, and Syria. Like the ordinary mole, to which it bears no little external resemblance, it passes its existence in the subterranean tunnels which it excavates by means of its powerful claws. As it but seldom ventures into the light of day, it stands in no need of visual organs, but is compensated for their absence by the very large development of the organs of hearing. The place of the eyes is taken by two little round black specks, which lie under the fur-covered skin, so that even if they were sensitive to light, they would be unable to perceive the brightest rays of the noontide sun. The ears, however,

SLEPEZ MOLE RAT.—(*Spalax Typhlus.*)

are extremely large, and the hearing is exceedingly sensitive, so that the animal receives earlier information of danger through its sense of hearing than through that of sight, which latter faculty would indeed be useless in its dark abode. Sometimes the Slepez leaves the burrow and lies basking in the warm sunshine, but upon the least alarm or unexpected sound it plunges into its tunnel, and will not again make its appearance until it feels perfectly assured of safety.

The general colour of the Slepez is a very light brown, slightly tinged with red in some parts, and fading into an ashen grey in others. Its total length is about ten or eleven inches, and the tail is wanting. The head is broad, flat on the crown, and terminates abruptly at the muzzle. The feet are short, and the claws small.

THE incisor teeth of the COAST RAT or SAND MOLE are even larger in proportion than those of the preceding animal, and those of the upper jaw are marked by a groove running throughout their length. The fore-feet are furnished with long and powerful claws, that of the second toe being the largest. The eyes are exceedingly small, the external ears are wanting, and the tail is extremely short.

The Coast Rat is an inhabitant of the Cape of Good Hope and the coasts of Southern Africa, where it is found in tolerable profusion, and drives such multitudes of shallow tunnels that the ground which it frequents is rather dangerous for horsemen, and not at all pleasant even to a man on foot. The

COAST RAT, OR SAND MOLE.—(*Bathyergus Maritimus.*)

burrows are made at so short a distance from the surface that the earth gives way under the tread of any moderately heavy animal. Mr. Burchell, the well-known African traveller, narrates that in traversing the great sand flats of

BAY BAMBOO RAT.—(*Rhizomys badius.*)

Southern Africa he was often endangered by his feet sinking into the burrows of the Coast Rat, which had undermined the light soil in every direction. The animal is rather slow of foot upon the surface of the ground, but drives its subterranean tunnels with marvellous rapidity, throwing up little sandy

M

hillocks at intervals, like those of the common mole. On account of this propensity it has received the name of *Zand Mall*, or Sand Mole, from the Dutch Boers who inhabit the Cape.

The colour of the Sand Mole is a uniformly light greyish-brown, rather variable in tinting. As it is very soft and full in texture, and can be obtained in great quantities, it might be profitably made a regular article of trade. The Sand Mole is as large as our ordinary wild rabbit, being about fifteen inches in total length, the tail measuring about three inches.

THE BAY BAMBOO RAT is one representative of the genus Rhizomys, of which there are several species.

This animal is a native of Nepal, Malacca, and China, and is very injurious to the bamboos, on the roots of which it feeds. In size it equals a rather small rabbit, and in colour it is of a uniform ruddy brown, slightly paler on the throat and abdomen. The long incisor teeth are faced with bright red enamel, which gives them a rather conspicuous appearance; the tail is short and marked, and the claws are rather small. The head is of a peculiar form, which will be better understood from the engraving than by description alone.

OXEN.

IN the large and important group of animals which now occupy our attention, the incisor teeth are entirely absent in the upper jaw, and are eight in number in the lower. There are six molars on each side of each jaw. The two middle toes of each foot are separate, and are furnished with hoofs instead of claws. From the frontal bones proceed two excrescences, which are generally armed with horns, particularly in the male animal. The structure of the stomach and gullet is very remarkable, and is employed in producing that peculiar action which is called "ruminating," or chewing the cud.

The DOMESTIC OX of Europe has been so modified in form, habits, and dimensions, by its long intercourse with mankind, that it has developed into as many permanent varieties as the dog, the pigeon, or the rabbit, and would in many cases be thought to belong to different species. Among the principal varieties of this animal may be noticed the Long-horned, the Short-horned, and the Polled or hornless breeds, and the Alderney cow, so celebrated for the quantity and quality of the milk which it daily furnishes. In almost every part of the world are found examples of the Ox, variously modified in order to suit the peculiar circumstances amid which they are placed, but in all instances they are susceptible of domestication, and are employed in the service of mankind.

There are few animals which are more thoroughly useful to man than the Ox, or whose loss we should feel more deeply in the privation of so many comforts. Putting aside the two obvious benefits of its flesh and its milk—both of which are so needful for our comfort that we almost forget to think about them at all—we derive very great benefit from its powers while living, and from many portions of its body when dead.

In many parts of England, Oxen are still employed in agricultural labour, drawing the plough or waggon with a slow but steady ploughing gait. The carpenter would find himself sadly at a loss were his supply of glue to be suddenly checked by the disappearance of the animal, from whose hoofs, ears, and hide-parings the greater part of that useful material is manufactured. The harness-maker, carriage-builder, and shoemaker would in that case be deprived of a most valuable article in their trade; the cutler and ivory turner would lose a considerable portion of the rough material upon which they work; the builder would find his best plaster sadly impaired without a proper admixture of cow's hair ; and the practical chemist would be greatly at a loss for some of his most valuable productions if the entire Ox tribe were swept from the earth. Not even the very intestines are allowed to be wasted, but are employed for a variety of purposes and in a variety of trades. Sometimes the bones are subjected to a process which extracts every nutritious particle out of them, and even in that case, the remaining innutritious portions of the bones are made useful by being calcined, and manufactured into the animal charcoal which has lately been so largely employed in many of the arts and sciences.

THE OX.—(*Bos.*)

The Domestic Cow is too well known to need any detailed description of form and colour. Few persons, however, except those who have been personally conversant with these animals, have any idea of their intelligent and affectionate natures.

As the Oxen, in common with the sheep, camel, giraffe, and deer, require a large amount of vegetable food, and are, while in their native regions, subject to innumerable disturbing causes that would effectually prevent them from satisfying their hunger in an ordinary manner, they are furnished with a peculiar arrangement of the stomach and digestive organs, by means of which they are enabled to gather hastily a large amount of food in any spot where the vegetation is luxuriant, and to postpone the business of mastication and digestion to a time when they may be less likely to be disturbed. The peculiarity of structure lies chiefly in the stomach and gullet, which are formed so as to act as an internal food-pouch, analogous in its use to the cheek-pouches of certain monkeys and rodents, together with an arrangement for regurgitating the food into the mouth at the will of the animal, previous to its mastication and digestion.

The domestic cattle of India are commonly known by the name of ZEBU, and are conspicuous for the curious fatty hump which projects from the withers. These animals are further remarkable for the heavy dewlap which falls in thick folds from the throat, and which gives to the forepart of the animal a very characteristic aspect. The limbs are slender, and the back, after rising towards the haunches, falls suddenly at the tail.

THE Zebu is a quiet and intelligent animal, and is capable of being trained in various modes for the service of mankind. It is a good draught animal, and is harnessed either to carriages or to ploughs, which it can draw with great steadiness, though with but little speed. Sometimes it is used for riding, and is possessed of considerable endurance, being capable of carrying a rider for fifteen hours a day, at an average rate of five or six miles per hour.

The Zebu race has a very wide range of locality, being found in India, China, Madagascar, and the eastern coast of Africa. It is believed, however, that its native land is India, and that it must have been imported from thence into other countries.

There are various breeds of Zebu, some being about the size of our ordinary cattle, and others varying in dimensions from a large Ox to a small Newfoundland dog. One of the most familiar of these varieties is the well-known Brahmin Bull, so called because it is considered to be sacred to Bramah.

The more religious among the Hindoos, scrupulously observant of the letter of a law which was intended to be universal in its application, but to which they give only a partial interpretation, indulge this animal in the most absurd manner. They place the sacred mark of Siva on its body, and permit it to wander about at its own sweet will, pampered by every luxury, and never opposed in any wish or caprice which it may form. A Brahmin Bull will walk along the street with a quaintly dignified air, inspect anything and anybody that may excite his curiosity, force any one to make way for himself, and if he should happen to take a fancy to the contents of a fruiterer's or greengrocer's shop, will deliberately make his choice and satisfy his wishes, none daring to cross him. The indulgence which is extended to this animal is carried to so great a height, that if a Brahmin Bull choses to lie down in a narrow lane, no one can pass until he gets up of his own accord.

ZEBU.—(*Bos Indicus.*)

THE BUFFALO is spread over a wide range of country, being found in Southern Europe, North Africa, India, and a few other localities.

This animal is subject to considerable modifications in external aspect, according to the climate or the particular locality in which it resides, and has in consequence been mentioned under very different names. In all cases the wild animals are larger and more powerful than their domesticated relations, and in many instances the slightly different shape and greater or lesser length of the horns, or the skin denuded of hairs, have been considered as sufficient evidence of separate species.

In India, the long, smooth-horned variety chiefly prevails, and is found in tolerable profusion. This animal frequents wet and marshy localities, being sometimes called the Water Buffalo on account of its aquatic predilections. It is a most fierce and dangerous animal, savage to a marvellous degree, and not hesitating to charge any animal that may arouse its ready ire. An angry buffalo has been known to attack a tolerably sized elephant, and by a vigorous charge in the ribs to prostrate its huge foe. Even the tiger is found to quail before the Buffalo, and displays the greatest uneasiness in its presence.

THE ARNEE lives in large herds, arranged after the manner of all bovine animals, the females and young being always placed in the safest spots, while the males post themselves in all positions of danger. These herds are never seen on elevated ground, preferring the low marshy districts where water and mud are abundant. In this mud they love to wallow, and when suddenly roused from their strange pastime, present a most terrible appearance, their eyes glaring fiercely from amid the mud-covered dripping masses of hair. Sometimes the Buffalo is said to fall a victim to its propensity for wallowing in the mud, and to be stuck so firmly in the oozy slime, as it dries under the scorching sunbeams of that burning climate, that it can be killed without danger. They generally chew the cud while they are lying immersed in mud or water.

The CAPE BUFFALO is quite as formidable an animal as its Indian relative, and much more terrible in outward aspect. The heavy bases of the horns, that nearly unite over the forehead, and under which the little fierce eyes twinkle with sullen rays, give to the

BUFFALO.—(*Bubalus buffelus.*)

creature's countenance an appearance of morose, lowering ill-temper, which is in perfect accordance with its real character.

Owing to the enormous heavy mass which is situated on the forehead, the Cape Buffalo does not see very well in a straight line, so that a man may sometimes cross the track of a Buffalo within a hundred yards, and not be seen by the animal, provided that he walks quietly, and does not attract attention by the sound of his footsteps. This animal is ever a dangerous neighbour, but when it leads a solitary life among the thickets and marshy places, it is a worse antagonist to a casual passenger than even the lion himself. In such a case, it has an unpleasant habit of remaining quietly in its lair until the unsuspecting traveller passes closely to its place of concealment, and then leaping suddenly upon him like some terrible monster of the waters, dripping with mud, and filled with rage.

Many such tragical incidents have occurred,

CAPE BUFFALO.—(*Bubalus Caffer.*)

chiefly, it must be acknowledged, owing to the imprudence of the sufferer : and there are few coverts in Southern Africa which are not celebrated for some such terrible incident. Sometimes the animal is so recklessly furious in its unreasoning anger, that it absolutely blinds itself by its heedless rush through the formidable thorn-bushes which are so common in Southern

Africa. Even when in company with others of their own species, they are liable to sudden bursts of emotion, and will rush blindly on, heedless of everything but the impulse that drives them forward. In one instance, the leader of the herd, being wounded, dropped on his knees, and was instantly crushed by the trampling hoofs of his comrades, as they rushed over the prostrate body of their chief.

The Cape Buffalo, although so terrible an animal, is not so large as the arnee, being little larger than an ordinary ox, but possessed of much greater strength. The strangely shaped horns are black in colour, and so large that the distance between their points is not unfrequently from four or five feet. On account of their great width at their bases, they form a kind of bony helmet, which is impenetrable to an ordinary musket ball, and effectually defend their owner against the severe shocks which are frequently suffered by these testy animals.

THE BISON is only found in Northern America, never appearing north of lat. 33°. It gathers together in enormous herds, consisting of many thousand in number, and in spite of the continual persecution to which it is subjected by man and beast, its multitudes are even now hardly diminished. The Bison is one of the most valuable of animals to the white hunter as well as to the aboriginal Red Indian, as its body supplies him with almost every necessary of life.

The flesh of the fat cow Bison is in great repute, being juicy, tender, and well-savoured, and possesses the invaluable quality of not cloying the appetite, even though it be eaten with the fierce hunger that is generated by a day's hunting. The fat is peculiarly excellent, and is said to bear some resemblance to the celebrated green fat of the turtle. The most delicate portion of the Bison is the flesh that composes the "hump," which gives to the animal's back so strange an aspect ; and the hunters are so fond of this delicacy that they will often slay a magnificent Bison merely for the sake of the hump, the tongue, and the marrow-bones, leaving the remainder of the body to the wolves and birds.

BISON.—(*Bison Americanus.*)

The hide is greatly valued both by Indians and civilised men, for the many purposes which it fulfils. From this hide the Indian makes his tents, many parts of his dress, his bed, and his shield. For nearly the whole of these uses the skin is deprived of hair, and is so dressed as to be impervious to water, and yet soft and pliable. The shield is very ingeniously made by pegging out the hide upon the ground with a multitude of little wooden skewers round its edge, imbuing it with a kind of glue, and gradually removing the pegs in proportion to the consequent shrinking and thickening of the skin. One of these shields, although still pliable, is sufficiently strong to resist an arrow, and will often turn a bullet that does not strike it fairly.

Vast quantities of Bisons are killed annually, whole herds being sometimes destroyed by the cunning of their human foes. The hunters, having discovered a herd of Bisons at no very great distance from one of the precipices which abound in the prairie lands, quietly surround the doomed animals, and drive them ever nearer and nearer to the precipice. When they have come within half a mile or so of the edge, they suddenly dash towards the Bisons, shouting, firing, waving hats in the air, and using every means to terrify the intended victims. The Bisons are timid creatures, and easily take alarm, so that on being startled by the unexpected sights and sounds, they dash off, panic-struck, in the only direction left open to them, and which leads directly to the precipice. When the leaders arrive at the edge they attempt to recoil, but they are so closely pressed upon by those behind them, that they are carried forward and forced into the gulf below. Many hundreds of Bisons are thus destroyed in the space of a few minutes.

The Bison is remarkably fond of wallowing in the mud, and when he cannot find a mud-hole ready excavated, sets busily to work to make one for himself. Choosing some wet and marshy spot, he flings himself down on his side, and whirls round and round until he wears away the soil, and forms a circular and rather shallow pit, into which the water rapidly drains from the surrounding earth. He now redoubles his efforts, and in a very short time succeeds in covering himself with a thick coating of mud, which is probably of very great service in defending him from the stings of the gnats and other noxious insects which swarm in such localities.

The Bison is a marvellously active animal, and displays powers of running and activity which would hardly be anticipated by one who had merely seen a stuffed specimen. The body is so loaded with hair that it appears to be of greater dimensions than is really the case, and seems out of all proportion to the slender legs that appear from under it and seem to bend beneath its weight. Yet the Bison is an enduring as well as a swift animal, and is also remarkably sure of foot, going at full speed over localities where a horse would be soon brought to a halt.

THE YAK, or GRUNTING OX, derives its name from its very peculiar voice, which sounds much like the grunt of a pig. It is a native of the mountains of Thibet, and according to Hodson, it inhabits all the loftiest plateaus of High Asia, between the Altai and the Himalayas.

The heavy fringes of hair that decorate the sides of

YAK.—(*Poephagus grunniens.*)

the Yak do not make their appearance until the animal has attained three month! of age, the calves being covered with rough curling hair, not unlike that of a black Newfoundland dog. The beautiful white bushy tail of the Yak is in great request for various ornamental purposes, and forms quite an

important article of commerce.   Dyed red, it is formed into those curious
tufts that decorate the caps of the Chinese, and when properly mounted in
a silver handle, it is used as a fly-flapper in India under the name of
"chowrie."   These tails are carried before certain officers of state, their
number indicating his rank.

THE curiously-shaped horns of the MUSK OX, its long woolly hair falling
almost to the ground in every direction, so as nearly to conceal its legs,
together with the peculiar form of the head and snout, are unfailing characteristics, whereby it can be dis-
criminated from any other
animal.   The horns of the
Musk Ox are extremely large
at their base, and form a kind
of helmet upon the summit
of the forehead.   They then
sweep boldly downwards, and
are again hooked upwards
towards the tips.   This
curious form of the horns is
only noticed in the male, as
the horns of the female are
set very widely apart from
the sides of the forehead, and
are simply curved.   The
muzzle is covered with hair,
with the exception of a very
slight line round the nostrils.

MUSK OX.—(*Ovibos moschatus.*)

This animal is an inhabit-
ant of the extreme north of America, being seldom seen south of the sixty-
first degree of latitude, and ascending as high as the seventy-fifth.   It
lives, in fact, in the same country which is inhabited by the Esquimaux,
and is known to them under the name of Oomingnoak.   It is a fleet and
active animal, and traverses with such ease the rocky and precipitous
ground on which it loves to dwell, that it cannot be overtaken by any pursuer
less swift than an arrow or bullet.   It is rather an irritable animal, and
becomes a dangerous foe to the hunters, by its habit of charging upon them
while they are perplexed amid the cliffs and crevices of its rocky home,
thus often escaping unharmed by the aid of its quick eye and agile limbs.
The flesh of this animal is very strongly perfumed with a musky odour, very
variable in its amount and strength.   Excepting, however, a few weeks in
the year, it is perfectly fit for food, and is fat and well flavoured.

The Musk Ox is a little animal, but, owing to the huge mass of woolly hair
with which it is thickly covered, appears to be of considerable dimensions.
The colour of this animal is a yellowish brown, deepening upon the sides.

ANTELOPES.

THE ANTELOPES form a large and important group of animals, finding
representatives in many portions of the globe.   Resembling the deer in many
respects, they are easily to be distinguished from those animals by the
character of the horns, which are hollow at the base, set upon a solid core
like those of the oxen, and are permanently retained throughout the life of
the animal.   Indeed, the Antelopes are allied very closely to the sheep and
goats, and in some instances, are very goat-like in external form.   In all cases

the Antelopes are light and elegant of body, their limbs are gracefully slender, and are furnished with small cloven hoofs. The tail is never of any great length, and in many species is very short. The horns, set above the eyebrows, are either simply conical or are bent so as to resemble the two horns of the ancient lyre, and are therefore termed "lyrate" in technical language.

The well-known GAZELLE is found in great numbers in Northern Africa, where it lives in herds of considerable size, and is largely hunted by man and beast.

Trusting to its swift limbs for its safety, the Gazelle will seldom, if ever, attempt to resist a foe, unless it be actually driven to bay in some spot from

GAZELLE.—(*Gazella Dorcas.*)

whence it cannot escape ; but prefers to flee across the sandy plains, in which it loves to dwell, with the marvellous speed for which it has long been proverbial. The herd seems to be actuated by a strong spirit of mutual attachment, which preserves its members from being isolated from their companions, and which, in many instances, is their only safeguard against the attacks of the smaller predaceous animals. The lion and the leopard can always find a meal whenever they can steal upon a band of Gazelles without being discovered by the sentries which watch the neighbourhood with jealous precaution ; for the Gazelles are too weak to withstand the attack of such terrible assailants, and do not even attempt resistance.

The eye of the Gazelle is large, soft, and lustrous, and has been long celebrated by the poets of its own land as the most flattering simile of a woman's eye. The colour of this pretty little animal is a light fawn upon the back, deepening into dark brown in a wide band which edges the flanks, and forms a line of demarcation between the yellow-brown of the upper portions of the body and the pure white of the abdomen. The face is rather curiously marked with two stripes of contrasting colours, one a dark black-brown line that passes from the eye to the curves of the mouth, and the other a white streak that begins at the horns and extends as far as the muzzle. The hinder quarters, too, are marked with white, which is very preceptible when the animal is walking directly from the spectator.

THE SPRING-BOK derives its very appropriate title from the extraordinary leaps which it is in the constant habit of making whenever it is alarmed.

As soon as it is frightened at any real or fancied danger, or whenever it desires to accelerate its pace suddenly, it leaps high into the air with a curiously easy movement, rising to a height of seven or eight feet without any difficulty, and being capable on occasions of reaching the height of twelve or thirteen feet. When leaping, the back is greatly curved, and the creature presents a very curious aspect, owing to the sudden exhibition of the long white hairs that cover the croup, and are nearly hidden by the folds of skin when the creature is at rest, but which come boldly into view as soon as the protecting skin-fold is obliterated by the tension of the muscles

SPRING-BOK—(*Antidorcas Euchore.*)

that serve to propel the animal in its aërial course.

The Spring-bok is a marvellously timid animal, and will never cross a road if it can avoid the necessity. When it is forced to do so, it often compromises the difficulty by leaping over the spot which has been tainted by the foot of man. The colour of the Spring-bok is very pleasing, the ground tinting being a warm cinnamon-brown upon the upper surface of the body, and pure white upon the abdomen, the two colours being separated from each other by a broad band of reddish brown. The flesh of the Spring-bok is held in some estimation, and the hide is in great request for many useful purposes.

Inhabiting the vast plains of Southern Africa, the Spring-bok is accustomed to make pilgrimages from one spot to another, vast herds being led by their chiefs, and ravaging the country over which they pass as if they were locusts.

THE GEMS-BOK, or KOOKAAM, is a large and powerful member of the Antelope tribe, equalling the domestic ass in size, and measuring about three feet ten inches at the shoulder. The manner in which the hide is decorated with boldly contrasted tints gives it a very peculiar aspect. The general hue is grey, but along the back, upon the hinder quarters, and along the flanks, the colour is deep black. A black streak also crosses the face, and, passing under the chin, gives it the appearance of wearing harness. It has a short, erect mane, and long, sweeping, black tail, and its heavy horns are nearly straight from base to tip.

The long and sharply-pointed horns with which its head is armed are terrible weapons of offence, and can be wielded with marvellous skill. Striking right and left with these natural bayonets, the adult Gems-bok is a match for most of the smaller carnivora, and has even been known to wage a successful duel with the lordly lion, and fairly to beat off its antagonist. Even when the lion has overcome the Gems-bok, the battle may sometimes be equally claimed by both sides, for in one instance the dead bodies of a lion and a Gems-bok were found lying on the plain, the horns of the Antelope being driven so firmly into the lion's body, that they could not be extracted by the efforts of a single man. The lion had evidently sprung upon the Gems-bok, which had re-ceived its foe upon the points of its horns, and had sacri-ficed its own life in destroy-ing that of its adversary.

As is the case with many long-horned animals, one of the horns, usually the left, is shorter than the other. In a fine pair of Gems-bok horns in my collection, the left horn is nearly three inches shorter than the right.

Although the Gems-bok is nearly independent of water, it stands as much in need of moisture as any other animal, and would speedily perish in the arid deserts were it not directed by its instincts towards certain suc-culent plants which are placed in those regions, and which possess the useful power of attracting and retaining every particle of moisture which

GEMS-BOK.—(*Oryx Gazella.*)

may happen to settle in their vicinity. The most common and most valuable of these plants is a bulbous root, belonging to the Liliacea, called, from its peculiar property of retaining the moisture, the Water-root. Only a very small portion of the valuable plant appears above the ground, and the water-bearing bulb is so encrusted with hardened soil that it must be dug out with a knife. Several other succulent plants also possess similar qualities, among which may be noticed a kind of little melon which is spread over the whole of the great Kalahari desert.

RESEMBLING the Gems-bok in many particulars, the ORYX can be easily distinguished from its predecessor by the shape of the horns, which, instead of being nearly straight, are considerably bent, and sweep towards the back in a noble curve.

It uses these horns with as much address as its near relative the gems-bok, and if it should be lying wounded on the ground, the hunter must beware of approaching the seemingly quiescent animal, lest it should suddenly strike at him with its long and keenly pointed horns, while its body lies prostrate on the earth. Should it be standing at bay, it is a very dangerous opponent, having a habit of suddenly lowering its head and charging forward with a quick, lightning-like speed, from which its antagonist cannot escape with-out difficulty.

The colour of this animal is greyish white upon the greater part of its person, and is diversified by sundry bold markings of black and ruddy brown, which are spread over the head and body in a manner that can be readily comprehended from the illustration. The height of the Oryx is rather more than three feet six inches, and the long curved horns are upwards of three feet in length. These horns are set closely together upon the head, from whence they diverge gradually to their extremities. These weapons are covered with rings at their bases, but at their tips they are smooth and exceedingly sharp. Their colour is black.

GOAT-LIKE in aspect, and very hircine in many of its habits, the CHAMOIS is often supposed to belong to the goats rather than to the Antelopes.

It is, however, a true Antelope, and may be readily distinguished from any of its relations by the peculiar form of the horns, which rise straight from the top of the head for some inches, and then suddenly curve backwards, so as to form a pair of sharp hooks. In descending a precipitous rock, the Chamois is greatly aided by the false hoofs of the hinder feet, which it hitches upon

CHAMOIS.—(*Rupicapra Tragus.*)

every little irregularity in the stony surface, and which seem to retard its progress as it slides downwards, guided by the sharp hoofs of the fore-feet, which are placed closely together, and pushed well in advance of the body. Thus flattened against the rock, the Chamois slides downwards until it comes to a ledge broad enough to permit it to repose itself for a while before descending further. In this manner the active creature will not hesitate to descend some twenty or thirty yards along the face of an almost perpendicular cliff, being sure to make good its footing on the first broad ledge that may present itself.

The Chamois is one of the most wary of Antelopes, and possesses the power of scenting mankind at an almost incredible distance. Even the old and half-obliterated footmarks which a man has made in the snow are sufficient to startle the sensitive fears of this animal, which has been observed to stop in mid career down a mountain side, and to bound away at right angles to its former course, merely because it had come across the track which had been left by the steps of some mountain traveller. Like all animals which live in herds, however small, they always depute one of their number to act as sentinel. They are not, however, entirely dependent on the vigilance of their picket, but are always on the alert to take alarm at the least suspicious scent, sight, or sound, and to communicate their fears to their comrades by a peculiar warning whistle. As soon as this sound is heard, the entire herd take to flight.

Their ears are as acute as their nostrils, so that there are few animals which are more difficult of approach than the Chamois. Only those who have been trained to climb the giddy heights of the Alpine mountains, to traverse the most fearful precipices with a quiet pulse and steady head, to exist for days amid the terrible solitudes of ice, rock, and snow, and to sustain almost every imaginable hardship in the pursuit of their game,—only these, or in very rare instances those who have a natural aptitude for the sport, and are, in consequence, soon initiated into its requisite accomplishments, can hope even to come within long rifle range of a Chamois when the animal is at large upon its native cliffs.

The food of the Chamois consists of the various herbs which grow upon the mountains, and in the winter season it finds its nourishment on the buds of sundry trees, mostly of an aromatic nature, such as the fir, pine, and juniper. In consequence of this diet the flesh assumes a rather powerful odour, which is decidedly repulsive to the palates of some persons, while others seem to appreciate the peculiar flavour, and to value it as highly as the modern gourmand appreciates the "gamey" flavour of long-kept venison. The skin is largely employed in the manufacture of a certain leather, which is widely famous for its soft though tough character. The colour of the Chamois is yellowish brown upon the greater portion of the body, the spinal line being marked with a black streak. In the winter months the fur darkens and becomes blackish brown. The face, cheeks, and throat are of a yellowish white hue, diversified by a dark brownish black band which passes from the corner of the mouth to the eyes, when it suddenly dilates and forms a nearly perfect ring round the eyes. The horns are jetty black and highly polished, especially towards the tips, which are extremely sharp. There are several obscure rings on the basal portions, and their entire surface is marked with longitudinal lines.

Several varieties of the Chamois are recorded, but the distinctions between them lie only in the comparative length of the horns and the hue of the coat. The full-grown Chamois is rather more than two feet in height, and the horns are from six to eight inches long.

OF all the Antelopes, the GNOO presents the most extraordinary conformation. At the first sight of this curious animal the spectator seems to doubt whether it is a horse, a bull, or an Antelope, as it appears to partake nearly equally of the nature of these three animals.

The Gnoos, of which there are several species, may be easily recognized by the fierce-looking head, their peculiarly shaped horns, which are bent downwards and then upwards again with a sharp curve, by their broad nose, and long hair-clad tail. They live together in considerable herds, often mixing with zebras, ostriches, and giraffes, in one huge army of living beings. In their habits they are not unlike the wild cattle which have already been described. Suspicious, timid, curious of disposition, and irritable of temper, they display these mingled qualities in a very ludicrous manner whenever they are alarmed by a strange object.

"They commence whisking their long white tails," says Cumming, "in a most eccentric manner ; then, springing suddenly into the air, they begin pawing and capering, and pursue each other in circles at their utmost speed. Suddenly they all pull up together to overhaul the intruder, when some of the bulls will often commence fighting in the most violent manner, dropping on their knees at every shock ; then, quickly wheeling about, they kick up their heels, whirl their tails with a fantastic flourish, and scour across the plain, enveloped in a cloud of dust." On account of these extraordinary manœuvres, the Gnoo is called *Wildebeest* by the Dutch settlers. The Gnoos in the Zoological Gardens may often be seen at their gambols.

The colour of the ordinary Gnoo (*Connochetes Gnu*) is brownish black, sometimes with a blue-grey wash. The mane is black, with the exception of the lower part, which is often greyish-white, as is the lower part of the tail. The nose is covered with a tuft of reversed hair, and there is a mane upon the chest.

THE GNOO.—(*Connochetes Gnu.*)

The Gnoo is about three feet nine inches high at the shoulders, and measures about six feet six inches from the nose to the root of the tail.

By far the most striking and imposing of all South African Antelopes, the Koodoo, now claims our attention.

This truly magnificent creature is about four feet in height at the shoulder, and its body is rather heavily made, so that it is really a large animal. The curiously twisted horns are nearly three feet in length, and are furnished with a strong ridge or keel, which extends throughout their entire length. It is not so swift or enduring as many Antelopes, and can be run down without difficulty, provided that the hunter be mounted on a good horse, and the ground be tolerably fair and open. Its leaping powers are very great, for one of these animals has been known to leap to a height of nearly ten feet without the advantage of a run.

The flesh of the Koodoo is remarkably good, and the marrow of the principal bones is thought to be one of Africa's best luxuries. So fond are the natives of this dainty, that they will break the bones and suck out the marrow without even cooking it in any way whatever. The skin of this animal is extremely valuable, and for some purposes is almost priceless. There is no skin that will make nearly so good a "fore-slock," or whip-lash, as that of the Koodoo ; for its thin, tough substance is absolutely required for such a purpose. Shoes, thongs, certain parts of harness, and other similar objects are manufactured from the Koodoo's skin, which, when properly prepared, is worth a sovereign or thirty shillings even in its own land.

The Koodoo is very retiring in disposition, and is seldom seen except by those who come to look for it. It lives in little herds or families of five or six in number, but it is not uncommon to find a solitary hermit here or there, probably an animal which has been expelled from some family, and is awaiting the time for setting up a family of his own. As it is in the habit of frequenting brushwood, the heavy spiral horns would appear to be great hindrances to their owner's progress; such is not, however, the case, for when the Koodoo runs, it lays its horns upon its back, and is thus enabled to thread the tangled bush without difficulty. Some writers say that the old males will sometimes establish a bachelors' club, and live harmoniously together, without admitting any of the opposite sex into their society.

The colour of the Koodoo is a reddish grey, marked with several white streaks running boldly over the back and down the sides. The females are destitute of horns.

THE ELAND, IMPOOFO, or CANNA, is the largest of the South African Antelopes, being equal in dimensions to a very large ox.

A fine specimen of an adult bull Eland will measure nearly six feet in height at the shoulders, and is more than proportionately ponderous in his build, being heavily burdened with fat as well as with flesh. Owing to this great weigth of body, the Eland is not so enduring as the generality of Antelopes, and can usually be ridden down without much trouble. Indeed the chase of the animal is so simple a matter, that the hunters generally contrive to drive it towards their encampment, and will not kill it until it has approached the waggon so closely that the hunters will have but little trouble in conveying its flesh and hide to their wheeled treasure-house.

The flesh of the Eland is peculiarly excellent; and as it possesses the valuable quality of being tender immediately after the animal is killed, it is highly appreciated in the interior of South Africa, where usually all the meat is as tough as shoe-leather, and nearly as dry. In some strange manner, the Eland contrives to live for

KOODOO.—(*Strepsiceros Kudu.*)

ELAND.—(*Oreas Canna.*)

months together without drinking, and even when the herbage is so dry that it crumbles into powder in the hand, the Eland preserves its good condition, and is, moreover, found to contain water in its stomach if opened. For its abstinence from liquids the Eland compensates by its ravenous appetite for solid food, and is so large a feeder that the expense of keeping the animal would be almost too great for any one who endeavoured to domesticate the animal in England with any hope of profit.

PASSING from Africa to Asia, we find a curious and handsome Antelope, partaking of many of the characteristics which are found in the Koodoo and Bosch-bok. This is the NYLGHAU, an inhabitant of the thickly-wooded districts of India.

This magnificent Antelope is rather more than four feet high at the shoulders, and its general colour is a slate-blue. The face is marked with brown or sepia; the long neck is furnished with a bold dark mane, and a long tuft of course hair hangs from the throat. The female is smaller than her mate, and hornless. Her coat is generally a reddish grey, instead of partaking of the slate-blue tint which colours the form of the male. The hind legs of this animal are rather shorter than the fore legs. Its name, Nylghau, is of Persian origin, and signifies " Blue Ox."

It does not seem to be of a social disposition, and is generally found in pairs inhabiting the borders of the jungle. There are, however, many examples of solitary males. It is a shy and wary animal, and the hunter who desires to shoot one of these animals is obliged to exert his bush-craft to the utmost in order to attain his purpose. To secure a Nylghau requires a good marksman as well as a good stalker, for the animal is very tenacious of life, and if not struck in the proper spot will carry off a heavy bullet without seeming to be much the worse at the time. The native chiefs are fond of hunting the Nylghau, and employ in the chase a whole army of beaters and trackers, so that the poor animal has no chance of fair play. These hunts are not without their excitement, for the Nylghau's temper is of the shortest, and when it feels itself aggrieved, it suddenly turns upon its opponent, drops on its knees, and leaps forward with such astounding rapidity that the attack can hardly be avoided, even when the intended victim is aware of the animal's intentions.

NYLGHAU.—(*Portax tragocamelus.*)

The Nylghau is not of very great value either to individual hunters or for commercial purposes. The hide is employed in the manufacture of shields, but the flesh is coarse and without flavour. There are, however, exceptions to be found in the " hump " of the male, the tongue, and the marrow bones : which are thought to be rather delicate articles of diet. Its gait is rather clumsy, but very rapid, and generally consists of a peculiar long swinging canter, which is not easily overtaken.

GOATS AND SHEEP.

CLOSELY allied to each other, the GOATS and SHEEP can be easily separated by a short examination. In the Goats, which will first come under consideration, the horns are erect, decidedly compressed, curved backwards and outwards, and are supplied with a ridge or heel of horny substance in front. The males generally possess a thickly-bearded chin, and are all notable for a powerful and very rank odour which is not present in the male sheep.

Of the genus Capra, which includes several species, the IBEX, or STEIN-BOCK, is a familiar and excellent example.

IBEX.—(*Capra Ibex.*)

This animal, an inhabitant of the Alps, is remarkable for the exceeding development of the horns, which are sometimes more than three feet in length, and of such extraordinary dimensions that they appear to a casual observer to be peculiarly unsuitable for an animal that traverses the craggy regions of Alpine precipices.

To hunt the Ibex successfully is as hard a matter as hunting the chamois, for the Ibex is to the full as wary and active an animal, and is sometimes

N

apt to turn the tables on its pursuer, and assume an offensive deportment. Should the hunter approach too near the Ibex, the animal will, as if suddenly urged by the reckless courage of despair, dash boldly forward at its foe, and strike him from the precipitous rock over which he is forced to pass. The difficulty of the chase is further increased by the fact that the Ibex is a remarkably endurant animal, and is capable of abstaining from food or water for a considerable time.

It lives in little bands of five or ten in number, each troop being under the command of an old male, and preserving admirable order among themselves. Their sentinel is ever on the watch, and at the slightest suspicious sound, scent, or object, the warning whistle is blown, and the whole troop make instantly for the highest attainable point. Their instinct always leads them upwards, an inborn "excelsior" being woven into their very natures, and as soon as they perceive danger they invariably begin to mount towards the line of perpetual snow. The young of this animal are produced in April, and in a few hours after their birth they are strong enough to follow their parent.

The colour of the Ibex is a reddish brown in summer, and grey brown in winter ; a dark stripe passes along the spine and over the face, and the

GOAT.—(*Hircus Ægágrus.*)

abdomen and interior faces of the limbs are washed with whitish grey. The horns are covered from base to point with strongly-marked transverse ridges, the number of which is variable, and is thought by some persons to denote the age of the animal. In the females the horns are not nearly so large nor so heavily ridged as in the male. The Ibex is also known under the name of BOUQUETIN.

THERE are an enormous number of varieties of the common domestic GOAT, many of them being so unlike the original stock from which they sprang as to appear like different species. For the present, we will turn to the common Goat of Europe, with which we are all so familiar. This animal is often seen domesticated, especially in and about stables, as there is a prevalent idea that the rank smell of the Goat is beneficial to horses. Be this as it may, the animal seems quite at home in a stable, and a very firm friendship often arises between the Goat and one of the horses. Sometimes it gets so petted by the frequenters of the stables, that it becomes presumptuous, and assaults any one whom it may not happen to recognise as a friend. Happily, a Goat, however belligerent he may be, is easily conquered if his beard can only be grasped, and when he is thus captured, he yields at once to his conqueror, assumes a downcast air and bleats in a very pitiful tone, as if asking for mercy.

In its wild state the Goat is a fleet and agile animal, delighting in rocks and precipitous localities, and treading their giddy heights with a foot as sure and an eye as steady as that of the chamois or ibex. Even in domesti-

CASHMIR GOAT.

cated life, this love of clambering is never eradicated, and wherever may be an accessible roof, or rock, or even a hill, there the Goat may be generally found.

The varieties of the Goat are almost numberless, and it will be impossible to engrave, or even to notice, more than one or two of the most prominent examples. One of the most valuable of these varieties is the celebrated Cashmir Goat, whose soft silky hair furnishes material for the soft and costly fabrics which are so highly valued in all civilized lands.

This animal is a native of Thibet and the neighbouring locality, but the Cashmir shawls are not manufactured in the same land which supplies the

N 2

material. The fur of the Cashmir Goat is of two sorts : a soft, woolly under-coat of greyish hair, and a covering of long silken hairs that serve to defend the interior coat from the effects of winter. The woolly under-coat is the substance from which the Cashmir shawls are woven, and in order to make a single shawl a yard and a half square, at least ten Goats are robbed of their natural covering. Beautiful as are these fabrics, they would be sold at a very much lower price but for the heavy and numerous taxes which are laid upon the material in all the stages of its manufacture, and after its completion upon the finished article. Indeed, the English buyer of a Cashmir shawl is forced to pay at least a thousand per cent. on his purchase.

FROM time immemorial the SHEEP has been subjected to the ways of mankind, and has provided him with meat and clothing, as well as with many articles of domestic use. The whole carcass of the Sheep is as useful as that of the ox, and there is not a single portion of its body that is not converted

HIGHLAND SHEEP.

to some beneficial purpose. The animal as we now possess it, and which has diverged into such innumerable varieties, is never found in a state of ab-solute wildness, and has evidently derived its origin from some hitherto undomesticated species. In the opinion of many naturalists, the mouflon may lay claim to the parentage of our domestic Sheep, but other writers have separated the mouflons from the Sheep, and placed them in a different genus.

Although the Sheep is generally considered to be a timid animal, and is really so when forced into adverse circumstances and deprived of its wonted liberty, it is truly as bold an animal as can well be seen, and even in this country gives many proofs of its courage. If, for example, a traveller comes unexpectedly upon a flock of the little Sheep that range the Welsh mountains, they will not flee from his presence, but draw together into a compact body, and watch him with stern and unyielding gaze. Should he attempt to advance, he would be instantly assailed by the rams, which form the first line in such cases, and would fare but badly in the encounter. A dog, if it should happen to accompany the intruder, would probably be at once charged and driven from the spot.

Even a single ram is no mean antagonist when he is thoroughly irritated, and his charge is really formidable. Sheep differ from Goats in their manner of fighting ; the latter animals rear themselves on their hind legs, and then plunge sideways upon their adversary, while the former animals hurl themselves forward, and strike their opponent with the whole weight as well as impetus of the body. So terrible is the shock of a ram's charge, that it has been known to prostrate a bull at the first blow. Nor is the sheep only combative when irritated by opposition, or when danger threatens itself. A Sheep that had been led into a slaughter-house has been known to turn fiercely upon the butcher as he was about to kill one of its companions, and to butt him severely in order to make him relinquish his grasp of its friend.

In the British Isles the Sheep breeds freely, producing generally one or two lambs every year, and sometimes presenting its owner with three lambs at a birth. One instance is on record of a wonderfully prolific ewe. She had hardly passed her second year when she produced four lambs. The next year she had five ; the year after that she bore twins ; and the next year five again. On two successive years she bore twins. Two out of the four and three out of the five were necessarily fed by hand.

We will now advert shortly to some of the principal breeds or varieties of the Sheep.

Of all the domestic varieties of this useful animal, the SPANISH, or MERINO SHEEP, has attracted the greatest attention.

Originally, this animal is a native of Spain, a country which has been for many centuries celebrated for the quantity and quality of its wool. The Merino Sheep, from whom the long and fine Spanish wool was obtained, were greatly improved by an admixture with the Cotswold Sheep of England, some of which were sent to Spain in 1464, and the fleece was so improved by the crossing, that famous English wool was surpassed by that which was supplied by Spain.

In Spain, the Merinos are kept in vast flocks, and divided into two general heads, the stationary and the migratory. The former animals remain in the same locality during the whole of their lives, but the latter are accustomed to undertake regular annual migrations. The summer months they spend in the cool mountainous districts, but as soon as the weather begins to grow cold, the flocks pass into the warmer regions of Andalusia, where they remain until April. The flocks are sometimes ten thousand in number, and the organization by which they are managed is very complex and perfect. Over each great flock is set one experienced shepherd, who is called the "mayoral," and who exercises despotic sway over his subordinates. Fifty shepherds are placed under his orders, and are supplied with boys and intelligent dogs.

Under the guardianship of their shepherds, the Merino Sheep, which have spent the summer in the mountains, begin their downward journey about the month of September ; and after a long and leisurely march they arrive at the pasture-grounds, which are recognized instinctively by the Sheep. In these pasturages the winter folds are prepared, and here are born the young Merinos, which generally enter the world in March or the beginning of April. Towards the end of that month the Sheep begin to be restless, and unless

they are at once removed, will often decamp of their own accord.   Sometimes a whole flock will thus escape, and, guided by some marvellous
instinct, will make their way to their old quarters, unharmed, except perchance by some prowling wolf, who takes advantage of the shepherd's absence.

MERINO, OR SPANISH SHEEP.

The very young lambs are not without their value, although they furnish
no wool, for their skins are prepared and sent to France and England, where
they are manufactured into gloves, and called by the name of "kid."

GIRAFFES.

TALLEST of all earthly dwellers, the GIRAFFE erects its stately head far
above any animal that walks the face of the globe.   It is an inhabitant of
various parts of Africa, and is evidently a unique being, comprising in itself
an entire tribe.   The colour of the coat is slightly different in the specimens
which inhabit the northern and the southern portions of Africa, the southern
animal being rather darker than its northern relative.

The height of a full-grown male Giraffe is from eighteen to twenty feet, the
female being somewhat less in her dimensions.   The greater part of this
enormous stature is obtained by the extraordinarily long neck, which is
nevertheless possessed of only seven vertebræ, as in ordinary animals. Those
bones are, however, extremely elongated, and their articulation is admirably

adapted to the purpose which they are called upon to fulfil. The back of the Giraffe slopes considerably from the shoulders to the tail, and at first sight the fore-legs of the animal appear to be longer than the hinder limbs. The legs themselves are, however, of equal length, and the elevation of the shoulders is due to the very great elongation of the shoulder-blades. Upon the head are two excrescences which resemble horns, and are popularly called by that name. They are merely growths or developments of certain bones of the skull, somewhat similar to the bony cores on which the hollow horns

GIRAFFE.—(*Giraffa Camelopárdalis.*)

of the oxen and antelopes are set. These quasi-horns are covered with skin, and have on their summits a tuft of dark hair. On the forehead, and nearly between the eyes, a third bony projection is seen occupying the same position that was traditionally accredited to the horn of the unicorn.

The singular height of this animal is entirely in accordance with its habits and its mode of acquiring food. As the creature is accustomed to feed upon the leaves of trees, it must necessarily be of very considerable stature to be able to reach the leaves on which it browses, and must also be possessed of organs by means of which it can select and gather such portions of the foliage as may suit its palate. The former object is gained by the great length of the neck and legs, and the latter by the wonderful development of the tongue, which is so marvellously formed that it is capable of a considerable amount of prehensile power, and can be elongated or contracted in a very wonderful manner. Large as is the animal, it can contract the tip of its tongue into so small a compass that it can pass into the pipe of an ordinary pocket-key, while its prehensile powers enable its owner to pluck any selected leaf with perfect ease. In captivity the Giraffe is rather apt to make too free a use of its tongue, such as twitching the artificial flowers and' foliage from ladies' bonnets, or any similar freak.

For grazing upon level ground the Giraffe is peculiarly unfitted, and never attempts that feat excepting when urged by hunger or some very pressing cause. It is, however, perfectly capable of bringing its mouth to the ground, although with considerable effort and much straddling of the fore-legs. By placing a lump of sugar on the ground, the Giraffe may be induced to lower its head to the earth, and to exhibit some of that curious mixture of grace and awkwardness which characterises this singular animal.

In its native country its usual food consists of the leaves of a kind of acacia, named the Kameel-dorn, or Camel-thorn (*Acacia giraffe*). The animal is exceedingly fastidious in its appetite, and carefully rejects every thorn, scrupulously plucking only the freshest and greenest leaves. When supplied with cut grass, the Giraffe takes each blade daintily between its lips, and nibbles gradually from the top to the stem, after the manner in which we eat asparagus. As soon as it has eaten the tender and green portion of the grass, it rejects the remainder as unfit for consumption. Hay, carrots, onions, and different vegetables form its principal diet while it is kept in a state of captivity.

As far as is at present known, the Giraffe is a silent animal, like the eland and the kangaroo, and has never been heard to utter a sound, even when struggling in the agonies of death. When in its native land, it is so strongly perfumed with the foliage on which it chiefly feeds, that it exhales a powerful odour, which is compared by Captain Cumming to the scent of a hive of heather honey.

To man it falls an easy prey, especially if it can be kept upon level ground, where a horse can run without danger. On rough soil, however, the Giraffe has by far the advantage, as it leaps easily over the various obstacles that lie in its way, and gets over the ground in a curiously agile manner. It is not a very swift animal, as it can easily be overtaken by a horse of ordinary speed, and is frequently run down by native hunters on foot. When running, it progresses in a very awkward and almost ludicrous manner, by a series of frog-like leaps, its tail switching and twisting about at regular intervals, and its long neck rocking stiffly up and down in a manner that irresistibly reminds the observer of those toy birds whose head and tail perform alternate obeisances by the swinging of a weight below. As the tail is switched sharply hither and thither, the tuft of the bristly hairs at the extremity makes a hissing sound as it passes through the air.

Besides the usual mode of hunting and stalking, the natives employ the pitfall for the purpose of destroying this large and valuable animal. For this purpose a very curiously-constructed pit is dug, being about ten feet in depth, proportionately wide, and having a wall or bank of earth extending from one side to the other, and about six or seven feet in height. When the Giraffe is caught in one of these pits, its fore limbs fall on one side of the wall and its hind legs on the other, the edge of the wall passing under the abdomen. The poor creature is thus balanced, as it were, upon its belly across the wall, and, in spite of all its plunging, is unable to obtain a foothold sufficiently firm to enable it to leap out of the treacherous cavity into which it has fallen. The pitfalls which are intended for the capture of the hippopotamus and the rhinoceros are furnished with a sharp stake at the bottom, which impales the luckless animal as it falls ; but it is found by experience that, in the capture of the Giraffe, the transverse wall is even more deadly than the sharpened spike.

The Giraffe is generally found in little herds, sometimes only five or six in number, and sometimes containing thirty or forty members, the average being about sixteen. These herds are always found either in or very close to forests, where they can obtain their daily food, and where they can be concealed from their enemies among the tree-trunks, to which they bear so close a resemblance.

The flesh of the Giraffe is considered to be good when rightly prepared, and its marrow is thought to be so great a delicacy that the natives eagerly suck it from the bones as they are taken from the animal. When cooked it is worthy of a place on a royal table. The flesh is well fitted for being made into jerked meat. The thick, strong hide is employed in the manufacture of shoe-soles, shields, and similar articles.

DEER.

FROM the Antelope the DEER are readily distinguished by the character of the horns ; which only belong to the male animals, are composed of solid bony substances, and are shed and renewed annually during the life of the animal. The process by which the horns are developed, die, and are shed, is a very curious one, and deserves a short notice before we proceed to consider the various species of Deer which will be noticed in the present work. For a familiar instance, we will take the common Stag, or Red Deer of Europe.

In the beginning of the month of March he is lurking in the sequestered spots of his forest home, harmless as his mate and as timorous. Soon a pair of prominences make their appearance on his forehead, covered with a velvety skin. In a few days these little prominences have attained some length, and give the first indication of their true form. Grasp one of these in the hand and it will be found burning hot to the touch, for the blood runs fiercely through the velvety skin, depositing at every touch a minute portion of bony matter. More and more rapidly grow the horns, the carotid arteries enlarging in order to supply a sufficiency of nourishment, and in the short period of ten weeks the enormous mass of bony matter has been completed. Such a process is almost, if not entirely, without parallel in the history of the animal kingdom.

When the horns have reached their due development, the bony rings at their bases, through which the arteries pass, begin to thicken, and by gradually filling up the holes, compress the blood-vessels, and ultimately obliterate them. The velvet now having no more nourishment, loses its vitality, and is soon rubbed off in shreds against tree-trunks, branches, or any inanimate object. The horns fall off in February, and in a very short time begin to be renewed. These ornaments are very variable at the different periods of the animal's life, the age of the Stag being well indicated by the number of "tines" upon its horns.

THE MOOSE or ELK is the largest of all the Deer tribe, attaining the extraordinary height of seven feet at the shoulders, thus equalling many an ordinary elephant in dimensions. The horns of this animal are very large, and widely palmated at their extremities, their united weight being so great as to excite a feeling of wonder at the ability of the animal to carry so heavy a burden. They do not reach their full development until the fourteenth year. The muzzle is very large and is much lengthened in front, so as to impart a most unique expression to the Elk's countenance. The colour of the animal is a dark brown, the legs being washed with a yellow hue. It is a native of Northern Europe and America, the Moose of the latter continent and the Elk of the former being one and the same species.

As the flesh of the Elk is palatable, and the skin and the horns extremely useful, the animal is much persecuted by hunters. It is a swift and enduring animal, although its gait is clumsy and awkward in the extreme. The only pace of the Elk is a long, swinging trot ; but its legs are so long and its paces so considerable, that its speed is much greater than it appears to be. Obstacles that are almost impassable to a horse are passed over easily by the Elk, which has been known to trot uninterruptedly over a number of fallen tree-trunks, some of them five feet in thickness. When the ground is hard and will bear the weight of so large an animal, the hunters are led a very long and severe chase before they come up with their prey ; but when the snow lies soft and thick on the ground, the creature soon succumbs to its lighter antagonists, who invest themselves in snow

MOOSE, OR ELK.—(*Alces Malchis.*)

shoes and scud over the soft snow with a speed that speedily overcomes that of the poor Elk, which sinks floundering into the deep snow-drifts at every step, and is soon worn out by its useless efforts.

It is as wary as any of the Deer tribe, being alarmed by the slightest sound or the faintest scent that gives warning of an enemy. As the Elk trots along, its course is marked by a succession of sharp sounds, which are produced by the snapping of the cloven hoofs, which separate at every step, and fall together as the animal raises its foot from the ground.

The enormous horns form no barrier to his progress through the woods, for when the Elk runs, he always throws his horns well back upon his shoulders, so that they rather assist than impede him in traversing the forest glades. The Elk is a capital swimmer, proceeding with great rapidity, and often taking to the water for its own amusement. During the summer months of the year it spends a considerable portion of its time under water, its nose and horn being the only parts of its form which appear above the surface. Even the very young Moose is a strong and fearless swimmer.

The skin of the Elk is extremely thick, and has been manufactured into clothing that would resist a sword-blow and repel an ordinary pistol-ball. The flesh is sometimes dressed fresh, but is generally smoked like hams, and is much esteemed. The large muzzle or upper lip is, however, the principal object of admiration to the lovers of Elk-flesh, and is said to be rich and gelatinous when boiled, resembling the celebrated green fat of the turtle.

Two. varieties of the REINDEER inhabit the earth ; the one, called the Reindeer, being placed upon the northern portions of Europe and Asia, and the other, termed the Caribou, being restricted to North America.

This animal is very variable in dimensions, specimens of very different height being in the British Museum. The colour is also variable, according to the season of the year. In winter the fur is long, and of a greyish brown tint, with the exception of the neck, hinder quarters, abdomen, and end of nose, which are white. In the summer, the grey-brown hair darkens into a sooty brown, and the white portions become grey.

The Laplanders place their chief happiness in the possession of many Reindeer, which are to them the only representatives of wealth. Those who possess a herd of a thousand or more are reckoned among the wealthy of their country ; those who own only a few hundreds are considered as persons of respectability ; while those who only possess forty or fifty are content to act as servants to their richer countrymen, and to merge their little herd in that of their employers. In the waste, dry parts of Lapland, grows a kind of white lichen, which forms the principal food of the Reindeer during winter, and is therefore highly prized by the natives. Although this lichen may be deeply covered with snow, the Reindeer is taught by instinct to scrape away the superincumbent snow with its head, hoofs, and snout, and to lay bare the welcome food that lies beneath. Sometimes the surface of the snow is frozen so firmly that the animal can make no impression ; and under these circumstances it is in very poor case, many of the unfortunate creatures dying of starvation, and the others being much reduced in condition.

REINDEER.—(*Tarandus Rångifer.*)

The Reindeer is extensively employed as a beast of draught and carriage, being taught to draw sledges and to carry men or packages upon its back. Each Reindeer can draw a weight of two hundred and fifty or even three hundred pounds, its pace being between nine and ten miles per hour. There is, however, a humane law which prohibits a weight of more than one hundred and ninety pounds upon a sledge, or one hundred and thirty upon the back. It is a very enduring animal, as it is able to keep up this rate of progress for twelve or more hours together.

The eyes of the Reindeer are very quick, and his hearing is also acute ; but his sense of smell is more wonderfully developed than either of the other senses.

WE now come to the Deer which inhabit the warm or temperate regions of the world, and which include the greater portion of the family.

The STAG, or RED DEER is spread over many parts of Europe and Asia, and is indigenous to the British Islands, where it still lingers, though in vastly reduced numbers.

In the olden days of chivalry and Robin Hood, the Red Deer were plentiful in every forest ; and especially in that sylvan chase which was made by the exercise of royal tyranny at the expense of such sorrow and suffering. Even in the New Forest itself the Red Deer is seldom seen, and those few survivors that still serve as relics of a bygone age are scarcely to be reckoned as living in a wild state, and approach nearly to the semi-domesticated condition of the Fallow Deer. Many of these splendid animals are preserved in parks or paddocks, but they no more roam the wide forests in unquestioned freedom. In Scotland, however, the Red Deer are still to be found, as can be testified by many a keen hunter of the present day, who has had his strength, craft, and coolness thoroughly tested before he could lay low in the dust the magnificent animal whose head with its forest of horns now graces his residence.

The great speed of the Stag is proverbial, and needs no mention. It is an admirable swimmer, having been known to swim for a distance of six or seven miles, and in one instance a Stag landed in the night upon a beach which he could not have reached without having swum for a distance of ten miles. The gallant beast was discovered by some dogs as he landed, and being chased by them immediately after his fatiguing aquatic exploit, was overcome by exhaustion, and found dead on the following morning.

The colour of the Stag varies slightly according to the time of the year. In the summer the coat is a warm reddish brown, but in the winter the ruddy hue becomes grey. The hind-quarters are paler than the rest of the fur. The young Red Deer are born about April, and are remarkable for the variegated appearance of their fur, which is mottled with white upon the back and sides. As the little creatures increase in dimensions, the white marking gradually fades, and the fur assumes the uniform reddish brown of the adult animal.

THE FALLOW DEER may readily be distinguished from the stag by the spotted coat, the smaller size, and the spreading, palmated horns.

The colour of the Fallow Deer is generally of a reddish brown, spotted with white, and with two or three white lines upon the body. There is, however, another variety which scarcely exhibits any of the white spots, and is of a deep blackish brown.

It is from the Fallow Deer that the best venison is procured, that of the stag being comparatively hard and dry. The skin is well known as furnishing a valuable leather, and the horns are manufactured into knife-handles and other articles of common use. The shavings of the horns are employed for the purpose of making ammonia, which has therefore been long popularly known under the name of hartshorn. The height of the adult Fallow Deer is three feet at the shoulders. It is a docile animal, and can be readily tamed. Indeed, it often needs no taming, but becomes quite familiar with strangers in a very short time, especially if they should happen to have any fruit, bread, or biscuit, and be willing to impart some of their provisions to their dappled friends.

THE well-known AXIS, CHITTRA, or SPOTTED HOG DEER, of India and Ceylon, belong to the Rusine Deer.

FALLOW DEER.—(*Dama Vulgaris.*)

The horns, like those of the sambur, a common Indian deer, are placed on long footstalks and simply forked at their tips. The colour of this pretty

AXIS DEER.—(*Axis maculata.*)

animal is rather various, but is generally a rich golden brown, with a dark brown stripe along the back, accompanied by two series of white spots, which at first sight appear to be scattered irregularly, but are seen on a careful inspection to be arranged in oblique curved lines. There is also a white streak across the haunches. There are, however, many varieties of the Axis Deer, which differ in size as well as in colour. The height of the adult Axis is almost equal to that of the Fallow Deer.

THE ROEBUCK is smaller than the Fallow Deer, being only two feet and three or four inches in length at the shoulder, but, although so small, can be really a formidable animal, on account of its rapid movements and great comparative strength.

It is not found in large herds like the Fallow Deer, but is strictly monogamous, the single pair living together, contented with each other's society. The horns of this animal have no basal snag, and rise straight from the forehead, throwing out one antler in front, and one or two behind, according to the age of the individual. From the base of the horn to the first antler the horn is thickly covered with wrinkles. It is a most active little Deer, always preferring the highest grounds, thence forming a contrast to the Fallow Deer, which loves the plains. It is seldom seen in England in a wild state, but may still be met with in many parts of Scotland.

ROEBUCK.—(*Capréolus Capræa.*)

The colour of the Roebuck is very variable, but is generally as follows. The body is always of a brown tint as a ground hue, worked with either red or grey, or remaining simply brown. Round the root of the tail is a patch of pure white hair, and the abdomen and inside of the limbs are greyish white. The chin is also white, and there is a white spot on each side of the lips.

THE MOSCHINE DEER are readily known by the absence of horns in both sexes, the extremely long canine teeth of the upper jaw in the males, and the powerfully odorous secretion in one of the species, from which they derive their popular as well as their scientific title. There are at least eight or nine species of these curious animals.

THE most celebrated of these little Deer is the common MUSK DEER, which is a native of the northern parts of India, and is found spread throughout a very large range of country, always preferring the cold and elevated mountainous regions. The height of the adult Musk Deer is about two feet three inches at the

MUSK DEER.—(*Moschus Moschiferus.*)

shoulders ; the colour is a light brown, marked with a shade of greyish yellow. Inhabiting the rocky and mountainous locations of its native home, it is remarkably active and sure-footed, rivalling even the chamois or the goat in the agility with which it can ascend or descend the most fearful precipices. The great length of the false hoofs adds much to the security of the Musk Deer's footing upon the crags.

It is only in the male that the long tusks are seen, and that the perfume called musk is secreted. The tusks are sometimes as much as three inches in length, and therefore project considerably beyond the jaw. In shape they are compressed, pointed, and rather sharp-edged. The natives say that their principal use is in digging up the kastoree plant, a kind of subterranean bulb on which the Musk Deer feeds, and which imparts the peculiar perfume to the odorous secretion. The musk is produced in a glandular pouch placed in the abdomen, and when the animal is killed for the sake of this treasure, the musk-bag is carefully removed, so as to defend its precious contents from exposure to the air. When recently taken from the animal, the musk is of so powerful an odour as to cause headache to those who inhale its overpowering fragrance. The affluence of perfume that resides in the musk is almost incredible, for a small piece of this wonderful secretion may remain in a room for many years, and at the end of that time will give forth an odour which is apparently not the least diminished by time.

SKULL OF MUSK DEER.

ANOTHER member of the Moschine group is the KANCHIL, or PIGMY MUSK (*Trágulus pygmæus*), a Deer which is found in the Asiatic islands, and which is as celebrated for its cunning as is the fox among ourselves.

KANCHIL, OR PIGMY MUSK.—(*Trágulus pygmæus.*)

This animal is not nearly so large as the Musk Deer, and although somewhat similar in colour, may be distinguished by a broad black stripe which runs along the back of the neck, and forms a wide band across the chest.

Instead of living in the cold and lofty mountain ranges which are inhabited by the Musk Deer, the Kanchil prefers the thickly wooded districts of the Javanese forests.   Like many other animals, the Kanchil is given to "possuming," or feigning death when it is taken in a noose or trap, and as soon as the successful hunter releases the clever actor from the retaining cord, it leaps upon its feet and darts away before he has recovered from his surprise.

FROM the earliest times that are recorded in history, the CAMEL is mentioned as one of the animals which are totally subject to the sway of man, and which in Eastern countries contribute so much to the wealth and influence of their owners.

There are two species of Camel acknowledged by zoologists, namely, the common Camel of Arabia, which has but one hump, and the Mecheri, or Bactrian Camel, which possesses two of these curious appendages.

As the animal is intended to traverse the parched sand-plains, and to pass several consecutive days without the possibility of obtaining liquid nourish-

ment, there is an internal structure which permits the animal to store up a considerable amount of water for future use. For this purpose the honey-comb cells of the "reticulum" are largely developed, and are enabled to receive and to retain the water which is received into the stomach after the natural thirst of the animal has been supplied. After a Camel has been accustomed to journeying across the hot and arid sand wastes, it learns wisdom by experience, and contrives to lay by a much greater supply of water than would be accumulated

CAMEL.—(*Camélus Arábicus.*)

by a young and untried animal.   It is supposed that the Camel is, in some way, able to dilate the honeycomb cells, and to force them to receive a larger quantity of the priceless liquid.

A large and experienced Camel will receive five or six quarts of water into its stomach, and is enabled to exist for as many days without needing to drink.   Aided by this internal supply of water, the Camel can satiate its hunger by browsing on the hard and withered thorns that are found scattered thinly through the deserts, and suffers no injury to its palate from their iron-like spears, that would direfully wound the mouth of any less sensitive creature.   The Camel has even been known to eat pieces of dry wood, and to derive apparent satisfaction from its strange meal.

The feet of the Camel are well adapted for walking upon the loose, dry sand, than which substance there is no more uncertain footing.   The toes are very broad, and are furnished with soft, wide cushions, that present a considerable surface to the loose soil, and enable the animal to maintain a firm hold upon the shifting sands. As the Camel is constantly forced to kneel in order to be loaded or relieved of its burden, it is furnished upon the knees and breast with thick callous pads, which support its weight without injuring the skin. Thus fitted by nature for its strange life, the Camel faces the desert sands with boldness, and traverses the arid regions with an ease and quiet celerity that has gained for the creature the title of " Ship of the Desert."

The Camel is invariably employed as an animal of carriage when in its native land, and is able to support a load of five or six hundred pounds' weight without being overtaxed.

The pace of the Camel is not nearly so rapid as is generally supposed, and even the speed of the Heirie, or swift Camel, has been greatly exaggerated.

The speed of the Heirie is seldom more than eight or ten miles per hour, but the endurance of the animal is so wonderful, that it is able to keep up this pace for twenty hours without stopping.

The " hump" of the Camel is a very curious part of its structure, and is of great importance in the eyes of the Arabs, who judge of the condition of their beasts by the size, shape, and firmness of the hump. They say, and truly, that the Camel feeds upon his hump, for in proportion as the animal traverses the sandy wastes of its desert lands, and suffers from privation and fatigue, the hump diminishes. At the end of a long and painful journey, the hump will often nearly vanish, and it cannot be restored to its pristine form until the animal has undergone a long course of good feeding. When an Arab is about to set forth on a desert journey, he pays great attention to the humps of his Camels, and watches them with jealous care.

Independently of its value as a beast of burden, the Camel is most precious to its owners, as it supplies them with food and clothing. The milk mixed with meal is a favourite dish among the children of the desert, and is sometimes purposely kept until it is sour, in which state it is very grateful to the Arab palate, but especially nauseous to that of a European. The Arabs think that any man is sadly devoid of taste who prefers the sweet new milk to that which has been mellowed by time. A kind of very rancid butter is churned from the cream by a remarkably simple process, consisting of pouring the cream into a goatskin sack, and shaking it constantly until the butter is formed.

The long hair of the Camel is spun into a coarse thread, and is employed in the manufacture of broad-cloths and similar articles. At certain times of the year, the Camel sheds its hair, in order to replace its old coat by a new one, and the Arabs avail themselves of the looseness with which the hair is at these times adherent to the skin, to pluck it away without injuring the animal.

The height of an ordinary Camel at the shoulder is about six or seven feet, and its colour is a light brown, of various depths in different individuals, some specimens being nearly black, and others almost white. The Dromedary is the lighter breed of Camel, and is chiefly used for riding, while the ordinary Camel is employed as a beast of burden. Between the two animals there is about the same difference as between a dray-horse and a hunter, the Heirie being analogous to the racehorse.

THE BACTRIAN CAMEL is readily to be distinguished from the ordinary Camel by the double hump which it bears on its back, and which is precisely analogous in its structure and office to that of the Arabian Camel.

The general formation of this animal ; its lofty neck, raising its head high above the solar radiations from the heated ground ; its valve-like nostrils, that close involuntarily if a grain of drifting sand should invade their precincts ; its wide cushion-like feet, and its powers of abstinence, prove that, like its

Arabian relative, it is intended for the purpose of traversing vast deserts without needing refreshment on the way. This species is spread through Central Asia, Thibet, and China, and is domesticated through a large portion of the world. It is not so enduring an animal as the Arabian species, requiring a fresh supply of liquid every three days ; while the Arabian Camel can exist without water for five or even six days.

The height of the Bactrian Camel is rather more than that of the Arabian species, and its colour is generally brown, which sometimes deepens into sooty black, and sometimes fades into a dirty white.

THE true camels are exclusively confined to the Old World, but find representatives in the New World in four acknowledged species of the genus LLAMA.

These animals are comparatively small in their dimensions, and possess no hump, so that they may easily be distinguished from the camels. Their hair is very woolly, and their countenance has a very sheep-like expression, so that a full-haired Llama instantly reminds the spectator of a long-legged, long-necked sheep. The feet of the Llamas are very different from those of the camels, as their haunts are always found to be upon rocky ground, and their feet must of necessity be accommodated to the ground on which they are accustomed to tread. The toes of the Llama are completely divided, and are each furnished with a rough cushion beneath, and a strong, claw-like hoof above, so that the member may take a firm hold of rocky and uneven ground.

BACTRIAN CAMEL.—(*Camélus Bactriánus.*)

Four species of Llamas are now acknowledged ; namely, the Vicugna, the Guanaco, the Yamma, and the Alpaca.

THE VICUGNA is found in the most elevated localities of Batavia and Northern Chili, and is a very wild and untameable animal, having resisted all the attempts of the patient natives to reduce it to a state of domestication. It is extremely active and sure-footed in its mountain home, and being equally timid and wary, is seldom captured in a living state. It lives in herds near the region of perpetual snow, and in its habits bears some resemblance to the chamois. The short, soft, silken fur of this animal is very valuable. The colour of the Vicugna is a nearly uniform brown, tinged with yellow on the back, and fading into grey on the abdomen. Its height at the shoulder is about two feet six inches.

THE GUANACO is spread over a very wide range of country extending over the whole of the temperate regions of Patagonia. The colour of this species is a reddish brown, the ears and hind legs grey. The neck is long in comparison with the size of the body, and the height at the shoulder is about three feet six inches.

The Guanaco lives in herds varying in number from ten to thirty or forty, but is sometimes seen in flocks of much greater numbers, resembling sheep, not only in their gregarious habits, but in the implicit obedience with which they rely upon their leader. Should they be deprived of his guardianship, they become so bewildered that they run aimlessly from spot to spot, and can be easily destroyed by experienced hunters.

The Guanaco is wonderfully sure-footed upon rocky ground, and is also a good swimmer, taking voluntarily to the water, and swimming from one island to another. When near the sea it will drink the salt water, and has often been observed in the act of drinking the briny waters of certain salt springs.

ALPACA LLAMA.—(*Llama Pacos.*)

THE YAMMA, or LLAMA, is of a brown or variegated colour, and its legs are long and slender. In former days this animal was the only beast of burden which was possessed by the natives, and it was largely used by the Spaniards (who described it as a sheep) for the same purpose. It is able to carry a weight of one hundred pounds, and to traverse about fourteen or fifteen miles per diem. As a beast of burden it is now being rapidly supplanted by the ass, while the European sheep is gradually taking its place as a wool-bearer. The flesh of the Llama is dark and coarse, and is accordingly held in bad repute.

THE ALPACA, or PACO, is, together with the last animal, supposed by several zoologists to be only a domesticated variety of the Guanaco. Its colour is generally black, but is often variegated with brown and white. The wool of this species is long, soft, silky, and extremely valuable in the commercial world.

HORSES.

THE HORSE has from time immemorial been made the companion and servant of man, and its original progenitors are unknown. It is supposed, however, that the Horse must have derived its origin from Central Asia, and from thence have spread to almost every portion of the globe.

The elegant, swift, and withal powerful Horses of which England is so proud, and which are employed in the chase or the course, owe their best qualities to the judicious admixture of the Arabian blood. The ARAB HORSE has long been celebrated for its swift limbs, exquisite form, and affectionate disposition.

There are several breeds of Arab Horses, only one of which is of very great value. This variety, termed the Kochlani, is so highly prized, that a mare of the pure breed can hardly be procured at any cost, and even the male animal is not easy of attainment. The pedigree of these Horses is carefully preserved, and written in most florid terms upon parchment. In some cases the genealogy is said to extend for nearly two thousand years. The body of the Arab Horse is very light, its neck long and arched, its eye full and soft, and its limbs delicate and slender. The temper of the animal is remarkably sweet, for as it has been born and bred among the family of its owner, it avoids injuring even the little children that roll about among its legs as carefully as if they were its own offspring. So attached to its owner is this beautiful Horse, that if he should be thrown from its back, the animal will stand quietly by its prostrate master, and wait until he gains strength to remount.

THE HORSE.

The training of the Kochlani is not so severe as is generally imagined, for the presence of water and abundant pasturage is absolutely necessary, in order to rear the animal in a proper manner. Not until the strength and muscles of the animal are developed is a trial permitted, and then it is truly a terrible one. When the mare—for the male animal is never ridden by the Arabs—has attained her full development, she is mounted for the first time, and ridden at full speed for fifty or sixty miles without respite. Hot and fainting, she is then forced into deep water, which compels her to swim, and if she does not feed freely immediately after this terrific trial, she is rejected as unworthy of being reckoned among the true Kochlani.

For the animals which will stand this terrible test the Arab has almost an idolatrous regard, and will oftentimes spare an enemy merely on account of his steed.

The RACEHORSE of England is, perhaps, with the exception of the fox-hound, the most admirable example of the perfection to which a domesticated animal can be brought by careful breeding and training.

Whatever may have been its original source, the Racer has been greatly improved by the mixture of Arab blood, through the means of the Godolphin and Darley Arabians. The celebrated horse Eclipse was a descendant, on the mother's side, of the Godolphin Arabian, that wonderful animal which was rescued from drawing a cart in Paris, and which was afterwards destined to play so important a part in regenerating the breed of English Racers. He was also descended, on his father's side, from the Darley Arabian.

The best bred Horses are generally the most affectionate and docile, although their spirit is very high and their temper hot and quick. There are few animals which are more affectionate than a Horse, which seems to feel a necessity for attachment, and if his sympathies be not aroused by human means, he will make friends with the nearest living being. Cats are great favourites with Horses, and even the famous Chillaby, called, from his ferocity, the Mad Arabian, had his little friend in the shape of a lamb, which would take any liberties with him, and was accustomed to butt at the flies as they came too near his strange ally. The Godolphin Arabian was also strongly attached to a cat, which usually sat on his back or nestled in the manger. When he died, the cat pined away and soon followed her loved friend.

These examples are sufficient to show that the ferocity of these animals was caused by the neglect or ignorance of their human associates, who either did not know how to arouse the affectionate feelings of the animal, or brutally despised and crushed them. The Horse is a much more intellectual animal than is generally supposed, as will be acknowledged by anyone who has possessed a favourite Horse and treated it with uniform kindness.

There is no need for whip or spur when the rider and steed understand each other, and the bridle is reduced almost to a mere form, as the touch of a finger, or the tone of a voice, is sufficient to direct the animal. We are all familiar with the elephantine dray-horses that march so majestically along with their load of casks, and which instantaneously obey the singular sounds which continually issue from the throats of their conductors, and back, stop, advance, or turn to the right or left, without requiring the touch of a rein or the blow of a whip. The infliction of pain is a clumsy and a barbarous manner of guiding a Horse, and we shall never reap the full value of the animal until we have learned to respect its feelings, and to shun the infliction of torture as a brutal, a cowardly, and an unnecessary act. To maltreat a child is always held to be a cowardly and unmanly act, and it is equally cowardly and unworthy of the human character to maltreat a poor animal which has no possibility of revenge, no hope of redress, and no words to make its wrongs known. Pain is pain, whether inflicted on man or beast, and we are equally responsible in either case.

As an unprejudiced observer, with no purpose to serve, and without bias in either direction, I cannot here refrain from observing, that Mr. Rarey's method of bringing the Horse under subjection is a considerable step in the right direction, and a very great improvement on the cruel and savage method which is so often employed by coarse and ignorant men, and truly called "breaking." Having repeatedly witnessed the successful operations of that gentleman in subduing Horses that had previously defied all efforts, I cannot be persuaded that it is a cruel process. The method by which it is achieved is now sufficiently familiar, and I will only observe that the idea is a true and philosophical one. The Horse is mostly fierce because it is nervous; and bites and kicks, not because it is enraged, but because it is alarmed. Restore confidence, and the creature becomes quiet, without any desire to use its hoofs and teeth in an aggressive manner. It is clearly impossible to do so as long as the animal is at liberty to annihilate its teacher, and the strap is only used until the Horse is convinced that the presence of a human form, or the touch of a human hand, has nothing of the terrible in it. Confidence soon takes the place of fear, and the animal seems to receive its teacher at once into its good graces, following him like a dog, and rubbing its nose against his shoulder.

SEVERAL breeds of partially wild Horses are still found in the British Islands, the best known of which is the SHETLAND PONY.

This odd, quaint, spirited little animal is an inhabitant of the islands at the northern extremity of Scotland, where it runs wild, and may be owned

SHETLAND PONY.

by anyone who can catch and hold it. Considering its diminutive proportions, which only average seven or eight hands in height, the Sheltie is wonderfully strong, and can trot away quite easily with a tolerably heavy man on its back. One of these little creatures carried a man of twelve stone weight for a distance of forty miles in a single day. The head of this little animal is small, the neck short and well arched, and covered with an abundance of heavy mane, that falls over the face and irresistibly reminds the spectator of a Skye terrier. It is an admirable draught-horse when harnessed to a carriage of proportionate size; and a pair of these spirited little creatures, when attached to a low lady's carriage, have a remarkably piquant and pretty appearance.

MAN has so long held the DOMESTIC ASS under his control, that its original progenitors have entirely disappeared from the face of the earth.

There are, as it is well known, abundant examples of wild Asses found in various lands, but it seems that these animals are either the descendants of domesticated Asses which have escaped from captivity, or the offspring of wild and domesticated animals. In size and general appearance the Ass varies greatly, according to the country which it inhabits and the treatment to which it is subjected. The Spanish kind, for example, is double the size of the ordinary English Ass, and even the latter animal is extremely variable in stature and general dimensions.

As a rule, the Ass is large and sleek-haired in warm countries, and small and woolly-haired in the colder parts of the globe.

Strong, sure-footed, hardy, and easily maintained, the Ass is of infinite use to the poorer classes of the community, who need the services of a beast of burden and cannot afford to purchase or keep so expensive an animal as a horse.

It is a very great mistake to employ the name of Ass or donkey as a metaphor for stupidity, for the Ass is truly one of the cleverest of our domesticated animals, and will lose no opportunity of displaying his capability whenever his intelligence

ASS.—(*Asinus vulgaris.*)

is allowed to expand by being freed from the crushing toil and constant pain that are too often the concomitants of a donkey's life.

Everyone who has petted a favourite donkey will remember many traits of its mental capacities ; for, as in the case of the domestic fool of the olden days, there is far more knavery than folly about the creature.

In the East, the Ass is used even more extensively than in Europe, and is generally employed for carrying burdens or for the saddle, the horse being used more for ostentation or for warfare than for the mere conveyance of human beings from one spot to another.

The colour of the Ass is

DZIGGETAI, OR KOULAN.—(*Asinus Onager.*)

a uniform grey, a dark streak passing along the spine, and another stripe being drawn transversely across the shoulders. In the quagga and zebra these stripes are much more extended.

THE Wild Asses are all celebrated for their extreme fleetness and sureness of foot, and among them the DZIGGETAI, KHUR, or KOULAN, deserves especial mention.

This animal is so wonderfully swift that it cannot be overtaken even by a fleet Arabian horse, and if it can get upon hilly or rocky ground, it bids defiance to all wingless enemies. Not even the greyhound can follow it with any hope of success when it once leaves level ground. This great speed renders it a favourite object of chase with the natives of the country which it inhabits ; and whether in Persia or India, it is held to be the noblest of game. Sometimes the falcon is trained to aid in the chase of the Wild Ass, but the usual method of securing this animal is to drive it towards rocky ground, and to kill it with a rifle bullet as it stands in fancied security upon some lofty crag.

QUAGGA.—(*Asinus Quagga.*)

It lives in troops, descending to the plains during the winter months, and returning to the cooler hills as soon as the summer begins to be unpleasantly warm.

It is very common in Mesopotamia, and is always a most shy and wary, as well as swift animal. Each troop is under the command of a leader, who sways his subjects with unlimited authority, and takes upon himself to make all needful arrangements for their welfare.

The colour of this animal is pale reddish brown in the summer, fading into a grey-brown in the winter, and marked with a black stripe along the spine, becoming wider upon the middle of the back.

ZEBRA.—(*Asinus Zebra.*)

Another species of Wild Ass is the KIANG, or Wild Ass of Thibet, sometimes, but erroneously, called the Wild Horse of Thibet, because its noise resembles the neighing of that animal rather than the braying of the Ass.

AFRICA produces some most beautiful examples of the Wild Asses, equalling the Asiatic species in speed and beauty of form, and far surpassing them in richness of colour and boldness of marking.

THE QUAGGA looks at first sight like a cross between the common wild ass and the zebra, as it only partially possesses the characteristic zebra stripes, and is decorated merely upon the hind and fore parts of the body. The streaks are not so deep as they are in the zebra, and the remainder of the body is brown, with the exception of the abdomen, legs, and part of the tail, which are whitish grey.   The Quagga lives in large herds, and is much persecuted by the natives of Southern Africa, who pursue it for the sake of its skin and its flesh, both of which are in high estimation.

AMONG all the species of the Ass tribe, the ZEBRA is by far the most conspicuous and the most beautiful.

The general colour of the Zebra is a creamy white, marked regularly with velvety black stripes that cover the entire head, neck, body, and limbs, and extend down to the very feet.   It is worthy of note that the stripes are drawn nearly at right angles to the part of the body on which they occur, so that the stripes of the legs are horizontal, while those of the body are vertical.   The abdomen and inside faces of the thighs are cream-white, and the end of the tail is nearly black.   This arrangement of colouring is strangely similar to that of the tiger, and has earned for the animal the name of " Hippotigris," or Horse-tiger, among some zoologists ancient and modern.   The skin of the neck is developed into a kind of dewlap, and the tail is sparingly covered with coarse black hair.   By the Cape colonists it is called *Wilde Paard,* or Wild Horse.

At the best of times the flesh of the Zebra is not very inviting, being rather tough, coarse, and of a very peculiar flavour.   The Boers, who call themselves by the title of " baptized men," think they would be derogating from their dignity to partake of the flesh of the zebra, and generously leave the animal to be consumed by their Hottentot servants.   When wounded, the Zebra gives a kind of groan, which is said to resemble that of a dying man.

In disposition the Zebra is fierce, obstinate, and nearly untameable.   The efforts used by Mr. Rarey in reducing to obedience the Zebra of the Zoological Gardens are now matter of history.   The little brindled animal gave him more trouble than the huge savages on whom he had so successfully operated, and it overset some of his calculations by the fact that it was able to kick as fiercely from three legs as a horse from four.

In its habits the Zebra resembles the Dziggetai, as it is always found in hilly districts, and inhabits the high craggy mountain ranges in preference to the plains.   It is a mild and very timid animal, fleeting instinctively to its mountain home as soon as it is alarmed by the sight of a strange object.

PACHYDERMATA;

OR, THICK-SKINNED ANIMALS.

THE important family of the Elephantidæ includes, according to the catalogue of the British Museum, the Elephants, Tapirs, Swine, Hyrax, Rhinoceros, and Hippopotamus.   All these animals, however different their aspect, are nearly related to each other by means of certain members of the family, which, although now extinct, have been recovered through the assistance of geological researches.

Of Elephants, two distinct species are found in different continents, the one inhabiting Asia, and the other taking up its residence in Africa. According to some zoologists these animals belong to different genera, but the distinctions between the two creatures are not sufficiently determined to warrant such a suggestion.   Although the Asiatic and African Elephants are very similar in external form, they may at once be distinguished from each other by the size of the ear.   In the Asiatic animal, the ears are of moderate size, while in the African Elephant they are of enormous magnitude, nearly meeting on the back of the head, and hanging with their tips below the neck.

The molar teeth also afford excellent indications of the country to which their owner has belonged, for the enamel upon the surface of the teeth of the Asiatic Elephant is moulded into a number of narrow bands like folded ribands, while that of the African species is formed into five or six diamond or lozenge-shaped folds. Indeed, each molar tooth seems to be composed of a number of flat, broad teeth, which are fastened closely together, so as to form a single large mass. Only a portion of each tooth is externally visible, the remainder being hidden in the jaw, and moving forward as the exposed portion is worn away. When the whole tooth is thus worn out, it falls from the jaw, and its place is taken by another which has been forming behind it. In this manner the Elephant sheds its molar teeth six or seven times in the course of its life. The tusks, however, are permanent, and are retained during the whole of the animal's existence. In the Indian Elephant, only the males are furnished with tusks, and not every individual of that sex, whereas in the African species both sexes are supplied with these valuable appendages, those of the male being much larger and heavier than those of his mate.

The strangest portion of the Elephant's form is the trunk or proboscis. This wonderful appendage is in fact a development of the upper lips and the nose, and is perforated through its entire length by the nostrils, and furnished at its extremity with a kind of finger-like appendage, which enables the animal to pluck a single blade of grass, or to pick a minute object from the ground. The value of the proboscis to the Elephant is incredible ; without its aid the creature would soon starve. The short, thick neck would prevent it from stooping to graze, while the projecting tusks would effectually hinder it from reaching any vegetables which might grow at the level of its mouth. And as it would be unable to draw water into its mouth without the use of the trunk, thirst would in a very short time end its existence.

In order to support the enormous weight of the teeth, tusks, and proboscis, the head is required to be of very large dimensions, so as to afford support for the powerful muscles and tendons which are requisite for such a task. It is also needful that lightness should be combined with magnitude, and this double condition is very beautifully fulfilled. The skull of the Elephant, instead of being a mere bony shell round the brain, is enormously enlarged by the separation of its bony plates, the intervening space being filled with a vast number of honeycomb-like bony cells, their walls being hardly thicker than strong paper, and their hollows filled during the life of the animal with a kind of semi-liquid fat or oil. The brain lies in a comparatively small cavity within this cellular structure, and is therefore defended from the severe concussions which it would otherwise experience from the frequency with which the animal employs its head as a battering ram.

In order to support the enormous weight which rests upon them, the legs are very stout, and are set perpendicularly, without that bend in the hinder leg which is found in most animals. There is an elongated cannon bone in the Elephant, so that the hind legs are without the so-called knee-joint. This structure, however, is of infinite use to the animal when it climbs or descends steep acclivities, a feat which it can perform with marvellous ease. It may seem strange, but it is nevertheless true, that localities which would be totally inaccessible to a horse are traversed by the Elephant with perfect ease.

In descending from a height, the animal performs a very curious series of manœuvres. Kneeling down with its fore-feet stretched out in front, and its hinder legs bent backward, as is their wont, the Elephant hitches one of its fore-feet upon some projection or in some crevice, and bearing firmly upon this support, lowers itself for a short distance. It then advances the other foot, secures it in like manner, and slides still farther, never losing its hold

of one place of vantage until another is gained.   Should no suitable projection be found, the Elephant scrapes a hole in the ground with its advanced
foot, and makes use of this artificial depression in its descent.   If the declivity
be very steep, the animal will not descend in a direct line, but make an
oblique track along the face of the hill.   Although the description of this
curious process occupies some time, the actual feat is performed with extreme
rapidity.

Though the foot of an Elephant is extremely large, it is most admirably
formed for the purpose which it is destined to fulfil, and does not, as might
be supposed, fall heavily upon the ground.   The hoof that incloses the foot is
composed of a vast number of horny plates, that are arranged on the principle
of the common carriage-spring, and seem to guard the animal from the jarring

ASIATIC ELEPHANT.—(*Elephas Indicus.*)

shock of the heavy limb upon the soil.   Those who for the first time witness
the walk or the run of the Elephant are always surprised at the silent ease of
the creature's free, sweeping step.   As there is no short ligament in the head
of the thigh-bone, the hind foot is swung forward at each step, clearing the
ground easily, but being scarcely raised above the surface of the earth.

Having thus given a short sketch of the characteristics which are common
to both species of Elephants, I will proceed to a short account of the Asiatic
animal.

THE ASIATIC ELEPHANT bears a world-wide fame for its capabilities as a
servant and companion of man, and for the extraordinary development of

its intellectual faculties. Hundreds of these animals are annually captured, and in a very short period of time become wholly subjected to their owners, and learn to obey their commands with implicit submission. Indeed, the power of the human intellect is never so conspicuous as in the supremacy which man maintains over so gigantic and clever an animal as the Elephant. In all work which requires the application of great strength combined with singular judgment, the Elephant is supreme ; but as a mere puller and hauler it is of no very great value. In piling logs, for example, the Elephant soon learns the proper mode of arrangement, and will place them upon each other with a regularity that would not be surpassed by human workmen. Sir Emerson Tennent mentions a pair of Elephants that were accustomed to labour conjointly, and which had been taught to raise their wood piles to a considerable height by constructing an inclined plane of sloping beams, and rolling the logs up the beams.

There are two modes of capturing the Asiatic Elephant, the one by pursuing solitary individuals and binding them with ropes as they wander at will through the forests, and the other by driving a herd of Elephants into a previously prepared pound, and securing the entrance so as to prevent their escape.

In the former method, the hunters are aided by certain trained females, termed " koomkies," which enter into the spirit of the chase with wonderful animation, and help their riders in every possible manner. When the koomkies see a fine male Elephant, they advance carelessly towards him, plucking leaves and grass, as if they were perfectly indifferent to his presence. He soon becomes attracted to them, when they overwhelm him with endearing feminine blandishments, and occupy his attention so fully that he does not observe the proceedings of the " mahouts," or riders. These men, seeing the Elephant engaged with the " koomkies," slip quietly to the ground and attach their rope nooses to his legs, fastening the ends of the cords to some neighbouring tree. Should no suitable tree be at hand, the koomkies are sagacious enough to comprehend the dilemma, and to urge their victim towards some large tree which is sufficiently strong to withstand his struggles. As soon as the preparations are complete the mahouts give the word of command to the koomkies, who move away, leaving the captive Elephant to his fate.

Finding himself deserted and bound, he becomes mad with rage, and struggles with all his force to get free. In these furious efforts the Elephant displays a flexibility and activity of body that are quite surprising, and are by no means in accordance with the clumsy, stiff aspect of its body and limbs. It rolls on the ground in despair, it rends the air with furious cries of rage, it butts at the fatal tree with all its force, in hope of bringing it to the ground, and has been known to stand with its hind legs fairly off the ground, in its furious endeavours to break the rope. After a while, however, it finds its exertions to be totally useless, and yields to its conquerors.

The second mode of capturing Elephants is more complicated. The inclosure into which the Elephants are driven is termed a " keddah," and is ingeniously constructed of stout logs and posts, which are supported by strong buttresses, and are so arranged that a man can pass through the interstices between the logs. When the keddah is set in good order, a vast number of hunters form themselves into a huge circle, inclosing one or more herds of Elephants, and moving gradually towards the inclosure of the keddah, and arranging themselves in such a manner as to leave the entrance towards the keddah always open. When they have thus brought the herd to the proper spot, a business which will often consume several weeks, the Elephants are excited by shouts, the waving of hands and spears, &c., to move towards the inclosure, which is cunningly concealed by the trees among which it is built. If the operation should take place at night, the surrounding hunters are supplied with burning torches, while the keddah is carefully kept in darkness. Being alarmed by the noise and the flames, the Elephants rush instinctively to the only open space, and are thus fairly brought within the precincts of the keddah, from which they never emerge again save as captives.

The terrified animals run round and round the inclosure, and often attempt a desperate charge, but are always driven back by the torch-bearers, who wave their flaming weapons and discourage the captured animals from their meditated assault. At last the poor creatures are so bewildered and fatigued that they gather together in the centre of the keddah, and are then considered to be ready for the professional elephant hunters. These courageous men enter the keddah either on foot or upon the back of their koomkies, and contrive to tie every one of the captives to some spot from whence it cannot move. Most ingenious stratagems are employed by the hunters in this perilous task, the details of which may be found in many works on the subject.

The Elephant is always guided by a mahout, who sits astride upon its neck and directs the movements of the animal by means of his voice, aided by a kind of spiked hook, called the haunkus, which is applied to the animal's head in such a manner as to convey the driver's wishes to the Elephant. The persons who ride upon the Elephant are either placed in the howdah, a kind of wheelless carriage strapped on the animal's back, or sit upon a large pad, which is furnished with cross ropes in order to give a firm hold. The latter plan is generally preferred, as the rider is able to change his position at will, and even to recline upon the Elephant's back if he should be fatigued by the heavy rolling gait of the animal. The Elephant generally kneels in order to permit the riders to mount, and then rises from the ground with a peculiar swinging motion. Very small Elephants are furnished with a saddle like that which is used upon horses, and is fitted with stirrups. The saddle, however, cannot be conveniently used on animals that are more than six feet in height.

The size of Elephants has been greatly exaggerated, as sundry writers have given fourteen or sixteen feet as an ordinary height, and have even mentioned instances where Elephants have attained to the height of twenty feet. It is true that the enormous bulk of the animal makes its height appear much greater than is really the case. Eight feet is about the average height of a large Elephant, and scarcely any Elephant measures much more than ten feet high at the shoulder.

The general colour of the Elephant is brown, of a lighter tint when the animal is at liberty, and considerably deeper when its hide is subjected to rubbing with a cocoa-nut brush and plenty of oil. Sometimes an albino or White Elephant is seen in the forests, the colour of the animal being a pinky white, and aptly compared to the nose of a white horse. The king of Ava, one of whose titles is " Lord of the White Elephants," generally contrives to monopolize every White Elephant, and employs them for purposes of state, decorating them with strings of priceless gems, pearls, and gold coins, and lodging them in the most magnificent of houses, where their very eating-troughs are of silver.

The AFRICAN ELEPHANT is spread over a very wide range of country, extending from Senegal and Abyssinia to the borders of the Cape Colony. Several conditions are required for its existence, such as water, dense forests, and the absence of human habitations.

Although it is very abundant in the locality which it inhabits, it is not often seen by casual travellers, owing to its great vigilance and its wonderful power of moving through the tangled forests without noise and without causing any perceptible agitation of the foliage. In spite of its enormous dimensions, it is one of the most invisible of forest creatures, and a herd of Elephants, of eight or nine feet in height, may stand within a few yards of a hunter without being detected by him, even though he is aware of their presence.

The Kaffirs are persevering elephant hunters, and are wonderfully expert in tracking any individual by the "spoor" or track, which is made by its footsteps. The foot of a male is easily to be distinguished by the roundness of its form, while that of the female is more oval, and the height of the animal is also ascertained by measurement of the footmarks, twice the circumference of the foot being equal to the height at the shoulder.

AFRICAN ELEPHANT. — (*Loxodonta Africana.*)

The death of a large Elephant is great matter of congratulation among the natives, who rejoice at the abundant supply of food which will fall to their share. Almost every portion of the animal is used by the Kaffirs, whose strong jaws are not to be daunted by the toughest meat, and whose accommodating palates are satisfied with various portions which would be rejected by any civilized being.

Some portions of the Elephant are, however, grateful even to European palates, and the foot, when baked, is really delicious. This part of the animal is cooked by being laid in a hole in the earth, over which a large fire has been suffered to burn itself out, and then covered over with the hot earth. Another fire is then built on the spot, and permitted to burn itself out as before, and when the place is thoroughly cool, the foot is properly cooked. The flesh of the boiled foot is quite soft and gelatinous, something resembling calf's head, and is so tender that it can be scooped away with a spoon,

The trunk and the skin around the eye are also enumerated as delicacies, but have been compared by one who has had practical experience as bearing a close resemblance to shoe-leather both in toughness and evil flavour.

The natives employ many methods of capturing Elephants, the pitfall being the most deadly. Even this insidious snare is often rendered useless by the sagacity of the crafty old leaders of the herds, who precede their little troops to the water, as they advance by night to drink, and carefully beating the ground with their trunks as they proceed, unmask the pitfalls that have been dug in their course. They then tear away the covers of the pits and render them harmless. These pits are terrible affairs when an animal gets into them, for a sharp stake is set perpendicularly at the bottom, so that the poor Elephant is transfixed by its own weight, and dies miserably. Each pit is about eight feet long by four in width.

The ivory of the African Elephant is extremely valuable, and vast quantities are imported annually into this country. The slaughter of an Elephant is therefore a matter of congratulation to the white hunter, who knows that he can obtain a good price for the tusks and teeth of the animal which he has slain. A pair of tusks weighing about a hundred and fifty pounds will fetch nearly forty pounds when sold, so that the produce of a successful chase is extremely valuable. One officer contrived to purchase every step in the army by the sale of the ivory which he had thus obtained. On an average, each pair of tusks, taking the small with the great, will weigh about a hundred and twenty pounds.

ONE of the links which unite the elephants to the swine and rhinoceros is to be found in the genus *Tapirus.* The animals which belong to this genus are remarkable for the prolonged upper lip, which is formed into a kind of small proboscis, not unlike that of the elephant, but upon a smaller scale, and devoid of the finger-like appendage at the extremity. Only two species are at present existing.

The common or American TAPIR, sometimes called the Mbórebi, is a native of tropical America, where it is found in great numbers, inhabiting the densely wooded regions that fringe the banks of rivers. It is a great water-lover, and can swim or dive with perfect ease. The tough, thick hide with which the Tapir is covered is of great service in enabling the animal to pursue its headlong course through the forest without suffering injury from the branches. When it runs, it carries its head very low, as does the wild boar under similar circumstances.

The colour of the adult Tapir is a uniform brown, but the young is beautifully variegated with yellowish fawn spots and stripes upon a rich brown black ground, reminding the observer of the peculiar tinting of the

KUDA-AYER, OR MALAYAN TAPIR.—(*Tapirus Malayánus.*)

Hood's marmot. The neck is adorned with a short and erect black mane.

The Tapir can easily be brought under the subjection of man, and is

readily tamed, becoming unpleasantly familiar with those persons whom it knows, and taking all kinds of liberties with them, which would be well enough in a little dog or a kitten, but are quite out of place with an animal as large as a donkey.

THE second species of Tapir is found in Malacca and Sumatra, and is a most conspicuous animal, in consequence of the broad band of white that encircles the body, and which at a little distance gives it the aspect of being muffled up in a white sheet.

The ground colour of the adult MALAYAN TAPIR is a deep sooty black, contrasting most strongly with the greyish white of the back and flanks. The young animal is as beautifully variegated as that of the preceding species, being striped and spotted with yellow fawn upon the upper parts of the body, and with white below. There is no mane upon the neck of the Malayan Tapir, and the proboscis is even longer in proportion. In size it rather exceeds the preceding animal. In many of its habits the Malayan animal is exactly similar to the species which inhabits America, but it is said that although the KUDA-AYER is very fond of the water, it does not attempt to swim, but contents itself with walking on the bed of the stream. Although a sufficiently common animal in its native country, it is but seldom seen, owing to its extremely shy habits, and its custom of concealing itself in the thickest underwood.

The hide of the Tapir is employed by the natives for several useful purposes, but the flesh is dry, tasteless, and not worth the trouble of cooking. The term Kuda-Ayer is a Malayan word, signifying "river-horse," and it is also known by the name of Tennu.

IN the SWINE, the snout is far less elephantine than in the preceding animals, and, though capable of considerable mobility, cannot be curled round any object so as to raise it from the ground. Nor, indeed, is such a power needed, as the Swine employ the snout for the purpose of rooting in the earth, and of distinguishing, by its tactile powers and the delicate sense of smell which is possessed by these animals, those substances which are suitable for food.

THE BOAR.—(*Sus scrofa.*)

There are many species as well as varieties of Swine, which are found in different parts of the earth, the first and most familiar of which is the DOMESTIC HOG of Europe.

This species is spread over the greater portion of the habitable globe, and was in former days common in a wild state even in England, from whence it has only been expelled within a comparatively late period. The chase of the wild boar was a favourite amusement of the upper classes, and the animal was one of those which were protected by the terribly severe forest laws which were then in vogue.

At the present time the wild Swine have ceased from out of England, in spite of several efforts that have been made to restore the breed by importing specimens from the Continent and turning them into the forests. There are, however, traces of the old wild boars still to be found in the forest pigs of Hampshire, with their high crests, broad shoulders, and thick bristling manes. These animals are very active, and are much fiercer than the ordinary Swine.

In this country the Hog is used not only for food, but for the sake of the hide, which, when prepared after a peculiar fashion, is found to make the best leather for saddles. The bristles which are so largely used in the manufacture of brushes are almost exclusively imported from the Continent.

In its wild and domesticated state the Hog is a most prolific animal, producing from eight to twelve pigs twice in each year, when it is in full vigour and in good health. Gilbert White records a sow which, when she died, was the parent of no less than three hundred pigs.

There is a prevalent idea, that whenever the Hog takes to the water he cuts his own throat with the sharp hoofs of his fore-feet. This, however, is by no means the case, for the animal is an admirable swimmer, and will often take to the water intuitively. In one of the Moray Islands, three domestic pigs belonging to the same litter swam a distance of five miles ; and it is said that if they had belonged to a wild family they would have swum to a much greater distance.

The flesh and fat of the Hog is especially valuable on account of its aptitude for taking salt without being rendered hard and indigestible by the process ; and the various breeds of domesticated Swine are noted for their adaptation to form pork or bacon in the shortest time and of the best quality. A full account of the various English varieties, together with the mode of breeding them and developing their peculiar characteristics, may be found in many books which are devoted especially to the subject.

ONE of the most formidable-looking of Swine is the BABYROUSSA of Malacca. This strange creature is notable for the curious manner in which the tusks are arranged, four of these weapons being seen to project above the snout. The tusks of the lower jaw project upward on each side of the upper, as is the case with the ordinary boar of Europe, but those of the upper jaw are directed in a very strange manner. Their sockets, instead of pointing downwards, are curved upwards, so that the tooth, in filling the curvatures of the socket, passes through a hole in the upper lip, and curls boldly

BABYROUSSA. —(*Babirussa Alfurus.*)

over the face. The curve, as well as the comparative size of these weapons, is extremely variable, and is seldom precisely the same in any two individuals. The upper tusks do not seem to be employed as offensive weapons ; indeed, in many instances they would be quite useless for such a purpose, as they are so strongly curved that their points nearly reach the skin of the forehead. The female is devoid of these curious appendages.

The skin of the Babyroussa is rather smooth, being sparsely covered with short bristly hairs. The object of the upper tusks is at present unknown, although certain old writers asserted that the animal was accustomed to suspend himself to branches by means of the appendage. The Babyroussa lives in herds of considerable size, and is found inhabiting the marshy parts of its native land.

The BOSCH VARK, or Bush Hog, of Southern Africa, is a very formidable animal in aspect, as well as in character, the heavy, lowering look, the projecting tusks, and the callous protuberance on the cheek, giving it a ferocious expression which is no way belied by the savage and sullen temper of the animal. The Bosch Vark inhabits the forests, and is generally found lying in excavations or hollows in the ground, from which it is apt to rush if suddenly disturbed, and to work dire vengeance upon its foe. In colour it is extremely variable, some species being of a uniform dark brown, others of a brown variegated with white, while others are tinged with bright chestnut. The young is richly mottled with yellow and brown. For the following account of the habits of the Bosch Vark I am indebted to Colonel Drayson's MS.

"Where the locality is sufficiently retired and wooded to afford shelter to the bush bucks which I have mentioned, we may generally expect to find traces of the Bush Pig. His spoor is like the letter M without the horizontal marks, the extremities of the toes forming two separate points, which is not the case with the Antelopes, at least very rarely so, the general impression of their feet being like the letter A with a division down the centre, thus Λ.

"The Bush Pig is about two feet six inches in height and five feet in length; his canine teeth are very large and strong, those in the upper jaw projecting horizontally; those in the lower upwards. He is covered with

BOSCH VARK.—(*Choiropotamus Africanus.*)

long bristles, and, taking him all in all, he is about as formidable-looking an animal, for his size, as can be seen.

"The Bosch Varks traverse the forests in herds, and subsist on roots and young shrubs. A large hard-shelled sort of orange, with an interior filled with seeds, grows in great quantities on the flats near the Natal forests; this is a favourite fruit of the wild pigs, and they will come out of the bush of an evening and roam over the plains in search of windfalls from these fruit-trees.

"The Kaffir tribes, although they refuse to eat the flesh of the domestic pig, will still feast without compunction on that of its bush brother.

"In the bush I always found the Kaffirs disinclined to encounter a herd of these wild Swine, stating as their reason for doing so that the animals were very dangerous; they also said that the wounds given by the tusks of this wild pig would not readily heal. The Berea bush of Natal was a favourite resort of these wild pigs, but although their spoor could be seen in all directions, the animals themselves were not so frequently encountered.

"The Kaffirs are much annoyed by these wild pigs, which force a passage through the imperfectly made fences, and root up the seeds, or destroy the pumpkins in the various gardens. As a defence, the Kaffirs leave nice enticing little openings in different parts of their fences, and the pigs, taking advantage of these ready-made doorways, frequently walk through them, and are then engulfed in a deep pit in which is a pointed stake, and they are assagaied with great delight by the expecting Kaffirs, who are on the alert, and who hear the cries of distress from piggy himself.

"The tusks are considered great ornaments, and are arranged on a piece of string and worn round the neck."

AMERICA possesses a representative of the porcine group in the PECCARIES, two species of which animals inhabit the Brazils.

The common PECCARY, or TAJACU, although it is of no very great dimensions, resembling a small pig in size, is yet a terrible animal. Ever fierce and irritable of temper, the Peccary is as formidable an antagonist as can be

PECCARY. —(*Dicotyles Tajacu.*)

seen in any land, for it knows no fear, and will attack any foe without hesitation. Although the Peccary is a very harmless animal to outward view, being only three feet long and weighing fifty or sixty pounds, and its armature consists of some short tusks that are barely seen beyond the lips, yet these little tusks are as fearful weapons as the longer teeth of the Bosch Vark, for they are shaped like a lancet, being acutely pointed and double-edged, so that they cut like knives and inflict very terrible wounds.

No animal seems to be capable of withstanding the united attacks of the Peccary, even the jaguar being forced to abandon the contest and to shrink from encountering the circular mass of Peccaries as they stand with angry eyes and gnashing teeth ready to do their worst on the foe.

The usual resting-place of the Peccary is in the hollow of a fallen tree, or in some burrow that has been dug by an armadillo and forsaken by the original inhabitant. The hollow tree, however, is the favourite resort, and into one of these curious habitations a party of Peccaries will retreat, each backing into the aperture as far as he can penetrate the trunk, until the entire hollow is filled with the odd little creatures. The one who last enters becomes the sentinel, and keeps a sharp watch on the neighbourhood.

The colour of the Peccary is a grizzled brown, with the exception of a white strip that is drawn over the neck, and has earned for the animal the name of the Collared Peccary.

INDIAN RHINOCEROS.—(*Rhinoceros unicornis.*)

SEVERAL species of the RHINOCEROS are still inhabitants of the earth. Of the existing species, two or three are found in various parts of Asia and its islands, and the remainder inhabit several portions of Africa. Before examining the separate species we will glance at some of the characteristics which are common to all the members of this very conspicuous group.

The so-called horn which projects from the nose of the Rhinoceros is a very remarkable structure, and worthy of a brief notice. It is in no way connected with the skull, but is simply a growth from the skin, and may take rank with hairs, spines, or quills, being indeed formed after a similar manner. If a Rhinoceros' horn be examined—the species of its owner is quite imma-terial—it will be seen to be polished and smooth at the tip, but rough and split into numerous filaments at the base. These filaments, which have a very close resemblance to those which terminate the plates of whalebone, can be stripped upwards for some length, and if the substance of the horn be cut across, it will be seen to be composed of a vast number of hairy filaments lying side by side.

The skin of the Rhinoceros is of very great thickness and strength, bidding defiance to ordinary bullets, and forcing the hunter to provide himself with balls which have been hardened with tin or solder. The extreme strength of the skin is well known both to the Asiatic and African natives, who manufacture it into shields, and set a high value on these weapons of defence.

In every species of Rhinoceros the sight appears to be rather imperfect, the animal being unable to see objects which are exactly in its front. The scent and hearing, however, are very acute, and seem to warn the animal of the approach of danger.

The Asiatic species of Rhinoceros are remarkable for the heavy folds into which the skin is gathered, and which hang massively over the shoulders, throat, flanks, and hind-quarters. Upon the abdomen the skin is comparatively soft, and can be pierced by a spear which would be harmlessly repelled from the thick folds of hide upon the upper portions of the body. In the INDIAN RHINOCEROS this weight of hide is especially conspicuous, the skin forming great flaps that can be easily lifted up by the hand. In a tamed state the Rhinoceros is pleased to be caressed on the softer skin under the thick hide, and in the wild state it suffers sadly from the parasitic insects that creep beneath the flaps, and lead the poor animal a miserable life, until they are stifled in the muddy compost with which the Rhinoceros loves to envelop its body. The horn of the Indian species is large in width, but inconsiderable in height, being often scarcely higher than its diameter. Yet with his short heavy weapon the animal can do terrible execution, and is said, upon the authority of Captain Williamson, to repel the attack of an adult male Elephant.

The height of this animal when fully grown is rather more than five feet, but the average height seems scarcely to exceed four feet. In colour it is a deep brown black, tinged with a purple hue, which is most perceptible when the animal has recently left its bath. The colour of the young animal is much paler than that of the mother, and partakes of a pinky hue.

OF African Rhinoceroses four species are clearly ascertained, and it is very probable that others may yet be in existence. Two of the known species are black, and the other two white; the animals differing from each other not only in colour, but in form, dimensions, habits, and disposition. The commonest of the African species is the BORELE, RHINASTER, or LITTLE BLACK RHINOCEROS, of Southern Africa; an animal which may be easily distinguished from its relations by the shape of the horns and the upper lip. In the Borele the foremost horn is of considerable length, and bent rather backward, while the second horn is short, conical, and much resembles the weapon of the Indian animal. The head is rather rounded, and the pointed upper lip overlaps the lower, and is capable of considerable extension.

The skin of this animal does not fall in heavy folds, like that of the Asiatic species, but is nevertheless extremely thick and hard, and will resist an ordinary leaden bullet, unless it be fired from a small distance. The skin is employed largely in the manufacture of whips, or jamboks.

The food of the Black Rhinoceros, whether the Borele or the Keitloa, is composed of roots, which the animal ploughs out of the ground with its horn, and of the young branches and shoots of the wait-a-bit thorn. It is rather remarkable that the black species is poisoned by one of the Euphorbiaceæ, which is eaten with impunity by the two white animals.

When wounded, the Black Rhinoceros is a truly fearful opponent, and it is generally considered very unsafe to fire at the animal unless the hunter be mounted on a good horse or provided with an accessible place of refuge. An old experienced hunter said that he would rather face fifty lions than one wounded Borele ; but Mr. Oswell, the well-known African sportsman, always preferred to shoot the Rhinoceros on foot. The best place to aim is just behind the shoulder, as if the lungs are wounded the animal very soon dies. There is but little blood externally, as the thick loose skin covers the bullet-hole, and prevents any outward effusion. When mortally wounded the Rhinoceros generally drops on its knees.

THE KEITLOA can readily be recognized by the horns, which are of considerable length, and nearly equal to each other in measurement. This is always a morose and ill-tempered animal, and is even more to be dreaded than the borele, on account of its greater size, strength, and length of horn. The upper lip of the Keitloa overlaps the lower even more than that of the borele ; the neck is longer in proportion, and the head is not so thickly covered with wrinkles. At its birth the horns of this animal are only indicated by a prominence on the nose, and at the age of two years the horn is hardly more than an inch in length. At six years of age it is nine or ten inches long, and does not reach its full measurement until the lapse of considerable time.

KEITLOA, OR SLOAN'S RHINOCEROS.—(*Rhinoceros Keitloa.*)

THE common WHITE RHINOCEROS (*Rhinoceros Simus*) is considerably larger than the two preceding animals, and, together with the kobaoba, or long-horned white Rhinoceros, is remarkable for its square muzzle and elongated head. The foremost horn of this animal is of very considerable length, attaining a measurement of more than three feet when fully grown. The second horn is short and conical, like that of the borele.

ONE of the most curious little animals in existence is the HYRAX, interesting not so much from its imposing external appearance, as for its importance in filling up a link in the chain of creation.

About as large as a tolerably-sized rabbit, covered with thick soft fur, inhabiting holes in the banks, possessing incisor-like teeth, and, in fine, being a very rabbit in habits, manners, and appearance, it was long classed among the rodents, and placed among the rabbits and hares. It has, however, been discovered in later years that this little rabbit-like animal is no rodent at all, but is one of the pachydermata, and that it forms a natural transition from the rhinoceros to the hippopotamus. On a close examination of the teeth, they are seen to be wonderfully like those of the hippopotamus, their edges being bevelled off in a similar manner, and therefore bearing some resemblance to the chisel-edged incisors of the rodents. There are several species of Hyrax, one of which inhabits Northern Africa and Syria, while the other two are found in Abyssinia and South Africa.

The South African Hyrax is termed by the colonists KLIP DAS, or ROCK RABBIT, and is found in considerable plenty among the mountainous districts of its native land, being especially common on the sides of the Table Mountain. It is largely eaten by the natives, who succeed in killing it in spite of its extreme wariness and activity. Among the crevices and fissures in the rock the Hyrax takes up its abode, and may often be seen sitting in the warm

HYRAX, OR KLIP DAS.—(*Hyrax Capensis.*)

rays of the sun, or feeding with apparent carelessness on the aromatic herbage of the mountain side. It is, however, perfectly secure, in spite of its apparent negligence, for a sentinel is always on guard, ready by a peculiar shrill cry to warn his companions of the approach of danger. Sometimes the Hyrax is seen at a considerable height, but is often observed near the sea-shore, seated on rocks which are barely above high-water mark.

Besides mankind, the Hyrax has many foes, such as the birds of prey and carnivorous quadrupeds, and is destroyed in considerable numbers. The fore-feet of this animal are apparently furnished with claws like those of the rabbit, but on a closer inspection the supposed claws are seen to be veritable hoofs, black in colour, and very similar to those of the rhinoceros in form. The Hyrax is an agile little creature, and can climb a rugged tree-trunk with great ease. It is rather hot in its temper, and if irritated becomes highly excited, and moves its teeth and feet with remarkable activity and force.

THE SYRIAN HYRAX is the animal which is mentioned under the name of "coney" in the Old Testament, and is found inhabiting the clefts and caverns of rocks. In its habits and general appearance it is very similar to the Cape Hyrax, and needs no further description.

THE last on the list of the pachydermatous animals is the well-known HIPPOPOTAMUS, or RIVER HORSE.

HIPPOPOTAMUS, OR ZEEKOE.—(*Hippopotamus amphibius.*)

This enormous quadruped is a native of various parts of Africa, and is always found either in water or in its near vicinity. In absolute height it is not very remarkable, as its legs are extremely short, but the actual bulk of its body is very great indeed. The average height of a full-grown Hippopotamus is about five feet. Its naked skin is dark brown, curiously marked with innumerable lines like those on "crackle" china or old oil-paintings, and is also dappled with a number of sooty black spots, which cannot be seen except on a close inspection. A vast number of pores penetrate the skin, and exude a thick, oily liquid, which effectually serves to protect the animal from the injurious effects of the water in which it is so constantly immersed. Some years ago, when the male Hippopotamus in the Zoological Gardens was young and gentle, I patted his back, and entirely spoiled a pair of new kid gloves. The mouth is enormous, and its size is greatly increased by the odd manner in which the jaw is set in the head.

Within the mouth is an array of white gleaming tusks, which have a terrific appearance, but they are solely intended for cutting grass and other vegetable substances, and are seldom employed as weapons of offence, except when the animal is wounded or otherwise irritated. The incisor teeth of the lower jaw lie almost horizontally, with their points directed forwards, and are said to be employed as crow-bars in tearing up the various aquatic plants on which the animal feeds. The canines are very large and curved, and are worn obliquely, in a manner very similar to the rodent type of teeth. Their shape is a bold curve, forming nearly the half of a circle, and their surface is deeply channeled and ridged on the outer line of the curve, and smoother on the face.

Possessed of an enormous appetite, having a stomach that is capable of containing five or six bushels of nutriment, and furnished with such powerful instruments, the Hippopotamus is a terrible nuisance to the owners of culti-vated lands that happen to be near the river in which the animal has taken up his abode.

The Hippopotamus is, as the import of its name, River Horse, implies, most aquatic in its habits. It generally prefers fresh water, but is not at all averse to the sea, and will sometimes prefer salt water to fresh. It is an admirable swimmer and diver, and is able to remain below the surface for a very considerable length of time. In common with the elephant, it possesses the power of sinking at will, which is the more extraordinary when the huge size of the animal is taken into consideration. Perhaps it may be enabled to contract itself by an exertion of the muscles whenever it desires to sink, and to return to its former dimensions when it wishes to return to the surface. It mostly affects the stillest reaches of the river, as it is less exposed to the current, and not so liable to be swept down the stream while asleep. The young Hippopotamus is not able to bear submersion so long as its parent, and is therefore carefully brought to the surface at short intervals for the purpose of breathing. During the first few months of the little animal's life, it takes its stand on its mother's neck, and is borne by her above or through the water as experience may dictate or necessity require.

The Hippopotamus is a gregarious animal, collecting in herds of twenty or thirty in number, and making the air resound with their resonant snorts. The snort of this creature is a most extraordinary sound, and one that is well calculated to disturb the nerves of sensitive persons, especially if heard un-expectedly. The animals at the Zoological Gardens make the very roof ring with the strange unearthly sounds which they emit. In their native state it is very difficult to ascertain even approximately the number of a herd, as the animals are continually diving and rising, and never appear simultaneously above the surface of the water.

DASYPIDÆ.

THIS small but important family includes the Manis, the Armadillo, the Ant-eater, and the Platypus, or Duck-bill.

THE PHATAGIN is one of the numerous species that compose the strange genus of MANIS. All these animals are covered with a series of horny plates, sharp-pointed and keen-edged, that lie with their points directed towards the tail, and overlap each other like the tiles upon the roof of a house.

The fore-claws of the Phatagin are very large, and are employed for the purpose of tearing down the nests of the termite, or white ant, as it is more popularly called, so as to enable it to feed upon the inmates, as they run about in confusion at the destruction of their premises. Ants, termites, and various insects are the favourite food of the Phatagin, which sweeps them up by means of its long and extensile tongue, caring nothing for their formidable jaws, which are powerful enough to drive a human being almost distracted with pain. The claws are not only employed in destroying the nest of the termite, but in digging burrows for its own residence, a task for which they are well adapted by reason of their great size and strength, and the vigour of the limbs to which they are attached.

The Phatagin is a native of Western Africa, and is of considerable dimensions, reaching five feet in average length, of which the tail occupies three feet. From the great length of the tail, it is sometimes called the LONG-TAILED MANIS.

PHATAGIN.—(*Manis tetradactyla.*)

THE BAJJERKEIT, or SHORT-TAILED MANIS, is a native of various parts of India, and is also found in Ceylon. Of this species Sir Emerson Tennent gives the following short account : " Of the Edentates, the only example in Ceylon is the scaly ant-eater, called by the Singalese, Caballaya, but usually known by its Malay name of Pengolin, a word indicative of its faculty of 'rolling itself up' into a compact ball, by bending its head towards its stomach, arching its back into a circle, and securing all by a powerful hold of its mail-covered tail. When at liberty, they burrow in the dry ground to a depth of seven or eight feet, where they reside in pairs, and produce annually two or three young.

" Of two specimens which I kept alive at different times, one from the vicinity of Kandy, about two feet in length, was a gentle and affectionate creature, which, after wandering over the house in search of ants, would attract attention to its wants by climbing up my knee, laying hold of my leg by its prehensile tail. The other, more than double that length, was caught in the jungle near Chilaw, and brought to me in Colombo. I had always understood that the Pengolin was unable to climb trees, but the one last mentioned frequently ascended a tree in my garden in search of ants, and this is effected by means of its hooked feet, aided by an oblique grasp of the tail. The ants it seized by extending its round and glutinous tongue along their tracks. Generally speaking, they were quiet during the day, and grew restless as evening and night approached."

BAJJERKEIT.—(*Manis pentadactyla.*)

The ARMADILLOS are inhabitants of Central and Southern America, and are tolerably common throughout the whole of the land in which they live. The general structure of the armour is similar in all the species, and consists of three large plates of horny covering ; one being placed on the head, another on the shoulders, and the third on the hind-quarters. These plates are connected by a series of bony rings, variable in number, overlapping each other, and permitting the animal to move freely.

THE common ARMADILLO, or POYOU, is about twenty inches in total length, the tail occupying some six or seven inches. It is very common in Paraguay, but is not easily captured, owing to its remarkable agility, perseverance, and wariness. Encumbered as it appears to be with its load of plate-armour, it runs with such speed that it can hardly be overtaken by a quick-footed man, and if it should contrive to reach its burrow, it can never be got out except by dint of hard work.

The food of the Armadillo is nearly as varied as that of the swine, for there are few eatable substances, whether vegetable or animal, which the Armadillo

ARMADILLO. — (*Dasypus sexcinctus.*)

will not devour, provided they are not too hard for its little teeth. Various roots, potatoes, and maize are among its articles of vegetable diet, and it also will eat eggs, worms, insects, and small reptiles of every description. Whenever wild cattle are slain, the Armadillo is sure to make its appearance in a short time, for the purpose of devouring the offal which the hunter leaves on the ground. It is not at all particular in taste, and devours the half-putrid remains with great eagerness, becoming quite fat on the revolting diet.

As the Armadillo is a nocturnal animal, its eyes are more fitted for the dark than for the bright glare of sunlight, which dazzles the creature and sadly bewilders it. If it should be detected on the

TAMANOIR, OR ANT BEAR.—(*Myrmecophaga jubata.*)

surface of the ground, and its retreat intercepted before it can regain its hole, the Armadillo rolls itself up as best it can, and, tucking its head under the chest, draws in its legs and awaits the result. Even when taken in hand it is not without a last resource, for it kicks so violently with its powerful legs that it can inflict severe lacerations with the sharp claws.

THE ANT-EATERS, as their name imports, feed very largely on ants, as well as on termites and various other insects, their long flexible tongue acting as a hand for the purpose of conveying food into the mouth. The tongue of the Ant-eater, when protruded to its fullest extent, bears some resemblance to a great earth-worm, and as it is employed in its food-collecting task, it coils and twists about as if it possessed a separate vitality of its own.

THE TAMANOIR, or GREAT ANT-EATER, or ANT BEAR, is entirely destitute of teeth, possesses a wonderfully elongated and narrow head, and is thickly covered with long coarse hair, which on the tail forms a heavy plume. The colour of this animal is brown, washed with grey on the head and face, and interspersed with pure white hairs on the head, body, and tail. The throat is black, and a long triangular black mark arises from the throat, and passes obliquely over the shoulders. There are four toes on the fore-feet, and five on the hinder. In total length it measures between six and seven feet, the tail being about two feet six inches long.

The claws of the fore-feet are extremely long and curved, and are totally unfitted for locomotion. When the animal is not employing these instruments in destroying, it folds the long claws upon a thick rough pad which is placed in the palm, and seems to render the exertion of walking less difficult. As, however, the Ant-bear is forced to walk upon the outer edge of its fore-feet, its progress is a peculiarly awkward one, and cannot be kept up for any long time. The creature seems to possess considerable grasping power in the toes of the fore-limbs, being able to pick up a small object in its paws. Though not a fighter, it can defend itself right well by means of these powerful instruments, and can not only strike with considerable violence, but when attacked by a dog or similar enemy, it clasps him in such a terrific grip, that the half-suffocated animal

THE MIDDLE ANT-EATER.

is only too glad to be able to escape.

The Ant Bear is said to make no burrow, but to content itself with the shade of its own plumy tail whenever it retires to rest. While sleeping, the creature looks very like a rough bundle of hay thrown loosely on the ground, for the hair of the main and tail is so long and so harsh that it can hardly be recognized at the first glance for the veritable coat of a living animal. The eye of this creature has a peculiar and indescribably cunning expression. The Tamanoir is a native of Guinea, Brazil, and Paraguay.

THE MIDDLE ANT-EATER, or Tamandua, is not so large as the preceding animal, from which it is readily distinguished by the tail, which is long and tapering, and almost devoid of hair except at the base. The tail indeed is used as an organ of prehension, to assist it in climbing trees, a feat which it sometimes performs, although not so often as the Little Ant-eater.

This animal produces a strong scent of musk, which is generally excited when it is enraged. The scent is not pleasant, like that of the musk deer, but very disagreeable, and can be perceived at a considerable distance.

The LITTLE ANT-EATER is a truly curious animal. The head of this creature is comparatively short ; its body is covered with fine silken fur, and its entire length does not exceed twenty or twenty-one inches.　The tail is well furred, excepting three inches of the under surface at the extremity, which is employed as the prehensile portion of that member, and is capable of sustaining the weight of the body as it swings from a branch.　On looking

LITTLE ANT-EATER.—(*Cyclothurus didactylus.*)

at the skeleton, a most curious structure presents itself.　On a side view, the cavity of the chest is completely hidden by the ribs, which are greatly flattened, and overlap each other so that on a hasty glance the ribs appear to be formed of one solid piece of bone.　There are only two claws on the fore-feet, and four on the hinder limbs.

The Little Ant-eater is a native of tropical America, and is always to be found on trees, where it generally takes up its residence, and where it finds its sustenance. It possesses many squirrel-like customs, using its fore-claws with great dexterity, and hooking the smaller insects out of the bark crevices in which they have taken unavailing refuge. While thus employed it sits upon its hind limbs, supporting itself with its prehensile tail. The claws are compressed, curved, and very sharp, and the little animal can use these instruments with some force as offensive weapons, and can strike smart blows with them. It is a bold little creature, attacking the nests of wasps, putting its little paw into the combs, and dragging the grubs from their cells.

Like its larger relations, it is nocturnal in its habits, and sleeps during the day with its tail safely twisted round the branch on which it sits. The generic name, *Cyclothurus*, signifies "twisted tail," and is very appropriate to the animal.

THERE are few animals which have attracted such universal attention, both from scientific men and the reading world in general, as the MULLINGONG, DUCK-BILL, or PLATYPUS, of Australia. This little creature, the largest being but twenty-two inches in length, has excited more interest than animals

DUCK-BILL, OR MULLINGONG. —(*Platypus Anatinus.*)

of a thousand times its dimensions, on account of its extraordinary shape and singular habits. It is most appropriately called the Duck-bill, on account of the curious development of the intermaxillary bones, which are very much flattened and elongated, and their ends turned inwards in a kind of angular hook. The lower jaw is also lengthened and flattened, although not to such an extent as the upper, and the bones are covered with a naked skin.

In the stuffed and dried specimens the "beak" appears as if it were composed of the black leather taken from an old shoe, but in the living animal it presents a very different aspect, being soft, rounded, and of a pinky hue at its tip, mottled with a number of little spots. Dr. Bennett, to whom the zoological world is so much indebted for his researches into the habits of this curious animal, kindly showed me some excellent drawings, which gave a very different idea of the animal from that which is obtained by the examination of stuffed skins. The beak is well supplied with nerves, and appears to be a sensitive organ of touch, by means of which the animal is enabled to feel as well as to smell the insects and other creatures on which it feeds.

The Mullingong is an essentially aquatic and burrowing animal, and is formed expressly for its residence in the water or under the earth. The fur is thick and soft, and is readily dried while the animal enjoys good health, although it becomes wet and draggled when the creature is weakly. The opening of the ears is small and can be closed at will, and the feet are furnished with large and complete webs, extending beyond the claws in the fore limbs, and to their base in the hind legs. The fore-feet are employed for digging as well as for swimming, and are therefore armed

PORCUPINE ANT-EATER, OR ECHIDNA.—(*Echidna Hystrix.*)

with powerful claws rather more than half an inch in length, and rounded as their extremities. With such force can these natural tools be used, that the Duck-bill has been seen to make a burrow two feet in length through hard gravelly soil in the space of ten minutes. While digging, the animal employs its beak as well as its feet, and the webbed membrane contracts between the joints so as not to be seen. The hind-feet of the male are furnished with a spur, about an inch in length, curved, perforated, and connected with a gland situated near the ankle. It was once supposed that this spur conveyed a poisonous liquid into the wound which it made, but this opinion has been disproved by Dr. Bennett, who frequently permitted, and even forced, the animal to wound him with its spurs, and experienced no ill consequences beyond the actual wound. The animal has the power of folding back the spur so as to conceal it entirely, and is then sometimes mistaken for a female.

The colour of the adult animal is a soft dark brown, interspersed with a number of glistening points which are produced by the long and shining hairs which protrude through the inner fur.

The food of the Mullingong consists of worms, water insects, and little molluscs, which it gathers in its cheek-pouches as long as it is engaged in its search for food, and then eats quietly when it rests from its labours. The teeth, if teeth they may be called, of this animal are very peculiar, consisting of four horny channeled plates, two in each jaw, which serve to crush the fragile shells and coverings of the animals on which it feeds. It seems seldom to feed during the day, or in the depth of night, preferring for that purpose the first dusk of evening or the dawn of morning. During the rest of the day it is generally asleep. While sleeping, it curls itself into a round ball, the tail shutting down over the head and serving to protect it.

The young Mullingongs are curious little creatures, with soft, short, flexible beaks, naked skins, and almost unrecognizable as the children of their long-nosed parents. When they attain to the honour of their first coat,

SLOTH.—(*Cholæpus didactylus.*)

they are most playful little things, knocking each other about like kittens, and rolling on the ground in the exuberance of their mirth. Their little twinkling eyes are not well adapted for daylight, nor from their position can they see spots directly in their front, so that a pair of these little creatures that were kept by Dr. Bennett used to bump themselves against the chairs, tables, or any other object that might be in their way. They bear a further similitude to the cat in their scrupulous cleanliness and the continual washing and pecking of their fur.

THE ECHIDNA is found in several parts of Australia, where it is popularly called the hedgehog, on account of the hedgehog-like spines with which the body is so thickly covered, and its custom of rolling itself up when alarmed. A number of coarse hairs are intermingled with the spines, and the head is devoid of these weapons. The head is strangely lengthened, in a manner somewhat similar to that of the Ant-eater, and there are no teeth of any kind in the jaws.

The food of the Echidna consists of ants and other insects, which it gathers into its mouth by means of the long extensile tongue. It is a burrowing animal, and is therefore furnished with limbs and claws of proportionate strength. Indeed, Lieutenant Breton, who kept one of these animals for some time, considers it as the strongest quadruped in existence in proportion to its size. On moderately soft ground it can hardly be captured, for it gathers all its legs under its body, and employs its digging claws with such extraordinary vigour that it sinks into the ground as if by magic. The Echidna is tolerably widely spread over the sandy wastes of Australia, but has not been seen in the more northern portions of that country.

IN the last group of the mammalia we find a very remarkable structure, adapted to serve a particular end, and misunderstood by zoologists. The common SLOTH, sometimes called the TWO-TOED SLOTH, is a native of the West Indies, where it is not very often seen, although it is not a very uncommon animal.

The peculiarity to be noticed in all the Sloths, of which there are several species, is, that they pass the whole of their lives suspended, with their backs downwards, from the branches of trees. The Sloth never gets upon a bough, but simply hooks his curved talons over it, and hangs in perfect security. In order to enable the animal to suspend itself without danger of falling, the limbs are enormously strong, the fore-legs are remarkable for their length, and the toes of all four feet are furnished with strong curved claws. Upon the ground the Sloth is entirely out of its element, as its limbs

AI, OR THREE-TOED SLOTH.—(*Bradypus tridactylus.*)

are wholly unadapted for supporting the weight of the body, and its long claws cannot be employed as adjuncts to the feet. The only manner in which a Sloth can advance, when he is unfortunately placed in such a position, is by hitching his claws into any depression that may afford him a hold, and so dragging himself slowly and painfully forward. On the trees, however, he is quite a different creature, full of life and animation, and traversing the branches at a speed which is anything but slothful. The

Q

Sloth travels best in windy weather, because the branches of trees are blown against each other, and permit the animal to pass from one tree to another without descending to the ground.

The food of the Sloth consists of leaves, buds, and young shoots.   It appears to stand in no need of water, being satisfied with the moisture which clings to the herbage on which it feeds.   In gathering the leaves and drawing the branches within reach, the Sloth makes great use of its fore-paws, which, however helpless upon the ground, can be managed with great dexterity.   It is very tenacious of life, and is protected from any injury which it might receive from falls by the peculiar structure of its skull.   In length it is about two feet.

THE AI, or THREE-TOED SLOTH, is an inhabitant of South America, and is more common than the preceding animal, from which it can easily be distin· guished by the third toe on its feet.   The colour of this animal is rather variable, but is generally of a brownish grey, slightly variegated by differ- ently tinted hairs, and the head and face being darker than the body and limbs.   The hair has a curious hay-like aspect, being coarse, flat, and harsh towards the extremity, although it is very fine towards the root.   Owing to the colour and structure of the hair, the Ai can hardly be distinguished from the bough under which it hangs, and owes much of its safety to this happy resemblance ; for its flesh is very good, and, in consequence, the poor creature is dreadfully persecuted by the natives, as well as by the white hunters.   The cry of this creature is low and plaintive, and is thought to resemble the sound *Ai.*   The head is short and round, the eyes deeply sunk in the head, and nose large and very moist.

The young of the Ai, as well as those of the other Sloths, cling to their mother as soon as they are born, and are carried about by her until they are able to transfer their weight from their parent to the branches.   Several other species of Sloths are known to exist, but all are similar in appearance and habits.

BIRDS.

BIRDS.

THE most conspicuous external characteristic by which the BIRDS are distinguished from all other inhabitants of earth is the feathery robe which invests their bodies, and which serves the double purpose of clothing and progression.

The fuller and more technical description of the Birds runs as follows : They are vertebrate animals, but do not suckle their young. The young are not produced in an actively animated state, but inclosed in the egg, from which they do not emerge until they have been warmed into independent life by the effects of constant warmth. Generally, the eggs are hatched by means of the natural warmth which proceeds from the mother bird; but in some instances, such as that of the Tallegalla of Australia, the eggs are placed in a vast heap of dead leaves and grass, and developed by means of the heat which is exhaled from decaying vegetable substances.

LAMMERGEYER.—(*Gypaetos barbatus.*)

BIRDS OF PREY.—VULTURES.

BY common consent VULTURES take the first rank among Birds, and in the catalogue of the British Museum the LAMMERGEYER, or BEARDED VULTURE, stands first upon the list.

THE CONDOR.

This magnificent bird is a native of Southern Europe and Western Asia, and often attains a very great size, the expanse of its wings being sometimes as much as ten feet, and its length nearly four feet.

The name of Bearded Vulture has been given to the Lammergeyer on account of the tufts of long and stiff bristle-like hairs which take rise at the nostrils and beneath the bill, and form a very prominent characteristic of the species.

The colour of the Lammergeyer is a grey-brown, curiously dashed with white upon the upper surface, in consequence of a white streak which runs along the centre of each feather. The under surface of the body, together with the neck, are nearly white, tinged with a wash of reddish brown, which is variable in depth in different individuals. In the earlier stages of its existence, the Lammergeyer is of a much darker hue, and the white dashes upon the back are not so purely white nor so clearly defined. The head and neck are dark brown, and the brown hue of the back is of so deep a tint that the young bird has been classed as a separate species, under the title of *Vultur niger*, or Black Vulture.

ALTHOUGH not exceeding the lammergeyer in dimensions, the CONDOR has been long celebrated as a giant among birds, the expanse of its wings being set down at eighteen or twenty feet, and its length exaggerated in the same proportion. In reality, the expanse of a large Condor's wings will very seldom reach eleven feet, and the average extent is from eight to nine feet.

CONDOR.—(*Sarcorhamphus Gryphus.*)

The general colour of the Condor is a greyish black, variable in depth and glossiness in different individuals. The upper wing-coverts are marked with white, which take a greyer tint in the female, and the exterior edges of the secondaries are also white. The adult male bird may easily be distinguished by the amount of white upon the feathers, so that the wings are marked with a large white patch. Around the neck is set a beautifully white downy collar of soft feathers, which does not entirely inclose the neck, but leaves a small naked band in front. This featherless band is, however, so small, that it is not perceptible except by a close examination.

The crest of the male Condor is of considerable size, occupying the top of the head and extending over a fourth of the basal portion of the beak.

The Condor is an inhabitant of the mountain chain of the Andes, and is celebrated not only for its strength and dimensions, but for its love of elevated localities. When enjoying the unrestricted advantages of its native home, it is seldom found lower than the line of perpetual snow, and only seems to seek lower and more temperate regions when driven by hunger to make a raid on the flocks or the wild quadrupeds of its native country. Although preferring carrion to the flesh of recently-killed animals, the Condor is a terrible pest to the cattle-keeper, for it will frequently make an attack upon a cow or a bull, and by dint of constant worrying, force the poor beast to succumb to its winged pursuers. Two of these birds will attack a vicugna, a deer, or even the formidable puma, and as they direct their assaults chiefly upon the eyes, they soon succeed in blinding their prey, who rapidly falls under the terrible blows which are delivered by the beaks of its assailants.

The Condor deposits its eggs, for it makes no nest whatever, upon a bare shelf of some lofty rock. The eggs are two in number, greyish white in colour, and are laid about November or December. When the young Condor is hatched it is nearly naked, but is furnished with a scanty covering of down, which in a short time becomes very plentiful, enveloping the body in a complete vestment of soft black plumage. The deep black grey of the adult bird is not attained until a lapse of three years, the colour of the plumage being a yellowish brown.

THE KING VULTURE has gained its regal title from a supposition which is prevalent among the natives of the country which it inhabits, that it wields royal sway over the aura, or zopilote Vultures, and that the latter birds will not venture to touch a dead carcass until the King Vulture has taken his share. There is some truth for this supposition, for the King Vulture will not permit any other bird to begin its meal until its own hunger is satisfied. The same habit may be seen in many other creatures, the more powerful lording it over the weaker, and leaving them only the remains of the feast instead of permitting them to partake of it on equal terms. But if the King Vulture should not happen to be present when the dead animal has reached a state of decomposition which renders it palatable to vulturine tastes, the subject Vultures would pay but little regard to the privileges of their absent monarch, and would leave him but a slight prospect of getting a meal on the remains of the feast.

Waterton, who often mentions this species in his interesting works, gives several curious instances of the sway which the King Vulture exercises over the inferior birds. "When I had done with the carcass of the large snake, it was conveyed into the forest, as I expected that it would attract the king of the Vultures, as soon as time should have rendered it sufficiently savoury. In a few days it sent forth the odour which a carcass should send forth, and about twenty of the common Vultures came and perched on the neighbouring trees. The king of the Vultures came too, and I observed that none of the common ones seemed inclined to begin breakfast until his majesty had finished. When he had consumed as much snake as nature informed him would do him good, he retired to the top of a high mora-tree, and then all the common Vultures fell to and made a hearty meal."

The King Vulture is a native of tropical America, and is most common near the equator, though it is found as far as the thirtieth degree of south

KING VULTURE.—(*Sarcorhamphus Papa.*)

latitude, and the thirty-second of north latitude. Peru, Brazil, Guiana, Paraguay and Mexico are the chosen residences of this fine species. It is a forest-loving bird, caring nothing for the lofty home of the condor, but taking up its residence upon the low and heavily-wooded regions in close proximity to swampy and marshy places, where it is most likely to find abundance of dead and putrefying animal substances. Its nest, or rather the spot on which it deposits its eggs, is within the hollow of some decaying tree. The eggs are two in number.

In its adult state the King Vulture is a most gorgeously decorated bird, though its general aspect and the whole expression of its demeanour are

FULVOUS, OR GRIFFIN VULTURE.—(*Gyps fulvus.*)

rather repulsive than otherwise. The greater part of the feathers upon the back are of a beautiful satiny white, tinged more or less deeply with fawn, and the abdomen is of a pure white. On account of its colour, the bird is termed the White Crow by the Spaniards of Paraguay. The long pinions of the wing and tail are deep black, and the base of the neck is surrounded with a thick ruff or collar of downy grey feathers.

The most brilliant tints are, however, those of the naked skin of the head and neck. " The throat and back of the neck," says Waterton, " are of a fine lemon colour ; both sides of the neck, from the ears downwards, of a rich scarlet ; behind the corrugated part there is a white spot. The crown of the head is scarlet, betwixt the lower mandible and the eye, and close by the ear, there is a part which has a fine silvery-blue appearance. Just above the white spot a portion of the skin is blue, and the rest scarlet ; the skin which juts out behind the neck, and appears like an oblong carbuncle, is blue in part, and part orange. The bill is orange and black, the caruncles on the forehead orange, and the cere orange, the orbits scarlet, and the irides white."

These gorgeous tints belong only to the adult bird of four years old, and in the previous years of its life the colours are very obscure. In the first year, for example, the general colour is deep blue-grey, the abdomen white, and the crest hardly distinguishable either for its colour or its size. In the second year of its age the plumage of the bird is nearly black, diversified with white spots, and the naked portions of the head and neck are violet-black interspersed with a few dashes of yellow. The third year gives the bird a very near approach to the beautiful satin fawn of the adult plumage, the back being nearly of the same hue as that of the four-year-old bird, but marked with many of the blue-black feathers of the second year. When full grown, the King Vulture is about the size of an ordinary goose.

THE FULVOUS or GRIFFIN VULTURE is one of the most familiar of these useful birds, being spread widely over nearly the whole of the Old World, and found in very many portions of Europe, Asia, and Africa.

It is one of the large Vultures, measuring four feet in length, and its expanse of wing being exceedingly wide. Like many of its relations, it is a high-roving bird, loving to rise out of the ken of ordinary eyes, and from that vast elevation to view the panorama which lies beneath its gaze ; not, however, tor the purpose of admiring the beauty of the prospect but for the more sensual object of seeking for food. Whenever it has discovered a dead or dying animal, the Vulture takes its stand on some adjoining tree or rock, and there patiently awaits the time when decomposition shall render the skin sufficiently soft to permit the entrance of the eager beak. As soon as its olfactory organs tell of that desired change, the Vulture descends upon its prey, and will not retire until it is so gorged with food that it can hardly stir. If it be suddenly attacked while in this condition, it can easily be overtaken and killed ; but if a pause of a few minutes only be allowed, the bird ejects by a spasmodic effort the load of food which it has taken into its interior, and is then ready for flight.

A controversy has long raged concerning the manner in which the Vulture obtains knowledge of the presence of food. Some naturalists assert that the wonderful powers of food-finding which are possessed by the Vulture are owing wholly to the eyes, while others as warmly attribute to the nose this currious capability. Others again, desirous of steering a middle course, believe that the eyes and the nostrils give equal aid in this never-ending duty of finding food, and many experiments have been made with a view to extracting the real truth of the matter. The following account has been kindly transmitted to me by Colonel Drayson, R.A., who has already contributed much original information to the present work :—

" Having shot an ourebi early in the morning, and when about three miles from home, I was not desirous of carrying the animal behind my saddle during the day's shooting, and I therefore sought for some method of concealment by which to preserve the dead quarry from jackals and Vultures. An ant-bear's hole offered a very convenient hiding-place, into which the buck was pushed, and the carcass was covered over with some grass cut for the purpose. As usual in South Africa there were some Vultures wheeling round at an enormous height above the horizon; these I believed would soon come down and push aside the grass and tear off the most assailable parts of the buck. There was, however, no, better means of protection, so I left the animal and rode away. When at about a quarter of a mile from the antbear's hole, I thought that it might be interesting to watch how the Vultures would approach and commence operations, so I 'off-saddled,' and kept watch.

"After about half an hour, I saw a Vulture coming down from the sky, followed by two or three others. They came down to the spot where the buck had been killed, and flew past this. They then returned and again overshot the mark. After circling several times within a radius of four hundred yards, they flew away. Other Vultures then came and performed similar manœuvres, but not one appeared to know where the buck was concealed. I then rode off to a greater distance, but the same results occurred.

"In the evening I returned for my buck, which

ARABIAN VULTURE.—(*Vultur Monachus.*)

however was totally useless in consequence of the intense heat of the sun, but which had not been touched by the Vultures."

ONE of the best known of the Vultures is the TURKEY BUZZARD, more rightly termed the CARRION VULTURE. Its name of Turkey Buzzard is earned from the strange resemblance which a Carrion Vulture bears to a turkey as it walks slowly and with a dignified air, stretching its long bare neck, and exhibiting the fleshly appendages which bear some likeness to the

wattles of the turkey. This bird is chiefly found in North America, but is also an inhabitant of Jamaica, where it is popularly known as the John Crow.

The nest of the Turkey Buzzard is a very inartistical affair, consisting merely of some suitable hollow tree or decayed log, in which there may be a depression of sufficient depth to contain the eggs. In this simple cradle the female deposits from two to four eggs, which are of a dull cream-white, blotched with irregular chocolate splashes, which seem to congregate towards the largest end. The young birds are covered with a plentiful supply of white down.

The adult Turkey Buzzard is rather a large bird, measuring two feet six inches in length, and six feet ten inches across the expanded wings. The weight is about five pounds. The general colour of the plumage is black, mingled with brown, the secondaries being slightly tipped with white, and a few of the coverts edged with the same tint. On the neck, the back, the shoulders, and the scapularies, the black hue is shot with bronze, green, and purple. Beneath the thick plumage is a light coating of soft white down, which apparently serves to preserve the creature at a proper temperature. The bare skin of the neck is not as wrinkled as in several Vultures, and the feathers make a complete ring round the neck. There is but little difference in the plumage of the two sexes, but the bill of the male is pure white.

TURKEY BUZZARD.— (*Catharista Aura.*)

WE now arrive at the true Vultures, the best known of which is the common ARABIAN VULTURE, a bird which is spread over a very large portion of the globe, being found in various parts of Europe, Asia and Africa.

It is a large bird, measuring nearly four feet in length, and the expansion of its wings being proportionately wide. The general colour of this species is a chocolate brown, the naked portions of the neck and head are of a bluish hue, and it is specially notable for a tuft of long soft feathers which spring from the insertion of the wings. In spite of its large size and great muscular powers, the Arabian Vulture is not a dangerous neighbour even to the farmer, for unless it is pressed by severe hunger, it seems rather to have a dread of living animals, and contents itself with feeding on any carrion which may come in its way. Sometimes, however, after a protracted fast, its fears are overruled by its hunger, and the bird makes a raid upon the sheepfolds or the goat-flocks, in the hope cf carrying off a tender lamb or kid.

The usual haunts of this species are situated on the mountain tops, and the bird does not descend into the valleys except when pressed by hunger.

The specific title of *Monachus*, or Monk, has been given to this species on account of the hood-like ruff around its neck, which is thought to bear a fanciful resemblance to the hood of a monk.

ALPINE, OR EGYPTIAN VULTURE.—(*Neophron percnopterus.*)

The ALPINE, or EGYPTIAN VULTURE, is, as its name imports, an inhabitant of Egypt and Southern Europe. It is also found in many parts of Asia, and as it has once been captured on our shores, has been placed among the list of British birds.

The general colour of the adult bird is nearly white, with the exception of the quill feathers of the wing, which are dark brown. The face, bill, and legs are bright yellow, so that the aspect of the bird is sufficiently curious. The sexes are clothed alike when adult. On account of the colour of its plumage, the Egyptian Vulture, is popularly termed the WHITE CROW by the Dutch colonists, and AKBOBAS, or White Father, by the Turks. It is also familiarly known by the name of PHARAOH'S CHICKEN, because it is so frequently represented in the hieroglyphical inscriptions of Egypt. When young the colour of its plumage is a chocolate brown, the neck and shoulders are covered with grey-tipped feathers, and the beak and feet are a very dull ochry yellow. The white plumage of the adult state is not attained until the bird has completed its third year.

As is the case with the Vultures in general, the Egyptian Vulture is protected from injury by the strictest laws, a heavy penalty being laid upon any one who should wilfully destroy one of these useful birds Secure under its human protection, the bird walks fearlessly about the streets of its native land, perches upon the houses, and, in common with the pariah dogs, soon clears away any refuse substances that are thrown into the open streets in those evil-smelling and undrained localities. This bird will eat almost anything which is not too hard for its beak, and renders great service to the husbandman by devouring myriads of lizards, rats, and mice, which would render all cultivation useless were not their numbers kept within limits by the exertion of this useful Vulture. It has been also seen to feed on the nara, a rough water-bearing melon, in common with cats, leopards, mice, ostriches, and many other creatures. The eggs of the ostrich are said to be a favourite food with the Egyptian Vulture, who is unable to break their strong shells with his beak, but attains his object by carrying a great pebble in the air, and letting it drop upon the eggs.

The wings of this species are extremely long in proportion to the size of the bird, and their lofty soaring flight is peculiarly graceful. It is but a small bird in comparison with many of those which have already been mentioned, being not much larger than the common rook of Europe. The nest of the Egyptian Vulture is made upon the shelf or in the cleft of a lofty rock, and the grey-white eggs are three or four in number. It is a curious fact, that during the season of reproduction the male bird slightly changes his aspect, the yellow bill becoming orange, and retaining that tint until the breeding season is over. Like many rapacious animals and birds, the Egyptian Vulture does not disdain to feed on insects, and has been observed in the act of following a ploughman along his furrows, picking up the worms and grubs after the fashion of the common rook.

EAGLES.

NEXT in order to the Vultures are placed the splendid birds which are so familiar to us under the general title of Eagles, and which form the first group of the great family Falconidæ, which includes the Eagles, Falcons, and Hawks.

The first and one of the finest of these grand birds is the well-known GOLDEN EAGLE. This magnificent bird is spread over a large portion of the world, being found in the British Islands, and in various parts of Europe, Asia, Africa, and America. The colour of this bird is a rich blackish brown on the greater part of the body, the head and neck being covered with feathers of a rich golden red, which have earned for the bird its popular name. The legs and sides of the thighs are grey-brown, and the tail is a deep grey, diversified with several regular dark-brown bars. In its immature state the plumage of the Golden Eagle is differently tinged, the whole of the feathers being reddish brown, the legs and sides of the thighs nearly white, and the tail white for the first three quarters of its length. So different an

GOLDEN EAGLE. —(*Aquila chrysaetos.*)

aspect does the immature bird present, that it has been often reckoned as a separate species, and named accordingly. It is a truly magnificent bird in point of size, for an adult female measures about three feet six inches in length, and the expanse of her wings is nine feet. The male is less by nearly six inches.

In England the Golden Eagle has long been extinct ; but it is still found in some plenty in the highlands of Scotland and Ireland, where it is observed to frequent certain favourite haunts, and to breed regularly in the same spot for a long series of years. The nest is always made upon some elevated spot, generally upon a ledge of rock, and is most inartistically constructed of sticks, which are thrown apparently at random, and rudely arranged for the purpose of containing the eggs and young. A neighbouring ledge of rock is generally reserved for a larder where the parent Eagles store up the food which they bring from the plains below.

In hunting for their prey, the Eagle and his mate mutually assist each other. It may here be mentioned that the Eagles are all monogamous, keeping themselves to a single mate, and living together in perfect harmony through their lives. As the rabbits and hares are generally under cover during the day, the Eagle is forced to drive them from their place of concealment, and manages the matter in a very clever and sportsmanlike manner. One of the Eagles conceals itself near the cover which is to be beaten, and its companion then dashes among the bushes, screaming and making such a disturbance that the terrified inmates rush out in hopes of escape, and are immediately pounced upon by the watchful confederate. The prey is immediately taken to the nest, and distributed to the young, if there should be any eaglets in the lofty cradle.

Owing to the expanse of the wings and the great power of the muscles, the flight of this bird is peculiarly bold, striking, and graceful. It sweeps through the air in a succession of spiral curves, rising with every spire, and making no perceptible motion with its wings, until it has attained an altitude at which it is hardly visible. From that post of vantage the Eagle marks the ground below, and swoops down with lightning rapidity upon bird or beast that may happen to take its fancy. It is not, however, so active at rising from the ground as might be imagined, and can be disabled by a comparatively slight injury on the wing. One of these birds, that was detected by a young shepherd boy in the act of devouring some dead sheep, was disabled by a pebble hurled at him from a sling and was at last ignominiously stoned to death.

The Eagle is supposed to be a very long-lived bird, and is thought to compass a century of existence when it is living wild and unrestrained in its native land. Even in captivity it has been known to attain a good old age, one of these birds which lived at Vienna being rather more than a hundred years old when it died.

ONE of the most interesting of the predaceous birds which belong to Great Britain is the celebrated OSPREY, or FISHING HAWK. This fine bird was formerly very common in England, but is now but rarely seen within the confines of the British Isles, although isolated species are now and then seen.

As the bird is a fish-eater, it is generally observed on the sea-coast or on the banks of some large river, but has occasionally been observed in some comparatively waterless situation, where it has probably been driven by stress of weather. In some parts of Scotland the Osprey still holds its own, and breeds year after year on the same spot, generally choosing the summit of an old ruined building or the top of a large tree for that purpose. The nest is a very large one, composed almost wholly of sticks, and contains two or three whitish eggs, largely blotched with reddish brown, the dark patches being collected towards the large end of the egg. As is the case with the Eagles, the Osprey is monogamous ; but on the death of either of the pair, the survivor soon finds another mate, and is straightway consoled by a new alliance. From all accounts it is an affectionate and domestic bird, paying the greatest attention to its mate and home, and displaying a constancy which is not to be surpassed by that of the turtle-dove, so celebrated for matrimonial felicity.

OSPREY.—(*Pandion haliaetus.*)

The flight of the Osprey is peculiarly easy and elegant, as might be expected from a bird the length of whose body is only twenty-two inches and the expanse of wing nearly five feet and a half. Living almost wholly on fish, the Osprey sails in wide undulating circles, hovering over the water and intently watching for its prey. No sooner does a fish come into view than the Osprey shoots through the air like a meteor, descends upon the

R

luckless fish with such force that it drives a shower of **spray in every**
direction, and soon emerging, flies away to its nest, bearing **its prey in its**
grasp.   In order to enable it to seize and retain so slippery **a creature as a**
fish, the claws of the Osprey are long, curved, and very sharp, the **soles of**

BALD, OR WHITE-HEADED EAGLE.—(*Haliaetus Leucocephalus.*)

the feet are rough, and the outer toe is capable of great versatility. **When**
the bird has settled upon its nest, or upon any spot where it intends to
eat its prey, it does not relinquish its hold, but, as if fearful that the fish
should escape, continues its grasp, and daintily picks away the flesh from
between its toes.

Harmless though the Osprey be—except to the fish—it is a much persecuted bird, being not only annoyed by rooks and crows, but robbed by
the more powerful white-headed Eagle. Mr. Thompson records an instance
where an Osprey, which had been fishing in Loch Ruthven, was greatly

BUZZARD.—(*Buteo vulgaris.*)

harassed by an impertinent Royston crow, which attacked the noble bird as
soon as it had caught a fish, and, as if knowing that it was incapable of
retaliation, actually struck it while on the wing. The Osprey kept quietly
on its way, but was so wearied by the repeated attacks of the crow, that

when pursued and pursuer had vanished out of sight, the poor Osprey had not been able to commence his repast.

The general colour of the Osprey is dark brown, but it is pleasingly variegated with various shades of black, grey, and white. The crown of

KITE.—(*Milvus regalis.*)

the head and the nape of the neck are covered with long grey-white feathers, streaked with dark brown. The under surface of the body is white, with the exception of a light-brown band which extends across the chest. The primaries are brown tipped with black, and the tail is barred above with a light and a deep brown, and below with brown and white. The legs, toes, and cere are blue, the eyes golden yellow, and the beak and claws black.

THE noble bird which is represented on page 242 is celebrated as being the type which has been chosen by the Americans as the emblem of their nation.

The name of BALD or WHITE-HEADED EAGLE has been applied to this bird on account of the snowy white colour of the head and neck, a peculiarity which renders it a most conspicuous bird when at large in its native land. The remainder of the body is a deep chocolate brown, inclining to black along the back. The tail and upper tail-coverts are of the same white hue as the head and neck. In its earlier stages of existence the creature is of more sombre tints not obtaining the beautifully white head and tail until it is four years of age.

The nest of the Bald Eagle is generally made upon some lofty tree, and in the course of years becomes of very great size, as the bird is in the habit of laying her eggs year after year in the same nest, and making additions of fresh building materials at every fresh breeding season. She commences this task at a very early period of the year, depositing her eggs in January, and hatching her young by the middle of February.

It is always a very affectionate bird, tends its young as long as they are helpless and unfledged, and will not forsake them even if the tree on which they rest be enveloped in flames.

The Bald Eagle often takes advantage of the fishing talents of the Osprey by robbing the lesser bird of its prey. The Eagle is, in truth, no very great fisher, but is very fond of fish, and finds that the easiest mode of obtaining the desired dainty is to rob those who are better qualified than himself for the sport.

The Bald Eagle is very accommodating in his appetite, and will eat almost anything that has ever possessed animal life. He is by no means averse to carrion, and has been seen seated regally upon a dead horse, keeping at a distance a horde of vultures which were collected round the carcase, and not permitting them to approach until he had gorged himself to the full. Another individual was seen by Wilson in a similar state of things. He had taken possession of a heap of dead squirrels that had been accidentally drowned, and prevented any other bird, or beast of prey, from approaching his treasure. He is especially fond of lambs, and is more than suspected of aiding the death of many a sickly sheep by the dexterous use of his beak and claws.

The Bald Eagle is found throughout the whole of North America, and may be seen haunting the greater part of the sea-coasts, as well as the mouths of the large rivers.

The common BUZZARD is one of our handsomest Falconidæ, and is one which, although banished from the greater part of England, is still found plentifully in many parts of Scotland and Ireland.

The plumage of this bird is looser and more downy than is seen in the generality of the hawk tribe, and bears a certain resemblance to that of the owl. This peculiarity is explained by the habits of the bird, which will presently be narrated. The average length of a Buzzard is from twenty to twenty-two inches, and the tinting of its plumage is extremely variable, even in adult birds. The usual colouring is as follows. The back and whole of the upper surface is a rich brown, becoming lighter on the head and neck, and diversified with longitudinal streaks of the darker hue. The tail is also dark brown, but is varied with stripes of a lighter colour, and the primary feathers of the wings are nearly black. The under portions of the body are grey-white, marked on the neck, chest, and abdomen with spots and streaks

of brown. The claws are black, the bill is a deep blue-black, and the legs,
toes, and ears are yellow.

The nest of the Buzzard is made either in some suitable tree or upon the
rocks, according to the locality, and is generally composed of grass and

SWALLOW-TAILED KITE.—(*Elanoides furcatus.*)

heather stems, intermingled with long, soft roots, and lined with wool,
heather, leaves, and other substances.

The flight of the Buzzard is rather variable. At times the bird seems
inspired with the very soul of laziness, and contents itself with pouncing

leisurely upon its prey, and returning to the branch on which it has been perched. Sometimes, however, and especially in the breeding season, it rises high in the air, and displays a power of wing and an easy grace of flight which would hardly be anticipated from its formerly sluggish movements. This fine bird may still be seen in the New Forest, where I have often watched its airy circling flight.

THE KITE may be known, even on the wing, from all other British birds of prey, by its beautifully easy flight and the long forked tail. Indeed, while flying, the Kite bears no small resemblance to a very large swallow, excepting that the flight is more gliding, and the wings are seldom flapped.

It was in former days one of the commonest of the British birds, swarming in every forest, building its nest near every village, and being the greatest pest of the farmer and poultry-keeper, on account of its voracity, craft, and swiftness. Even the Metropolis was filled with these birds, which acted the same part that is played by vultures in more eastern lands, and were accustomed to haunt the streets for the purpose of eating the offal which was so liberally flung out of doors in the good old times, and which, but for the providential instincts of the Kites, would have been permitted to decompose in the open streets.

In the present day, however, the Kite is comparatively seldom seen in England, and when observed, is of sufficient rarity to be mentioned in the floating records of natural history.

The flight of this bird is peculiarly easy and graceful, as the wings are seldom flapped, and the Kite sails through the air as by the mere power of volition. From the gliding movements of the Kite when on the wing it has derived the name of Gled, from the old Saxon word *glida*. When in pursuit of prey, the Kite sails in circles, at a considerable height from the ground, watching with its penetrating gaze the ground beneath, and sweeping with unerring aim upon any bird, quadruped, or reptile that may take its fancy.

The food of the Kite is rather general in its nature, consisting of various quadrupeds, young rabbits, hares, rats, mice, and moles, of which latter animals no less than twenty-two were discovered in the nest of a single Kite, showing how rapid and noiseless must be its movements when it can secure so wary and keen-eared an animal as a mole. It does not chase the swift-winged birds through the air, but pounces on many a partridge as it sits on the ground, and is remarkably fond of taking young and unfledged birds from their nests; reptiles of different kinds, such as snakes, frogs, lizards, and newts, also form part of its food, and it will not disdain to pick up a bee or a grasshopper when it can find no larger prey. The Kite is also a good fisher, waging nearly as successful war against the finny inhabitants of the rivers or ponds as the osprey itself; sweeping suddenly down upon the fish as they rise to the surface in search of food, or in their accustomed gambollings, and bearing them away to the shore, where it settles down and eats them in peace.

The nest of the Kite is chiefly built with sticks as a foundation, upon which is placed a layer of moss, wool, hair, and other soft and warm articles. The locality which is chosen for the nest is generally in some thick wood, and the bird prefers a strong, forked branch for the resting-place. The eggs are generally two in number, and sometimes three, of a greyish or light-brownish white colour, speckled with reddish chestnut blotches, which, as is the case with so many hawk's eggs, are gathered towards the larger end.

The ordinary length of the common Kite is about twenty inches, but the sexes are rather variable in that respect, the females being always larger than the males. The colouring of the bird is very elegant, although composed of few tints, and is remarkable more on account of the delicate gradations and contrasts of hue than for any peculiar brilliancy of the feathers. The general aspect of the Kite is reddish brown, which on a close inspection is resolved into the following tints. The back and upper portions are dark brown, relieved by a reddish tinge upon the edges of the feathers; the primaries are black, and the upper tail-coverts chestnut. There is a little white upon the edges of the tertiaries, and the head and back of the neck are covered with greyish white feathers, the centre of each feather being streaked with brown. The forked tail is reddish brown, barred on the under surface with dark brown stripes, the centre feathers being the darkest. The chin and throat are coloured like the head, and the abdomen and under portions are reddish brown. The under tail-coverts are white, with a slight reddish tinge, and the under surface of the rectrices are also white, but washed with grey.

THE beautiful bird which is so well known under the appropriate title of the SWALLOW-TAILED KITE is an inhabitant of various parts of America, though it has occasionally been noticed on the British shores.

This bird bears so strong an external resemblance to the swallow, that it might easily be taken for a common swallow or swift, as it flies circling in the air in search of the insect prey on which it usually feeds. Even the flight is very much of the same character in both birds, and the mode of feeding very similar. The usual food of the Swallow-tailed Kite consists of the larger insects, which it either catches on the wing or snatches from the leaves as it shoots past the bushes. Reptiles, such as small snakes, lizards, and frogs, also form part of the food of this elegant bird. While it is engaged in the pursuit of such prey, or in catching the large insects upon the branches, it may be approached and shot without much difficulty, as it is so intent upon its prey that it fails to notice its human foe.

The nest of the Swallow-tailed Kite is generally found on the very summit of some lofty rock or pine, and is almost invariably in the near vicinity of water. It is composed of small sticks externally, and is lined with grasses, moss, and feathers. The eggs are rather more numerous than is generally the case with the Hawks, being from four to six in number. Their colour is white with a greenish tinge, and they are marked with some dark brown blotches which are gathered towards the larger end. There is only one brood in the year, and when the young birds are first hatched, they are covered with a uniformly buff-coloured downy coat. The colour of the adult bird is variable, consisting mostly of white and black, but, on account of the bold manner in which their hues are contrasted, is remarkably pleasing in its effect. The back, the upper part of the wings, with the exception of the inner webs of the tertiaries, upper tail-coverts and rectrices, are a deep purple-black, the head, neck, and all other parts of the plumage being pure white. The legs and toes are blue with a green tinge, the cere is blue, and the beak blue-black. The claws are orange-brown. The length of this bird averages twenty inches.

THE true FALCONS are known by their strong, thick, and curved beak, the upper mandible having a projecting tooth near the curve, which fits into a corresponding socket in the under mandible. The talons are strongly curved, sharp-pointed, and are either flat or grooved in their under sides. The Falcons all obtain their prey by striking it while on the wing; and with such terrible force is the attack made, that a Peregrine Falcon has been known to strike the head completely from the shoulders of its quarry, while the mere force of its stroke, without the use of its claws, is sufficient to kill a pigeon or a partridge, and send it dead to the ground.

In striking their prey the Falcons make no use of the beak, reserving that weapon for the purpose of completing the slaughter when they and the wounded quarry are struggling on the ground.

Among the true Falcons the JER-FALCON is the most conspicuous on account of the superior dimensions of its body and the striking power of its wing.

This splendid bird is a native of Northern Europe, being mostly found in Iceland and Norway, and it also inhabits parts of both Americas. Some naturalists believe that the Norwegian and Icelandic birds ought to be reckoned as different species, but others think that any differences between them are occasioned by age and sex. It is said that of the two birds the Iceland variety is the more powerful, of bolder flight, and greater age, and therefore better adapted for the purpose of falconry. Sometimes it is seen in the northern parts of the British Islands, having evidently flown over the five hundred miles or so of sea that divides Scotland frcm Iceland; this journey, however, is no difficult task for the Jer-falcon, who is quite capable of paying a morning visit to these islands and returning to its home on the same day. In 1859 one of these birds was shot in Northumberland, and others have been observed in the more southern counties. Towards the south, however, it has seldom if ever been observed.

JER-FALCON.—(*Falco Gyrfalco.*)

The colour of the adult Jer-falcon is nearly white, being purely white on the under surface and flecked with narrow transverse bars of greyish brown upon the upper parts. The sharp claws are black, the beak of a bluish tint, and the cere, tarsus, and toes yellow. When young, however, the bird presents a very different aspect, and would hardly be recognized as belonging to the same species. In its earlier stages of life it is almost wholly of a greyish brown tint, the feathers being slightly marked with a little white upon their edges. As the bird increases in age the white edges become wider and by degrees the entire feather is of a snowy whiteness. The name Jer-falcon is supposed to be a corruption of " Geier-falcon," or Vulture Falcon.

LESS powerful, but more graceful than the Jer-falcon, the PEREGRINE FALCON has ever held the first place among the hawks that are trained for the chase.

When thoroughly tamed, the Peregrine Falcon displays a very considerable amount of attachment to its owner, and even while flying at perfect liberty will single him out from a large company, fly voluntarily towards him, and perch lovingly on his hand or shoulder.

It will chase and kill many of the coast birds, such as the dunlin, the gull, and the plover. The curlew is a very favourite prey, and, being a strong-winged bird, affords great sport. It is rather remarkable that the dunlin, together with birds of similar habits, fly instinctively to the sea, lake, or river, when attacked by the Peregrine Falcon, as if knowing that the winged hunter is very unwilling to swoop upon any object that is flying upon the surface of the water. The Falcon has been seen to drive a dunlin repeatedly into the sea before it could intercept the poor bird between the dry land and its watery refuge. It will also strike at the grey crows, or at herons, but unless specially trained to the pursuit, will not trouble itself further about them.

PEREGRINE FALCON. (*Falco peregrinus.*)

The full speed of the Peregrine Falcon has been computed at a hundred and fifty miles per hour, and a single chase will often occupy a space of eight or ten miles. Its power of wing is not only useful in enabling it to wage successful pursuit of swift-winged birds, but in giving it sufficient buoyancy to carry off the prey which it has secured. So strong is the

Peregrine's wing, that it has often been observed to bear in its talons a bird larger than itself, and to carry it to the nest without difficulty. Even a guillemot has been struck and carried off by the Peregrine.

The eggs of this bird are generally two or three in number, although a fourth is sometimes known to be laid in the same nest. The colour of the egg is a very pale reddish brown, usually mottled with a darker tint.

In its adult state, the Peregrine Falcon is very elegantly coloured. The top of the head, the back of the neck, the primaries, and a stripe beneath the eye, are of a deep black-brown ; the upper parts of the body are ashy brown, the latter tint becoming fainter in each successive moult, and being always marked with a series of dark bars upon its back, tail, and wing-coverts : the breast is white, deepening into a chestnut hue, and being barred transversely with reddish brown upon the breast, and marked on the front of the throat with longitudinal dashes of very dark brown. The remainder of the under plumage is greyish white, profusely barred with dark brown. When young the plumage is altogether of a more ruddy hue, and the birds are termed in the language of falconry, Red Tercels, or Red Falcons, according to their sex.

THE small but exquisitely shaped HOBBY is found spread over the greater part of the old world, specimens having been taken in Northern Africa, and in many portions of Asia, as well as in Europe, which seems to be its chief residence. It was formerly very common in England, but is year by year less seldom seen in our island, as is the case with all its predaceous relations. From all accounts, it seems to be rather a local bird, being partially influenced by the nature of the ground and the quantity of food which it is able to procure.

This bird appears to favour inland and well-wooded lands rather than the sea-shore or the barren rocks ; thus presenting a strong contrast to the Peregrine Falcon. We may find an obvious reason for this preference in the fact that a considerable proportion of its food is composed of the larger insects, especially of the fat-bodied beetles, which it seizes on the wing. Chafers of various kinds are a favourite prey with the Hobby, and in several cases the stomachs of Hobbies that had been shot were found to contain nothing but the shelly portions of the larger dung-chafer (*geotrupes sterco-rarius*). As therefore the common cock-chafer is a leaf-eating insect and frequents forest lands for the purpose of attaining its food, the Hobby will constantly be found in the same locality for the object of feeding on the cock-chafer. And as the dung-chafer swarms wherever cattle are most abundantly nourished, the Hobby is attracted to the same spot for the sake of the plentiful supply of food which it can obtain.

ALTHOUGH of the smallest of the British Falconidæ, being only from ten to thirteen inches in length, according to the sex of the individual, the MERLIN is one of the most dashing and brilliant of all the hawks which frequent our island.

This beautiful little bird is almost invaluable to the young falconer, as it is so docile in disposition and so remarkably intelligent in character, that it repays his instructions much sooner than any of the more showy but less teachable falcons. Every movement of this admirable little hawk is full of life and vivacity ; its head turns sharply from side to side as it sits on its master's hand ; its eyes almost flame with fiery eagerness, and it ever and anon gives vent to its impatience by a volley of ear-piercing shrieks.

Before the young bird is able to tear to pieces its winged prey, it should always be accustomed to have its food placed upon the stuffed skin of a partridge, and when it has attained sufficient strength, the breast of a real partridge should be cut open, and a small portion of its ordinary food placed within the aperture, so as to encourage the bird to tear away the flesh in order to satiate its hunger. The next step is to substitute an entire partridge for the ordinary diet, and by degrees to teach it to pounce upon the dead bird as it is flung to a daily increasing distance. It is a good

MERLIN.—(*Hypotriorchis æsalon.*)

pigeon-hunter, and if the owner choose to train it for smaller game, it is unrivalled as a chaser of thrushes, larks, and similar birds, owing to the pertinacity with which it carries on the pursuit, and the resolutely agile manner with which it will thread the mazes of branch and leaf in chase of a bird which seeks for refuge in the covert.

The Merlin frequently breeds in England, and makes its nest on the ground, generally choosing for that purpose some spot where large stones are tolerably plentiful, and may serve as a protection to the nest, as well as for a perch on which the Merlin, like the harrier, loves to sit and survey the prospect. From this habit of perching on pieces of stone it has derived the name of STONE FALCON, a title which has been applied to this bird in Germany and France as well as in England. Sometimes, but not often, the nest is made on some rocky shelf on a precipice. The eggs are four or five in number, of a light reddish brown hue, covered with mottlings and splashings of a deeper tint.

The colour of the Merlin is very pleasing, but not very easy to describe, as it is not so conspicuous as in many of the hawks, and moreover is rather different in the two sexes.

The top of the head is a slaty grey, marked with dark streaks running along the line of the head; the beak and upper portions of the body are of a similar slaty grey, but without the dark lines. The shafts of each feather are, however, of a dark brown, and give a very rich and peculiar colouring to those portions of the plumage. The pinions are black, the upper surface of the tail is nearly grey, with the exception of three faint dark bands, the last being the broadest, and the tip white. The chin and throat are white, and the under parts of the body are reddish fawn, thickly marked with patches of a darker colour and streaks of deep brown. The cere, legs, and toes are yellow, the claws black, and the beak a slaty grey, deepening towards the point, and slightly marked with longitudinal dark lines. Round the neck runs a band of pale reddish brown, which also extends to the cheeks, and there forms a patch on each side.

This description belongs to the male bird, the colouring of the female being of a rather different nature. The beautiful blue-grey which tints the upper parts of the male bird is in the female of a dark reddish brown, marked with slender longitudinal streaks covered by the black-brown shafts of each feather. The secondaries and the wing-coverts are of the same hue as the back. The tail is brown, varied with five narrow streaks of dark brown, and the under surface of the body is a very pale brown, marked with longitudinal dashes of a darker hue. The young of both sexes are nearly alike for the first year, after which time the males assume their peculiar colouring, and the females retain the same tints.

THE common KESTREL is one of the most familiar of the British hawks, being seen in almost every part of the country where a mouse, a lizard, or a beetle may be found.

It may be easily distinguished while on the wing from any other hawk, by the peculiar manner in which it remains poised in air in a single spot, its head invariably pointing towards the wind, its tail spread, and its wings widely extended, almost as if it were a toy kite raised in the air by artificial means, and preserved in the same spot by the trammels of a string. While hanging thus strangely suspended in the air, its head is bent downwards, and its keen eyes glance restlessly in every direction, watching every blade of grass beneath its ken, and shooting down with unerring certainty of aim upon any unhappy field-mouse that may be foolish enough to poke his red face out of his hole while the Kestrel is on the watch.

The number of field-mice consumed by this hawk is very great, for it is hardly possible to open the stomach of a Kestrel without finding the remains of one or more of these destructive little animals. On account of its mouse-eating propensities, the Kestrel is a most useful bird to the farmer, who in his ignorance confounds all hawks together, and now shoots the Kestrel which catches mice because kites used formerly to steal chickens.

In the use of its claws the Kestrel is remarkably quick **and** ready, and being also a swift-winged bird, it is in the habit of chasing cock-chaffers and other large beetles on the wing, and catching them neatly with its claws as it shoots past their course. Without pausing in its flight, the bird transfers

KESTREL.—(*Tinnunculus Alaudarius.*)

the insect from the foot to the mouth, and eats it without taking the trouble to alight. With such eagerness does it pursue this kind of prey, which we may suppose to be taken as a dessert after a more substantial meal upon mouse-flesh, that it continues its chase far into the evening, and may be seen in hot pursuit of the high-flying beetles long after dusk. Caterpillars and other larvæ are also eaten by the Kestrel, which does not disdain to alight on the ground and draw the earthworms out of their holes.

Mice, however, are always its favourite diet, and as the multiplication of these little pests is much increased by the abundant food which they find in cultivated grounds and stacks and barns, the Kestrel has learnt to attach itself to human residences, instead of becoming self-banished, as is the case with almost every other hawk. There is hardly a village where the Kestrel may not be seen hovering with outspread wings, and surveying the fields below.

With the aid of a good telescope, every movement of the bird may be discovered as it hangs in the air, and the sight is a very interesting one. Its wings keep up a continual shivering, its widely-spread tail is occasionally moved so as to suit the slight changes of the breeze, the spirited little head is in perpetual motion, and the dark brown eyes gleam with animation as they keep their restless watch. It seems from various observations that each Kestrel has its regular beat or hunting-grounds, and may be observed punctually repairing to the same spot at the same hour, much after the manner of the golden eagle.

The Kestrel is known by various names in different parts of the country. Its most common name is Windhover, in allusion to its peculiar mode of flight. For the same reason it is termed Stannel, Stand-gall, or Stand-gale, and has also obtained the title of Vanner Hawk.

The nest of the Kestrel is generally placed upon the topmost bough of some lofty tree, although it is sometimes found upon a ledge of some precipitous cliff, should the bird have taken up its residence in a mountainous country. Many of these birds have built their nests upon the rocky heights of Dovedale in Derbyshire, and may be seen hovering in mid-air near the spot where their young are nourished. The nest itself is a very simple construction of sticks and moss ; and the bird is so averse to trouble that it often takes possession of the deserted nest of the carrion crow. I have several times been greatly surprised in my nest-hunting expeditions, by finding the ruddy eggs of the Kestrel lying in the nest which I thought only to be that of the crow. This bird also deposits its eggs in the crannies of old ruined buildings and lofty towers, but I have never as yet been fortunate enough to find them in such a situation.

The colour of the male Kestrel is briefly as follows. The head, cheeks, and back of the neck are ashen grey, marked with narrow longitudinal streaks of deeper grey. The back and upper portions of the body, together with the tertiaries and wing-coverts, are bright ruddy fawn, dotted with little triangular black spots, caused by the extreme tips of the feathers being black. The larger quill feathers of the wing are black-grey, marked with a paler hue ; the under portions of the body are pale reddish fawn, marked with dark streaks on the chest and spotted on the abdomen ; the thighs and under tail-coverts are of the same hue as the abdomen, but without the spots. The upper surface of the tail is of the same hue as the head, marked with a single broad band of black near its extremity and tipped with white, while its under surface is grey-white, marked with a number of narrow irregular bars of a darker hue, in addition to the black band and white tip, which are the same as on the upper surface. The legs, toes, cere, and orbits of the eyes are yellow, the claws are black, and the beak is slaty blue, deepening towards the point.

The females and young males are differently marked, and are altogether of a darker and more ruddy hue. The head and neck are ruddy fawn, marked with many transverse darker stripes, and the back, upper portions, and tail are red-brown covered with numerous irregular blue-black bars. The males do not assume their appropriate plumage until they have completed their first year. The length of the male bird is about thirteen inches, and that of the female fifteen inches.

GOSHAWK.—(*Astur palumbarius.*)

We now come to a large and important genus of Hawks, which is represented in England by the GOSHAWK.

This handsome bird is even larger than the jer-falcon, the length of an adult male being eighteen inches, and that of his mate rather more than two feet. It is not, however, so powerful or so swift-winged a bird as the jer-falcon, and its mode of taking prey is entirely different.

When trained, the Goshawk is best employed at hares, rabbits, and other furred game, and in this particular sport is unrivalled. Its mode of hunting is singularly like that of the chetah, which has already been mentioned on page 50. Like that animal, it is not nearly so swift as its prey, and therefore is obliged to steal upon them, and seize its victim by a sudden and unexpected pounce. When it has once grasped its prey it is rarely found to loose its hold, even after the most violent struggles or the most furious attack.

This species is found spread over nearly the whole of Europe and Asia, and has also been seen in Northern Africa. The nest of this bird is generally placed on the topmost boughs of some lofty tree, and the eggs are of a uniform spotless blue-white. Their number is from three to four, and the young are hatched about May or the beginning of June.

In colour, the adult birds of both sexes are very similar to each other, the tinting of the plumage being briefly as follows. The top of the head and the entire upper portions of the body and wings are grey-brown, and the under portions of the body, together with a band over the cheeks and the back of the neck, are nearly white, diversified with numerous irregular spots, splashes, and partial bars of black. The cheeks and ear-coverts are dark greyish brown, the upper surface of the tail is the same hue as the back, and barred with dark brown ; the under tail-coverts are white. The cere, legs, and toes are yellow, the claws black, and the beak blue-black. In the female the grey-brown of the back is a more ruddy hue, and in the young the plumage is curiously diversified with reddish white, buff, and grey.

THE well-known SPARROW HAWK is almost as familiar to us as the kestrel, the two birds being, indeed, often confounded with each other by those who ought to know better. This fine and active little bird is an inhabitant of many portions of the world, being very common in nearly all parts of Europe, equally so in Egypt and Northern Africa, and being very frequently found in India and other Asiatic countries.

The courage of the Sparrow Hawk is of the most reckless character, for the bird will fly unhesitatingly at almost any other inhabitant of air, no matter what its size may be.

In consequence of the headlong courage possessed by this handsome little hawk, it is very valuable to the falconer if properly trained, for it will dash at any quarry which may be pointed out to it. Unfortunately, however, the Sparrow Hawk is one of the most difficult and refractory of pupils, being shy to a singular degree, slow at receiving a lesson and quick at forgetting it. Besides, its temper is of a very crabbed and uncertain nature, and it is so quarrelsome, that if several of these birds should be fastened to the same perch, or placed in the same cage, they will certainly fight each other, and, in all probability, the conqueror will eat his vanquished foe. Such an event has actually occurred ; the victrix—for it was a female— killing and devouring her intended spouse.

One of these birds afforded an excellent example of the shyness above mentioned. Although he was most kindly treated and liberally fed, he used to scream in the most ear-piercing manner when approached, even by the person who generally carried his food. The only companion whose presence he would tolerate was a little Skye terrier, named Rosy, and the two strangely matched comrades used to execute the most singular gambols together, the dog generally taking the initiative and persecuting the Hawk, until she forced him to fly.

The nest of the Sparrow Hawk is placed in some elevated spot, and contains three or four eggs, rather variable in their marking, but always

S

possessing a certain unmistakable character. The ground tint of the egg is a greyish white, slightly tinged with blue, and a number of bold blotches of a very dark brown are placed upon the surface, sometimes scattered rather irregularly, but generally forming a broad ring round the larger end. The bird seldom troubles itself to build a new nest, but takes possession of the deserted tenement of a crow or rook.

SPARROW HAWK.—(*Accipiter Nisus.*)

THE very remarkable SECRETARY BIRD derives its name from the curious feathery plumes which project from each side of its head, and bear a fanciful resemblance to pens carried behind the ear by human secretaries.

It is an inhabitant of Southern Africa, and is most valuable in destroying the serpent race, on which creatures it most exclusively feeds. Undaunted by the deadly teeth of the cobra, the Secretary Bird comes boldly to the attack, and in spite of all the efforts of the infuriated and desperate reptile, is sure to come off victorious. Many other creatures fall victims to the ravenous appetite of the Secretary, and in the stomach of one of these birds which was found by Le Vaillant, were discovered eleven rather large lizards, eleven small tortoises, a great number of insects nearly entire, and three snakes as thick as a man's arm.

The nest of the Secretary is built on the summit of a lofty tree, and contains two or three large white eggs.

The ordinary length of the adult Secretary Bird is about three feet, and its colour is almost wholly a slaty grey. The peculiar feathers which form the crest are black, as are the primaries and the feathers of the thigh. There is a lighter patch towards the abdomen. The tail is black with the exception of the two central rectrices, which are grey, with a white tip and a broad black bar towards their extremities.

We now arrive at the HARRIERS, probably so called because they "harry" and persecute the game. Several species of this genus are found in England, the most common of which is the HEN HARRIER.

The Harrier may be readily distinguished from the other hawks by the manner in which the feathers radiate around the eyes, forming a kind of funnel-shaped depression, somewhat similar to but not so perfect as that of the owl. The flight of the Harrier is very low, seldom being more than a few yards above the ground, and as the bird flies along it beats every bush and pries

THE SECRETARY BIRD.—(*Serpentarius Secretarius.*)

into every little covert in search of prey. There are few of the smaller animals that do not fall victims to the Hen Harrier, which is always ready to pick up a field-mouse, a lizard, a small snake, a newt, or a bird, and will even pounce upon so large a bird as a partridge or a pheasant. Sometimes it sits on a stone or small hillock, and from that post keeps up a vigilant watch on the surrounding country, swooping off as soon as it observes indications of any creature on which it may feed.

The flight of the Hen Harrier, although it is not remarkable for its power, is yet very swift, easy, and gliding, and, as the bird quarters the ground after its prey, is remarkably graceful. The Harriers prefer to live on moors and similiar localities, where they can pursue their rather peculiar mode of hunting, and where they may find a secluded spot for a secure home. Like the kestrel, the Hen Harrier appears to have regular hunting-grounds, and is very punctual in its visits. The nest of this bird is generally placed under the shadow of some convenient furze-bush, and is composed of a few sticks thrown loosely together, in which are deposited four or five very pale blue eggs.   The young are hatched about the middle of June.

The two sexes differ very greatly in colour, and until comparatively recent times were recorded as distinct species.   The general colour of the adult

HEN HARRIER.—(*Circus cyaneus.*)

male is ashen grey from the beak and upper parts, the only exception being the primaries, which are black.   The throat and chin are nearly of the same hue as the beak, but the chest and abdomen are white, with a slight blue tinge, which is lost upon the plumage of the thigh.   On the under surface of the tail are several indistinct dark bars, and the hair-like feathers between the eye and the base of the beak are black. The legs, toes, and cere are yellow, the claws black, and the beak nearly black, with a bluish tinge. The length of the male bird is about eighteen inches.

The female is a much darker bird, the back and upper portions being of a deep dusky brown, and the primaries being but a little darker than the plumage of the back. The feathers of the under parts are lighter brown, with pale margins, so as to present a kind of mottled buff and chestnut aspect ; the upper surface of the tail is marked with partial dark bands, and its under surface is very distinctly bound with broad bands of black and greyish white. The funnel-shaped depression round the eyes, technically called the concha, or shell, is brown towards the base of the feathers, but merges into a white eyebrow above, reaching to the cere. Her length is about two inches more than that of the male, and her spread of wing is about three feet six inches.

OWLS.

THERE are few groups of birds which are so decidedly marked as the OWLS, and so easy of recognition. The round, puffy head, the little hooked beak just appearing from the downy plumage with which it is surrounded, the large, soft, blinking eyes, and the curious disk of feathers which radiate from the eye, and form a funnel-shaped depression, are such characteristic distinc-tions, that an Owl, even of the least owl-like aspect, can at once be detected and referred to its proper place in the animal kingdom.

These birds are, almost without an exception, noc-turnal in their habits, and are fitted for their peculiar life by a most wonderfully adapted form and structure. The eyes are made so as to take in every ray of light, and are so sensitive to its in-fluence that they are unable to endure the glare of day-light, being formed expressly for the dim light of evening or earliest dawn. An ordinary Owl of almost any species, when brought into the full light of day, becomes quite bewildered with the unwont-ed glare, and sits blinking uncomfortably, in a pitiable manner.

THE SNOWY OWL.—(*Nyctea nivea.*)

The SNOWY OWL is one of the handsomest of this group, not so much on account of its dimen-sions, which are not very considerable, but by reason of the beautiful white mantle with which it is clothed, and the large orange eyeballs.

This bird is properly a native of the north of Europe and America, but has also a few domains in the more northern parts of England, being constantly seen, though rather a scarce bird, in the Shetland and Orkney

Islands, where it builds and rears its young.   Like the Hawk Owl, it is a day-flying bird, and is a terrible foe to the smaller mammalia and to various birds.

In proportion to its size the Snowy Owl is a mighty hunter, having been detected in chasing the American hare, and carrying of wounded grouse before the sportsman can secure his prey.   According to Yarrell, the Swedish name of *Harfang*, which has been given to this bird, is derived from its habit of feeding on hares.   It is also a good fisherman, posting itself on some convenient spot overhanging the water, and securing its finny prey with a lightning-like grasp of the claw.   Sometimes it will sail over the surface of the stream, and snatch the fish as they rise for food, but its general mode of angling is that which has just been mentioned.   It is also a great eater of lemmings ; and in the destruction of these quadrupedal pests does infinite service to the agriculturist and the population in general.

COQUIMBO, OR BURROWING OWL.—(*Athene cunicularia.*)

The colour of an old snowy Owl is pure white without any markings whatever ; but in the earlier years of its life its plumage is covered with numerous dark brown spots and bars caused by a dark tip to each feather.   Upon the breast and abdomen these markings form short abrupt curves, but on the back and upper surface they are nearly straight.   The beak and claws are black.   The length of the male Snowy Owl is about twenty-two inches, and that of the female twenty-six or twenty-seven.

The quaint, long-legged little Owl which is represented in the accompanying illustration is a native of many parts of America, where it inhabits the same locality with the prairie dog.   The description of that curious marmot and its peculiar burrow may be found on pages 157, 158.

The prairie dogs and Burrowing Owls live together very harmoniously ;

and this strange society is said also to be augmented by a third member, namely, the rattlesnake. It is now, however, ascertained with tolerable accuracy that the rattlesnake is nothing but a very unwelcome intruder upon the marmot, and, as has been shown by the Hon. G. F. Berkeley's experiments, is liable to be attacked and destroyed by the legal owner of the burrow. If all had their rights, it would seem that the Owl is nearly as much an intruder as the snake, and that it only takes possession of the burrow excavated by the prairie dog in order to save itself the trouble of making a subterranean abode for itself. Indeed, there are some parts of the country where the Owl is perforce obliged to be its own workman, and, in default of convenient "dog" burrows, is fain to employ its claws and bill in excavating a home for itself.

The tunnel which is made by the Owl is not nearly so deep or so neatly constructed as that which is dug by the marmot, being only eighteen inches or two feet in depth, and very rough in the interior. At the bottom of this burrow is placed a tolerably seized heap of dried grass, moss, leaves, and other soft substances, upon which are deposited its white-shelled eggs.

The Coquimbo Owl is by no means a nocturnal bird, facing the glare of the mid-day sun without inconvenience, and standing at all times in the day or evening on the little heaps of earth which are thrown up at the entrance of the burrow. It is a lively little bird, moving about among the burrows with considerable vivacity, rising on the wing if suddenly disturbed, and making a short undulating aërial journey before it again settles upon the ground. When it has alighted from one of these little flights it turns round and earnestly regards the pursuer. Sometimes it will dive into one of the burrows, heedless of prior occupants, and thus it is that marmot, owl, and snake come to be found in the same burrow.

VIRGINIAN EARED-OWL.—(*Bubo Virginianus.*)

The colour of the Burrowing Owl is a rather rich brown upon the upper parts of the body, diversified with a number of small grey-white spots, and altogether darker upon the upper surface of the wings. The under parts are greyish white. The length of the bird is not quite eleven inches. The cry

of this curious bird is unlike that of any other Owl, and bears a very great resemblance to the short, sharp bark of the prairie dog.

We now arrive at a large group of Owls which are remarkable for two tufts of feathers which rise from the head, and occupy nearly the same relative position as the ears of quadrupeds. These "ears," as they are called, have, however nothing to do with the organs of hearing, but are simply tufts of feathers, which can be raised or depressed at the will of the bird, and give a most singular expression to the countenance.

THE VIRGINIAN-EARED OWL is found spread over the greater portion of North America, and in former days did great damage among the poultry of the agriculturists, being a bold as well as a voracious bird. Now, however, the ever-ready rifle of the farmer has thinned its numbers greatly, and has inspired the survivors with such awe that they mostly keep clear of cultivated lands, and confine themselves to seeking after their legitimate prey.

BROWN OWL.—(*Surnium Aluco.*)

It is a terrible destroyer of game, snatching up grouse, partridges, hares, ducks, sparrows, squirrels, and many other furred and feathered creatures, and not unfrequently striving after larger quarry. The wild turkey is a favourite article of diet with this Owl; but on account of the extreme wariness of the turkey nature, the depredator finds an unseen approach to be no easy matter. The usual mode in which the Owl catches the turkey is, to find out a spot where its intended prey is quietly sleeping at night, and then to swoop down suddenly upon the slumbering bird before it awakes. Some-

times, however, the Owl is baffled in a very curious manner. When the turkey happens to be roused by the rush of the winged foe, it instinctively ducks its head and spreads its tail flatly over its back. The Owl, impinging upon the slippery plane of stiff tail-feathers, finds no hold for its claws, and glides off the back of its intended victim, which immediately dives into the brushwood before the Owl can recover from the surprise of its unexpected failure.

The flight of this bird is remarkably powerful, easy, and graceful, as may be gathered from the enormous expanse of wing in comparison with the weight and dimensions of the body. Its voice is of a hollow and weird-like character, and when heard by night from some spot on which the Owl has silently settled, is apt to cause many a manly cheek to pale. As Wilson well observes, the loud and sudden cry of "Waugh O! Waugh O!" is sufficient to alarm a whole garrison of soldiers. Probably on account of the peculiar sounds which are uttered by this bird, the Cṛee Indians know it by the name of *Ottowuck-oho !*

The Virginian Horned Owl takes up its residence in the deep swampy forests, where it remains hidden during the day, and comes out at night and morning, heralding its approach with its loud unearthly cries, as of an unquiet wandering spirit. Sometimes, according to Wilson, "he has other nocturnal solos, one of which very strikingly resembles the half-suppressed screams of a person suffocating or throttled."

THE common BROWN OWL, or TAWNY OWL, as it is often named, is, with the exception of the Barn Owl, one of the best known of the British Owls.

Although rather a small bird, being only about fifteen inches in total length it is possessed of a powerful pounce and an audacious spirit, and, when roused to anger or urged by despair, is a remarkably unchancy antagonist.

The food of this Owl is of a very varied nature, consisting of all the smaller mammalia, many reptiles, some birds, fishes when it can get them, and insects. It seems to be a good fisherman, and catches its finny prey by waiting on the stones that project a little above the water, and adroitly snatching the fish from the stream by a rapid movement of the foot. Sometimes it flies at much higher game, especially when it has a young family to maintain, and will then attack birds and quadrupeds of very great size when compared with its own dimensions. In a single nest of this bird have been found according to a writer in the *Field*, three young Owls, five leverets, four young rabbits, three thrushes, and one trout weighing nearly half a pound.

The voice of the Brown Owl is a loud monotonous hoot, that may be often heard in the evening in localities where the bird has made its home.

The nest is usually placed in the hollow of a tree, and contains several white eggs. The colour of the Brown or Tawny Owl is an ashen grey upon the upper parts of the body, variegated with choco-

SCOPS (Gr. Σκώψ, an Owl), *the Scops Eared-Owl.*

late and wood-brown. Several whitish grey bars are seen upon the primaries, and there are several rows of whitish spots upon the wings and scapularies. The facial disc is nearly white, edged with brown, and the under surface of the body is of the same hue, covered with longitudinal mottlings of variously tinted brown. The claws are nearly white at their base, darkening towards their extremities, and the beak is nearly of the same colour. The eyes are of a very dark black-blue.

This species is found in many parts of Europe, and is said to be one of the indigenous birds of Japan.

We now come to an example of the British Owls, a bird that has attracted great notice on account of its singular aspect.

The SCOPS EARED-OWL has been once or twice found in Yorkshire, but usually resides in the southern parts of the Continent. It is remarkable for the regularity with which it utters its monotonous cry, as if a person were constantly repeating the letter Q at regular intervals of two seconds. It does not seem to prey upon mice and other animals, like most of its relations, but feeds on large insects, such as beetles and grasshoppers. The size of this owl is very small, as it only.measures seven inches in length ; the third primary feather is the longest. It lays from two to four white eggs in a simple nest made in a hollow tree or in a cleft in the rock

THE best known of the British Owls is the WHITE, BARN, or SCREECH OWL, by either of which appellations the bird is familiarly known over the whole of England.

This delicately-coloured and soft-plumed bird is always found near human habitations, and is generally in the vicinity of farmyards, where it loves to dwell, not for the sake of devouring the young poultry, but of eating the various mice which make such havoc in the ricks, fields, and barns. The "feathered cat," as this bird has happily been termed, is a terrible foe to mice, especially to the common field-mouse, great numbers of which are killed daily by a single pair of Owls when they are bringing up their young family. In the evening dusk, when the mice begin to stir abroad in search of a mole, the Owl starts in search of the mice, and with noiseless flight quarters the ground in a sportsmanlike and systematic manner, watching with

WHITE, OR BARN OWL.—(*Strix flammea.*)

its great round eyes every movement of a grass-blade, and catching with its sensitive ears every sound that issues from behind. Never a field-mouse can come within ken of the bird's eye, or make the least rustling among the leaves within hearing of the Owl's ear, that is not detected and captured. The claws are the instruments by which the Owl seizes its victim, and it does not employ the beak until it desires to devour the prey.

This bird is easily tamed when taken young, and is a very amusing pet. If properly treated, and fed with appropriate diet, it will live for a consider-able time without requiring very close attendance. Even if it be set at liberty, and its wings permitted to reach their full growth, it will voluntarily remain with its owner, whom it recognizes with evident pleasure, evincing its dislike of strangers by a sharp hiss and an impatient snap of the bill.

The nest of this species is placed either in a hollow tree or in a crevice of some old building, where it deposits its white, rough-surfaced eggs upon a soft layer of dried "castings." These nests have a most ill-conditioned and penetrating odour, which taints the hand which is introduced, and cannot be removed without considerable care and several lavations. The young are curious little puffs of white down, and the Barn Owl is so prolific, that it has been known to be sitting on one brood of eggs while it is feeding the young of a previous hatching.

As may be supposed from its popular title of White Owl, this species is very light in its colouring. The general colour of this bird is buff of different tints, with grey, white, and black variegations. The head and neck are light buff, speckled slightly with black and white spots, and the back and wings are of a deeper buff, spotted with grey, black, and white. The tail is also buff, with several broad bars of grey. The facial disc is nearly white, becoming rusty brown towards the eye, and a deeper brown round the edge.

The under surface of the male bird is beautifully white, the claws are brown, the beak nearly white, and the eyes blue-black. The sexes are very similar in their colouring, but the females and young males may be distinguished by the under surface of the body, which is fawn instead of white.

GOAT-SUCKERS.

WITH the owls closes the history of these birds which are called predaceous, although to a considerable extent nearly all birds are somewhat predaceous, even if they prey upon smaller victims than do the vultures, eagles, falcons, or owls. Next to the predaceous birds come the Passeres, distinguished by their cereless and pointed beak, their legs feathered as far as the heel, their tarsus covered in front with shield-like scales, and their sligthly-curved and sharply-pointed claws. This order is a very large one, and embraces a vast variety of birds.

First among the Passerine birds are placed the Fissirostres, or cleft-beaked birds, so called from the enormous gape of the mouth, a structure which is intended to aid them in the capture of the agile prey on which they feed.

The GOAT-SUCKERS, as they are familiarly termed, from a stupid notion that was formerly in great vogue among farmers, and is not even yet quite extinct, that these birds were in the habit of sucking the wild goats, cows, and sheep, are placed first among the Fissirostres on account of the wonderfully perfect manner in which their structure is adapted to the chasing and securing of the swift-winged insects on which they feed. The colour of all these birds is sombre ; black, brown, and grey being the prevailing tints. The gape of the mouth is so large that when the bird opens its beak to its fullest extent, it seems to have been severely wounded across the mouth, and the plumage is lax and soft like that of the owl.

There are many well-known proverbs relating to the power of calumny, and the readiness with which an evil report is received and retained, notwithstanding that it has been repeatedly proved to be false and libellous. The common GOAT-SUCKER is a good instance of the truth of this remark, for it was called *Aigothéles,* or Goat-sucker, by Aristotle in the days of old, and has been religiously supposed to have sucked goats ever afterwards. The Latin word *caprimulgus* bears the same signification. It was even supposed that after the bird had succeeded in sucking some unfortunate goat, the fount of nature was immediately dried up, and the poor beast also lost its sight. Starting from this report, all kinds of strange rumours flew about the world, and the poor Goat-sucker, or NIGHTJAR, as it ought more rightly to be called, has been invariably hated as a bird of ill omen to man and beast.

As usual, mankind reviles its best benefactors, for there are very few creatures which do such service to mankind as the Nightjar. Arriving in this country in the month of May or June, it reaches our shores just in time to catch the cockchafers, as they fly about during the night in search of their food, and does not leave us until it has done its best to eat every chafer that comes across its path.

The Nightjar also feeds on moths of various kinds, and catches them by sweeping quickly and silently among the branches of the trees near which the moth tribes most love to congregate. While engaged in their sport, they will occasionally settle on a bank, a wall, a post, or other convenient perch, crouch downward until they bring their head almost on a level with their feet, and utter the peculiar churning note which has earned for them the name of Churn-Owls, Jar-Owls, and Spinners. Their cry has been rather well compared to that sound which is produced by the larger beetles of the night, but of course much louder, and with the addition of the characteristic " chur-r-r !—chur-r-r ! " Sometimes, although but seldom, the Nightjar utters its cry while on the wing. When it settles, it always seats itself along a branch, and almost invariably with its head pointing towards the trunk of the tree.

There is also a strange squeaking sound which is emitted by the Nightjar while playing round the trees at night, and which is supposed to be a cry of playfulness or a call to its mate.

Unlike the Falconidæ, the Goat-sucker catches its prey, not with its claws, but with its mouth, and is aided in retaining them in that very wide receptacle, by the glutinous secretion with which it is lined, and the " vibrissæ " or hair-like feathers which surround its margin. On an examination of the foot of this bird, the claw of the middle toe is seen to be serrated like the teeth of a comb, a structure which has never yet been satisfactorily explained, notwith-

EUROPEAN GOATSUCKER.—(*Caprimulgus Europæus.*)

standing the various theories which have been put forward concerning its use. The hind toe of each foot is very mobile, and can be brought round to the remaining toes, so that all the claws take their hold in the same direction. Apparently, this structure is intended to enable the bird to run along the branches of trees, in its nocturnal chase after beetles and other insects.

The Nightjar makes no nest, but choosing some sheltered hollow under the shade of a grass tuft, a bunch of fern, bramble, or other defence, there lays two eggs on the bare ground. The colour of the egg is greyish white, plentifully mottled with pale buff and grey. The young are very similar to those of the cuckoo. The plumage of the nightjar is very rich in its colouring, the tints of buff, grey, black, white, brown, and chestnut being arranged in pleasing but most intricate patterns, and easier to be understood from a pencil illustration than a description of the pen.

SWALLOWS.

THE close-set plumage of the SWALLOW tribe, their long sickle-like wings, their stiff, firm tail, forked in most of the species, and their slight legs and toes, are characteristics which mark them out as birds which spend the greater part of their existence in the air, and exercise their wings far more than their feet.

They all feed upon insects, and capture their prey in the air, ascending at one time to such a height that they are hardly perceptible to the naked eye, and look merely like tiny dots moving upon the sky ; while at other seasons they skim the earth and play for hours together over the surface of the water, in chase of the gnats that emerge in myriads from the streams. The gape of the mouth is therefore exceedingly great in these birds, reaching as far as a point below the eyes. The bill itself is very short, flattened, pointed, slightly curved downwards, and broad at the base.

SWIFT.—(*Cypselus apus.*)

The group which is scientifically termed the Hirundinidæ is a very large one, and is divided into two lesser groups, the members of one being classed together under the title of Swifts, while the others are known by the name of Swallows. With the former birds we have first to deal.

The SWIFTS are readily distinguished from the Swallows by the very great comparative length of the two first primary feathers of the wing, which are either equal to each other, or have the second feather longer than the first. The secondaries are remarkably small, being nearly concealed under the coverts. There are ten primaries in the wing, and the same number of quill feathers in the tail.

The true SWIFTS, of which England affords two examples, one very familiarly known, and the other a very rare and almost unnoticed species, are remarkable for the feathered tarsus, the long wings, and the peculiar form of the feet. In this member, all the toes are directed forward, a structure which is admirably adapted to the purpose which it fulfils. The Swifts build their nests, or rather lay their eggs, for the nest is hardly worthy of the name, in holes under the eaves of houses, or in similar localities, and would find themselves greatly inconvenienced when seeking admission into their domiciles, but for the shape of the feet, which enables them to cling to the slightest projection, and to clamber up a perpendicular surface with perfect ease and safety.

DEVOID of all pretensions to the brilliantly-tinted plumage which decorates so many of its relations, and clad only in sober black and grey, the common SWIFT is, nevertheless, one of the most pleasing and interesting of the British birds ; resting its claims to favourable notice upon its graceful form and its unrivalled powers of wing.

There are very few birds which are so essentially inhabitants of the air as our common Swift, which cuts the atmosphere with its sabre-like wings with such marvellous ease and rapidity that at times its form is hardly discernible as it shoots along, and it leaves the impression of a dark black streak upon the eyes of the observer. The plumage of this bird is constructed especially with a view to securing great speed, as may be seen by an inspection of the closely-set and firmly-webbed feathers with which the entire body and limbs are clad. The muscles which move the wings are enormously developed, and in consequence the breast-bone is furnished with a remarkably strong and deep " keel."

The flight of the Swift is quite peculiar to the bird, and cannot be mistaken even for that of the swallow by any one who has a practical acquaintance with the habits of the two species. The Swift does not flap its wings so often as the swallow, and has a curious mode of shooting through the air as if hurled from some invisible bow, and guiding itself in its headlong course by means of its wings and tail.

This indefatigable bird is an early riser, and very late in returning to rest, later indeed than any of the diurnal birds. Though engaged in flight during the live-long day, the Swift appears to be proof against fatigue, and will, during the long summer days, remain upon the wing until after nine in the evening. As the days become shorter, the Swift is found to retire earlier, but during its st· y in this country it is almost invariably later than other birds, sometimes being on the wing together with the owl. Indeed, the air seems to the Swift even a more familiar element than the earth, and the bird is able to pass the whole of its life, and to perform all the bodily functions except those of sleep, while upborne on the untiring pinions with which it is furnished. The Swift that has a nest to take care of is forced to descend at intervals for the purpose of supplying its family with food, but except when urged by such considerations, it is able to remain in the air for many successive hours without needing to rest.

The Swifts may generally be found near buildings, rocks, and cliffs, for in such localities they build their nest, and from their home they seldom wander to any great distance, as long as they remain in the country.

In general, the Swift loves to build its nest in a hole under a roof, whether slated, tiled, or thatched, preferring, however, the warm, thick straw-thatch to the tile or slate. Sometimes it makes a hole in the thatch, through which it gains access to the nest, but in most instances it makes use of some already existing crevice for that purpose. In all cases the nest is placed above the entrance, and generally may be found about eighteen inches or two feet from the orifice. Even by the touch, the eggs of the Swift may be discerned from those of any other bird, as their length is singularly disproportionate to their width.

The sound which these birds utter is of the most piercing description, and can be heard at a very great distance, thus betraying them when they are hawking after the high-flying insects at such an altitude that their forms are hardly perceptible to the unassisted eye.

The nest is a very firmly made but yet rude and inartificial structure. The materials of which it is made are generally straw, hay, and feathers, pieces of rag, or any soft and warm substance which the bird may find in its rambles, and, when woven into a kind of nest, are firmly cemented together with a kind of glutinous substance secreted by certain glands. In Norway and Sweden the Swift builds in hollow trees. The eggs are from two to five in number, not often, however, exceeding three, and in colour they are pure white. In this country the Swift pays but a very short visit, as the bird evidently requires a very high temperature, and is forced to depart as soon as the weather becomes chilly. Generally the Swifts leave England by the end of August, but there are often instances where a solitary bird has delayed its voyages for some good reason.

AMONG the many "travellers' tales" which called forth such repudiation and ridicule from the sceptical readers of the earlier voyagers, the accounts of the Chinese cuisine were held to be amongst the most extravagant.

That civilized beings should condescend to eat dogs and rats specially fattened for the table was an idea from which their

ESCULENT SWALLOW.—(*Collocalia nidifica.*)

own better sense revolted, that the same nation should reckon sharks' fins and sea-slugs among their delicacies was clearly an invention of the writer; but that the Chinese should make soup out of birds' nests was an absurdity so self-evident that it destroyed all possibility of faith in the writers' previous assertions.

The birds that make these remarkable nests belong to several species, four of which have been acknowledged. There are the ESCULENT SWALLOW, the Linchi (*Collocalia fuciphaga*), the White-backed Swallow (*Collocalia troglodytes*), and the Grey-backed Swallow (*Collocalia Francica*).

These nests could hardly be recognised as specimens of bird architecture by any one who had not previously seen them, as they look much more like a set of sponges, corals, or fungi, than nests of birds. They are most irregular in shape, are adherent to each other, and are so rudely made that the hollow in which the eggs and young are intended to live is barely perceptible. They are always placed against the face of a perpendicular rock, generally upon the side of one of the tremendous caverns in Java and other places where these strange birds love to dwell. The men who procure the nests are lowered by ropes from above, and their occupation is always considered as perilous in the extreme.

The nests are of very different value, those which have been used in rearing a brood of young being comparatively low in price, while those which are quite new and nearly white are held in such esteem that they are worth their weight in silver.

In the British Museum may be seen a very fine specimen of the nest of the Esculent Swallow, comprehending a mass of the nests still adhering to the rock. It is rather remarkable that the birds have a habit of building these curious nests in horizontal layers.

The Esculent Swallow is a small bird, and its colour is brown on the upper parts of the body, and white beneath. The extremity of the tail is greyish white. The British Museum possesses specimens of all the Swallows which are known to make these curious edible nests.

THE elegantly-shaped and beautifully-coloured SWALLOW may be readily distinguished from any of its British relations by the very great elongation of the feathers which edge its tale, and which form nearly two-thirds of the bird's entire length.

SWALLOW.—(*Hirundo rustica.*)

It is the most familiar of all the Hirundinidæ of England, and from its great familiarity with man, and the trustfulness with which it fixes its domicile under the shelter of human habitations, is generally held as an almost sacred bird, in common with the robin and the wren.

The Swallow wages a never-ceasing war against many species of insects, and seems to be as capricious in its feeding as are the roach and other river fish.

The nest of the Swallow is always placed in some locality where it is effectually sheltered from wind and rain. Generally it is constructed under the eaves of houses, but as it is frequently built within disused chimneys, it has given to the species the popular title of Chimney Swallow. The bird is probably attracted to the chimney by the warmth of some neighbour fire.

The nest is composed externally of mud or clay, which is brought by the bird in small lumps and stuck in irregular rows so as to build up the sides of its little edifice. There is an attempt at smoothing the surface of the nest, but each lump of clay is easily distinguishable upon the spot where it has been stuck. While engaged at the commencement of its labours, the Swallow clings perpendicularly to the wall of the house or chimney, clinging with its sharp little claws to any small projection, and sticking itself by the pressure of its tail against the wall. The interior of the nest is lined with grass and other soft substances.

There are sometimes two broods in the year, and when the second brood has been hatched at a very late period of the year, the young are frequently deserted and left to starve by their parents, who are unable to resist the innate impulse that urges them to seek a warmer climate. When fully fledged, and before they are forced to migrate, the young birds generally roost for the night in osiers and other water-loving trees.

Except in confinement, the Swallow knows not the existence of frost, nor the extreme of heat, passing from Europe to Africa as soon as the cold weather begins to draw in, and migrating again to the cooler climes as soon as the temperature of its second home becomes inconvenient to its sensitive

FAIRY MARTIN.—(*Hirundo Ariel.*)

existence. The time of its arrival in England is various, and depends almost entirely on the state of the weather. Solitary individuals are now and then seen in very early months, but, as a general fact, the Swallow does not arrive until the second week in April ; the time of its departure is generally about the middle of September, although some few lingerers remain in the country for more than a month after the departure of their fellows.

Guided by some wondrous instinct, the Swallow always finds its way back to the nest which it had made, or in which it had been reared, as has frequently been proved by affixing certain marks to individual birds, and watching for their return. Sometimes it happens that the house on which they had built has been taken down during their absence, and in that case the distress of the poor birds is quite pitiable. They fly to and fro over the

T

spot in vain search after their lost homes, and fill the air with the mournful cries that tell of their sorrows.

The Swallow is widely spread over various parts of the world, being familiarly known throughout the whole of Europe, not excepting Norway, Sweden, and the northern portions of the Continent. It is also seen in Western Airica, and Mr. Yarrell mentions an instance where it was observed in the Island of St. Thomas, which is situated upon the equator. The martin and the swift were seen at the same place.

The colour of the Swallow is very beautiful. Upon the forehead the feathers are of a light chestnut, which gives place to deep glossy steel-blue upon the upper portions of the body and wings. The primaries and secondaries are black, as are the tail feathers, with the exception of a few white patches. The throat is chestnut, and a very dark-blue band crosses the upper part of the chest. The under parts are white, and the beak, legs, and toes black. The female is distinguished by the smaller chestnut on the

SAND MARTIN.—(*Cotile riparia.*)

forehead, the smaller tint of the feathers, and the narrowness of the dark band across the chest.

MANY examples of white Swallows are on record, and specimens may be seen in almost every collection of British birds.

Among the most ingenious of bird architects, the FAIRY MARTIN holds a very high place in virtue of the singular nest which it constructs.

The nest of the Fairy Martin has a very close resemblance to a common oil-flask, and reminds the observer of the flask-shaped nests which are constructed by the Pensile Oriole and similar birds, although made of harder materials. The Fairy Martin builds its curious house of mud and clay, which it kneads thoroughly in its beak before bringing it to the spot where it will be required. Six or seven birds work amicably at each nest, one remaining in the interior enacting the part of chief architect, while others act as hodsmen, and bring material as fast as it is required. Except upon wet days, this bird only works in the evening and early morning, as the heat of midday seems to dry the mud so rapidly that it cannot be rightly kneaded together. The mouths, or "spouts," of these nests vary from eight to ten inches in length, and point indifferently in all directions. The diameter of the widest portion of the nest is very variable, and ranges between four and seven inches.

The exterior of the nests is as rough as that of the common swallow of England, but the interior is comparatively smooth, and is lined with feathers and fine grass. The eggs are generally four or five in number, and the bird rears two broods in the course of the year.

THE pretty little SAND MARTIN is, in spite of its sober plumage and diminutive form, a very interesting bird, and one which adds much to the liveliness of any spot where it may take up its abode.

In size it is less than any other British Hirundinidæ, being less than five inches in total length. The colour of this bird is very simple, the general tint of the entire upper surface of the head and body being a soft brown,

HOUSE MARTIN.—(*Chelidon urbica.*)

relieved from too great uniformity by the sooty black quill feathers of the wings and tail. The under surface is pure white, with the exception of a band of brown across the upper part of the chest. The young bird possesses a lighter plumage than the adult, owing to the yellowish white tips of the back, tertiaries, and upper coverts. The beak is dark brown, and the eyes hazel.

Although its little beak and slender claws would seem at first sight to be utterly inadequate for the performance of miner's work, the Sand Martin is in its way as good a tunnel driver as the mole or the rat, and can manage to dig a burrow of considerable depth. The soil which it most loves is light sandstone, because the labour which is expended in the tunnelling is very little more than that which would be required for softer soils, and the sides of its burrow are sufficiently firm to escape the likelihood of breaking down.

The depth of the burrow is extremely variable, some tunnels being only eighteen inches or two feet deep, while others run to a depth of nearly five feet. During some five years' experience and constant watching of these birds in Derbyshire, I generally found that the hand could reach to the end of the burrows, and remove the eggs, provided that the birds had not been forced to change the direction of the tunnel by the intervention of a stone or a piece of rock too hard for their bills to penetrate.

As is generally the case with burrowing birds, the Sand Martin takes very little trouble about the construction of its nest, but contents itself with laying down a small handful of various soft substances, such as moss, hay, and feathers. The eggs are very small and fragile, and are not easily removed from the burrow without being fractured. Their colour is, when freshly laid, a delicate semi-transparent pink, which darkens to a dull opaque grey when incubation has proceeded to some extent, and changes to a beautiful white when the contents are removed from the shell. Their number is from four to six.

The food of this bird is composed of insects, and, in spite of the small dimensions of the little creature, it will pursue, capture, and eat insects of considerable dimensions and strength of wing, such as wasps and dragon-flies. Gnats and similar insects, however, form the staple of its diet.

This bird generally makes its appearance in England about the beginning of April, and has even been noticed before the end of March, so that its arrival is earlier than that of the swallow or martin. It departs about the beginning of September, and, like other British Hirundinidæ, makes its way to Africa, where it remains until the succeeding year.

RESEMBLING the common swallow in habits and general appearance, the HOUSE MARTIN may easily be distinguished from that bird by the large white patch upon the upper tail-coverts, a peculiarity which is even more notable when the bird is engaged in flight than when it is seated on the ground or clinging to its nest. In the dusk of evening the Martins may often be seen flying about at so late an hour that their bodies are almost invisible in the dim and fading twilight, and their presence is only indicated by the white patches upon their backs, which reflect every fading ray, and bear a singular resemblance to white moths or butterflies darting through the air.

This beautiful little bird is found in all parts of England, and is equally familiar with the swallow and sand martin. It places its clay-built nest principally under the shelter afforded by human habitations, and becomes so trustful and fearless that it will often fix its nest close to a window, and will rear its young without being dismayed at the near presence of human beings.

The nest of this species is extremely variable in shape and size, no two being precisely similar in both respects. Generally the edifice is cup-shaped, with the rim closely pressed against the eaves of some friendy house, and having a small semicircular aperture cut out of the edge in order to permit the ingress and egress of the birds. Sometimes, however, the nest is supported on a kind of solid pedestal, composed of mud, and often containing nearly as much material as would have made an ordinary nest. These pedestals are generally constructed in spots where the Martin finds that her nest does not find adequate support from the wall.

There are generally several broods in the course of the year, two being the usual number, three or even four being sometimes noticed. In such cases, however, the young birds seldom reach maturity, for they are hatched at such a late period of the year that the parents are unable to withstand the instinct which leads them to migrate, and in obeying the promptings of this principle, leave their unfortunate family to perish miserably of hunger. The parents do not seem to grieve over their dead children, and when they return to the nest in the succeeding season, they unconcernedly pull the dry and shrivelled bodies out of the nest, and rearrange it in readiness for the next brood.

The general colouring of this bird is composed of rich blue-black and white, arranged in bold masses, so as to present a fine contrast of two very opposite tints. The head and upper portions of the body are of a very deep glossy blue, with the exception of the quill-feathers of the wings and tail, which are sooty black, and the upper tail-coverts, which are snowy white. The chin, breast, and abdomen are of the same pure white as the upper tail-coverts, except in the young birds, which are greyish white beneath. The female bird is rather grey on the under portions of the body. A number of tiny white feathers are spread over the legs and toes, and the beak is black and the eyes brown. The total length of the Martin is rather more than five inches.

ROLLERS.

THE ROLLERS evidently form one of the connecting links between the swallows and the bee-eaters, as may be seen by the shape of their feet, which have the two hinder toes partially joined together, while those of the bee-eaters are wholly connected, or, as it were, soldered together. The Rollers, as is evident from their long pointed wings, stiff tail, and comparatively feeble legs and feet, are to a great extent feeders on the wing, although they do not depend wholly on their powers of flight for subsistence, but take many insects, worms, and grubs, from the ground.

GARRULOUS ROLLER.—(*Coracias garrula.*)

Although tolerably common on several parts of the Continent, GARRU-LOUS ROLLER is at the present time a very rare visitant to this country. There seems, however, to be reason to believe that in former days, when England was less cultivated and more covered with pathless woods, the Roller was frequently seen in the ancient forests, and that it probably built its nest in the hollows of trees, as it does in the German forests at the present day.

Africa is the legitimate home of the Roller, which passes from that land in the early spring, and makes its way to Europe, *viâ* Malta and the Mediterranean islands, which afford it resting-places during its long journey.

Accordingly, in those islands the Rollers are found in great plenty, and as they are considered a great delicacy when fat and in good condition, they are killed in considerable numbers, and exposed for sale like pigeons, whose flesh they are said greatly to resemble. Even in its flight it possesses something of the pigeon character, having often been observed while flying at a considerable elevation to "tumble" after the manner of the well-known tumbler pigeons. It is rather curious that throughout Asia Minor the Rollers and magpies were always found in close proximity to each other.

The food of the Roller is almost wholly of an insect nature, but is diversified with a few berries and other vegetable productions. It has even been known to become carnivorous in its habits, for, according to Temminck, it sometimes feeds on the smaller mammalia.

Worms, slugs, millipedes, and similar creatures also fall victims to its voracity.

In the colouring of its plumage it is truly a gorgeous bird. The general tint of the head, neck, breast, and abdomen is that peculiar green blue termed "verditer" by

GREEN TODY.—(*Todus viridis.*)

artists, changing into pale green in certain lights, and deepening into rich azure upon the shoulders. The back is a warm chestnut-brown, changing to purple upon the upper tail-coverts. The tail is of the same verditer hue as the head and neck, with the exception of the exterior feathers, which are furnished with black tips. The quill feathers of the wings are of a dark blue-black, becoming lighter at their edges, and the legs are covered with chestnut-brown feathers like those of the back. These gorgeous hues are not attained until the bird has passed through the moult of its second year. Both male and female are nearly equally decorated, the latter being slightly less brilliant than her mate. It is not a very large bird, scarcely exceeding a foot in total length.

THE curious little birds which are termed TODIES bear a considerable resemblance to the kingfishers, from which they may be easily distinguished by the flattened bill.

The Todies are natives of tropical America, and are very conspicuous among the brilliant-plumaged and strangely-shaped birds of that part of the world.

The GREEN TODY is a very small bird, being hardly larger than the common wren of England, but yet very conspicuous on account of the brilliant hues with which its plumage is decorated. The whole of the upper surface is a light green, the flanks are rose-coloured, deepening into scarlet upon the throat and fading into a pale yellow upon the abdomen and under the tail-coverts.

TROGONS.

FOR our systematic knowledge of the magnificent tribe of the TROGONS we are now almost wholly indebted to Mr. Gould, who by the most persevering labour and the most careful investigations has reduced to order this most perplexing group of birds, and brought into one volume a mass of information that is rarely found in similar compass. There are few groups of birds which are more attractive to the eye than the Trogons, with all their glowing hues of carmine, orange, green, and gold; and few there are which presented greater difficulties to the ornithologist until their various characteristics were thoroughly sifted and compared together. The two sexes are so different from each other, both in the colour and shape of the feathers, that they would hardly be recognizable as belonging to a single species, and even the young bird is very differently coloured from his older relatives.

These beautiful birds are found in the Old and the New Worlds, those which inhabit the latter locality being easily distinguishable by their deeply-barred tails. Those of the Old World are generally found in Ceylon,

RESPLENDENT TROGON.—(*Calurus resplendens.*)

Sumatra, Java, and Borneo, while only a single species, the Narina Trogon, is as yet known to inhabit Africa.

The Trogons are mostly silent birds, the only cry used being that of the male during the season of pairing. It is not a very agreeable sound, being of a sombre and melancholy cast, and thought to resemble the word "courou-courou."

Several of the Trogons are distinguished from their relatives by the length and downy looseness of many of the feathers, more especially the lance-shaped feathers of the shoulders and the elongated upper tail-coverts. On account of this structure of the plumage, they are gathered into a separate genus under the appropriate title of Calurus, or Beautiful-tailed Trogons.

The RESPLENDENT TROGON is a native of Central America, and was in former days one of the most honoured by the ancient Mexican monarchs, who assumed the sole right of wearing the long plumes, and permitted none but the members of the royal family to decorate themselves with the flowing feathers of this beautiful bird.

This species is fond of inhabiting the densest forests of Southern Mexico, and generally haunts the topmost branches of the loftiest trees, where it clings to the boughs like a parrot, and traverses their ramifications with much address.

The colour of the adult male bird is generally of a rich golden green on the upper parts of the body, including the graceful rounded crest, the head, neck, throat, chest, and long lancet-shaped plumes of the shoulders. The breast and under-parts are brilliant scarlet, the central feathers of the tail are black, and the exterior white, with black bars. The wonderful plumes which hang over the tail are generally about three feet in length, and in particularly fine specimens have been known to exceed that measurement by four inches, so that the entire length of the bird may be reckoned at four feet. The bill is light yellow.

As is often the case with birds, where the male is remarkable for the beauty of his plumage, the female is altogether an ordinary and comparatively insignificant bird.

KINGFISHERS.

THE KINGFISHERS form a tolerably well-marked group of birds all of which are remarkable for their long bills and the comparative shortness of their bodies, which give them a peculiar bearing that is not to be mistaken.

The bills of these birds are all long and sharp, and in most cases are straight. Their front toes are always joined together more or less, and the number of the toes is very variable in form and arrangement; some species possessing them in pairs, like those of the parrots; others having them arranged three in front and one behind, as is usually the case with birds; while a few species have only three toes altogether, two in front and one behind. The wings are rounded. As may be gathered from their popular name, they mostly feed upon fish, which they capture by pouncing upon the finny prey.

Our first example of the Kingfisher is the LAUGHING JACKASS, or GIANT KINGFISHER, its former title being derived from the strange character of its cry.

This bird is an inhabitant of Australia, being found chiefly in the south-eastern district of that country, and in New South Wales. In Van Diemen's Land Mr. Gould believes that it does not exist. In no place is it found in any great numbers; for although it is sufficiently common, it is but thinly dispersed over the country. It is rather a large bird, being eighteen inches in total length, and is powerful in proportion, being able to wage successful war against creatures of considerable size.

Although one of the true Kingfishers, it so far departs from the habits of the family as to be comparatively careless about catching fish, and often resides in the vast arid plains where it can find no stream sufficiently large to harbour fish in their waters. Crabs of various kinds are a favourite food with this bird, which also eats insects, small mammalia, and reptiles. Mr. Gould mentions an instance where he shot one of these birds for the sake of possessing a rare and valuable species of rat which it was carrying off in its bill. It is also known to eat snakes, catching them with great dexterity by the tail, and crushing their heads with its powerful beak. Sometimes it is known to pounce upon fish, but it usually adheres to the above-mentioned diet.

The cry of this bird is a singular, dissonant, abrupt laugh, even more startling than that of the hyæna, and raising strange panics in the heart of the novice who first hears it while bivouacking in the "bush." Being of a mightily inquisitive nature, the Laughing Jackass seems to find great

THE KINGFISHER. Page 282.

The Popular Natural History

attraction in the glare of a fire, and in the evening is apt to glide silently through the branches towards the blaze, and, perching upon a neighbouring bough, to pour forth its loud yelling cry. The " old hands " are in no wise disconcerted at the sudden disturbance, but shoot the intruder on the spot,

LAUGHING JACKASS.—(*Dacelo gigas.*)

and in a very few minutes convert him into a savoury broil over the fire which he had come to inspect.

At the rising and the setting of the sun the Laughing Jackass becomes very lively, and is the first to welcome the approach of dawn, and to chant its strange exulting pæans at the return of darkness. From this peculiarity it has been called the Settler's clock. In allusion to the cry of this bird, which has been compared by Sturt to the yelling chorus of unquiet demons, the natives call it by the name Gogobera.

The home of the Laughing Jackass is usually made in the hole of a gum-tree (*Eucalyptus*), where it makes no sort of nest, but simply lays its eggs upon the soft decaying wood. The eggs are pearly white, and the bird keeps a vigilant watch over the burrow which holds its treasures, fiercely combating any creature that may approach the entrance, and aiming the most desperate blows with its long pointed and powerful beak.

It is a really handsome bird, and, although not possessing such an array of brilliant plumage as falls to the lot of many Kingfishers, is yet very richly coloured. The bird is decorated with a dark brown crest, and the general tint of the back and upper surface is olive brown. The wings are brown-black, a few of the feathers being slightly tipped with verditer and the breast and under portions are white, washed with pale brown, which forms a series of faint bars across the breast. The tail is rather long, and rounded at the extremity, and is of a rich chestnut colour, banded with deep black and tipped with white.

THE common KINGFISHER is by far the most gorgeously decorated of all our indigenous birds, and can bear comparison with many of the gaily decorated inhabitants of tropical climates.

It is a sufficiently common bird, although distributed very thinly over the whole country ; and considering the great number of eggs which it lays, and the large proportion of young which it rears, is probably more plentiful than is generally supposed to be the case. The straight, glancing flight of the Kingfisher, as it shoots along the river-bank, its azure back gleaming in the sunlight with meteoric splendour, is a sight familiar to all those who have been accustomed to wander by the sides of rivers, whether for the purpose of angling

KINGFISHER.—(*Alcedo Ipsida.*)

or merely to study the beauties of nature. So swift is the flight of this bird, and with such wonderful rapidity does it move its short wings, that its shape is hardly perceptible as it passes through the air, and it leaves upon the eye of the observer the impression of a blue streak of light.

The food of this bird consists chiefly, though not exclusively, of fish, which it takes, kills, and eats in the following manner :—

Seated upon a convenient bough or rail that overhangs a stream where the smaller fish love to pass, the Kingfisher waits very patiently until he sees an unsuspecting minnow or stickleback pass below his perch, and then, with a rapid movement, drops into the water like a stone and secures his prey. Should it be a small fish, he swallows it at once ; but if it should be of rather large dimensions, he carries it to a stone or stump, beats it two or three times against the hard substance, and then swallows it without any trouble.

With the fish it generally feeds its young, being able to disgorge at will the semi-digested food which it has swallowed, after the manner of most birds of prey. Fish, however, do not constitute its sole nourishment, as it is known to eat various insects, such as dragon-flies and water-beetles, and will often in cold weather pay a visit to the sea-shore for the purpose of feeding upon the little crabs, shrimps, and sandhoppers that are found upon the edge of the tide.

The nest of the Kingfisher is always made in some convenient bank, at the extremity of a hole which has previously been occupied and deserted by the water-rat or other mining quadrupeds, and been enlarged and adapted for use by the Kingfisher. Now and then the nest of this bird has been found built in the deserted hole of a rabbit warren. It is always found that the tunnel slopes gently upward, and that the bird has shaped the extremity into a globular form in order to contain the parent bird, the nest, and eggs. Sometimes the nest is placed in the natural crevices formed by the roots of trees growing on the water's edge. In many cases it is easily detected, for the birds are very careless about the concealment of their nest even before the eggs are hatched ; and after the young have made their appearance in the world, they are so clamorous for food and so insatiable in their appetite, that their noisy voices can be heard for some distance, and indicate with great precision the direction of their home.

Some writers say that the interior of the burrow is kept so scrupulously clean that it is free from all evil scents. My own experience, however, contradicts this assertion, for after introducing the hand into a Kingfisher's nest, I have always found it imbued with so offensive an odour that I was fain to wash it repeatedly in the nearest stream. As the Kingfisher is so piscatorial in its habits it would naturally be imagined that the nest would be placed in close connection with the stream from which the parent birds obtained their daily food. I have, however, several times seen a Kingfisher's nest, and obtained the eggs, in spots that were not within half a mile of a fish-inhabited stream. The bird is greatly attached to the burrow in which it has once made its nest, and will make use of the same spot year after year, even though the nest be plundered and the eggs stolen.

The eggs are from six to eight in number, rather globular in form, and of an exquisitely delicate pink in colour while fresh, changing to a pearly white when the contents are removed. As soon as the young are able to exert themselves, they perch on a neighbouring twig or other convenient resting-place, and squall incessantly for food. In a very short time they assume their yearling plumage, which is very nearly the same as that of the adult bird, and soon learn to fish on their own account.

The nest of the Kingfisher has long been known to consist of the bones, scales, and other indigestible portions of the food, which are ejected from the mouth by "castings," like those of the hawk or owl ; but until Mr. Gould recently procured a perfect Kingfisher's nest, its shape and the manner of construction were entirely unknown. His account of its discovery, and the ingenious manner in which it was procured, is so interesting that it must be given in his own words.

" Ornithologists are divided in opinion as to whether the fish-bones found in the cavity in which the Kingfisher deposits its eggs are to be considered in the light of a nest, or as merely the castings from the bird during the period of incubation. Some are disposed to consider these bones as entirely the castings and faeces of the young brood of the year before they quit the nest, and that the same hole being frequented for a succession of years, a great mass is at length formed ; while others believe that they are deposited by the parents as a platform for the eggs, constituting, in fact, a nest ; in which latter view I fully concur, and the following are my reason for so doing :—

" On the 18th of the past month of April, during one of my fishing excursions on the Thames, I saw a hole in a precipitous bank, which I felt sure was a nesting-place of the Kingfisher, and on passing a spare top of my fly-rod to the extremity of the hole, a distance of nearly three feet, I brought out some freshly-cast bones of fish, convincing me that I was right in my surmise. On a subsequent day, the 9th of May, I again visited the spot with a spade, and after moving nearly two feet square of the turf, dug down to the nest without disturbing the entrance hole or the passage which led to it. Here I found four eggs placed on the usual layers of fish-bones ; all of these I removed with care, and then filled up the hole, beating the earth down as hard as the bank itself, and replacing the sod on the top in order that barge-horses passing to and fro might not put a foot in the hole. A fortnight afterwards the bird was seen to leave the hole again, and my suspicion was awakened that she had taken to her old breeding-quarters a second time.

"The first opportunity I had of again visiting this place, which was exactly twenty-one days from the date of my former exploration and taking the eggs, I again passed the top of my fly-rod up the hole, and found not only that the hole was of the former length, but that the female was within. I then took a large mass of cotton-wool from my collecting-box, and stuffed it to the extremity of the hole, in order to preserve the eggs and nest from damage during my again laying it open from above. On removing the sod and digging down as before, I came upon the cotton-wool, and beneath it a well-formed nest of fish-bones, the size of a small saucer, the walls of which were fully half an inch thick, together with eight beautiful eggs and the old female herself. This nest and eggs I removed with the greatest care, and I now have the pleasure of exhibiting it to the Society, before its transmission to the British Museum, the proper resting-place of so interesting a bird's nest. This mass of bones then, weighing 700 grains, had been cast up and deposited by the bird, or the bird and its mate, besides the unusual number of eight eggs, in the short space of twenty-one days.

"To gain anything like an approximate idea of the number of fish that had been taken to form this mass, the skeleton of a minnow, their usual food, must be carefully made and weighed, and this I may probably do upon some future occasion. I think we may now conclude, from what I have adduced, that the bird purposely deposits these bones as a nest ; and nothing can be better adapted, as a platform, to defend the eggs from the damp earth."

The voice of the Kingfisher is a peculiarly shrill and piping cry, that can be heard at some distance, and is not easily mistaken for any other sound.

The colour of this bird is very gorgeous, and rather complicated in its arrangement. The top of the head and back of the neck are dark green, flecked with many spots of verditer blue upon the tips of the feathers. The upper part of the back is also dark green, and the lower part is light violet or blue, gleaming vividly under a strong light, and being very conspicuous as the bird is on the wing. The tail is deep indigo, and the quill feathers of the wing are dark blackish green, lightened by a brighter hue of green on the outer webs, and set off by the verditer blue spots of the tertiaries. A white patch or streak passes from the eye to the back of the neck, and a dark green streak is drawn immediately under the white patch. The throat and chin are yellowish white, and the whole of the under surface is chestnut. The eyes are crimson, and the bill is black, with the exception of the orange-tinted base of the lower mandible. The total length of the bird is about seven inches.

BEE-EATERS.

THE BEE-EATERS may at once be distinguished by the shape of the bill, which is curved, and by the formation of the wings, which are long and

>ointed, and give to their owners a wonderful command of the air, while :ngaged in chasing their winged prey.

The common BEE-EATER of Europe is very frequently found in many)arts of the Continent, and has been several times taken in England. t is, however, a scarce bird in Great Britain, and is of sufficient rarity o excite some curiosity whenever it is found within the confines of)ur shores.

The food of the Bee-eater consists wholly of insects, hive-bees and others)f the hymenopterous order being the favourite article of diet. In chasing hese insects, which are for the most part very active of flight, the Bee-eater lisplays very great command of wing, and while urging its pursuit, can twist .nd turn in the air with as much ease and skill as is exhibited by the swallow)r the roller.

To the apiarian, who resides in the same country with the Bee-eater, the)ird is a terrible foe, as it has an insatiable appetite for the honey-making nsects, and haunts every spot where it is likely to meet them. The hives .re constantly visited by the Bee-eaters, who are ingenious enough to resort o the turpentine pines for the sake of catching the bees that come to carry .way the exudations for the purpose of converting them into " propolis," or hat substance with which they har-len the edges of their cells, caulk the :revices of the hives, and perform nany other useful tasks. It does not, .owever, confine itself to the hymen-)pterous insects, but is fond of bee-les, cicadæ, grasshoppers, and similar :reatures.

The nest of the Bee-eater is not inlike that of the kingfisher, being)laced at the extremity of a burrow nade in some convenient bank. The)urrow is excavated by the bird itself, ind it often happens that the Bee-:aters are as gregarious in their nest-ng as in their flight, honeycombing he clay banks in a manner very simi-ar to that of the sand martin. The

BEE-EATER.—(*Merops apiaster*).

)urrows do not run to any great depth, seldom exceeding six or eight inches n length. The nest is composed of moss, and contains about five or six)eautifully white and pearly eggs.

The colours of the adult male bird are extremely varied and very beautiful. The top of the head is rich chestnut brown, extending to the neck, back, and wing-coverts. Over the rump the chestnut changes to light reddish yellow. The primaries and secondaries of the wing are bright blue-green, tipped with)lack, and their shafts painted with the same colour, and the tertiaries are green throughout their entire length. The upper tail-coverts are of the jame hue as the wings, and the tail is likewise green, tinted with a darker 1ue, graphically called by Mr. Yarrell " duck-green." The chin and throat ire a reddish yellow, and round the throat runs a band of deep blue-black. The under part of the body is green with a blue tinge, and the under jurface of the wings and tail is greyish brown. The ear-coverts are black, ind the eye is light scarlet, which contrasts beautifully with the chestnut)lack, and yellow of the head and neck.

The female may be distinguished from the male by the paler hue of the :eddish yellow on the throat, and the reddish tinge that runs throughout the green of the body and wings. In size the **Bee-eater is** nearly equal to the English starling.

SLENDER-BILLED BIRDS.

UPUPIDÆ, OR HOOPOES.

THE large group of birds which are termed TENUIROSTRAL, or Slender-billed, always possess a long and slender beak, sometimes curved, as in the creepers, hoopoes, and many humming birds ; and sometimes straight, as in the nuthatch and other humming birds. The feet are furnished with lengthened toes, and the outer toe is generally connected at the base with the middle toe.

The first family of the Tenuirostres is called after the hoopoe, and termed Upupidæ. In all these birds the bill is curved throughout its entire length, long, slender, and sharply pointed. The wings are rounded, showing that the birds are not intended for aerial feats, and the tail is rather long. The legs are short, and the claws strong and decidedly curved.

The common HOOPOE enjoys a very wide range of country, being found in Northern Africa, where its principal home is generally stationed, in several parts of Asia, and nearly the whole of Europe. On account of its very striking and remarkable form it has attracted much notice, and has been the subject of innumerable legends and strange tales, nearly all of which relate to its feathery crest.

The Turks call the Hoopoe *Tir-Chaous,* or Courier-Bird, because its feathery crown bears some resemblance to the plume of feathers which the *chaous,* or courier, wears as a token of his office. The Swedes are rather fearful of the Hoopoe, and dread its presence, which is rare in their country, as a presage of war, considering the plume as analogous to a helmet. Even in our own country the uneducated rustics think it an unlucky bird, most probably on account of some old legend which, although forgotten has not entirely lost its power of exciting prejudice.

HOOPOE. —(*Upupa epops*).

The food of the Hoopoe is almost entirely of an insect nature, although the bird will frequently vary its diet with tadpoles and other small creatures. Beetles and their larvæ, caterpillars and grubs of all kinds, are a favourite food with the Hoopoe, which displays much ingenuity in digging them out of the decayed wood in which they are often found. The jet-ant (*Formica fuliginosa*), which greatly haunts the centre of decaying trees, is also eaten by this bird.

The nest is made in hollow trees, and consists of dried grass-stems, feathers, and other soft substances. The eggs are of a light grey colour, and in number vary from four to seven. They are laid in May, and the young make their appearance in June. It is worthy of notice that the beak of the young Hoopoe is short and quite straight, not attaining its long curved form until the bird has attained its full growth. The nest of the Hoopoe has a very pungent and disgusting odour.

The general colours of the Hoopoe are white, buff, and black, distributed in the following manner :—The plumes of the crest, which is composed of a double row of feathers, are of a reddish buff, each feather being tipped with black. The remainder of the head, neck, and breast is purplish buff, and the upper part of the beak purple-grey. Three semicircular black bands are drawn across the back, and the quill feathers of the wings are marked with broad bands of black and white. The tail is also black, with the exception of a sharply-defined white semicircular band that runs across the centre.

The under portions of the body are pale yellowish buff, and the under tail-coverts are white. In their colours the two sexes are rather different from each other, the male being of a more ruddy hue than his mate, and having a larger crest. The total length of the adult Hoopoe is not quite thirteen inches.

SUN-BIRDS.

THE beautiful and glittering SUN-BIRDS evidently represent in the Old World the humming birds of the New. In their dimensions, colour, general form, and habits, they are very similar to their brilliant representatives in the western hemisphere, although not quite so gorgeous in plumage, nor so powerful and enduring of wing. They are termed Sun-birds because the hues with which their feathers are so lavishly embellished gleam out with peculiar brilliancy in the sunlight.

These exquisite little birds feed on the juice of flowers and the minute insects that are found in their interior, but are not in the habit of feeding while on the wing, hovering over a flower and sweeping up its nectar with the tongue, as is the case among the humming-birds.

The COLLARED SUN-BIRD is an inhabitant of many parts of Africa, stretching from the northern portions of that continent as far as the western coasts. It is extremely plentiful in the larger forests of the Cape and the interior, but there is very little information concerning its habits, saving that they resemble those of its relations. The nidification of this species differs according to the locality, for it places its nest in the interior of hollow trees wherein it resides in the forests, and is content with the shelter of a thick bough when there are no decaying trees within reach.

The male Collared Sun-bird is a most beautiful little creature, bedecked with glowing tints of wonderful intensity. The general colour of the upper parts of the body and breast is a rich golden green, the upper surface of the wings and tail being blackish brown with green reflections. Across the breast are drawn several coloured bands, which have earned for the bird its popular and expressive name, as all names should be. A narrow band of bright steel-blue runs across the upper part of the breast, being rather wide in the centre and narrowing rapidly towards the sides of the neck. Below this blue band runs a broad belt of rich carmine, and immediately below the carmine is a third narrow band of bright golden yellow. From the sides of the breast proceed several small feathery plumes of the same golden hue. The remainder of the abdomen is greyish brown, and the upper tail-coverts are violet-purple.

The female is rather less in dimensions than her mate, and is very sober in her attire, wearing a suit of uniform olive-brown, darker upon the wings and tail, and very pale behind. The total length of this species is rather more than four and a half inches.

The JAVANESE SUN-BIRD is a native of the country from which it derives its name. It is a very pretty little creature, although its colours are not so

COLLARED SUN-BIRD.—(*Nectarinia chalybeia.*)
JAVANESE SUN-BIRD.—(*Nectarinia Javanica.*)

resplendent as in several of the species. The upper parts of the body are shining steely-purple, and the under surface is olive-yellow. The throat is chestnut, and a bright violet streak runs from the angle of the mouth to the breast.

The beautiful little DICÆUM, although very common throughout the whole of Australia, and a remarkably interesting little bird, was, when Mr. Gould wrote his animated description, so little known among the colonists that there was no popular name for the bright little creature.

This tiny bird is fond of inhabiting the extreme summits of the tallest trees, and habitually dwells at so great an elevation that its minute form is hardly perceptible, and not even the bright scarlet hue of the throat and breast can betray its position to the unaccustomed eye of a passenger below.

The flight of the Dicæum is very quick and darting, and it makes more use of its wings and less of its feet than any of the insect-hunting birds. The nest is remarkably pretty, being woven as it were out of white cotton cloth, and suspended from a branch as if the twigs had been pushed through its substance. The peculiar purse-like shape of the nest is shown in the illustration. The material of which it is woven is the soft cottony down which is found in the seed-vessels of many plants. The eggs are four or five in number, and their colour is a dull greyish white, profusely covered with minute speckles of brown.

The head, back, and upper parts of the adult male are deep black, with a beatiful steely blue gloss, the sides are brownish grey, and the throat, breast, and under tail-coverts are a bright glaring scarlet. The abdomen is snowy white, with the exception of a tolerably large black patch on its centre. The female is more sombre in her apparel, the head and back being

AUSTRALIAN DICÆUM.—(*Dicæum hirundinaceum.*)

of a dull sooty black, and the steel-blue reflection only appearing on the upper surface of the wings and tail. The throat and centre of the abdomen are buff, the sides are pale greyish brown, and the under tail-coverts scarlet, of a less brilliant hue than in the male. In its dimensions the Dicæum is hardly so large as our common wren.

HONEY-EATERS.

THE true Honey-eaters form a very numerous group of birds, all of which are graceful in their forms and pleasing in the colour of their plumage, while

U

in some instances the hues with which they are decorated are so bright as to afford ground for classing them among the really beautiful birds. They all feed on similar substances, which, as indicated by their name, consist chiefly of honey and the sweet juices of flowers, although they also vary their diet by insects and other small living beings.

POË BIRD.—(*Prosthemadera Zeelandiæ.*)

Among this group of birds the POË BIRD, or TUE, or PARSON BIRD, is one of the most conspicuous, being nearly as remarkable for its peculiar colouring as the rifle bird itself, although the hues of its feathers are not quite so resplendently brilliant as in that creature.

The Poë Bird is a native of New Zealand, where it is far from uncommon, and is captured by the natives for the purpose of sale. Many individuals are brought over to Sydney, where, according to Dr. Bennett, they are kept in cages, and are very amusing in their habits, being easily domesticated and becoming very familiar with those who belong to the household. Independently of its handsome and rather peculiar colour, which makes it very effective in a room, it possesses several other qualifications which render it a very desirable inhabitant of an aviary. Its native notes are very fine, the bird being considered a remarkably fine songster, and it also possesses the p)wer of mimicking in a degree surpassing that of the common magpie or raven, and hardly yielding even to the famous mocking-bird himself. It learns to speak with great accuracy and fluency, and readily imitates any sound that may reach its ear, being especially successful in its reproduction of the song of other birds.

While at liberty in its native land it is remarkable for its quick, restless activity, as it flits rapidly about the branches, pecking here and there at a stray insect, diving into the recesses of a newly-opened flower, and continually uttering its shrill sharp whistle. Although one of the large group of Meliphagidæ, or Honey-eaters, the Poë Bird feeds less upon honey than upon insects, which it discovers with great sharpness of vision and catches in a particularly adroit manner. It will also feed upon worms, and sometimes varies its diet by fruits.

In New Zealand it is often killed for the sake of its flesh, which is said to be very delicate and well flavoured.

The general colour of the Poë Bird is a very deep metallic green, becoming black in certain lights, and having a decided bronze reflection in others. The back is deep brown, also with a bronze reflection, and upon the shoulders there is a patch of pure white. On the back of the neck the feathers are long and lancet shaped, each feather having a very narrow white streak along its centre. From each side of the neck depends a tuft of snowy curling downy feathers, spreading in fan-like fashion from their bases This creature is called the Parson Bird because these white tufts are thought to bear some resemblance to the absurd parallelograms of white lawn that are denominated " bands."

HUMMING-BIRDS, OR TROCHILIDÆ.

> "Bright Humming-bird of gem-like plumeletage,
> By western Indians 'Living Sun-beam' named."—BAILEY, *Mystic.*

THE wonderful little HUMMING-BIRDS are only found in America and the adjacent islands, where they take the place of the sun-birds of the Old World. It is rather remarkable that, as yet, no Humming-birds have been discovered in Australia.

These little winged gems are most capricious in their choice of locality some being spread over a vast range of country, while others are confined. within the limits of a narrow belt of earth hardly more than a few hundred yards in width, and some refuse to roam beyond the narrow precincts of a single mountain. Some of these birds are furnished with comparatively short and feeble wings, and, in consequence, are obliged to remain in the same land throughout the year, while others are strong of flight, and migrate over numerous tracts of country. They gather most thickly in Mexico and about the equator. the number of species diminishing rapidly as they recede from the equatorial line.

The name of Humming-birds is given to them on account of the humming or buzzing sound which they produce with their wings, especially while they are hovering in their curious fashion over a tempting blossom, and feeding on its contents while suspended in the air.

The legs of these birds are remarkably weak and delicate, and the wings are proportionately strong, a combination which shows that the creatures are intended to pass more of their time in the air than on foot. Even when feeding they very seldom trouble themselves to perch, but suspend themselves in the air before the flower on which they desire to operate, and with their long slender tongues are able to feed at ease without alighting. In the skeleton, especially in the shape of the breast-bone and wings, as well as in the comparative small size of the feet, the Humming-birds bear some analogy to the swifts, and, like those birds, never lay more than two eggs.

The flight of these birds is inconceivably rapid, so rapid indeed that the eye cannot follow it when the bird puts forth its full speed ; and with such wonderful rapidity do the little sharp-cut wings beat the air, that their form is quite lost, and while the bird is hovering near a single spot, the wings look like two filmy grey fans attached to the sides. While darting from one flower to another the bird can hardly be seen at all, and it seems to come suddenly into existence at some spot, and as suddenly to vanish from sight. Some Humming-birds are fond of towering to a great height in the air, and descending from thence to their nests or to feed, while others keep near the ground,

HUMMING-BIRDS.

and are seldom seen at an elevation of many yards.

The food of the Humming-bird is much the same as that of the honey-suckers, except, perhaps, that they consume more honey and fewer flies. Still, they are extremely fond of small insects, and if kept away from this kind of diet soon pine away, in spite of unlimited supplies of syrup and other sweet food.

In order to enable the Humming-bird to extract the various substances on which it feeds from the interior of the flowers, the beak is always long and delicate, and in shape is extremely variable, probably on account of the particular flowers on which the bird feeds. In some instances the bill is nearly straight, in others it takes a sharp sickle-like downward curve, while in some it possesses a double curve. The general form of the beak is however a very gently downward curve, and in all instances it is pointed at its extremity. At the base the upper mandible is wider than the lower, which is received into its hollow. Their nostrils are placed at the base of the beak, and defended by a little scale-like shield.

The plumage is very closely set on the body, and is possessed of a metallic brilliancy in every species, the males being always more gorgeously decorated than their mates.

The tongue is a very curious structure, being extremely long, filamentous, and double nearly to its base. At the throat it is taken up by that curious forked bony structure called the hyoid bone, the forks of which are enormously elongated, and, passing under the throat and round the head, are terminated upon the forehead. By means of this structure the Humming-bird is enabled to project the tongue to a great distance from the bill, and to probe the inmost recesses of the largest flowers. The common woodpecker has a very similar description of tongue and employs it in a similar manner.

In their habits the Humming-birds are mostly diurnal, although many species are only seen at dawn and just after sunset. Many, indeed, live in such dense recesses of their tropical woods, that the beams of the sun never fairly penetrate into their gloomy depths, and the Humming-bird dwells in a permanent twilight beneath the foliage. It is worthy of notice that the name Trochilidæ is not a very apt one, as the Trochilus was evidently a bird which had nothing in common with the Humming-bird, and was most probably the zic-zac of Egypt.

The upper figure in the illustration on page 292 represents the CORA HUMMING-BIRD, a native of Peru. The head and back are gold-green, the wings are purple, and the throat is violet or crimson according to the direction of the light.

SLENDER SHEAR-TAIL HUMMING-BIRD.--(*Thau-mastura encicura.*)

In the centre is the DOUBLE-CREST, a Brazilian species. It derives its name from the crest-like feathers that start from either side of the head. The top of the head is azure, and the throat fiery crimson. The sides of the face and the chin are velvet black.

The BAR-TAIL occupies the right of the illustration. It is a native of Bolivia, and derives its name from the black tips of the crimson tail feathers. The body is green.

On the left is GOULD'S HUMMING-BIRD, a lovely little creature, remarkable for the beautiful neck-tufts, with their pure white feathers tipped with green. It is found in the Amazon district.

The SLENDER SHEAR-TAIL is an inhabitant of Central America, and appears to be rather a local bird. It is supposed not to be found south of the Isthmus of Panama, nor to extend more than eighteen degrees northward. As its wings are rather short, and not remarkable for strength, it is conjectured to be a non-migratory bird. The country where it is seen in the greatest plenty is Guatemala.

The sexes of this creature are very different in their form and colour of their plumage, and could hardly be recognized as belonging to the same species. In the adult male bird the upper parts of the body are a deep shining green, becoming brown on the head, and changing into bronze on the back and wing-coverts. The wings are purple-brown. The long and deep forked tail is black, with the exception of a little brown upon the inner web of the two uttermost feathers. The chin is black glossed with green, the throat is deep metallic purple, and upon the upper art of the chest is placed a large crescent-shaped mark of buff. The abdomen is bronze, with a grey spot in its centre ; and there is a buff spot on each flank. The under tail-coverts are of a greenish hue.

COPPER-BELLIED PUFF-LEG HUMMING-BIRD.
(*Eriocnemis cupreiventris.*)

The female does not possess the long tail, and her colours are golden-green above and reddish buff below. The tail is very curiously marked. The central feathers are entirely gold-green ; the exterior feathers are rusty red at their base, black for a considerable portion of their length, and tipped with white.

Several of the Humming-birds are remarkable for a tuft of pure white downy feathers which envelop each leg, and which has obtained for them the popular title of Puff-legs, because the white tufts bear some resemblance to a powder-puff. The COPPER-BELLIED PUFF-LEG is an inhabitant of Santa Fé de Bogóta, and is a very common bird in that locality. It may easily be found, as it is a remarkably local bird, being confined to a narrow strip or belt of land, which possesses the requisite characteristics of temperature and vegetation.

It is a very beautiful little bird, and both the sexes are nearly similar in their colour and general appearance, except that in the female the puffs of white down are not so large nor so conspicuous as in her mate. In the adult male, the top of the head, the sides of the neck, and the back are green, washed with a decided tint of bronze, except upon the upper tail-coverts, where the green is very pure and of a metallic brilliancy. As is generally the case with Humming-birds, the fine and sharply-cut wings are brown washed with purple. The tail is black, with a purple gloss in a side light. The throat is of a beautiful shining metallic green, and the general colour of the breast and under portions of the body is green glossed with gold, with the exception of the abdomen, where the green takes a coppery hue, from which the bird has received its popular name. The "puffs" are of a snowy whiteness, and look like refined swan's-down.

WHITE-BOOTED RACKET-TAIL.—(*Spathura Underwoodii*—male and female.)

The female is very similar in colour, except that the hues of the throat are not possessed of so metallic a brilliancy, and, as has already been stated, the leg-tufts are comparatively small.

We have in the Racket-tailed Humming-birds one of those singular forms which are so often found among these strange little birds.

The WHITE-BOOTED RACKET-TAIL inhabits the Columbian Andes, and is very common near Santa Fé de Bogóta. It is a hill-loving bird, being generally found at an elevation of five or ten thousand feet above the level of the sea. It is thought to be confined within the third and tenth degrees of north latitude. This bird is remarkably swift of wing, its darting flight reminding the spectator of the passage of an arrow through the air. At one time it will hover close to the ground, hanging over some favourite flower and extracting the sweet contents of the blossoms ; and at the next moment it will shoot to the very summit of some lofty tree, as if impelled from a bow, and leave but the impression of an emerald-green line of light upon the observer's eye. While hovering over the flowers the long racket-shaped feathers of the tail are in constant motion, waving gently in the air, crossing each other, opening and closing in the most graceful manner. But when the bird darts off with its peculiar arrowy flight, the tail-feathers lie straight behind it.

The male of this species is bronze-green upon the greater part of the body, the green taking a richer and redder hue upon the upper tail-coverts. The throat and breast are brilliant emerald green. The wings are purple-brown, and the tail is brown, with the exception of the rackets, which are black "shot" with green. The feet are yellow, and upon the legs are placed two beautiful white puffs. The whole length of the bird is rather more than three inches. The female bird does not possess the racket-shaped tail-feathers, and is of a bronze - green upon the upper surface. The tail is brown, with the exception of the two middle feathers,

SPANGLED COQUETTE.—(*Lophornis Reginæ.*)

which are bronze-green like the body. The two exterior feathers are tipped with white, and the others with bronze-green. The under surface is white, diversified with bronze-green spots on the breast and flanks. The puffs are smaller than in the male.

THE accompanying illustration represents another remarkable little bird possessed of a most beautiful and graceful crest. The SPANGLED COQUETTE is an excellent example of the very remarkable genus to which it belongs All the Coquettes possess a well-defined crest upon the head and a series of projecting feathers from the neck, some being especially notable for the one ornament, and others for the other.

The crown of the head and the crest are light ruddy chestnut, each feather having a ball-like spot of dark bronze-green at the tip. The throat and face are shining metallic green, below which is a small tuft of pointed white feathers that have a very curious effect as they protude from beneath the gorget. The upper parts are bronze-green as far as the lower part of the back, where a band crosses from side to side, and the rest of the plumage is dark ruddy chestnut as far as the tail. The tail is also chestnut-brown, with a slight wash of metallic green. The female has no crest nor green gorget.

The RUBY AND TOPAZ HUMMING-BIRD derives its name from the colouring of its head and throat, the former being of a deep ruby tint, and the latter of a resplendent topaz. Sometimes it is called the Ruby-headed Humming-bird, and it is also known under the name of the Aurora. It is very common in Bahia, the Guianas, Trinidad, and the Caraccas, and, as it is in great request for the dealers, is killed by thousands annually. There is no species so common in ornamental cases of Humming-birds as the Ruby and Topaz. It makes a very beautiful nest, round, cuplike, and delicately woven of cotton and various fibres, and covered externally with little leaves and bits of lichen.

The plumage of this species is extremely variable, but may be described briefly as follows. The forehead, the crown, and the nape of the neck are metallic ruby-red; and the chin, throat, and chest are effulgent topaz. The upper parts of the body are velvety bronze-brown, and the wings are purple-brown. The tail is rich chestnut-red, tipped with black, and the abdomen is a dark olive-brown. The female has none of the ruby patches on the head, but retains a little of the topaz on the throat.

THE RUBY-THROATED HUMMING-BIRD inhabits North America, and derives its name from the brilliant ruby hue of the feathers that adorn its throat. It is one of the commonest as well as the most beautiful of this lovely group.

The beautiful little VERVAIN HUMMING-BIRD is one of the minutest examples of feathered life that are at present known to zoologists. In total length this bird does not measure three inches; while, as the tail occupies nearly an inch and the head

RUBY AND TOPAZ HUMMING-BIRD.—(*Chrysolampis moschitus.*)

half an inch, the actual length of the body will be seen to be not quite an inch and a half. It is a native of Jamaica, and has been admirably described by Mr. Gosse, while treating of the birds which inhabit that island.

The name of Vervain Humming-bird has been given to this tiny creature, because it is in the habit of feeding on the blossoms of the West Indian Vervain, but it is also known under a variety of other titles, and has been described by many scientific writers under different names.

The general colour of this beautiful little bird is a brilliant metallic green, the wings being, as usual, purple-brown, and the tail deep black. The throat and chin are white, sprinkled profusely with little black spots, and the breast is pure white. The abdomen is also white, but diversified with a slight green tip to each feather, and the flanks are bright metallic green, nearly as resplendent as upon the back. The under tail-coverts are white, with a few very pale green spots. The colours of the female are rather more dull than those of her mate, the green being tinged with yellow, and the under parts without the green spots. The first half of the tail is yellowish green, and all the feathers of the tail, with the exception of the two central feathers, are furnished with white tips.

The nest of the Vervain Humming-bird is very small, in accordance with the dimensions of the architect, is round and cup-like in shape, and beautifully constructed of cotton fibres and other soft and warm substances. As is the case with the nests of almost all the species of Trochilidæ, the rim is so made as to curve slightly inwards, and is, in all probability, constructed for the purpose of preventing the eggs from rolling out of the nest when the "procreant cradle" is rocked by the tempestuous winds of the tropics.

RUBY-THROATED HUMMING-BIRD.
(*Colubris.*)

VERVAIN HUMMING-BIRD.—(*Mellisuga minima.*)

We now arrive at the CERTHIDÆ or CREEPERS, the best known of which is our English CREEPER.

This little bird is one of the prettiest and most interesting of the feathered tribes that are found in this country. It is a very small bird, hardly so large as a sparrow, and beautifully slender in shape. The bill is rather long, ointed, and curved, and the tail-feathers are stiff and pointed at their extremities. The food of the Creeper consists chiefly of insects, although the ird will sometimes vary its diet by seeds and other vegetable substances. The insects on which it feeds live principally under the bark of various ough-skinned trees, and when it is engaged in running after its food, it runs pirally up the trunk with wonderful ease and celerity, probing every crevice vith ready adroitness, its whole frame instinct with sparkling eagerness, and ts little black eyes glancing with the exuberance of its delight. While unning on the side of the tree which is nearest to the spectator, it presents . very curious appearance, as its dark-brown back and quick tripping movements give it a great resemblance to a mouse, and ever and anon, as it comes gain into sight from the opposite side of the trunk, its beautifully white reast gleams suddenly in contrast with the sombre-coloured bark. Its eyes re wonderfully keen, as it vill discern insects of so minute a form that the human ye can hardly perceive them, nd it seems to possess some nstinctive mode of detecting the presence of its insect rey beneath moss or lichens, nd will perseveringly bore hrough the substance in vhich they are hidden, never ailing to secure them at ist.

The Creeper is a very timid ird, and if it is alarmed at he sight of a human being, will either fly off to a distant tree, or will quietly slip ound the trunk of the tree n which it is running, and eep itself carefully out of ght. It soon, however, gains onfidence, and, provided that he spectator remains perfectly quiet, the little head nd white breast may soon e seen peering anxiously ound the trunk, and in a few inutes the bird will resume

COMMON TREE CREEPER.—(*Certhia familiaris.*)

s progress upon the tree, and run cheerily up the bark, accompanying itself ith its faint trilling song. It seldom attempts a long flight, seeming to content itself with flitting from tree to tree.

The nest of the Creeper is usually made in the hollow of some decaying ee, and is made of grasses, leaves, and vegetable fibres, and lined with athers. The eggs are very small, about seven or eight in number, and of a ashen grey colour, sprinkled with little grey-brown spots. Sometimes it uilds in the hole of an old wall, and has been known to make its nest in a isused spout.

THE Nuthatches are represented in England by the common NUTHATCH f our woods. They are all remarkable for their peculiarly stout and sturdy uild, their strong, pointed, cylindrical beaks, and their very short tails.

The Nuthatch, although by no means a rare bird, is seldom seen except by those who are acquainted with its haunts, on account of its shy and retiring habits. As it feeds mostly on nuts, it is seldom seen except in woods or their immediate vicinity, although it will sometimes become rather bold, and frequent gardens and orchards where nuts are grown. The bird also feeds upon insects, which it procures from under the bark after the manner of the creepers, and it is not unlikely that many of the nuts which are eaten by the Nuthatch have been inhabited by the grub of the nut weevil. It will also feed upon the seeds of different plants, especially preferring those which it pecks off the fir-cones.

In order to extract the kernel of the nut, the bird fixes the fruit securely in some convenient crevice, and, by dint of repeated hammerings with its beak, breaks a large ragged hole in the shell, through which the kernel is readily extracted. The blows are not merely given by the stroke of the beak, but the bird grasps firmly with its strong claws, and, swinging its whole body upon its feet, delivers its stroke with the full weight and sway of the body.

NUTHATCH.—(*Sitta Europæa.*)

The nest of the Nuthatch is placed in the hollow of a decaying tree, and the bird always chooses some hole to which there is but a small entrance. Should the orifice be too large to please its taste, it ingeniously builds up the hole with clay and mud, probably to prevent the intrusion of any other bird. If any foe should venture too near the nest, the mother bird becomes exceedingly valiant, and dashing boldly at her enemy, bites and pecks so vigorously with her powerful beak, hissing and scolding the while, that she mostly succeeds in driving away the assailant. The nest is a very inartificial structure, made chiefly of dried leaves laid loosely upon the decaying wood, and rudely scraped into the form of a nest.

In its colour the Nuthatch is rather a pretty bird, of pleasing though not of brilliantly tinted plumage. The general colour of the upper parts is a delicate bluish grey, the throat is white, and the abdomen and under parts are reddish brown, warming into rich chestnut on the flanks. From the angle of the mouth a narrow black band passes towards the back of the neck, enveloping the eye in its course and terminating suddenly before it reaches the shoulders. The tail is black on the base and grey towards the tip, except the two outer tail-feathers, which have each a black spot near the extremity. The shafts are also black.

LYRE BIRD. —(*Menura superba.*)

WE now arrive at the family of the WRENS, in which group we find two birds so dissimilar in outward appearance as apparently to belong to different orders, the one being the common WREN of England, and the other the celebrated Lyre-bird of Australia.

This bird, which also goes under the name of NATIVE PHEASANT among the colonists, and is generally called BULLEN-BULLEN by the natives, on account of its peculiar cry, would, if it had been known to the ancients, have been consecrated to Apollo, its lyre-shaped tail and flexible voice giving it a double claim to such honours. The extraordinary tail of this bird is often upwards of two feet in length, and consists of sixteen feathers, formed and arranged in a very curious and graceful manner. The two outer feathers are broadly webbed, and, as may be seen in the illustration, are curved in a manner that gives to the widely-spread tail the appearance of an ancient lyre. When the tail is merely held erect and not spread, the two lyre-shaped feathers cross each other, and produce an entirely different outline. The two central tail-feathers are narrowly webbed, and all the others are modified with long slender shafts, bearded by alternate feathery filaments, and well representing the strings of the lyre.

The tail is seen in its greatest beauty between the months of June and September, after which time it is shed, to make its first reappearance in the ensuing February or March. The habits of this bird are very curious, and are so well and graphically related by Mr. Gould, that they must be given in his own words :—

"The great stronghold of the Lyre-bird is the colony of New South Wales, and, from what I could learn, its range does not extend so far to the eastward as Moreton Bay, neither have I been able to trace it to the westward of Port Phillip on the southern coast ; but further research can only determine these points. It inhabits equally the bushes on the coast and those that clothe the sides of the mountains in the interior. On the coast it is especially abundant at the Western Port, and Illawarra ; in the interior the cedar brushes of the Liverpool range, and, according to Mr. G. Bennett, the mountains of the Tumat country are among the places of which it is the denizen.

"Of all the birds I have ever met with, the Menura is far the most shy and difficult to procure. While among the mountains I have been surrounded by these birds, pouring forth their loud and liquid calls for days together, without being able to get a sight of them, and it was only by the most determined perseverance and extreme caution that I was enabled to effect this desirable object, which was rendered more difficult by their often frequenting the almost inaccessible and precipitous sides of gullies and ravines, covered with tangled masses of creepers and umbrageous trees ; the cracking of a stick, the rolling down of a small stone, or any other noise, however slight, is sufficient to alarm it ; and none but those who have traversed these rugged, hot, and suffocating bushes, can fully understand the anxious labour attendant on the pursuit of the Menura.

"At Illawarra it is sometimes successfully pursued by dogs trained to rush suddenly upon it, when it immediatey leaps upon the branch of a tree, and, its attention being attracted by the dog below barking, it is easily approached and shot. Another successful mode of procuring specimens is by wearing the tail of a full-plumaged male in the hat, keeping it constantly in motion, and concealing the person among the bushes, when, the attention of the bird being arrested by the apparent intrusion of another of its own sex, it will be attracted within the range of the gun. If the bird be hidden from view by surrounding objects, any unusual sound, such as a shrill whistle, will generally induce him to show himself for an instant, by causing him to leap with a gay and sprightly air upon some neighbouring branch to ascertain the cause of the disturbance ; advantage must be taken of this circumstance immediately, or the next moment it may be half-way down the gully.

"The Menura seldom, if ever, attempts to escape by flight, but easily eludes pursuit by its extraordinary powers of running. None are so efficient in obtaining specimens as the naked black, whose noiseless and gliding steps enable him to steal upon it unheard or unperceived, and with a gun in his hand he rarely allows it to escape, and in many instances he will even kill it with his own weapons.

"The food of the Menura appears to consist principally of insects, particularly of centipedes and coleoptera. I also found the remains of shelled snails in the gizzard, which is very strong and muscular."

The nest of the Lyre-bird is a large, loosely-built, domed structure, composed of small sticks, roots, and leaves, and of an oven-like shape, the entrance being in front. The lining is warm and soft, being composed of downy feathers.

The egg of this singular bird is quite as curious as its general form, and presents the curious anomaly of an egg as large as that of a common fowl, possessing all the characteristics of the insessorial egg. The general colour of the egg is a deep chocolate tint, marked with purple more or less deep in different specimens, and its surface is covered with a number of stains and blotches of a darker hue, which are gathered towards the larger end, as is usual in spotted eggs.

WE are all familiar with the WREN.

The long and harsh name of Troglodytes, which has been given to this bird, signifies a diver into caves, and has been attributed to the Wren on account of its shy and retiring habits, and its custom of hiding its nest in some hollow or crevice where it may escape observation. The Wren is seldom to be seen in the open country, and does not venture upon any lengthened flight, but con-

WREN.—(*Troglodytes vulgaris.*)

fines itself to the hedge-rows and brushwood, where it may often be observed hopping and skipping like a tiny feathered mouse among the branches. It especially haunts the hedges which are flanked by ditches, as it can easily hide itself in such localities, and can also obtain a plentiful supply of food. By remaining perfectly quiet, the observer can readily watch its movements, and it is really an interesting sight to see the little creature flitting about the brushwood, flirting its saucily expressive tail, and uttering its quick and cheering note.

The voice of the Wren is very sweet and melodious, and of a more powerful character than would be imagined from the dimensions of the bird. The Wren is a merry little creature, and chants its gay song on the slightest encouragement of weather. Even in winter there needs but the gleam of a few stray sunbeams to set the Wren a-singing, and the cold Christmas season is often cheered with its happy notes.

The nest of the Wren is rather an ambitious structure, being a completely domed edifice, and built in a singularly ingenious manner. If, however, the bird can find a suitable spot, such as the hole of a decaying tree, the gnarled and knotted branches of old ivy, or the overhanging eaves of a deserted building, where a natural dome is formed, it is sure to seize upon the opportunity and to make a dome of very slight workmanship. The dome, however, always exists in some form.

The materials of which the nest is composed are always leaves, moss, grass, and lichens, and it is almost always so neatly built that it can hardly be seen by one who was not previously aware of its position. The opening of the nest is always at the side, so that the eggs are securely shielded from the effects of weather.

As to the locality and position in which the nest is placed, no definite rule is observed, for the Wren is more capricious than the generality of birds in fixing upon a house for her young. Wrens' nests have been found in branches, hedges, hayricks, waterspouts, hollow trees, barns and outhouses. Sometimes the Wren becomes absolutely eccentric in its choice, and builds its nest in spots which no one would conjecture that a bird would select. A Wren has been known to make its nest in the body of a dead hawk which had been killed and nailed to the side of a barn. Another Wren chose to make her house in the throat of a dead calf, which had been hung upon a tree, and another of these curious little birds was seen to build in the interior of a pump, gaining access to her eggs and young through the spout.

The eggs of the Wren are very small, and are generally from six to eight in number.

During the winter, the Wren generally shelters itself from the weather in the same nest which it had inhabited during the breeding season, and in very cold seasons it is not an uncommon event to find six or seven Wrens all huddled into a heap for the sake of warmth, and presenting to the eye or hand of the spectator nothing but a shapeless mass of soft brown feathers. It is probable that these little gatherings may be composed of members of the same family.

The colour of the Wren is a rich reddish brown, paling considerably on the under surface of the body, and darkening into dusky brown upon the quill feathers of the wings and tail. The outer webs of the former are sprinkled with reddish brown spots, and the short tail-feathers are barred with the same hue. The bill is slender, and rather long in proportion to the general dimensions of the bird. The total length of the Wren is rather more than four inches. White and pied varieties are not uncommon.

WE now arrive at the very large family of the WARBLERS.

The first example of the Warblers is the celebrated TAILOR-BIRD of India and the Indian Archipelago.

The Tailor-bird is a sober little creature, not more conspicuous than a common sparrow, and is chiefly remarkable for its curious nest, which is made in a singular and most ingenious manner. Taking two leaves at the extremity of a slender twig, the bird literally sews them together at their edges, its bill taking the place of the needle, and vegetable fibres constituting the thread. A quantity of soft cottony down is then pushed between the leaves, and a convenient hollow scraped out in which the eggs may lie and the young birds may rest at their ease. Sometimes, if the leaf be large enough, its two edges are drawn together, but in general a pair of leaves are needed. A few feathers are sometimes mixed with the down.

This curious nest is evidently hung at the very extremity of the twigs, in order to keep it out of the way of the monkeys, snakes, and other enemies which might otherwise attack and devour mother and young together.

THE tiny GOLDEN-CRESTED WREN, as it is popularly called, is very common throughout England, and may be seen hopping and flitting merrily among the branches in copses, orchards, and plantations. Although from its diminutive size it has gained the title of Wren, it has no claim to that designation, and is more rightly termed the Kinglet, or Regulus.

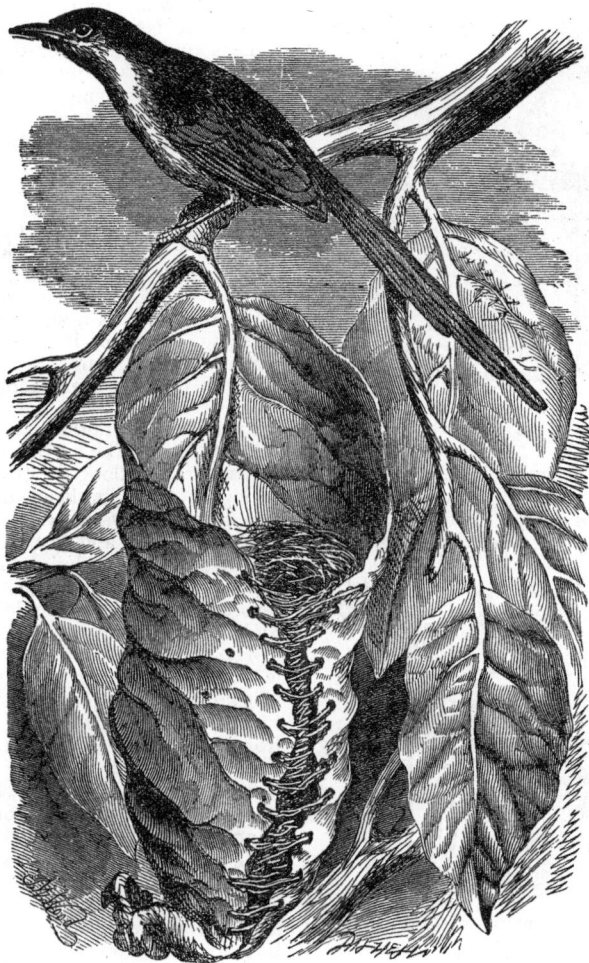

TAILOR-BIRD.—(*Orthotomus longicaudus.*)

The Golden-crested Wren is notable for the crest of golden-coloured feathers which is placed upon the crown of its head, which it can raise or depress at pleasure, and which gives so pert and changeful an expression to the little creature. But for this golden crest, which is not at all conspicuous when the feathers are lowered, the bird might easily be mistaken for a tree-

X

creeper as it runs up and down the branches, searching into the crevices of the bark for the little insects on which it feeds. The first specimen that I ever saw was traversing the branches of a fine "Blenheim Orange" apple-tree in an adjoining garden, and by my inexperienced eyes was at first taken for a very young creeper. Like the creeper, it can even run up a perpendicular wall, peering into every little crevice, and stocking up the moss and lichens for the purpose of obtaining the insects and their eggs that are lying concealed. It will also eat the chrysalides that are found so abundantly upon the walls.

GOLDEN-CRESTED WREN.—(*Regulus cristatus.*)

All the movements of the Golden-crested Wren are full of spring and fiery activity, and the manner in which it will launch itself from one tree to another, and then, without a pause, commence traversing the branches, is a sight well worth seeing. Perhaps it is seen to best advantage among the fir-trees, where it finds great scope for its active habits. Up one branch it scuds, down another, then whisks itself through the air to a fresh tree, and then flings itself back again to its former perch. Along the twigs it runs with astonishing rapidity, sometimes clinging with its head downward, sometimes running round and round them spirally, always twisting its pert little head in every direction, and probing each hole and crevice with its sharp, slender little bill. The roughest-barked trees are its favourite resort, because in such localities it finds its best supply of insect food.

The nest of this beautiful little bird is exquisitely woven of various soft substances, and is generally suspended to a trunk where it is well sheltered from the weather. I have often found their nests, and in every instance have noticed that they are shaded by leaves, the projecting portion of a banch, or some such protection. In one case the nest, which was suspended to a fir-branch, was almost invisible beneath a heavy bunch of large cones

BLACKCAP WARBLER.—(*Sylvia atricapilla.*)

that drooped over it, and forced the bird to gain admission by creeping along the branch to which the nest was suspended. The edifice is usually supported by three branches, one above and one at either side. The nest is usually lined with feathers, and contains a considerable number of eggs, generally from six to ten. These eggs are hardly bigger than peas, and, as may be supposed, their shells are so delicately thin, that to extract the interior without damaging them is a very difficult matter.

The entire length of this bird is about three inches and a half, and its general colour is brownish above, marked with olive-green, and flanked with white on the wing-coverts. The under surface is yellowish grey, the beak is black, and the eye hazel-brown. The forehead is marked with greyish white ; the crest is brilliant yellow tipped with orange, and on each side of the crest runs a black line. The female is not so brilliant in her colouring, and the crest is wholly of a pale yellow.

The FIRE-CRESTED WREN is very similar to the preceding species, but may be distinguished from it by the ruddy hue of the forehead, the fiery orange of the crest, and the decidedly yellow hue of the sides of the neck. It is an inhabitant of England, but it is a much rarer bird than the Golden-crest. Owing to the great resemblance between the two species, they have often been mistaken for each other, and it is only within a comparatively recent period that their diversity was established.

WITH the exception of the nightingale, the BLACKCAP WARBLER is the sweetest and richest of all the British song-birds, and in many points the voice of the Blackcap is even superior to that of the far-famed Philomel.

The Blackcap derives its name from the tuft of dark feathers which crown the head, and which in the males are coal black, but in the females are deep reddish brown. It is rather late in arriving, seldom being seen or heard until the end of April, and it remains with us until the middle of September. As several specimens of this pretty bird have been noticed in England in the months of December and January, it is probable that some individuals may not migrate at all, but remain in this country through-out the entire winter. Should it do so, it might easily escape notice, as it would not be likely to sing much during the cold months, and owing to its retiring habits it is at all times more likely to be heard than seen.

NIGHTINGALE.—(*Luscinia Philomela.*)

The food of the Blackcap consists chiefly of insects, but it also pays attention to the ripe fruit in the autumn, being especially fond of rasp-berries. Perhaps it may choose this fruit on account of the little white maggots that are so often found in the centre of the over-ripe raspberry.

The nest of the Blackcap is generally placed only a foot or so above the ground, within the shelter of a dense bush or tuft of rank herbage, and is composed of vegetable fibres and hairs rather loosely put together. The eggs are four or five in number, and are of a pale reddish brown, dappled with a deeper hue of brown. The general colour of the Blackcap is grey, with a wash of dark green upon the upper surface and ashen grey upon the lower surface. The total length of the bird is not quite six inches, its extent of wings nearly nine inches, and its weight not quite half an ounce.

THE well-known and far-famed NIGHTINGALE is, happily for us, an in-habitant of England, visiting us about the middle of April and remaining until the breeding season is over.

It seems to be rather a local bird, some parts of England appearing to be quite unsuited to its habits. The northern counties are seldom visited by this bird, and in Ireland and Scotland it is almost unknown.

The food of the Nightingale consists principally of various insects, and it

is so powerfully attracted by the common mealworm, that one of these creatures employed as a bait is sure to attract the bird to its destruction.   It
appears to make great havoc among the caterpillars, which come out to feed
at night, and are to be seen so abundantly on damp warm evenings.   In the
autumn it is somewhat of a fruit-eater, and has been seen in the act of eating
"blackheart" cherries, plucking them from the tree and carrying them to its
young.

   As is well known, the song of the Nightingale is mostly uttered after sunset, but the bird may be heard in full song throughout the day.   Towards
the end of June, when the young birds are hatched, the song changes into a
kind of rough croaking sound, which is uttered by way of warning, and
accompanied with a sharp snapping sound of the beak.   The time when the
Nightingales sing loudest and most constantly is during the week or two
after their arrival, for they are then engaged in attracting their mates, and
sing in fierce rivalry of each other, hoping to fascinate their brides by the

WHEATEAR. —(*Saxicola œnanthe.*)

splendour of their voices.   When once the bird has procured a partner, he
becomes deeply attached to her, and if he should be captured, soon pines
away and dies, full of sorrowful remembrances.   The bird-dealers are therefore anxious to catch the Nightingale before the first week has elapsed, as
they can then, by dint of care and attention, preserve the bird in full song to
a very late period.   Mr. Yarrel mentions an instance where a caged Nightingale sang upon a hundred and fourteen successive days.

   The nest of the Nightingale is always placed upon or very near the
ground, and is generally carefully hidden beneath heavy foliage.   One such
nest that I discovered in Wiltshire was placed among the knotted and
gnarled roots of an old ivy-covered thorn-stump that still maintained its
place within a yard of a footpath.   The nest is made of grass and leaves,
and is of exceedingly slight construction, so slight, indeed, that to remove it
without damage is a very difficult process, and requires the careful use of the
hands.   The eggs are generally four and sometimes five in number, and are

of a peculiar smooth olive-brown, which distinguishes them at once from the egg of any other British bird of the same size.

The colour of the Nightingale is a rich hair-brown upon the upper part of the body, and greyish white below, the throat being of a lighter hue than the breast and abdomen. The entire length of the bird rather exceeds six inches.

A SMALL but very interesting group of birds now claim our attention. These are the Erythacinæ, or Redbreast kind, including the Redbreast, the Wheatear, and other birds.

The WHEATEAR, or FALLOW CHAT, is a well-known visitant of the British Isles, and on account of the delicate flavour of its flesh when fat, is sadly persecuted throughout the whole time of its sojourn.

Being in great favour for the table, where it is popularly known as the English ortolan, and consequently fetching a good price in the market, it is caught in great numbers, and sold to the game-dealers of London. The trap by which it is captured is a remarkably simple affair, consisting merely of an oblong piece of turf cut from the soil, and arranged crosswise over the cavity from which it was taken. A horsehair noose is supported under the turf by means of a stick, and the trap is complete, needing no bait or super-vision. It is the nature of the Wheat-ear to run under shelter at the least alarm; a passing cloud sufficing to drive it under a stone or into a hole in a bank. Seeing, therefore, the sheltering turf, the Wheatear runs beneath it, and is caught in the noose. These simple traps are much used by the shepherds, who can make and attend to four or five hundred in a day, and have been known to catch upwards of a thousand Wheatears within twenty-four hours.

As a general rule, the nest of the Wheatear is hidden in the most per-fect manner, the bird ordinarily choos-ing to place its domicile within the recesses of large stone-heaps, in deep rocky crannies, and in similar loca-

REDSTART.—(*Ruticilla phœnicura.*)

lities; so that, even if it should be discovered, the work of obtaining it is very severe. In some pats of the cliff-bound sea-coast, the Wheatear's nest is so deeply buried in the rocky crevices, that the only mode of obtaining the eggs is to hook out the nest by means of a bent wire at the end of a long stick.

The upper part of the body is light silver-grey, and the quill feathers of the wings, together with their coverts, are deep black. The middle tail-feathers and the tips of the various rectrices, are of the same hue, and a black streak passes from the edge of the beak to the ear, enveloping the eye, and spreading widely upon the ear-coverts. The breast is buff, with a decided orange tinge, and the abdomen is beautifully white. The female is not quite so handsome; the wings, tail, and ear-coverts being dark brown, and the lighter portions of the body tinged with brown. The total length of the bird is about six inches and a half.

THE specific title of *phœnicura*, which is given to the REDSTART, signifies "ruddy-tail," and is attributed to the bird in consequence of the light ruddy chestnut feathers of the tail and upper tail-coverts.

It is a handsomely-coloured and elegantly-shaped bird, and is a great ornament to our fields and hedge-rows. The name of Redstart is a very appropriate one, and has been given to the bird in allusion to the peculiar character of its flight. While walking quietly along the hedge-rows, the observer may often see a bird flash suddenly out of the leafage, flirt its tail in the air, displaying strongly a bright gleam of ruddy hue, and after a sharp dash of a few yards, turn into the hedge again with as much suddenness as it had displayed in its exit. These manœuvres it will repeat frequently, always keeping well in front, and at last it will quietly slip through the hedge, double back on the opposite side, and return to the spot whence it had started.

No one need fancy, from seeing the bird in the hedge, that its nest is in close proximity, for the Redstart seldom builds in such localities, only haunting them for the sake of obtaining food for its young. The nest is almost invariably built in the hole of an old wall, in a crevice of rock, a heap of

REDBREAST.—(*Erythacus rubecula.*)

large stones, in a hollow tree, or in very thick ivy. The eggs are generally five in number, although they vary from four to seven, and are of a beautiful blue, with a slight tinge of green. They are not unlike those of the common hedge-sparrow, but are shorter and of a different contour.

The Redstart has a very sweet song, which, although not very powerful, is soft and melodious, bearing some resemblance to that of the nightingale.

The food of the Redstart is mostly of an insect nature, and is obtained in various ways. Sometimes the bird dashes from its perch upon a passing insect, after the manner of the flycatcher ; sometimes it chases beetles and other creeping insects upon the leaves and branches of the hedges ; sometimes it hunts for worms, grubs, and snails from the ground ; and it often picks maggots out of fungi, decaying wood, mosses, and lichens. Soft ripe fruit is also eaten by the Redstart, which, however, ought to be allowed its free range of the garden in recompense for the great service which it has performed in the earlier portion of the year, by devouring the myriad insects that feed upon the blossoms of fruit-trees. The softer berries form part of the Redstart's diet, but the bird does not seem to care about the hard seeds.

THERE are few birds which are more familiar to us than the REDBREAST, or ROBIN, a bird which is interwoven among our earliest recollections, through the medium of "The Children in the Wood" and the mournful ballad of "The Death and Burial of Cock Robin."

Although the Redbreast remains in England throughout the winter, it is very susceptible to cold, and one of the first birds to seek for shelter, its appearance among the outhouses being always an indication of coming inclemency. In cold weather, the Redbreast seldom perches upon twigs and branches, but crouches in holes or sits upon the ground. The bird seems strongly attached to man and his home, and will follow the ploughman over the fields, picking up the worms which he turns up with the ploughshare, or enter his house and partake of his evening meal.

The nest of this bird is generally placed near the ground in a thick leafy bush, or in a bank, and is composed of dry leaves, moss, grass, hair, and feathers. I have seen the nest very well concealed among the thick ivy that had wreathed round a tree-trunk, and placed about eight feet from the ground. The bird seldom flies directly to its nest, or leaves it directly, but alights at a little distance, and creeps through the leaves or branches until it enters its home.

The eggs of the Redbreast are generally five in number, as is the case with most of the song-birds, and their colour is greyish white, covered with variously-sized spots of pale rusty red. The song of this bird is very sweet and pleasing; and it is a pretty sight to observe two or more Redbreasts perched on different trees, and answering each other with their musical cries. Whenever the Redbreast perches on the top of a tree or other elevated spot, and begins to sing merrily, it is an unfailing indication that the weather of the coming day promises to be fair. The bird sings throughout the greater part of the year, beginning early in spring, and continuing it very late into the autumn. Even in the winter months, a bright sunny day is apt to excite the Robin to perch upon a twig and pour forth a sweet though broken melody.

The colour of the male Robin is bright olive-brown on the back, orange-red on the throat, chin, breast, forehead, and round the eye. A stripe of blue-grey runs round the red, and the abdomen and lower part of the breast are white. The bill and eyes are black. The female is coloured after the same manner, but the tints are not so vivid as in her mate. The total length of the bird is nearly six inches, and its weight about half an ounce.

OF the pretty though sober-plumaged Accentors, we have one or two British examples, that which is best known being the HEDGE ACCENTOR, or HEDGE SPARROW, as it is often, though wrongly, called, as it by no means belongs to the same group of birds.

The Hedge Accentor is very common through the whole of England, and may be heard in the gardens, copses, and hedge-rows, chanting its pleasing and plaintive melody without displaying much fear of its auditors.

It is especially adapted for living among hedges, as it possesses a singular facility in threading its way through the twigs, stems, and branches. It seems equally at home in dried brushwood, and may often be seen traversing the interior of a woodpile with perfect ease. The nest is one of the earliest to be built, and is frequently completed and the eggs laid before the genial warmth of spring has induced the green leaves to burst their inclosures.

The nest is generally placed at a very low elevation, seldom more than two or three feet from the ground, and it is rather large in proportion to the size of the bird. The materials of which the structure is made are various mosses, wool, and hair, and the eggs are usually five in number, of a bright bluish green colour. Sometimes, but very rarely, six eggs are found in a single nest.

The song of the Hedge Accentor is sweet, but not varied nor powerful, and has a peculiar plaintive air about it. The bird is a persevering songster, continuing to sing throughout a large portion of the year, and only ceasing during the time of the ordinary moult. Like many other warbling birds, it possesses considerable powers of imitation, and can mock with some success the greater number of British song-birds.

The colour of the Hedge Accentor is bluish grey, covered with small brown streaks upon the head and the back and sides of the neck. The back and wings are brown, streaked with a deeper tint of the same hue, and the quill feathers of the wings and tail are of a rather darker brown, and not quite so glossy. The chin, the throat, and upper part of the breast are grey, and the lower part of the breast and the abdomen are white with a wash of pale buff. The legs and toes are brown, with a decided orange tinge, and the beak is dark brown. The total length of the bird is nearly six inches.

HEDGE SPARROW, OR ACCENTOR.
(*Accentor modularius.*)

THE group of birds which are distinguished by the name of Parinæ or TITMICE, are easily recognizable, having all a kind of famliy resemblance which guards the observer from mistaking them for any other bird.

The first example of these birds is the GREAT TITMOUSE, an inhabitant of England and many parts of Europe.

It does not migrate, finding a sufficiency of winter food in its native land. During the summer it generally haunts the forests, gardens, or shrubberies, and may be seen hopping and running about the branches of the trees in a most adroit manner, searching for insects, and occasionally knocking them out of their hiding-places by sharp blows of the bill. The beak of the Great Titmouse is, although so small, a very

GREAT TITMOUSE —(*Parus major.*)

formidable one, for the creature has often been known to set upon the smaller birds and to kill them by repeated blows on the head, afterwards pulling the skull to pieces and picking out the brains. During the winter the Great Titmouse draws near to human habitations, and by foraging among the barns and outhouses, seldom fails in discovering an ample supply of food.

The nest is always made in some convenient hollow, generally that of a tree, but often in the holes of old walls, and in the cavities that are formed by thick gnarled roots in the sides of a bank. Hollow trees, however, are the favourite nesting-places of this bird, which is able to shape the hollow to its liking, by chiselling away the decaying wood with its sharp strong beak. The materials of which the nest is made vary according to the locality. There are generally from eight to twelve eggs in each

BLUE TITMOUSE. —(*Parus cæruleus.*)

nest, and their colour is whitish grey, covered with mottlings of a rusty red, which are thickly gathered towards the larger end.

The colouring of this species is very bold, and is briefly as follows : The top of the head and throat, as far as the middle of the neck, together with a rather broad streak down the centre of the chest and abdomen, are rich purple-black, relieved by a spot of pure white on the nape of the neck, and a large flask-shaped patch under each eye. The back and shoulders are ashy green, the greater wing-coverts are blue-black, each feather being tipped with white, so as to form a bar across the wings. The quill feathers are dark green-grey, the primaries being edged with greyish white. The tail-feathers are the same green-grey, except that the extreme feathers are white on their outer ends. The under parts are light sulphurous yellow, and the under tail-coverts are white. The total length of the bird is not quite six inches.

THE little BLUE TITMOUSE is one of the most familiar birds of England, as it is widely spread throughout the land, and is of so bold a nature that it exhibits itself fearlessly to any observer. In many of its habits it resembles the last-mentioned species, but it nevertheless possesses a very marked character, and has peculiarities which are all its own. As it trips glancingly over the branches, it hardly looks like a bird, for its quick limbs and strong claws carry it over the twigs with such rapidity that it resembles a blue mouse rather than one of the feathered tribe. Being almost exclusively an insect-eating bird, and a most voracious little creature, it renders invaluable service to the agriculturist and the gardener by discovering and destroying the insects which crowd upon the trees and plants in the early days of spring, and which, if not removed, would effectually injure a very large proportion of the fruit and produce.

The nest of this species may be found in the most extraordinary localities, such as hollow trees, holes in old walls, the interior of disused spouts, sides of gravel-pits, the hat of a scarecrow, the inside of a porcelain jar, or the cylinder of a pump. One bird had actually chosen a bee-hive as its residence, and had succeeded in building its nest and rearing its young while surrounded by the bees going to and returning from their work. Another Titmouse contrived to get into a weathercock on the summit of a spire, and there made its nest in security. The eggs are small and rather numerous, being generally about eight or ten, but sometimes exceeding the latter number.

THE LONG-TAILED TIT-MOUSE is familiarly known throughout England, and is designated under different titles, according to the locality in which it resides, some of its popular names being derived from its shape, and others from its crest. In some parts of the country it is called "Long Tom," while others it goes by the name of "Bottle-crested Tit," or "Poke-Pudding," the latter word being a provincial rendering of the useful culinary apparatus termed a pudding-bag.

This pretty little bird is a notable frequenter of trees, hedgerows, and orchards, and

LONG-TAILED TITMOUSE.—(*Parus caudatus.*)

is remarkable for its sociable habits, being generally seen in little troops of six or eight in number. It appears that the young birds always remain with their parents throughout the whole of the first year, so that when the brood happens to be a large one, as many as sixteen Long-tailed Titmice may be seen hopping and skipping about together.

As far as is known, the Long-tailed Titmouse feeds exclusively on insects, and on account of its microscopical eyes is able to see and to catch the very minutest. The service which is rendered to agriculture by even a single nest of these birds is almost invaluable, for at all seasons of the year they continue to obtain their food, catching the perfect insect in the summer, and feeding on the eggs, hidden larvæ, and chrysalides in the winter.

The nest of this species is undoubtedly the most wonderful example of bird architecture that is to be found in the British Islands, and is not exceeded in beauty by the home of any bird whatever. In form it somewhat resembles an egg, and it is built of moss, hair, a very little wool, the cocoon webs of spiders, and the silken hammocks of certain caterpillars, all woven into each other in the most admirable manner. The exterior of the nest is spangled with silvery lichens, which generally correspond in colour with the bark of the tree on which it is placed, and serve to render it as little conspicuous as possible. The interior of the nest is wonderfully soft and warm, being literally crammed with downy feathers to such an extent that the eggs are deeply buried in the feathery bed, and cannot be counted until the whole lining of the nest is removed. The nest is generally placed rather near the ground, and is so well concealed that it is not easily seen except by experienced eyes.

The number of eggs which this little bird lays is really surprising. Very seldom does it content itself with eight, and double that number has been frequently counted in a single nest. In consequence, the young birds are packed like so many herrings in a barrel, and the ingenuity which must be exerted by the parent birds in giving each little one its food in proper rotation must be very great indeed.

The colouring of this species is as follows. The upper part of the head, the cheeks, the throat, and the whole of the under surface are greyish white, warming into a rosy hue upon the sides, flanks, and under tail-coverts. A broad stripe of deep black passes over the eye and ear-coverts, and joins a large triangular patch of the same jetty hue, which extends from the shoulders as far as the upper tail-coverts. The shoulders, the scapularies, and the lower part of the back are washed with a decided tinge of a ruddy hue. The wings are mostly black, with the exception of the tertiary quill feathers, which are edged with white. The long central feathers of the tail are black, and the remainder are black on the inner webs and white on the outer. They are regularly graduated in length, each pair being about half an inch shorter than the preceding pair. Both sexes are similar in their colouring. The total length of the bird is about five inches and a half.

WAGTAILS.

WE now arrive at a small group of birds which is sufficiently familiar to every observer of nature through the different representatives which inhabit this country. The WAGTAILS, so called from their well-known habit of jerking their tails while running on the ground or on settling immediately after a flight, are found in both hemispheres, and are all well known by the habit from which they derive their popular title. N 'ess than nine species of this group occur in Britain, some of which are nearly as well known as the common sparrow, while others are less familiar to the casual observer.

The PIED WAGTAIL is the most common of all the British examples of this genus, and may be seen at the proper season of the year near almost every pond or brook, or even in the open road, tripping daintily over the ground, pecking away at the insects and wagging its tail with hearty good-will.

Mr. Yarrell mentions that this bird is an accomplished fisher, and excels in snapping up the smaller minnows and fry as they come to the surface of the water. It also haunts the fields where sheep, horses, or horned cattle are kept, and hovers confidingly close to their hoofs, pecking away briskly at the little insects which are disturbed by their tread. It also delights in newly mown lawns, and runs over the smooth surface with great agility, peering between every grass-blade in search of the insects which may be lying concealed in their green shelter. The flight of the Pied Wagtail is short and jerking, the bird rising and falling in a very peculiar manner with every stroke of the wings.

The Pied Wagtail remains in England throughout the year, but generally retires to the southern counties during the winter, as it would othewise be unable to obtain its food.

The nest of the Wagtail is generally placed at no great distance from water, and is always built in some retired situation. Holes in walls, the hollows of aged trees, or niches in old gravel-pits are favourite localities with this bird. Heaps of large stones are also in great favour with the Wagtail, and I have generally found that wherever a pile of rough stones has remained for some time in the vicinity of water, a Wagtail's nest is almost invariably somewhere within it. I have also found the nest in heaps of dry brushwood piled up for the purpose of being cut into faggots. In every case the nest is placed at a considerable depth. The eggs are generally four or five in number, and their colour is grey-white, speckled with a great number of very small brown spots.

The colouring of the Pied Wagtail is almost entirely black and white, very boldly disposed and distributed as follows. The top of the head, the nape of the neck, part of the shoulders, the chin, neck, and throat, are jetty black, contrasting boldly with the pure snowy white of the sides of the face and the white patch on the sides of the neck. The upper tail-coverts and the coverts of the wings are also black. The quill feathers of the wings are black, edged on the outer web with a lighter hue. The two exterior feathers of the tail are pure white, edged on the inner web with white, and the remainder jetty black. The under parts of the body are greyish white, taking a blue tint upon the flanks. The entire length of the bird is between seven and eight inches. This is the summer plumage of the male bird. In the winter the chin and throat exchange their jetty hue for a pure white, leaving only a collar of black round the throat. The female much resembles her mate in the general colouring of her plumage, but is about half an inch shorter.

PIED WAGTAIL.—(*Motacilla Yarrellii.*)

The PIPITS, or TITLARKS as they are sometimes called, form a well-marked group, which possesses the long hind toe of the hawk, together with very similar plumage, and also bears the long tail which is found in the Wagtails Several species of Pipit inhabit England.

The best-known is the common MEADOW PIPIT, or MEADOW TITLING, a bird which may be seen throughout the year upon moors, waste lands, and marshy ground, changing its locality according to the season of year. It is a pretty though rather sombre little bird, and is quick and active in its movements, often jerking its long tail in a fashion that reminds the observer of the Wagtail's habits. It moves with considerable celerity, tripping over the rough and rocky ground which it frequents, and picking up insects with the stroke of its unerring beak. Its food, however, is of a mixed description, as in the crops of several individuals were found seeds, insect and water-shells, some of the latter being entire.

The nest of this species is placed on the ground, and generally hidden in a large grass-tuft.

THE very large family of the THRUSHES now engages our attention. Many of these birds are renowned for their song, and some of them are remarkable for their imitative powers.

THE Ant Thrushes find an English representative in the well-known DIPPER, or WATER OUSEL, of our river banks.

Possessing neither brilliant plumage nor graceful shape, it is yet one of the most interesting of British birds when watched in its favourite haunts. It always frequents rapid streams and channels, and being a very shy and retiring bird, invariably prefers those spots where the banks overhang the water, and are clothed with thick brushwood. Should the bed of the stream be broken up with rocks or large stones, and the fall be sufficiently sharp to wear away an occasional pool, the Dipper is all the better pleased with its home, and in such a locality may generally be found by a patient observer.

All the movements of this little bird are quick, jerking, and wren-like, a similitude which is enhanced by its habit of continually flirting its apology for a tail. Caring nothing for the frosts of winter, so long as the water

MEADOW PIPIT. —(*Anthus pratensis.*)

remains free from ice, the Dipper may be seen throughout the winter months flitting from stone to stone with the most animated gestures, occasionally stopping to pick up some morsel of food, and ever and anon taking to the water, where it sometimes dives entirely out of sight, and at others merely walks into the shallows and there flaps about with great rapidity.

The food of the Dipper seems to be exclusively of an animal character, and, in the various specimens which have been examined, consists of insects in their different stages, small crustaceæ, and the spawn and fry of various fishes. Its fish-eating propensities have been questioned by some writers. but the matter has been entirely set at rest by the discovery of fish-bones and half-digested fish in the stomachs of Dippers that had been shot. Generally, however, the food consists of water-beetles, particularly of the genus known by the name of Hydrophilus, a flat, oval-shaped insect, with hard wing-cases and oar-like hind-legs. The bird has also been known to pick up the caddis worms, taking them on shore, pulling and knocking to pieces the tough case in which the fat white grub is enveloped, and swallowing the contents.

The song of the Dipper is a lively and cheerful performance, and is uttered most frequently in the bright frosty mornings. Sometimes it will stand upon a stone when singing, and accompany its song with the oddest imaginable gestures, hopping and skipping about, twisting its head in all directions and acting as if it were performing for the amusement of the spectator.

The nest is not unlike that of the wren, being chiefly composed of mosses built into a dome-like shape with a single aperture in the side. It is generally placed near the water, and always under some sort of cover, usually a hole in the bank.

The nest is not, however, always so close to the water, for I found one near Swindon, in the side of an old disused pit, at some little distance from the

DIPPER.—(*Hydrobates cinclus.*)

great Swindon reservoir. It was discovered more by accident than by intention, the touch having given the first intimation of its presence. The moss always remains in a green state, as it is placed in a damp locality, so that it can with great difficulty be distinguished from the vegetation of the spot whereon it is situated.

The eggs are pure white, and rather long in proportion to their breadth. Their full number is five, and the young remain with their parents for a considerable period, forming little companies of five or six of these curious birds.

The general colour of this bird is brown on the upper surface of the body; the throat and upper parts of the chest are white, and the abdomen is rusty red. The young birds possess a rather variegated plumage of black, brown, ash colour, and white. The total length of the adult bird is about seven inches.

THE MOCKING BIRD of America is universally allowed to be the most wonderful of all songsters, as it not only possesses a very fine and melodious voice,

MOCKING BIRD.—(*Mimus polyglottus.*)

but is also endowed with the capacity for imitating the notes of any other bird, and, indeed, of immediately reproducing with the most astonishing exactness any sound which it may hear.

All persons who come within the sound of the Mocking Bird's voice are fascinated with the thrilling strains that are poured without effort from the melodious throat, and every professed ornithologist who has heard this wonderful bird has exhausted the powers of his language in endeavouring to describe the varied and entrancing melody of the Mocking Bird. Within the compass of one single throat the whole feathered race seems to be comprised, for the Mocking Bird can with equal ease imitate, or rather reproduce, the sweet and gentle twittering of the blue-bird, the rich full song of the thrush, or the harsh ear-piercing scream of the eagle.

Let it but approach the habitation of man, and it straightway adds a new series of sounds to its already vast store, laying up in its most retentive memory the various noises that are produced by man and his surroundings, and introducing among its other imitations the barking of dogs, the harsh "setting" of saws, the whirring buzz of the millstone, the everlasting clack of the hoppers, the dull heavy blow of the mallet, and the cracking of splitting timbers, the fragments of songs whistled by the labourers, the creaking of ungreased wheels, the neighing of horses, the plaintive "baa" of the sheep, and the deep lowing of the oxen, together with all the innumerable and accidental sounds which are necessarily produced through human means. Unfortunately, the bird is rather apt to spoil his own wonderful song by a sudden introduction of one of these inharmonious sounds, so that the listener, whose ear is being delighted with a succession of the softest and richest toned vocalists, will suddenly be electrified with the loud shriek of the angry hawk or the grating whirr of the grindstone.

The nest of this bird is usually placed in some thick bush, and is in general very carefully concealed. Sometimes, however, when the bird builds in localities where it knows that it will be protected from human interference, it is quite indifferent about the concealment of its home, and trusts to its own prowess for the defence of its mate and young. The nest is always placed at a short distance from the ground, being seldom seen at an elevation of more than eight feet.

MISSEL THRUSH.—(*Turdus viscivorus.*)

The materials of which the nest is composed are generally dried weeds and very slender twigs as a foundation ; straw, hay, wool, dried leaves, and moss, as the main wall ; and fine vegetable fibres as the lining. The eggs are four or five in number, and there are often two broods in the course of the year. The colour of the eggs is greenish blue, spotted with amber-brown.

THE first example of the true Thrushes is the MISSEL THRUSH, one of the largest and handsomest of the species.

It is one of our resident birds, and on account of its great size, its combative nature, its brightly-feathered breast, its rich voice and gregarious habits, is one of the best known of the British birds. About the beginning of April the Missel Thrush sets about its nest, and in general builds a large weighty edifice, that can be seen through the leafless bushes from a great distance.

The materials of which the nest is composed are the most heterogeneous that can be imagined. Every substance that can be woven into a nest is pressed into the service. Moss, hay, straw, dead leaves, and grasses are among the ruling substances that are employed for the purpose, and the bird often adds manufactured products, such as scraps of rag, paper, or shavings. I once found one of these nests that was ingeniously placed in the crown of an old hat which had evidently been flung into the tree by some traveller. At first, it hardly looked like a nest, but there were a few bits of grass lying over the brim that had a very suspicious aspect, and on climbing the tree, the old hat was proved to have been made the basis of a warm nest, with the proper complement of eggs.

Towards the end of the summer the Missel Thrushes assemble in flocks of considerable size, and in the autumn often do great harm to gardens and plantations, by devouring the fruit. They are particularly fond of raspberries and cherries, and have been known entirely to ruin the crop of these fruits. They are also fond of the berries of the mountain ash and the arbutus, and are so partial to the viscid berries of the mistletoe plant that they have been called by its name. Insects of various kinds, caterpillars, and spiders also form part of the Missel Thrush's diet, and a partly-digested lizard has been found in the interior of one of these birds.

The song of the Missel Thrush is rich, loud, clear, and ringing, and is often uttered during the stormiest period of the year, the bird seeming to prefer the roughest and most inclement weather for the exercise of its voice.

ANOTHER large example of the British Thrushes is found in the FIELDFARE. This bird is one of the migratory species, making only a winter visit to this country, and often meeting a very inhospitable reception from the gun of the winter sports-boy. Very seldom is it seen in this country till November, and is often absent until the cold month of December, when it makes its appearance in great flocks, searching eagerly for food over the fields. When the snow lies heavily upon the fields, this bird betakes itself to the hedge-rows and outskirts of woods and copses, and there feeds on the various berries that have survived the autumn. During this inclement season the Fieldfare may be approached and shot without much difficulty. Their shyness, however, depends greatly on the amount of persecution which they have sustained.

In its colour the Fieldfare bears a decided resemblance to the generality of the Thrushes. The upper parts of the body as far as the shoulders are ashen grey, dotted with dark-brown spots upon the head ; the back and wings are rich brown, and the tail is dark blackish brown. The chin and throat are a peculiar golden hue, not unlike amber, and covered with numerous black streaks ; the breast is reddish brown, also spotted with black, and the abdomen and under parts are white, spotted on the flanks and under tail-coverts with brown of various shades. The Fieldfare is not quite so large a bird as the Missel Thrush, being about ten inches in total length.

AMONG the best-known and best-loved of our British songsters, the BLACKBIRD is one of the most conspicuous.

This well-known bird derives its popular name from the uniformly black hue of its plumage, which is only relieved by the bright orange-coloured bill of the male bird. The song of this creature is remarkable for its full mellowness of note, and is ever a welcome sound to the lover of nature and her vocal and visual harmonies.

The Blackbird feeds usually on insects, but it also possesses a great love of fruit, and in the autumn ravages the gardens and orchards in a most destructive manner, picking out all the best and ripest fruit, and wisely leaving the still immatured produce to ripen on the branches.

The nest of this bird is made very early in the spring, and is always carefully placed in the centre of some thick bush, a spreading holly-tree being a very favourite locality. It is a large, rough, but carefully-constructed habitation, being made externally of grass-stems and roots, plastered on the interior with a rather thick lining of coarse mud, which, when thoroughly dried, forms a kind of rude earthenware cup. A lining of fine grass is placed within the earthen cup, and upon this lining the five eggs are laid These eggs are of a light greyish blue ground colour, splashed, spotted and freckled over their entire surface with brown of various shades and intensity. The colouring of these eggs is extremely variable, even those of a single nest being very different in their appearance.

The Blackbird is very courageous in defence of its nest, and will attack almost any animal that threatens the security of its home. On one occasion a prowling cat was forced to retreat ignominiously from the united assaults of two Blackbirds near whose domicile she had ventured.

THE well-known SONG THRUSH, or THROSTLE as it is sometimes called, bears a deservedly high rank among our British birds of song.

It is plentifully found in most parts of England, and favours us with its vocal efforts throughout a considerable portion of the year. The song of the Thrush is peculiarly rich, mellow, and sustained, and is remarkable for the full purity of its intonation and the variety of its notes. The Thrush begins to sing as soon as incubation commences, and continues its song from the beginning of the spring until the middle of autumn. In many cases the bird sings to a very late period of the year, and has been heard in the months of November and December.

The food of the Thrush is mostly of an animal character, and consists largely of worms, snails, slugs, and similar creatures. In eating snails it is very dexterous, taking them in its bill, battering them against a stone until the shells are entirely crushed, and then swallowing the inclosed mollusc. When a Thrush has found

BLACKBIRD.—(*Turdus merula.*)

a stone that suits his purpose peculiarly well, he brings all his snails to the spot, and leaves quite a large heap of empty snail-shells under the stone. One of the best examples that I have ever seen was a large squared boulder-stone, forming part of a rustic stile in Wiltshire. There was a large pile of shells immediately under the stone, and the ground was strewed for some distance with the crushed fragments that had evidently been trodden upon and carried away by the feet of passengers.

The Thrush does not, however, confine itself wholly to this kind of diet, but in the autumn months feeds largely on berries and different fruits, being very fond of cherries, and often working great havoc in an orchard or fruit-garden. But in spite of its occasional inroads upon the gardens, it deserves the gratitude of the agriculturist on account of its service in destroying the snails and other garden pests, and may well be allowed to take its autumnal toll of a few of the fruits of which it has been such an efficient preserver.

The nest of the Thrush is rather large, and shaped like a basin. The

Y

shell of the nest is composed of roots and mosses, inside which is worked a rather thin but wonderfully compact layer of cow-dung and decayed wood, so strongly kneaded that when dry it will hold water almost as well as an earthenware vessel. There are usually five eggs, of a beautiful blue spotted with black. The spots are small, round, and well marked, and are extremely variable in size and number; they are always gathered towards the larger end of the egg.

THE GOLDEN ORIOLE is an extremely rare visitant of this country, having been but seldom observed within our coasts, but is far from uncommon in many parts of the Continent, especially the more southern portions of Europe, such as the shores of the Mediterranean and Southern Italy.

It derives its name from the bright golden yellow with which the feathers of the adult male bird are largely tinged; but as the full glory of its plumage is not displayed until the bird has entered its third year, it is possible that many specimens may have visited this country and again departed without having attracted particular attention.

It is rather gregarious in its habits, generally associating in little flocks, and frequenting lofty trees and orchards, where it can obtain abundance of food.

The nest of this bird is a very elegantly-formed and well-constructed edifice, of a shallow cup-like shape, and usually placed in a horizontal fork of a convenient branch. The materials of which it is made are mostly delicate grass-stems, interwoven with wool so firmly that the whole structure is strong and warm. The eggs are generally four or five in number, and their colour is purplish white, sparingly marked with blotches of a deep red and ashen grey. It is believed that there is but one brood in the year, so that the species does not multiply very rapidly. Sometimes the bird is said to build a deep and purse-like nest, which is suspended from a forked branch instead of being placed upon it.

THE interesting family of FLYCATCHERS is composed of a large number of

SONG THRUSH.—(*Turdus musicus.*)

GOLDEN ORIOLE.—(*Oriolus galbula.*)

species, extremely variable in size, form, and colour. The average dimensions of these birds are about equal to those of a large sparrow, and many are smaller than that bird, although two or three species nearly equal the thrush in size.

PIED FLYCATCHER. —(*Muscicapa atricapilla.*)
SPOTTED FLYCATCHER. —(*Muscicapa grisola.*)

The SPOTTED FLYCATCHER has received several local names in allusion to its habits, the titles WALL BIRD and BEAM BIRD being those by which it is most frequently designated. It is one of the migrating birds, arriving in this country at a rather late season, being seldom seen before the middle or even towards the end of May.

Y 2

This bird is fond of haunting parks, gardens, meadows, and shrubberies, always choosing those spots where flies are most common, and attaching itself to the same perch for many days in succession.   When the Flycatcher inhabits any place where it has been accustomed to live undisturbed, it is a remarkably trustful bird, and permits the near approach of man, even availing itself of his assistance.

The Spotted Flycatcher builds a very neatly made nest, and is in the habit of fixing its home in the most curious and unsuspected localities.   The hinge of a door has on more than one occasion been selected for the purpose, and in one instance the nest retained its position although the door was repeatedly opened and closed, until a more severe shock than ordinary shook the eggs out of the nest and broke them.   It is fond of selecting some human habitation for the locality in which it builds its nest, and its titles of Beam Bird and Wall Bird have been given to it because it is in the habit of making its home on beams or in the holes of walls.   The branches of a pear, apricot, vine, or honeysuckle are favourite resorts of the Spotted Flycatcher, when the tree has been trained against a wall.

The nest is generally round and cup-shaped, and is made of fine grasses, moss, roots, hair, and feathers, the harder materials forming the walls of the nest, and the softer being employed as lining.

The eggs of the Spotted Flycatcher are four or five in number, and their colour is a very pale bluish white, spotted with ruddy speckles.   As the nest is made at so late a period of the year, being but just begun when some birds have hatched their first brood, there is not often more than a single family in the course of the season.   Sometimes, however, it has been known to hatch and rear a second brood in safety.

The general colour of the Spotted Flycatcher is a delicate brown on the upper parts of the body, the quill feathers of the wings and tail being, as is usually the case, of a blacker hue than the feathers of the back.   There are a few dark spots on the top of the head, and the tertial feathers of the wings are edged with light brown.   The breast is white, with a patch of very light dull brown across its upper portion, and both the chin and breast are marked with dark brown longitudinal streaks.

THE other species of British Flycatchers is much more rare than the bird just described, and may easily be distinguished from it by the peculiarity of plumage, from which it derives its popular title.   The PIED FLYCATCHER has been observed in most parts of England, but seems to be of very rare occurrence, except in the counties of Cumberland and Westmoreland, where it is found in the vicinity of the lakes.

The colouring of this bird is as follows : In the adult male, the top of the head, back of the neck, back, and wings, are dark blackish brown, with the exception of a white patch upon the forehead, and a broad stripe of white on the tertiary and greater wing-coverts. The tail is black, except some bold white marks on some of the outer feathers, and the whole of the under surface is pure white.   The female is of a delicate brown on the upper parts of the body, and those portions which in the male are pure white are in the female of a dull whitish grey.   In dimensions the bird is not equal to the Spotted Flycatcher, barely exceeding five inches in total length.

A SMALL but interesting group of birds has been designated by the name of Ampelinæ, or Chatterers, in allusion to the loquacity for which some of the species are remarkable.   They all have a wide mouth, opening nearly as far as the eyes, but without the bristly appendages which so often accompany a large extent of gape.

One well-known species, the WAXEN CHATTERER, is a tolerably frequent visitor of England, though it cannot be reckoned among the common British birds. It is also known by the name of the BOHEMIAN CHATTERER, the latter name being singularly inappropriate, as the bird is quite as rare in Bohemia as in Engand.

It is a very gregarious bird, assembling in very large flocks, and congregating so closely together that great numbers have been killed at a single discharge of a gun.

The long, flat, scarlet appendages to the wings are usually confined to the secondaries and tertiaries, at whose extremities they dangle as if they had been formed separately, and fastened to the feathers as an after thought. Indeed they so precisely resemble red sealing-wax, that anyone on seeing the bird for the first time would probably suppose that a trick had been played upon him by some one who desired to tax his credulity to a very great extent.

BOHEMIAN WAXWING, OR WAXEN CHATTERER.—(*Ampelis garrula.*)

To this country it only comes in the winter months, although there has been an example of its appearance as early as August.

In its plumage the Bohemian Waxwing is a very pretty and striking bird, being as notable for the silken softness of its feathers as for its pleasingly blended colours and the remarkable appendage from which it derives its popular name. The colouring of the bird is very varied, but may briefly be described as follows. The top of the head and crest are a light soft brown, warming into ruddy chestnut on the forehead. A well-defined band of black passes over the upper base of the beak, and runs round the back of the head, developing the eyes on each side, and there is a patch of the same jetty hue on the chin. The general colour of the bird is grey-brown, the primary and secondary feathers of the wings and tail are black, tipped with yellow, the primary wing-coverts are tipped with white, and the tertiaries are purplish brown, also tipped with white. The under surface of the bird is sober grey, and the under tail-coverts are rich ruddy brown. The length of the Waxen Chatterer is about eight inches.

WE now arrive at the family of Lanidæ, or SHRIKES, or BUTCHER BIRDS, whose character is given in the names by which they are distinguished The scientific term "Lanidæ" is of Latin origin, and is derived from a word which signifies lacerating or tearing, in allusion to the habits of the bird. These birds are found in all parts of the globe, and in all countries are celebrated for their sanguinary and savage character. They are quite as rapacious as any of the hawk tribe, and in proportion to their size are much more destructive and bloodthirsty. They feed upon small and disabled mammalia, and birds of various kinds, especially preferring them while young and still unfledged, and upon several kinds of reptiles, and also find great part of their subsistence among the members of the insect world.

In order to fit them for these rapacious pursuits, the bill is strong, rather elongated, sharp-edged, curved at the tip, and armed on each side with a well-marked tooth. The wings are powerful, the plumage closely set, and the claws strong, curved, and sharp. The Shrikes are separated for convenience of reference into two groups or sub-families, namely, the true Shrikes, or Laninæ, and the Bush Shrikes, or Thamnophilinæ.

RED-BACKED SHRIKE.—(*Enneoctonus collurio*.)

The RED-BACKED SHRIKE is a summer visitant to this country, and is tolerably common. Its winter quarters seem to be situated in Africa, and it reaches us at the end of April, or the beginning of May, passing through Italy on its passage.

During the time of its residence it may often be seen flitting about the tops of hedges and small trees, evidently in search of its prey, and even at a considerable distance may be recognized by its habit of wagging its tail up and down whenever it settles, in a manner very similar to that of the wag-tails. Usually it is seen in pairs ; but when the eggs are laid, the male bird is generally engaged in procuring food while the mother bird stays at home and attends to her domestic affairs.

The food of the Red-Backed Shrike chiefly consists of the larger insects, such as grasshoppers, beetles, and chafers, and it is in the habit of impaling them on the thorns near its nest, probably to save the mother bird the trouble of going to look for her own meals.

The nest of this Shrike is situated in hedges or bushes, generally from five to ten feet from the ground, the average elevation being about seven feet. It is large, rather clumsy, and very easily seen through the foliage, being made of thick grass-stems, moss, and roots on the exterior, and lined with very fine grasses and hair. In some places the nests are quite common, and I have found three in a hedge surrounding a single field of no very great extent. The eggs are generally five in number, and are rather variable in colouring, their ground colour being always white, tinged in some cases with blue, in others with green, and in a few specimens with rusty red. The spots with which they are marked are quite as variable, sometimes being numerous, dark, and gathered into a ring at the large end of the egg, and sometimes being only grey and light brown scattered irregularly. In all cases, however, they are gathered upon the large end of the egg.

In the adult male, the head, neck, and upper parts of the shoulders are pearly grey, with a black stripe across the base of the beak and running through the eye. The back and wing-coverts are ruddy chestnut, fading into reddish grey upon the upper tail-coverts. The quill feathers of the wings are black, edged with red upon their outer webs, and the quill feathers of the tail are white at the basal half, and the remainder of each feather is black tipped with a very narrow line of white. The chin and under tail-coverts are white, and the rest of the under surface is pale rusty red. The strongly notched and hooked beak is deep shining black. The female bird may at once be known by the absence of the black streak across the eye, which in her case is replaced by a light-coloured stripe over the eye. The head and all the upper parts of the body are reddish brown, and the red edges of the wing-feathers are narrower than in the male. The under side of the body is wholly greyish white, covered with very numerous transverse lines of a darker hue.

WE now arrive at a very large and important group, called from the shape of their beaks the CONIROSTRES, or Cone-billed Birds. In these birds the bill varies in length and development, in some being exceedingly short, while in others it is much elongated ; in some being straight and simple, while in others it is curiously curved and furnished with singular appendages ; in some being toothless, while in others there is a small but perceptible tooth near the tip. In all, however, the bill is more or less conical in form, being very thick and rounded at the base, and diminishing to a point at the extremity. There are no less than eight recognized families of this large group, containing some of the most important and most remarkable members of the feathered race.

THE first family is that which is well known under the title of Corvidæ, or CROWS, containing the crows, rooks, magpies, starlings, and other familiar birds, together with the equally celebrated bower birds but less known paradise birds, troopials, and orioles. The beak of all these birds is long, powerful, and somewhat compressed—*i.e.*, flattened at the sides—curved more or less on the ridge of the upper mandible, and with a notch at the extremity.

THE best known of the Garruline, or talkative birds, is our common English JAY, one of the handsomest of our resident birds.

The ordinary note of the Jay is a rather soft cry, but the bird is a most adroit imitator of various sounds, particularly those of a harsh character. It has one especially harsh scream, which is its note of alarm, and serves to set on the alert not only its own kind but every other bird that happens to be within hearing. The sportsman is often baffled in his endeavours to get a shot at his game by the mingled curiosity and timidity of the Jay, which cannot hear a strange rustling or see an unaccustomed object without sneaking silently up to inspect it, and is so terribly frightened at the sight of a man a dog, and a gun, that it dashes off in alarm, uttering its loud " squawk," which indicates to every bird and beast that danger is abroad.

The Jay, like all the crow tribe, will eat animal or vegetable substances with equal zest, and will plunder the hoards of small quadrupeds or swallow the owner with perfect impartiality. Young birds are a favourite food of the Jay, which is wonderfully clever at discovering nests and devouring the fledgelings. Occasionally it even feeds upon birds, and has been seen to catch a full grown thrush. Eggs also are great dainties with this bird, particularly those of pheasants and partridges, so that it is ranked among the " vermin " by all gamekeepers or owners of preserves. So fond is it of eggs, that it can almost invariably be enticed into a trap by means of an egg or two placed as bait, and it is a curious fact that the Jay does not seem to be aware of the right season for eggs, and suspects no guile even when it finds a nest full of fine eggs in the depth of winter.

It also eats caterpillars, moths, beetles, and various similar insects preferring the soft, fat, and full-bodied species to those of a more slender shape. Fruits and berries form a considerable portion of the autumnal food of this bird, and it occasionally makes great havoc in the cherry orchards, slipping in quietly at the early dawn, accompanied by its mate and young family, and stripping the branches of the bark and finest fruit. The kitchen garden also suffers severely from the attacks of the Jay, which has a great liking for young peas and beans. It also eats chestnuts, nuts, and acorns, being so fond of the last-mentioned fruit as to have received the title of *glandarius,* meaning " a lover of acorns." Sometimes it becomes more re-fined in its taste, and eats the flower of several cruciferous plants, which, according to Mudie, it plucks slowly and carefully, petal by petal.

The nest of the Jay is a flattish kind of edifice, constructed of sticks, grass, and roots, the sticks acting as the foundations, and a rude superstructure of the softer substances being placed upon them. It is always situated at a con-siderable elevation from the ground. There are gene-rally four or five eggs, and the bird mostly brings up two broods in the year.

In size, the Jay equals a rather large pigeon; and the colouring of its plum-age is very attractive. The general tint of the upper part of the body is light reddish brown, with a perceptible purple tinge, varying in in-tensity in different specimens. The primary wing-coverts are bright azure, banded with jetty black, and form a most conspicuous orna-

JAY.—(*Garrulus glandarius.*)

ment on the sides as the bird sits with closed wings. The head is decorated with a crest, which can be raised or lowered at pleasure, and the feathers of which it is composed are whitish grey, spotted with black.

THE true CROWS are known by their beaks, which have no tooth in the upper mandible, and by their wings, which are tolerably long and ample. There are very many species spread over the world, and they are well repre-sented in our own country.

THE first of these birds on our list is the celebrated RAVEN, our finest representative of the family.

This truly handsome bird is spread over almost all portions of the habit-able globe, finding a livelihood wherever there are wide expanses of un-cultivated ground, and only being driven from its home by the advance of cultivation and consequent inhabitance of the soil by human beings. It is a solitary bird, living in the wildest district that it can find, and especially preferring those that are intersected with hills. In such localities the Raven reigns supreme, hardly the eagle himself daring to contest the supremacy with so powerful, crafty, and strong-beaked a bird.

The food of the Raven is almost entirely of an animal nature, and there are few living things which the Raven will not eat whenever it finds an oppor- tunity of so doing. Worms, grubs, caterpillars, and insects of all kinds are swallowed by hundreds, but the diet in which the Raven most delights is dead carrion. In conse- quence of this taste, the Raven may be found rather plentiful on the Scottish sheep-feeding grounds, where the flocks are of such im- mense size that the bird is sure to find a sufficiency of food among the daily dead ; for its wings are large and powerful, and its daily range of flight is so great, that many thousands of sheep pass daily under its ken, and it is tolerably sure in the course of the day to find at least one dead sheep or lamb. Sometimes the Raven ac- celerates matters, for if it should find an unfortunate sheep lying in a ditch, a mis- fortune to which these ani- mals are especially prone, it is sure to cause the speedy death of the poor creature by repeated attacks upon its eyes. Weakly or ailing sheep

RAVEN.—(*Corvus Corax.*)

are also favourite subjects with the Raven, who soon puts an end to their sufferings by the strokes of his long and powerful beak. Even the larger cattle are not free from the assaults of this voracious bird, which performs in every case the office of a vulture.

The cunning of the Raven is proverbial, and anecdotes of its extraordinary intellectual powers abound in various works.

The Raven is an excellent linguist, acquiring the art of conversation with wonderful rapidity, and retaining with a singularly powerful memory many sounds which it has once learned. Whole sentences are acquired by this strange bird, and repeated with great accuracy of intonation, the voice being a good imitation of human speech, but always sounding as if spoken from behind a thick woollen wrapper.

The Raven is celebrated for its longevity, many instances being known where it has attained the age of seventy or eighty years, without losing one jot of its activity, or the fading of one spark from its eyes. What may be the duration of a Raven's life in its wild state is quite unknown.

The colour of the Raven is a uniform blue-black, with green reflections in certain lights. The female is always larger than her mate.

THE common CARRION CROW, so plentiful in this country, much resembles in habits and appearance the bird which has just been described, and may almost be reckoned as a miniature raven. In many of its customs the Crow is very raven-like, especially in its love for carrion, and its propensity for attacking the eyes of any dead or dying animal. Like the raven, it has been known to attack game of various kinds, although its inferior size forces

it to call to its assistance the aid of one or more of its fellows, before it can successfully cope with the larger creatures. Rabbits and hares are frequently the prey of this bird, which pounces on them as they steal abroad to feed, and while they are young is able to kill and carry them off without difficulty. The Crow also eats reptiles of various sorts, frogs and lizards being common dainties, and is a confirmed plunderer of other birds' nests, even carrying away the eggs of game and poultry by the simple device of driving the beak through them and flying away with them when thus impaled. Even the large egg of the duck has thus been stolen by the Crow. Sometimes it goes to feed on the sea-shore, and there finds plenty of food among the crabs, shrimps, and shells that are found near low - water mark, and ingeni-ously cracks the harder shelled crea-tures by flying with

CROW.—(*Corvus Corone.*)

them to a great height and letting them fall upon a convenient rock.

The nest of the Crow is invariably placed in some tree remote from the habitations of other birds, and is a structure of considerable dimensions, and very conspicuous at a distance. It is always fixed on one of the topmost branches, so that to obtain the eggs safely requires a steady head, a practised foot, and a ready hand.

The materials of which the Crow's nest is made are very various, but always consist of a foundation of sticks, upon which the softer substances are laid. The interior of the nest is made of grasses, fibrous roots, the hair of cows and horses, which the Crow mostly obtains from trees and posts where the cattle are in the habit of rubbing themselves, mosses, and wool.

ROOK.—(*Corvus frugilegus.*)

The colour of the Crow is a uniform blue-black, like that of the raven, but varie-ties are known in which the feathers have been pied or even cream-white.

THE most familiar of all the British Corvidæ is the common ROOK, a bird which has attached itself to the habitations of mankind, and in course of time has partially domesticated itself in his dominions.

The habits of the Rook are very interesting, and easily watched. Its extreme caution is very remarkable, when combined with its attachment to human homes. A colony of a thousand birds may form a rookery in a park, placing themselves under the protection of its owner ; and yet, if they see a man with a gun, or even with a suspicious-looking stick, they fly off their nests with astounding clamour, and will not return until the cause of their alarm is dissipated. During the "rook-shooting" time, all the strong-winged birds leave their nests at the first report of the gun, and, rising to an enormous elevation, sail about like so many black midges over their deserted homes, and pour out their complaints in loud and doleful cries, which are plainly audible even from the great height at which they are soaring.

The nest of the Rook is large, and rather clumsily built ; consisting chiefly of sticks, upon which are laid sundry softer materials as a resting-place for the eggs. The Rook is a very gregarious bird, building in numbers on the boughs of contiguous trees, and having a kind of social compact that often rises into the dignity of law. For example, the elder Rooks will not permit the younger members of the community to build their nests upon an isolated tree at a distance from the general assemblage ; and, if they attempt to infringe this regulation, always attack the offending nest in a body, and tear it to pieces.

The number of birds which are to be found in such rookeries is enormously great, several thousands having been counted in a single assemblage. In such cases they do great damage to the upper branches of the trees, and in some in-stances have been known to kill the tree, by the con-tinual destruction of the growing boughs.

The colour of the Rook is a glossy deep blue-black, the blue being more conspicious on the wing-coverts and the sides of the head and neck. The length of an adult Rook is about eighteen or nineteen inches.

THE smallest of the British Corvidæ is the well-known JACKDAW, a bird of infinite wit and humour, and one that has an extraordinary attachment for man and his habitations.

JACKDAW.—(*Corvus monedula.*)

The Jackdaw may easily be distinguished from either the rook or the crow by the grey patch upon the crown of the head and back of the neck, which is very conspicuous, and can be seen at a considerable distance. The voice, too, is entirely different from the caw of the rook or the hoarse cry of the crow ; and as the bird is very loquacious, it soon announces itself by the tone of its voice. It generally takes up its home near houses, and is fond of nesting in old buildings, espe-cially preferring the steeples and towers of churches and similar edifices, where its nest and young are safe from the depredations of stoats, weasels, and other destroyers.

The grey patch on the head and neck is not seen until the bird attains maturity, the feathers being of the same black hue as on the remainder of the body until the first moult, when the juvenile plumage is shed and the adult garments assumed.

The nest of the Jackdaw is a very rude structure of sticks, lined, or rather covered with hay, wool, feathers, and all kinds of miscellaneous substances of a warm kind for the eggs and young. It is placed in various localities, generally in buildings or rocks, but has often been found in hollow trees and even in the holes of rabbit warrens, the last-mentioned locality being a very remarkable one, as the young birds must be in constant danger of marauding stoats and weasels. In one instance a quantity of broken glass was employed in the foundation of the nest. The Jackdaw is not choice in the selection of feathered neighbours, for I have found in the same tower the nests of pigeons, jackdaws, and starlings, in amicable proximity to each other. The eggs are smaller and much paler than those of the rook or crow, but have a similar general aspect. Their number is about five.

THE ROYSTON CROW, or HOODED CROW, or GREY CROW, is a very conspicuous bird, on account of the curiously pied plumage with which it is invested.

This bird is not very common in England, but is plentifully found both in Ireland, Scotland, and the Scottish isles, having been seen in large flocks of several hundreds in number on the east coast of Jura. Generally it is not very gregarious, the male and female only being found in company; but it some-times chooses to associate in little flocks of fifteen or sixteen in number. It seems to prefer the sea-coast to any inland locality, as it there finds a great variety of food, and is not much ex-posed to danger. I have often seen these birds in the Bay of Dublin, perched upon the rocks at low water, and searching for food among the dank seaweed and in the rock puddles that are left by the retreating tide. They seemed always extremely bold, and would permit a very close approach without exhibiting any alarm. The banks of the Medway between Rochester and Sheerness are much frequented by the Hooded Crow.

ROYSTON, OR HOODED CROW.—(*Corvus cornix.*)

The Hooded Crow never breeds in society, but always builds its nest at some distance from the homes of any other of the same species, so that, although a forest or a range of cliffs may be inhabited by these birds, the nests are scattered very sparingly over the whole extent. The structure of the nest is somewhat similar to that of the crows and rooks, being a mass of sticks and heather stalks as a foundation, upon which is placed a layer of wool, hair, and other soft substances. Sometir..es the bird builds a better and more compact nest with the bark of trees; and in all cases this species breeds very early in the season.

The Hooded Crow is boldly and conspicuously pied with grey and black, distributed as follows :—The head, back of the neck, and throat, together with the wings and tail, are a glossy bluish black, while the remainder of the body is a peculiar grey, with a slight blackish wash.

WHO does not know the MAGPIE, the pert, the gay, the mischievous ? What denizen of the country is not familiar with his many exploits in the way of barefaced and audacious theft, his dipping flight, and his ingenuity in baffling the devices of the fowler and the gunner ? What inhabitant of the town has not seen him cooped in his wicker dwelling, dull and begrimed with the daily smoke, but yet pert as ever, talkative, and a wonderful admirer of his dingy plumage and ragged tail ?

The food of the Magpie is as multifarious as that of the crow or raven, and consists of various animal and vegetable substances. It is a determined robber of other birds' nests, dragging the unfledged young out of their homes, or driving its bill through their eggs, and thus carrying them away.

The nest of the Magpie is a rather complicated edifice, domed, with an entrance at the side, and mostly formed at the juncture of three branches, so as to afford an effectual protection against any foe who endeavours to force admittance into so strong a fortress. Generally the nest is placed at the very summit of some lofty tree, the bird usually preferring those trees which run for many feet without a branch. The tops of tall pines are favourite localities for the Magpie's nest, as the trunk of these trees is bare of branches except at the summit, and the dark-green foliage of the spreading branches is so thick that it affords an effectual shelter to the large and conspicuous edifice which rests upon the boughs.

When tame, it is a most amusing bird, teaching itself all kinds of odd tricks, and learning to talk with an accuracy and volubility little inferior to that of the parrot. It is, however, a most incorrigibly mischievous bird, and unless subjected to the most careful supervision is capable of doing a very great amount of damage in a wonderfully

MAGPIE.—(*Pica caudata.*)

short space of time. I have witnessed a multitude of these exploits, but as I have already related many of them in " My Feathered Friends," the reader is referred to the pages of that little work for a tolerably long series of new and original Magpie anecdotes.

The plumage of this bird is remarkably handsome both in colour and form. The head, neck, back, and upper tail-coverts are deep black, with a light green gloss in certain lights ; and the same colour is found on the chin, the throat, the upper part of the breast, and the base, tips, and outer edges of the primary quill feathers. The secondaries are also black, but with a blue gloss, which becomes peculiarly rich on the tertials and wing-coverts. The inner web of the primaries is white for a considerable portion of its length, presenting a bold and conspicuous appearance when the bird spreads its wings. The central feathers of the tail are nearly eleven inches in length, and they decrease gradually in size ; those on the exterior being hardly five inches long. Their colour is a wonderfully rich mixture of the deepest blue, purple, and green, the green being towards the base, and the blue and purple towards the extremity. The under surface of the tail-feathers is dull black. The lower parts of the breast, abdomen, and flanks are snowy white.

OF the next little group of Corvidæ, named the Pyrrhocoracinæ, or Scarlet

Crows, in allusion to the red bill and legs of some of the species, England possesses a good example in the common CHOUGH. In all these birds the beak is long and slender, slightly curved downwards, and with a small notch at the extremity.

The Chough is essentially a coast bird, loving rocks and stones, and having a great dislike to grass or hedges of every kind. When in search of food it will venture for some little distance inland, and has been observed in the act of following the ploughman after the manner of the rook, busily engaged in picking up the grubs that are unearthed. Sometimes it will feed upon berries and grain, but evidently prefers animal food, pecking its prey out of the crevices among the rocks with great rapidity and certainty of aim, its long and curved beak aiding it in drawing the concealed insects out of their hiding-places.

As is the case with nearly all coast birds, the Chough builds its nest at no great distance from the sea, generally choosing some convenient crevice in a cliff, or an old ruin near the sea-shore. The nest is always placed at a considerable elevation from the ground, and is made of sticks lined with wool, hair, and other soft substances. The eggs are usally five in number, and in colour they are yellower than those of the crow or rook but are spotted with similar tints. The general colour of the Chough is black, with a rich blue gloss, contrasting well with the vermilion-red of the beak, legs, and toes. The claws are black, and the eyes are curiously coloured with red and blue in concentric circles. The total length of the adult male Chough is about seventeen inches, and the female is about three inches shorter.

CHOUGH.—(*Coracia gracula.*)

THE supremely glorious members of the feathered tribe which have by common consent been termed BIRDS OF PARADISE are not very numerous in species, but are so different in form and colour, according to the sex and age, that they have been considered far more numerous than is really the case. The plumage of these birds is wonderfully rich and varied, and not even the humming-birds themselves present such an inexhaustible treasury of form and colour as is found among the comparatively few species of the Birds of Paradise. In all, the feathers glow with resplendent radiance ; in nearly all there is some strange and altogether unique arrangement of the plumage ; and in many the feathers are modified into plumes, ribbons, and streamers, that produce the most surprising and lovely effects.

THE EMERALD BIRD OF PARADISE is the species which is most generally known, and is the one of which were related many absurd tales. The specific term, *apoda,* signifies "footless," and was given to the bird by Linnæus in allusion to those fables which were then current, but which he did not believe.

This most lovely bird is a native of New Guinea, where it is far from uncommon, and is annually killed in great numbers for the sake of its plumage, which always commands a high price in the market. It is a very retiring bird, concealing itself during the day in the thick foliage of the teak-tree, and only coming from the green shelter at the rising and setting of the sun, for the purpose of obtaining food. Almost the only successful method of shooting the Emerald Paradise Bird is to visit a teak or fig-tree before dawn, take up a position under the branches, and there wait patiently until one of the birds comes to settle upon the branches, or leaves the spot which has sheltered it during the night. This bird is rather tenacious of life, and unless killed instantly is sure to make its escape amid the dense brush-wood that grows luxuriantly beneath the trees, and if the sportsman ventured to chase a wounded bird amid the bushes, he would, in all probability, lose his way and perish of hunger. Those sportsmen, therefore, who desire to shoot this bird, always provide themselves with guns that will carry their charge to a great distance, and employ very large shot for the purpose, as the bird always perches on the summits of the loftiest trees of the neigh-bourhood, and would not be much damaged by the shot ordinarily used in shooting.

This species is very sus-picious, so that the sports-man must maintain a pro-found silence, or not a bird will show itself or utter its loud full cry, by which the hunter's attention is directed to his victim.

THE large and important family of the STARLINGS now claims our attention. These birds are seldom of great size, the common Star-ling of England being about an average example of their dimensions. The bill of the Starling tribe is straight until near its extremity, when it suddenly curves downward, and is generally armed with a slight notch. The first sub-family of these birds is that which is known by the name of Glossy Starlings, so called on account of the silken sheen of their plumage.

EMERALD BIRD OF PARADISE.—(*Paradiseapoda.*)

The best representative of this little group is the celebrated SATIN BOWER BIRD of Australia. This beautiful and remarkable bird is found in many parts of New South Wales, and although it is by no means uncommon, is so cautious in the concealment of its home, that even the hawk-eyed natives seem never to have discovered its nest.

The chief peculiarity for which this bird is famous is a kind of bower or arbour, which it constructs from twigs in a manner almost unique among the feathered tribes. The form of this bower may be seen in the illustration, and the mode of construction, together with the use to which the bird puts the building, may be learned from Mr. Gould's account :—

"On visiting the cedar brushes of the Liverpool range, I discovered several of these bowers or playing-places ; they are usually placed under the shelter of the branches of some overhanging tree in the most retired part of the forest ; they differ considerably in size, some being larger, while others are much smaller. The base consists of an exterior and rather convex platform of sticks, firmly interwoven, on the centre of which the bower itself is built. This, like the platform on which it is placed and with which it is interwoven, is formed of sticks and twigs, but of a more slender and flexible description, the tips of the twigs being so arranged as to curve inwards and nearly meet at the top ; in the interior of the bower, the materials are so placed that the forks of the twigs are always presented outwards, by which arrangement not the slightest obstruction is offered to the passage of the birds.

" For what purpose these curious bowers are made is not yet, perhaps, fully understood; they are certainly not used as a nest, but as a place of resort for many individuals of both sexes, who when there assembled run through and round the bower in a sportive and playful manner, and that so frequently that it is seldom entirely deserted.

" The interest of this curious bower is much enhanced by the manner in which it is decorated, at and near the entrance, with the most gaily coloured articles that can be collected, such as the blue tail-feathers of the Rose Hill and Lory Parrots, bleached bones, the shells of snails, &c. Some of the feathers are stuck in among the twigs, while others, with the bones and shells, are strewed about near the entrance. The propensity of these birds to fly off with any attractive object is so well known that the blacks always search the runs for any missing article."

SATIN BOWER BIRD.—(*Ptilonorhynchus holosericeus.*)

So persevering are these birds in carrying off anything that may strike their fancy, that they have been known to steal a stone tomahawk, some blue cotton rags, and an old tobacco-pipe. Two of these bowers are now in the nest-room of the British Museum, and at the Zoological Gardens the Bower Bird may be seen hard at work at its surface, fastening the twigs or adorning the entrances, and ever and anon running through the edifice with a curious loud full cry that always attracts the attention of a passer-by. The Satin Bower Bird bears confinement well, and although it will not breed in captivity, it is very industrious in building bowers for recreation.

The food of this bird seems to consist chiefly of fruit and berries, as the stomachs of several specimens were found to contain nothing but vegetable remains. Those which are caged in Australia are fed upon rice, fruit, moistened bread, and a very little meat at intervals, a diet on which they thrive well. The plumage of the adult male is a very glossy satin-like purple, so deep as to appear black in a faint light, but the young males and the females are almost entirely of an olive-green.

WE now come to the true Starlings. In these birds the bill is almost straight, tapering and elongated, slightly flattened at the top, and with a hardly perceptible notch.

The common STARLING is one of the handsomest of our British birds, the bright mottlings of its plumage, the vivacity of its movements, and the elegance of its form, rendering it a truly beautiful bird.

COMMON STARLING. —(*Sturnus vulgaris.*)

It is very common in all parts of the British Isles, as well as in many other countries, and assembles in vast flocks of many thousands in number, enormous accessions being made to their ranks after the breeding season. These vast assemblies are seen to best advantage in the fenny districts, where they couch for the night amid the osiers and aquatic plants, and often crush whole acres to the ground by their united weight. In their flight the Starlings are most wonderful birds, each flock, no matter how large its dimensions, seeming to be under the command of one single bird, and to obey his orders with an instantaneous action which appears little short of a miracle. A whole cloud of Starlings may often be seen flying along at a considerable elevation from the ground, darkening the sky as they pass overhead, when of a sudden the flock becomes momentarily indistinguishable, every bird having simultaneously turned itself on its side so as to present only the edge of its wings to the eye. The whole body will then separate into several divisions, each division wheeling with the most wonderful accuracy, and after again uniting their forces they will execute some singular manœuvre, and then resume their onward progress to the feeding-ground or resting-place.

The nest of the Starling is a very loose kind of affair, composed of straw, roots, and grasses, thrust carelessly together, and hardly deserving the name of a nest. In many cases the bird is so heedless that it allows bits of straw or grass to hang from the hole in which the nest is placed, just as if it had intentionally furnished the bird-nesting boy with a clue to the position of the nest. Although this bird makes its home in some retired spot, such as the cleft of a rock, a niche in some old ruin, a ledge in a church-tower, or a hole in a decaying tree, there are few nests more easy to discover ; for not only does the bird leave indications of its home in the manner already described, but is so very loquacious that it cannot resist the temptation of squalling loudly at intervals, especially when returning to its domicile laden with food for its young, and so betrays the position of its home. The eggs are generally five in number, and of the faintest imaginable blue.

The food of the Starling is very varied, but consists chiefly of insects. These birds have a habit of following cows, sheep, and horses, fluttering about them as they move for the purpose of preying upon the insects which are put to flight by their feet. The Starlings also perch upon the backs of the cattle, and rid them of the parasitic insects that infest them. From the sheep the Starling often takes toll, pulling out a beakful of wool now and then, and carrying it away to its nest. It is a voracious bird, the stomach of one of these birds having been found to contain more than twenty shells, some of no small size, and all nearly perfect ; a great number of insects, and some grain. Another Starling had eaten fifteen molluscs of different kinds, a number of perfect beetles, and many grubs.

The colour of this bird is very beautiful, and is briefly as follows : The general tint is an extremely dark purplish green, having an almost metallic glitter in a strong light. The feathers of the shoulders are tipped with buff, and the wing-coverts, together with the quill feathers of the tail and wings, are edged with pale reddish brown. The beak is a fine yellow. The feathers of the upper part of the breast are elongated and pointed. This is the plumage of the adult male, and is not brought to its perfection until three years have elapsed. The first year's bird, before its autumnal moult, is almost wholly of a brownish grey, and after its moult is partly brown and partly purple and green. In the second year the plumage is more decided in its tints, but is variegated with a great number of light-coloured spots on the under and upper surfaces, and the beak does not attain its beautiful yellow tinge.

WE now arrive at the large and important families of the FINCHES. In all these birds the bill is conical, short, and stout, sharp at the extremity and without any notch in the upper mandible.

The first group of the Finches is composed of a number of species which, although for the most part not conspicuous either for size, beauty of form, or brilliancy of colour, are yet among the most remarkable of the feathered tribe, on account of their architectural powers. Dissimilar in shape, form, and material, there is yet a nameless something in the construction of their edifices which at once points them out as the workmanship of the Weaver Birds. Some of them are huge, heavy, and massive, clustered together in vast multitudes, and bearing down the branches with their weight. Others are light, delicate and airy, woven so thinly as to permit the breeze to pass through their net-like interior, and dangling daintily from the extremity of some slender twig. Others, again, are so firmly built of flattened reeds and glass blades,

SOCIABLE WEAVER BIRD.—(*Philæœrus socius.*)

that they can be detached from their branches and subjected to very rough handling without losing their shape, while others are so curiously formed of stiff grass-stalks that their interior is studded with sharp points like the skin of a hedgehog.

The true Weaver Birds all inhabit the hotter portions of the Old World, the greater number of them being found in Africa, and the remainder in various parts of India.

The SOCIABLE WEAVER BIRD is found in several parts of Africa, and has always attracted the attention of travellers from the very remarkable edifice which it constructs. The large social nests of this bird are so conspicuous as to be notable objects at many miles' distance, and it is found that they generally build in the branches of the giraffe thorn or "kameeldorn," one of the acacia tribe.

The Sociable Weaver Bird, which is by some writers termed the "Sociable Grosbeak," in choosing a place for its residence, is careful to select a tree which grows in a retired and sheltered situation, secluded as far as may be from the fierce wind-storms which are so common in hot countries. When a pair of these birds have determined to make a new habitation, they proceed after the following fashion :—They gather a vast amount of dry grasses, the favourite being a long, tough, and wiry species, called "Booschmanees-grass," and by hanging the long stems over the branches and ingeniously interweaving them, they make a kind of roof, or thatch, which is destined to shelter the habitations of the community.

In the under sides of this thatch they fasten a number of separate nests, each being inhabited by a single pair of birds, and only divided by its walls from the neighbouring habitation. All these nests are placed with their mouths downwards, so that when an entire edifice is completed, it reminds the observer very strongly of a common wasp's nest. This curious resemblance is often further strengthened by the manner in which these birds will build one row of nests immediately above or below another, so that the nest-groups are arranged in layers precisely similar to those of the wasp or hornet. The number of habitations thus placed under a single roof is often very great. Le Vaillant mentions that in one nest which he examined there were three hundred and twenty inhabited cells, each of which was in the possession of a distinct pair of birds, and would at the close of the breeding season have quadrupled their numbers.

The number of eggs in each nest is usually from three to five, and their colour is bluish white, dotted towards the larger end with small brown spots. The food of this bird seems to consist mostly of insects, as, when the nests are pulled to pieces, wings, legs, and other hard portions of various insects are often found in the interior of the cells. It is said that the Sociable Weaver Birds have but one enemy to fear in the persons of the small parrots, who also delight in assembling together in society, and will sometimes make forcible entries into the Weaver Bird's nest and disperse the rightful inhabitants.

The colour of the Sociable Weaver Bird is brown, taking a pale buff tint on the under surface of the body, and mottled on the back with the same hue. It is quite a small bird, measuring only five inches in length.

There are several allied species, which are remarkable for the structure of their nests. There is the MAHALI WEAVER BIRD, whose nest is shaped like a Florence flask ; the TAHA, the RUFOUS-NECKED WEAVER, and others, all of which suspend their nests to the ends of branches that overhang water, thus ensuring safety against almost any enemy.

THE GROSBEAKS or HAWFINCHES now claim our attention. They are all remarkable for their very large, broad, and thick beaks, a peculiarity of construction which is intended to serve them in their seed-crushing habits.

England possesses a good example of this group in the well-known HAW-FINCH, or GROSBEAK.

This bird was once thought to be exceedingly scarce, but is now known to be anything but uncommon, although it is rarely seen, owing to its very shy and retiring habits, which lead it to eschew the vicinity of man and to bury itself in the recesses of forests. So extremely wary is the Hawfinch, that to approach within gunshot is a very difficult matter, and can seldom be accomplished without the assistance of a decoy-bird, or by imitating the call-note, which bears some resemblance to that of a robin. It feeds chiefly on the various wild berries, not rejecting even the hard stones of plums and the laurel berries. In the spring, it is apt to make inroads in the early dawn upon the cultivated grounds, and has an especial liking for peas, among which it often works dire havoc.

The nest of the Hawfinch is not remarkable either for elegance or peculiarity of form. It is very simply built of slender twigs, bits of dried creepers, grey lichens, roots, and hairs, and is so carelessly put together that it can hardly be moved entire. The eggs are from four to six in number, and their colour is very pale olive-green, streaked with grey and spotted with black dots. The birds pair in the middle of April, begin to build their nests about the end of that month, and the young are hatched about the third week in May.

THE true FINCHES are known by their rather short and conical beak, their long and pointed wings, and the absence of nostrils in the beak. England possesses many examples of these birds.

The CHAFFINCH is one of our commonest field birds, being spread over the whole of England in very great numbers.

The specific title of CŒLEBS which is given to the Chaffinch, signifies "a bachelor," and refers to the annual separation of the sexes, which takes place in the autumn, the females departing to some other region, and

HAWFINCH OR GROSBEAK. —(*Coccothraustes vulgaris.*)

the males congregating in vast multitudes, consoling themselves as they best can by the pleasures of society for the absence of the gentler portion of the community.

The note of this bird is a merry kind of whistle, and the call-note is very musical and ringing, somewhat resembling the word "pinck," which has therefore been often applied to the bird as its provincial name.

The nest of the Chaffinch is one of the prettiest and neatest among the British nests. It is deeply cup-shaped, and the materials of which it is composed are moss, wool, hair, and lichens, the latter substances being always stuck profusely over the surface, so as to give it a resemblance to the bough on which it has been built. The nest is almost invariably made in the upright fork of a branch, just at its junction with the main stem or bough from which it sprang, and is so beautifully worked into harmony with the bark of the particular tree on which it is placed, that it escapes the eye of any but a practised observer. Great pains are taken by the female in making her nest, and the structure occupies her about three weeks. The eggs are from four to five in number, and their colour is a pale brownish buff, decorated with several largish spots and streaks of very dark brown.

The colour of this pretty bird is as follows. At the base of the beak the feathers are jetty black, and the same hue, but with a slight dash of brown.

is found on the wings and the greater wing-coverts. The top of the head and back of the neck are slaty grey, the back is chestnut, and the sides of the head, the chin, throat, and breast are bright ruddy chestnut, fading into a colder tint upon the abdomen. The larger wing-coverts are tipped with white, the lesser coverts are entirely of the same hue, and the tertials are edged with yellowish white. The tail has the two central feathers greyish black, the next three pairs black, and the remaining feathers variegated with black and white. The total length of the bird is six inches. The female is coloured something like the male. but not so brilliantly.

CHAFFINCH.—(*Fringilla cœlebs.*)

OF all the British Finches, none is so truly handsome as the GOLDFINCH, a bird whose bright yellow-orange hues suffer but little even when it is placed in close proximity to the more gaudy Finches of tropical climates. Like the chaffinch, it is spread over the whole of England, and may be seen in great numbers feeding on the white thistledown. There are few prettier sights than to watch a cloud of Goldfinches fluttering along a hedge, chasing the thistledown as it is whirled away by the breeze, and uttering all the while their sweet merry notes.

The birds are not very shy, and by lying quietly in the hedge the observer may watch them as they come flying along, ever and anon perching upon the thistle tops, dragging out a beakful of down, and biting off the seeds with infinite satisfaction. Sometimes a Goldfinch will make a dart at a thistle or burdock, and without perching snatch several of the seeds from their bed, and then alighting on the stem, will run up it as nimbly as a squirrel, and peck away at the seeds, quite careless as to the attitude it may be forced to adopt. These beautiful little birds are most useful to the farmer, for they not only devour multitudes of insects during the spring months, but in the autumn they turn their attention to the thistle, burdock, groundsel, plantain, and other weeds, and work more effectual destruction than the farmer could hope to attain with all his labourers. Several Goldfinches may often be seen at one time on the stem and top of a single thistle, and two or three are frequently busily engaged on the same plant of groundsel.

GOLDFINCH.—(*Fringilla carduelis.*)

The nest of the Goldfinch is very neat and prettily made, sometimes built in a hedge or thick bush, but mostly placed towards the extremity of a thickly foliaged tree-branch, such being a favourite for this purpose. In this position the nest is so ingeniously concealed from the gaze of every one beneath, by the disposition of the branches and leaves and by the manner in which the exterior of the nest is made to harmonize in tint with the bark, that it can scarcely be discerned even when the observer has climbed the tree and is looking down upon the nest. The bird, too, seldom flies directly in or out of the nest, but alights at a little distance from her tree, and then slips quietly through the leaves until she reaches her eggs or young.

The materials of which the exterior of the nest is made differ according to the tree in which it is placed. In general, fine grasses, wool, hairs, and very slender twigs are employed in constructing the walls, and the interior is softly lined with feathers, down, and hairs. The eggs are generally four or five in number, and delicately marked with small dots and streaks of light purplish brown upon a white ground, having a tinge of blue, something like " sky-blue " milk.

THE GREENFINCH is one of our commonest birds, being resident in this country throughout the year, and not even requiring a partial migration. It is mostly found in hedges, bushes, and copses ; and as it is a bold and familiar bird, is in the habit of frequenting the habitations of men, and even building its nest within close proximity to houses or gardens. When young the bird is fed almost wholly upon caterpillars and various insects, and not until it has attained its full growth does it try upon the hard seeds the large bill which has obtained for it the title of Green Grosbeak.

The nest of this bird is generally built rather later than is usual with the Finches, and is seldom completed until May has fairly set in. Its substance is not unike that of the chaffinch, being composed of roots, wool, moss, and feathers. It is not, however, so neatly made, nor so finely woven together, as the nest of that bird. The eggs are from three to five in number, and the colour is bluish white,

GREENFINCH.—(*Fringilla chloris.*)

covered at the larger end with spots of brown and grey.

FEW birds are better known than the COMMON LINNET, although the change of plumage to which it is subject in the different seasons of the year has caused the same bird, while in its winter plumage, to be considered as distinct from the same individual in its summer dress. Except during the breeding season, the Linnets associate in flocks, flying from spot to spot, and feeding upon the seeds of various plants, evidently preferring those of the thistle, dandelion, and various cruciferous plants.

The nest of this bird is strangely variable in the positions which it occupies, sometimes being placed at a considerable height upon a tree, and at other times built in some bush quite close to the ground, the latter being the usual locality chosen by the bird. The full number of the eggs is five, and the colour is mostly blue, spotted with dark brown, and a rather faint and undecided purple.

The summer plumage of the male bird is as follows : On the top of the head the feathers are greyish brown at their base, but are tipped with bright vermilion, a tint which contrasts well with the ashen grey-brown of the face and back of the neck. The upper parts of the body are warm chestnut, and the wing is black, excepting the narrower exterior webs, which are white. The chin and throat are grey, the breast bright red slightly dappled with brown, and the under portions of the body are grey-brown, taking a yellowish tint on the flanks. The tail is rather forked, and the feathers are black, edged with white.

THE pretty little CANARY BIRD, so prized as a domestic pet, derives its name from the locality whence it was originally brought.

Rather more than three hundred years ago, a ship was partly laden with little green birds captured in the Canary Islands, and having been wrecked near Elba, the birds made their escape, flew to the island and there settled themselves. Numbers of them were caught by the inhabitants, and on account of their sprightly vivacity and the brilliancy of their voice, they soon became great favourites, and rapidly spread over Europe.

The original colour of the Canary is not the bright yellow with which its feathers are generally tinted, but a kind of dappled olive-green, black, and yellow, either colour predominating according to circumstances.

I have kept Canaries for many years, and could fill pages with anecdotes and histories of them and their habits, but as I have already written rather a long biography of my Canaries in " My Feathered Friends," together with instructions for the management and rearing of these pretty birds, there is no need to repeat the account in the present pages.

CANARY.—(*Carduelis canaria.*)

THE SISKIN is hardly to be considered more than an occasional visitor in England, but in Scotland it sometimes breeds, as may be seen from the following extract :—

" The Siskin is a common bird in all the high parts of Aberdeenshire, which abound in fir-woods. They build generally near the extremities of the branches of tall fir-trees, or near the summit of the tree. Sometimes the nest is found in plantations of young fir-wood. In one instance I met with a nest not three feet from the ground. I visited it every day until four or five eggs were deposited. During incubation the female showed no fear at my approach. On bringing my hand close to the nest, she showed some inclination to pugnacity, and tried to frighten me away with her open bill, following my hand round and round when I attempted to touch her. At last she would only look anxiously round to my finger without making any attack on me. The nest was formed of small twigs of birch or heath outside and neatly lined with hair." Its eggs are a bluish-white spotted with purplish red.

SISKIN.—(*Fringillus spinus.*)

THE noisy, familiar, impatient SPARROW, is one of those creatures that has attached itself to man, and follows him wherever he goes.

Nothing seems to daunt this bold little bird, which is equally at home in the fresh air of the country farm, in the midst of a crowded city, or among the strange sights and sounds of a large railway station; treating with equal indifference the slow-paced waggon-horses, as they deliberately drag their load over the country roads, the noisy cabs and omnibuses as they rattle over the city pavements, and the snorting, puffing engines, as they dash through the stations with a velocity that makes the earth tremble beneath their terrible rush.

Although its ordinary food consists of insects and grain, both of which articles it can only obtain in the open country, it accommodates itself to a town life with perfect ease, and picks up a plentiful subsistence upon the various refuse that is thrown daily out of town houses, and which, before it is handed over to the dustman, is made by the Sparrow to yield many a meal.

When in the country the Sparrow feeds almost wholly on insects and grain, the former being procured in the spring and early summer, and the latter in autumn and winter. As these birds assemble in large flocks and are always very plentiful, they devour great quantities of grain, and are consequently much persecuted by the farmer, and their numbers thinned by guns, traps, nets, and all kinds of devices. Yet their services in insect-killing are so great as to render them most useful birds to the agriculturist. A single pair of these birds have been watched during a whole day, and

SPARROW.—(*Passer domesticus.*)

were seen to convey to their young no less than forty grubs per hour, making an average exceeding three thousand in the course of the week. In every case where the Sparrows have been extirpated, there has been a proportional decrease in the crops from the ravages of insects. At Maine, for example, the total destruction of the Sparrows was ordered by Government, and the consequence was that in the succeeding year even the trees were killed by caterpillars, and a similar occurrence took place near Auxerre.

The nest of the Sparrow is a very inartificial structure, composed of hay, straw, leaves, and various similar substances, and always filled with a prodigious lining of feathers. For, although the Sparrow is as hardy a bird as can be seen, and appears to care little for snow or frost, it likes a warm bed to which it may retire after the toils of the day, and always stuffs its resting-place full of feathers, which it gets from all kinds of sources. Even their roosting-places are often crammed with feathers.

Generally the nest is built in some convenient crevice, such as an old wall, especially if it be covered with ivy; but the bird is by no means particular in the

YELLOW HAMMER.—(*Emberiza citrinella.*)

choice of a locality, and will build in many other situations. There are

generally five eggs, though they sometimes reach the number of six, and their colour is greyish white, profusely covered with spots and dashes of grey-brown. They are, however, extremely variable, and even in the same nest it is not uncommon to find some eggs that are almost black with the mottlings, while others have hardly a spot about them. The Sparrow is a very prolific bird, bringing up several broods in the course of a season, and has been known to rear no less than fourteen young in a single breeding season.

THE BUNTINGS are known by their sharp conical bills, with the edges of the upper mandible rounded and slightly turned inwards, and the knob on the palate. They are common in most parts of the world, are gregarious during the winter months, and in some cases become so fat upon the autumn grain that they are considered great dainties.

ONE of the most familiar of all these birds is the YELLOW BUNTING, or YELLOW HAMMER, as it is often called.

This lively bird frequents our fields and hedge-rows, and is remarkable for a curious mixture of wariness and curiosity, the latter feeling impelling it to observe a traveller with great attention, and the former to keep out of reach of any missile. So, in walking along a country lane, the passenger is often preceded by one or more of these birds, which always keeps about seventy or eighty yards in advance, and flutters in and out of the hedges or trees with a peculiar and unmistakable flirt of the wings and tail.

The song—if it may so be called—of the bird has a peculiar intonation, which is almost articulate, and is variously rendered in different parts of England. For example, among the southern counties it is well represented by the words, " A little bit of bread and no che-ee-ee-ee-se ! " In Scotland it assumes a sense quite in accordance with the character of its auditors, and is supposed to say, " De'il, de'il, de'il take ye-ee-ee-ee."

ORTOLAN.—(*Emberiza hortulana.*)

The nest of the Yellow Bunting is generally placed upon or very close to the earth, and the best place to seek for the structure is the bottom of a hedge where the grass has been allowed to grow freely, and the ground has been well drained by the ditch. In rustic parlance, a " rough gripe " is the place wherein to look for the Yellow Hammer's nest. It is a neatly built edifice, composed chiefly of grasses, and lined with hair. The eggs are five in number, and their colour is white, with a dash of very pale purple, and dotted and scribbled all over with dark purple-brown. Both dots and lines are most variable, and it also frequently happens that an egg appears with hardly a mark upon it, while others in the same nest are entirely covered with the quaint-looking decorations.

The general colour of this bird is bright yellow, variegated with patches of dark brown, and having a richly-mottled brownish yellow on the back, with a decided warm ruddy tinge. The primary feathers of the wing are black, edged with yellow, and the remainder of the feathers throughout, with all the wing-coverts, are deep brown-black, edged with ruddy brown. The chin, throat, and all the under parts of the body are bright pure yellow, sobering into rusty brown on the flanks. The female is similarly marked, but is not so brilliant in her hues. The total length of the bird is about seven inches.

THE ORTOLAN, or GARDEN BUNTING, is widely celebrated for the delicacy of its flesh, or rather for that of its fat; the fat of the Ortolan being somewhat analogous to the green fat of the turtle in the opinion of gourmands.

The Ortolan has occasionally been shot in England, but it is most frequently found on the Continent, where its advent is expected with great anxiety, and vast numbers are annually captured for the table. These birds are not killed at once, as they would not be in proper condition, but they are placed in a dark room, so as to prevent them from moving about, and are fed largely with oats and millet, until they become mere lumps of fat, weighing nearly three ounces, and are then killed and sent to table.

The colouring of this bird is as follows : The head is grey, with a green tinge, and the back is ruddy brown, beautifully mot-

SKYLARK.—(*Alauda arvensis.*)

tled with black. The wings are black, with brown edges to the feathers ; the chin, throat, and upper portions of the breast are greenish yellow ; and the abdomen is warm buff. The total length of the Ortolan is rather more than six inches.

THE LARKS may be readily recognized by the very great length of the claw of the hind toe, the short and conical bill, and the great length of the tertiary quill feathers of the wing, which are often as long as the primaries.

The first example of these birds is the well-known SKYLARK, so deservedly famous for its song and its aspiring character.

This most interesting bird is happily a native of our land, and has cheered many a sad heart by its blithe jubilant notes as it wings skyward on strong pinions, or flutters between cloud and earth, pouring out its very soul in its rich wild melody. Early in the spring the Lark begins its song, and continues its musical efforts for nearly eight months, so that on almost every warm day of the year on which a country walk is pacticable, the Skylark's happy notes may be heard ringing throughout the air, long after the bird which utters them has dwindled to a mere speck, hardly distinguishable from a midge floating in the sunbeams.

The nest of the Skylark is always placed on the ground, and generally in some little depression, such as the imprint of a horse's hoof, the side of a mole-hill, or the old furrow of a plough. It is very well concealed, the top of the nest being only just on a level with the surface of the ground, and sometimes below it. I have known several instances where the young Larks would suffer themselves to be fed by hand as they sat in their nests, but the parent birds always seemed distressed at the intrusion into their premises. The materials of which it is made are dry grasses, bents, leaves, and hair, the hair being generally used in the lining. It will be seen that the sober colouring of those substances renders the nest so uniform in tint with the surrounding soil, that to discover it is no easy matter. The eggs are four or five in number, and their colour is grey-yellow, washed with light brown, and speckled with brown of a darker hue. They are laid in May, and are hatched in about a fortnight.

Towards the end of autumn and throughout the winter the Lark becomes very gregarious, "packing" in flocks of thousands in number, and becoming very fat when snow should cover the ground, in which case they speedily lose their condition. These flocks are often augmented by the arrival of numerous little flocks from the Continent that come flying over the sea about the end of autumn, so that the bird-catchers generally reap a rich harvest in a sharp winter.

THE next group is that of the Pyrrhulinæ, of which our BULLFINCH is a familiar example.

It cares little for open country, preferring cultivated grounds, woods, and copses, and is very fond of orchards and fruit-gardens, finding there its greatest supply of food. This bird seems to feed almost wholly on buds during their season, and is consequently shot without mercy by the owners of fruit-gardens. The Bullfinch has a curious propensity for selecting those buds which would produce fruit, so that the leafage of the tree is not at all diminished. Although the general verdict of the garden-keeping public goes against the Bullfinch, there are, nevertheless, some owners of gardens who are willing to say a kind word for Bully, and who assert that its mischievous propensities have been much overated.

It is true that the bird will oftentimes set hard to work upon a fruit-tree, and ruthlessly strip off every single flower-bud, thereby destroying to all appearance the prospects of the crop for that season. Yet there are cases when a gooseberry-bush has thus been completely disbudded, and yet borne a heavy crop of fruit. The reason of this curious phenomenon may probably be, that some of the buds were attacked by insects, and that the kind of pruning process achieved by the Bullfinch was beneficial rather than hurtful to the plant.

BULLFINCH.—(*Pyrrhlua rubicilla.*)

The Bullfinch affords a curious instance of the change wrought by domestication. In its natural state its notes are by no means remarkable, but its memory is so good, and its powers of imitation so singular, that it can be taught to pipe tunes with a sweet and flute-like intonation, having some of that peculiar "woody" quality that is observable in the flute.

Those who desire to find the nest of the Bullfinch must search in the thickets and most retired parts of woods or copses, and they may perhaps find the nest hidden very carefully away in some leafy branch at no great

height from the ground. A thick bush is a very favourite spot for the nest; but I have more than once found them in hazel branches so slender that the mere weight of the nest has bent them aside. The eggs are very prettily marked with deep violet and purple-brown streaks and mottlings upon a greenish white ground, and are easily recognizable by the more or less perfect ring which they form round the larger end of the egg. The eggs are generally five in number.

THE CROSSBILLS, of which three species are known to inhabit England, are most remarkable birds, and have long been celebrated on account of the singular form of beak from which they derive their name.

In all these birds, the two mandibles completely cross one another, so that at first sight the structure appears to be a malformation, and to prohibit the bird from picking up seeds or feeding itself in any way. But when the Crossbill is seen feeding, it speedily proves itself to be favoured with all the ordinary faculties of birds, and to be as capable of obtaining its food as any of the straight-beaked birds.

The food of the Crossbill consists almost, if not wholly, of seeds, which it obtains in a very curious manner. It is very fond of apple-pips, and settling on the tree where ripe apples are to be found, attacks the fruit with its beak, and in a very few moments cuts a hole fairly into the " core," from which it picks out the seeds daintily and eats them, rejecting the ripe pulpy fruit in which they had been enveloped. As the Crossbill is rather a voracious bird, the havoc which it will make in an orchard may be imgined.

This bird is also very fond of the seeds of cone-bearing trees, and haunts the pine-forests in great numbers. While engaged in eating, it breaks the cones from branch-es, and, holding them firmly in its feet after the fashion of the parrots, inserts its beak below the scales, wrenches them away, and with its tongue scoops out the seed.

The Crossbill is not common in this country, although when it does make its appearance it generally comes over in flocks. Usually it consorts in little assemblies consisting of the parents and their young,

CROSSBILL. (*Loxia curvirostris.*)

but it has often been known to associate in considerable numbers. It is a very shy bird, and has a peculiar knack of concealing itself at a moment's notice, pressing itself closely upon the branches at the least alarm, and remaining without a movement or a sound to indicate its position until the danger has departed.

In Sweden and Norway the Crossbill is a very common bird; and the North of Europe seems to be their proper breeding-place.

The nests are always placed in rather close proximity, so that if one nest is found, others are sure to be at no great distance. The nest is made of little fir-twigs, mosses, and wool, and is of rather a loose texture. It is always found upon the part of the branch that is nearest the stem. The fir is the tree that is almost always if not invariably employed by this bird as its nesting-place. The eggs are generally three, but sometimes four in number, and are something like those of the greenfinch, but rather larger.

SCANSORES, OR CLIMBING BIRDS.

A LARGE group of birds is arranged by naturalists under the title ot SCANSORES, or CLIMBING BIRDS, and may be recognized by the structure of their feet. Two toes are directed forward, and the other two backward, so that the bird is able to take a very powerful hold of the substance on which it is sitting, and enables some species, as the woodpeckers, to run nimbly up tree-trunks and to hold themselves tightly on the bark while they hammer away with their beaks; and other species, of which the parrots are familiar examples, to clasp the bough as with a hand.

There are many strange and wonderful forms among the feathered tribes; but there are, perhaps, none which more astonish the beholder who sees them for the first time, than the group of birds known by the name of HORNBILLS.

They are all distinguished by a very large beak, to which is added a singular helmet-like appendage, equalling the beak itself in some species, while in others it is so small as to attract but little notice. On account of the enormous size of the beak and the helmet, which in some species recede to the crown of the head, the bird appears to be overweighted by the mass of horny substance which it has to carry; but on a closer investigation, the whole structure is found to be singularly light, and yet very strong.

On cutting asunder the beak and helmet of a Hornbill, we shall find that the outer shell of horny substance is very thin indeed,

RHINOCEROS HORNBILL.—(*Buceros Rhinoceros.*)

scarcely thicker than the paper on which this description is printed, and that the whole interior is composed of numerous honey-combed cells, with very thin walls and very wide spaces, the walls of the cells being so arranged as to give very great strength when the bill is used for biting, and with a very slight expenditure of material.

Perhaps the greatest development of beak and helmet is found in the RHINOCEROS HORNBILL.

As is the case with all the Hornbills, the beak varies greatly in proportion to the age of the individual, the helmet being almost imperceptible when it is first hatched, and the bill not very striking in its dimensions. But as the bird gains in strength, so does the beak gain in size, and when it is adult the helmet and beak attain their full proportions.

When at liberty in its native forests, the Hornbill is lively and active, leaping from bough to bough with great lightness, and appearing not to be in the least incommoded by its large beak. It ascends the tree by a succession of easy jumps, each of which brings it to a higher branch, and when it has attained the very summit of the tree, it stops and pours forth a succession of loud roaring sounds, which can be heard at a considerable distance.

The food of the Hornbill seems to consist both of animal and vegetable matters, and Lesson remarks that those species which inhabit Africa live on carrion, while those that are found in Asia feed on fruits, and that their flesh acquires thereby an agreeable and peculiar flavour. While on the ground, the movements of the Hornbill are rather peculiar, for instead of walking soberly along, as might be expected from a bird of its size, it hops along by a succession of jumps. It is but seldom seen on the ground, preferring the trunks of trees, which its powerful feet are well calculated to clasp firmly.

THE very curious birds that go by the name of TOUCANS are not one whit less remarkable than hornbills, their beak being often as extravagantly large, and their colours by far superior. They are inhabitants of America, the greater number of species being found in the tropical regions of that country.

Of these birds there are many species, of which no less than five were living in the Zoological Gardens in a single year. Mr. Gould, in his magnificent work, the "Monograph of the Rhamphastidæ," figures fifty-one species, and ranks them under six genera.

The most extraordinary part of these birds is the enormous beak, which in some species, such as the Toco Toucan, is of gigantic dimensions, seeming big enough to give its owner a perpetual headache ; while in others, such as the Toucanets, it is not so large as to attract much attention.

COMMON TOUCAN.—(*Rhamphastos Ariel.*)

As in the case of the hornbills, their beak is very thin and is strengthened by a vast number of honeycomb-cells, so that it is very light and does not incommode the bird in the least. The beak partakes of the brilliant colouring which decorates the plumage, but its beautiful hues are sadly evanescent, often disappearing or changing so thoroughly as to give no intimation of their former beauty.

The voice of the Toucan is hoarse and rather disagreeable, and is in many cases rather articulate. In one species the cry resembles the word " Tucano," which has given origin to the peculiar name by which the whole group is designated. They have a habit of sitting on the branches in flocks, having a sentinel to guard them, and are fond of lifting up their beaks, clattering them together, and shouting hoarsely, from which custom the natives term them Preacher-birds. Sometimes the whole party, including the sentinel, set up a simultaneous yell, which is so deafeningly loud that it can be heard at the distance of a mile. They are very loquacious birds, and are often discovered through their perpetual chattering.

When settling itself to sleep, the Toucan packs itself up in a very systematic manner, supporting its huge beak by resting it on its back, and tucking it completely among the feathers, while it doubles its tail across its back, just as if it moved on a spring hinge. So completely is the bill hidden among the feathers, that hardly a trace of it is visible in spite of its great size and bright colour, and the bird when sleeping looks like a great ball of loose feathers.

PARROTS.

THE general form of the PARROTS is too well known to need description. All birds belonging to this large and splendid group can be recognized by the shape of their beaks, which are large, and have the upper mandible extensively curved and hanging far over the lower; in some species the upper mandible is of extraordinary length. The tongue is short, thick, and fleshy, and the structure of this member aids the bird in no slight degree in its singular powers of articulation. The wings and tail are generally long, and in some species, such as the Macaws, the tail is of very great length, while in most of the Parrakeets it is longer than the body.

THE genus Palæornis, of which the RINGED PARRAKEET is an excellent example, is a very extensive one, and has representatives in almost every hot portion of the world, even including Australia.

The Ringed Parrakeet is found both in Africa and Asia, the only difference perceptible between the individuals brought from the two continents being that the Asiatic species is rather larger than its African relative. It has long been the favourite of man as a caged bird, and is one of the species to which such frequent reference is made by the ancient writers, the other species being the Alexandrine Ringed Parrakeet (*Palæornis Alexandri*).

RINGED PARRAKEET.
(*Palæornis torquatus.*)

This species of Parrakeet is not very good at talking, though it can learn to repeat a few words, and is very apt at communicating its own ideas by a language of gesture and information especially its own. It is, however, very docile, and will soon learn any lesson that may be imposed, even that most difficult task to a Parrot—remaining silent while anyone is speaking.

The general colour of this species is grass-green, variegated in the adult male as follows:—The feathers of the forehead are light green, which take a bluish tinge as they approach the crown and nape of the neck, where they are of a lovely purple blue. Just below the purple runs a narrow band of rose colour, and immediately below the rosy line is a streak of black, which is narrow towards the back of the neck but soon becomes broader, and envelops the cheek and chin. It does not go quite round the neck, as there is an interval of nearly half an inch on the back of the neck. The quill feathers of both wings and tail are darkish green; the wings are black beneath, and the tail yellowish.

ONE of the very prettiest and most interesting of the Parrot tribe is the GRASS or ZEBRA PARRAKEET, deriving its names from its habits and the markings of its plumage.

It is a native of Australia, and may be found in almost all the central portions of that land, whence it has been imported in such great numbers as an inhabitant of our aviaries, that when Dr. Bennett was last in England he found that he could purchase the birds at a cheaper rate in England than in New South Wales. This graceful little creature derives its name of Grass Parrakeet from its fondness for the grass lands, where it may be seen in great

WARBLING GRASS PARRAKEET.—(*Melopsittacus undulatus.*)

numbers, running amid the thick grass blades, clinging to their stems, or feeding on their seeds. It is always an inland bird, being very seldom seen between the mountain ranges and the coasts.

The voice of this bird is quite unlike the rough screeching sounds in which Parrots seem to delight, and is a gentle, soft, warbling kind of song, which seems to be contained within the body, and is not poured out with that

A A

decision which is usually found in birds that can sing, however small their efforts may be. This song, if it may be so called, belongs only to the male bird, who seems to have an idea that his voice must be very agreeable to his mate, for in light warm weather he will warble nearly all day long, and often pushes his beak almost into the ear of his mate, so as to give her the full benefit of his song.

The food of this Parrakeet consists almost chiefly of seeds, those of the grass plant being their constant food in their native country. In England they take well to canary-seed, and it is somewhat remarkable that they do not pick up food with their feet, but always with their beaks. It is a great mistake to confine these lively little birds in a small cage, as their wild habits are peculiarly lively and active, and require much space. The difference between a Grass Parrakeet when in a little cage and after it has been removed into a large house, where it has plenty of space to move about, is really wonderful.

BLUE AND YELLOW MACAW.—(*Ara Ararauna.*)

In its native land it is a migratory bird, assembling after the breeding season in enormous flocks as a preparation for their intended journey. The general number of the eggs is three or four, and they are merely laid in the holes of the gum-tree without requiring a nest.

THE MACAWS are mostly inhabitants of Southern America, in which country so many magnificent birds find their home.

They are all very splendid birds, and are remarkable for their great size, their very long tails, and the splendid hues of their plumage. The beak is also very large and powerful, and in some species the ring round the eyes and part of the face is devoid of covering. As their habits are all very similar, only one example has been figured. This is the great BLUE AND YELLOW MACAW, a bird which is mostly found in Demerara. It is a wood-loving bird, particularly haunting those places where the ground is wet and swampy, and where grows a certain palm on the fruit of which it chiefly feeds.

The wings of this species are strong, and the long tail is so firmly set that considerable powers of flight are manifested. The Macaws often fly at a very great elevation, in large flocks, and are fond of executing sundry aerial evolutions before they alight. With one or two exceptions they care little for the ground, and are generally seen on the summit of the highest trees.

THE true Parrots constitute a group which are easily recognized by their short squared tails, the absence of any crest upon the head, and the toothed edges of the upper mandible.

The GREY PARROT has long been celebrated for its wonderful powers of imitation and its excellent memory.

It is a native of Western Africa, and is one of the commonest inhabitants of our aviaries, being brought over in great numbers by sailors, and always finding a ready sale as soon as the vessel arrives in port.

Its power of imitating all kinds of sounds is really astonishing. I have heard a Parrot imitate, or rather reproduce, in rapid succession the most

GREY PARROT.—(*Psittacus erythacus.*)

dissimilar of sounds, without the least effort and with the most astonishing truthfulness. He could whistle lazily like a street idler, cry prawns and shrimps as well as any costermonger, creak like an ungreased "sheave" in the pulley that is set in the blocks through which ropes run for sundry nautical purposes, or keep up a quiet and gentle monologue about his own accomplishments with a simplicity of attitude that was most absurd.

A A 2

Even in the imitation of louder noises he was equally expert, and could sound the danger-whistle or blow off steam with astonishing accuracy. Until I came to understand the bird, I used to wonder why some invisible person was always turning an imperceptible capstan in my close vicinity, for the Parrot had also learned to imitate the grinding of the capstan bars and the metallic clink of the catch as it falls rapidly upon the cogs.

GREEN PARROT.—(*Chrysotis festivus.*)

As for the ordinary accomplishments of Parrots, he possessed them in perfection, but in my mind his most perfect performance was the imitation of a dog having his foot run over by a cart-wheel. First there came the sudden half-frightened bark, as the beast found itself in unexpected danger, and then the loud shriek of pain, followed by the series of howls that is popularly termed "pen and ink." Lastly the howls grew fainter, as the dog was supposed to be limping away, and you really seemed to hear him turn the corner and retreat into the distance. The memory of the bird must have been most tenacious, and its powers of observation far beyond the common order ; for he could not have been witness to such a canine accident more than once.

The food of this Parrot consists chiefly of seeds of various kinds, and in captivity may be varied to some extent. Hemp-seed, grain, canary-seed, and the cones of fir-trees are favourite articles of diet with this bird. Of the cones it is especially fond, nibbling them to pieces when they are young and tender ; but, when they are old and ripe, breaking away the hard scales and scooping out the seeds with its very useful tongue. Hawthorn berries are very good for the Parrot, as are several vegetables. These, however, should be given with great caution, as several, such as parsley and chick-weed, are exceedingly hurtful to the bird.

When proper precautions are taken, the Parrot is one of our hardiest cage-birds, and will live to a great age even in captivity. Some of these birds have been known to attain an age of sixty or seventy years, and one which was seen by Le Vaillant had attained the patriarchal age of ninety-three. At sixty its memory began to fail ; and at sixty-five the moult became very irregular, and the tail changed to yellow. At ninety it was a very decrepit creature, almost blind and quite silent, having forgotten its former abundant stock of words.

The general colour of this bird is a very pure ashen grey, except the tail, which is deep scarlet.

Two species of GREEN PARROT are tolerably common, the one being the Festive Green Parrot, and the other the Amazon Green Parrot.

The AMAZON GREEN PARROT is the species most commonly seen in England. It is a native of Southern America, and especially frequents the banks of the Amazon. It is not, however, so retiring in its habits as most Parrots, and will often leave the woods for the sake of preying upon the orange plantations, among which it works great havoc. Its nest is made in the decayed trunks of trees.

As a general fact, it is not so apt at learning and repeating phrases as the Grey Parrot, but I have known more than one instance where its powers of speech could hardly be exceeded, and very seldom rivalled. One of these birds, which used to live in a little garden into which my window looked, was, on our first entrance into the house, the cause of much perplexity to ourselves and the servants. The nurserymaid's name was Sarah, and the unfortunate girl was continually running up

SULPHUR-CRESTED COCKATOO.
(*Cacatua galerita.*)

and down stairs, fancying herself called by one of the children in distress. The voice of the Parrot was just that of a child, and it would call Sarah in every imaginable tone, varying from a mere enunciation of the name as if in conversation, to angry remonstrance, petulant peevishness, or sudden terror.

THE COCKATOOS are very familiar birds, as several species are common inhabitants of our aviaries, where they create much amusement by their grotesque movements, their exceeding love of approbation, and their repeated mention of their own name. Wherever two or three of these birds are found in the same apartment, however silent they may be when left alone, the presence of a visitor excites them to immediate conversation, and the air

LEADBEATER'S COCKATOO.—(*Cacatua Leadbeateri.*)

resounds with " Cockatoo ! " " Pretty Cocky ! " in all directions, diversified with an occasional yell, if the utterer be not immediately noticed.

They are confined to the Eastern Archipelago and Australia, in which latter country a considerable number of large and splendid species are found. The nesting-place of the Cockatoos is always in the holes of decaying trees, and by means of their very powerful beaks they tear away the wood until they have enlarged the hollow to their liking. Their food consists almost wholly of fruits and seeds, and they are often very great pests to the agriculturist, settling in large flocks upon the fields of maize and corn, and devouring the ripened ears or disinterring the newly-sown seeds with hearty good-will. The wrath of the farmer is naturally aroused by these frequent raids, and the Cockatoos perish annually in great numbers from the constant persecution to which they are subjected, their nests being destroyed, and themselves shot and trapped.

To those, however, who own no land and are anxious about no crops, a flock of Cockatoos is a most beautiful and welcome sight, as they flit among the heavy-leaved trees of the Australian forest, their pinky-white plumage relieved against the dark masses of umbrageous shade, as they appear and vanish among the branches like the bright visions of a dream.

THE remarkably handsome bird which is represented on the opposite page is a native of Australia. It is called by several names, such as the TRICOLOR CRESTED COCKATOO, and the PINK COCKATOO, by which latter name it is known to the colonists. The title of LEADBEATER'S COCKATOO was given to the bird in honour of the well-known naturalist, who possessed the first specimen brought to England.

It is not so noisy as the common species, and may possibly prove a favourite inhabitant of our aviaries, its soft blush-white plumage and splendid crest well meriting the attention of bird-fanciers. The crest is remarkable for its great development, and for the manner in which the bird can raise it like a fan over its head, or depress it upon the back of its neck at will. In either case it has a very fine effect, and especially so when it is elevated, and the bird is excited with anger or pleasure.

The general colour of this bird is white, with a slight pinkish flush. Round the base of the beak runs a very narrow crimson line, and the feathers of the crest are long and pointed, each feather being crimson at the base, then broadly barred with golden yellow, then with crimson, and the remainder is white. The neck, breast, flanks, and under-tail coverts are deeply stained with crimson, and the under surface of the wing is deep crimson-red. The beak is pale greyish white, the eyes brown, and the feet and legs dark grey, each scale being edged with a lighter tint. In size it is rather superior to the common white Cockatoo.

THE species of Cockatoo which is most common in England is the SULPHUR-CRESTED COCKATOO, a representation of which will be found on page 357.

This bird is an inhabitant of different parts of Australia, and is especially common in Van Diemen's Land, where it may be found in flocks of a thousand in number. Owing to the ease with which it is obtained, it is frequently brought to England and is held in much estimation as a pet.

The colour of this species is white, with the exception of the crest, which is of a bright sulphur-yellow, and the under surface of the wings and the basal portions of the inner webs of the tail-feathers, which are of the same colour, but much paler in hue. The total length of this species is about eighteen inches.

WE now take our leave of the Parrots, and come to a very interesting family of scansorial birds, known popularly as WOODPECKERS ; and scientifically as Picidæ.

As is well known, the name of Woodpecker is given to these birds from their habit of pecking among the decaying wood of trees in order to feed upon the insects that are found within. They also chip away the wood for the purpose of making the holes or tunnels wherein their eggs are deposited. In order to enable them to perform these duties, the structure of the Woodpecker is very curiously modified. The feet are extremely powerful, and the claws are strong and sharply hooked, so that the bird can retain a firm hold of the tree to which it is clinging while it works away at the bark or wood with its bill. The tail, too, is furnished with very stiff and pointed feathers, which are pressed against the bark, and form a kind of support on which the bird can rest a large proportion of its weight. The breast-bone is not so prominent

GREAT SPOTTED WOODPECKER.—(*Picus major.*)

as in the generality of flying birds, in order to enable the Woodpecker to press its breast closely to the tree; and the beak is long, strong, and sharp.

These modifications aid the bird in cutting away the wood, but there is yet a provision needful to render the Woodpecker capable of seizing the little insects on which it feeds, and which lurk in small holes and crannies into which the beak of the Woodpecker could not penetrate. This structure is shown when a Woodpecker's head is carefully dissected. The tongue-bones or "hyoid" bones are greatly lengthened, and pass over the top of the head, being fastened in the skull just above the right nostril. These long tendinous-looking bones are accompanied by a narrow strip of muscle by which they are moved.

The tongue is furnished at the tip with a long horny appendage covered with barbs and sharply pointed at the extremity, so that the bird is enabled to project this instrument to a considerable distance from the bill, transfix an insect, and draw it into the mouth. Those insects that are too small to be thus treated are captured by means of a glutinous liquid poured upon the tongue from certain glands within the mouth, and which cause the little insects to adhere to the weapon suddenly projected among them. Some authors deny the transfixion.

THE GREAT SPOTTED WOODPECKER is one of the five British species, and is also known by the names of Frenchpie and Woodpie.

It is found in many parts of England, and, like the other Woodpeckers, must be sought in the forests and woods rather than in orchards and gardens. Like other shy birds, however, it soon finds out where it may take up its abode unmolested, and will occasionally make its nest in some cultivated ground, where it has the instinctive assurance of safety, rather than entrust itself to the uncertain security of the forest.

Although the Woodpeckers were formerly much persecuted, under the idea that they killed the trees by pecking holes in them, they are most useful birds, cutting away the decaying wood, as a surgeon removes a gangrened spot, and eating the hosts of insects which encamp in dead or dying wood, and would soon bring the whole tree to the ground. They do not confine themselves to trees, but seek their food wherever they can find it, searching old posts and rails, and especially delighting in those trees that are much infested with the green-fly, or aphis, as the wood-ants swarm in such trees for the purpose of obtaining the "honey-dew," as it distils from the aphides, and then the Woodpeckers eat the ants. Those destructive creatures generally called wood-lice, and known to boys as "monkey-peas," are a favourite article of diet with the Woodpeckers, to whom our best thanks are therefore due.

But the Woodpeckers, although living mostly on insects, do not confine themselves wholly to that diet, but are very fond of fruits, always choosing the ripest.

As is the case with all its congeners, the Great Spotted Woodpecker lays its eggs in the hollow of a tree.

The locality chosen for this purpose is carefully selected, and is a tunnel excavated, or at all events altered, by the bird for the special purpose of nidi-fication. Before commencing the operation, the Woodpeckers always find out whether the tree is sound or rotten, and they can ascertain the latter fact even through several layers of sound wood. When they have fixed upon a site for their domicile, they set determinately to work, and speedily cut out a circular tunnel just large enough to admit their bodies, but no larger. Sometimes this tunnel is tolerably straight, but it generally turns off in another direction.

At the bottom of the hole the female bird collects the little chips of decayed wood that have been cut off during the boring process, and deposits her eggs upon them without any attempt at nest-making. Some excellent examples of these nests are in the British Museum. The eggs are generally five in number, but six have been taken from the nest of this species.

Generally the nests of birds are kept scrupulously clean, but that of the Woodpecker is a sad exception to the rule, the amount of filth and potency of stench being quite beyond human endurance. The colour of the eggs is white, and their surface glossy, and they are remarkable, when fresh, for some very faint and very narrow lines, which run longitudinally down the shell towards the small end.

The general colour of this species is black and white, curiously disposed, with the exception of the back of the head, which is light scarlet, and contrasts strongly with the sober hues of the body. Taking the black to be

GREEN WOODPECKER.—(*Gecinus viridis.*)

the ground colour, the white is thus arranged. The forehead and **ear-coverts,** a patch on each side of the neck, the scapularies, and part of the wing-coverts, several little squared spots on the wings, and large patches on the tail, are pure white. The throat and the whole of the under surface are also white, but with a greyish cast, and the under tail-coverts are red. The total length of the adult male is rather more than nine inches. The female has no red on the head and the young birds of the first year are remarkable for having the back of the head black and the top of the head red, often mixed with a few little black feathers.

THE commonest of the British Woodpeckers is that which is generally known by the name of the GREEN WOODPECKER. It has, however, many popular titles, such as Rain-bird, Wood-spite, Hew-hole, and Wood-wall. This bird is our representative of the Gecinæ, or Green Woodpeckers.

Although the Green Woodpecker is a haunter of woods and forests, it will sometimes leave those favoured localities and visit the neighbourhood of man. The grounds between the Isis and Merton College, Oxford, are rather favourite resorts of this pretty bird.

The name of Rain-bird has been given to this species because it becomes very vociferous at the approach of wet weather, and is, as Mr. Yarrell well observes, "a living barometer to good observers." Most birds, however, will answer the same purpose to those who know how and where to look for them. The other titles are equally appropriate, Wood-spite being clearly a corruption of the German term *specht*. Hew-hole speaks for itself ; and Wood-wall is an ancient name for the bird, occurring in the old English poets.

The other British species are the Great Black Woodpecker (*Dryocopus Martius*), the Northern Three-toed Woodpecker (*Picoides tridactylus*), and the Lesser Spotted Woodpecker (*Picus minor*).

WRYNECK.—(*Yunx torquilla.*)

THE curious bird known under the popular and appropriate name of the WRYNECK is by some considered to be closely allied to the woodpeckers.

The Wryneck is a summer visitor to this country, appearing just before the cuckoo, and therefore known in some parts of England as the cuckoo's footman. There is a Welsh name for this bird, signifying " Cuckoo's knave," —*Gwas-y-gôg*.

The tongue of this bird is long, slender, and capable of being projected to the distance of an inch or so from the extremity of the beak, and its construction is almost exactly the same as that of the woodpecker. As might be supposed, it is employed for the same purpose, being used in capturing little insects, of which ants form its favourite diet. So, fond, indeed, is the Wryneck of these insects, that in some parts of England it is popularly known by the name of Emmet-hunter. In pursuit of ants it trips nimbly about the trunks and branches of trees, picking them off neatly with its tongue as they run their untiring course. It also frequents ant-hills, especially when the insects are bringing out their pupæ to lie in the sun, and swallows ants and pupæ at a great pace. When the ants remain within their fortress, the Wryneck pecks briskly at the hillock until it breaks its way through the fragile walls of the nest, and as the warlike insects come rushing out to attack the intruder of their home and to repair damages, it makes an excellent meal of them in spite of their anger and their stings.

When ants are scarce and scantily spread over the ground, the Wryneck runs after them in a very agile fashion ; but when it comes upon a well-stocked spot, it stands motionless, with the exception of the head, which is darted rapidly in every direction, the neck and central line of the back twisting in a manner that reminds the observer of a snake. When captured or wounded, it will lie on its back, ruffle up its feathers, erect its neck, and hiss so like an angry serpent that it is in some places known by the name of the snake-bird.

The nest of the Wryneck is hardly deserving of that name, being merely composed of chips of decaying wood. The eggs are laid in the hollow of a tree, not wholly excavated by the bird, as is the case with the woodpeckers, its beak not being sufficiently strong for such a task, but adapted to the purpose from some already existing hole.

The number of eggs laid by the Wryneck is rather great, as many as ten having often been found in a single nest. In one instance no less than twenty-two eggs were taken at four intervals. Their colour is beautiful white with a pinky tinge, not unlike those of the kingfisher; and as this pink colour is produced by the yolk showing itself through the delicate shell, it is, of course, lost when the egg is emptied of its contents. The plumage of this little bird, although devoid of brilliant hues, and decked only with brown, black, and grey, is really handsome, from the manner in which those apparently sombre tints are disposed.

THE CUCKOOS constitute a large family, containing several smaller groups and many species. Two representatives of the groups will be found in the following pages. All these birds have a rather long, slender, and somewhat curved beak, which in

CUCKOO.—(*Cuculus Canorus.*)

some species takes a curve so decided, that it gives quite a predaceous air to its owner. Examples of the Cuckoo tribe are to be found in almost every portion of the globe, and are most plentiful about the tropics.

THERE are few birds which are more widely known by good and evil report than the common CUCKOO.

It is well known that the female Cuckoo does not make any nest, but places her egg in the nest of some small bird, and leaves it to the care of its unwitting foster-parents. Various birds are burdened with this charge, such as the hedge-warbler, the pied-wagtail, the meadow-pipit, the red-backed shrike, the blackbird, and various finches. Generally, however, the three first are those preferred. Considering the size of the mother-bird, the egg of the Cuckoo is remarkably small, being about the same size as that of the skylark, although the latter bird has barely one-fourth the dimensions of the former. The little birds, therefore, which are always careless about the colour or form of an egg, provided that it be nearly the size of their own productions, do not detect the imposition, and hatch the interloper together with their own young.

The general colour of the Cuckoo's egg is mottled reddish grey, but the tint is very variable in different individuals, as I can testify from personal experience. It has also been noted that the colour of the egg varies with the species in whose nest it is to be placed, so that the egg which is intended to be hatched by the hedge-warbler is not precisely of the same colour as that which is destined for the nest of the pipit.

The mode by which the Cuckoo contrives to deposit her eggs in the nest of sundry birds was extremely dubious, until a key was found to the problem by a chance discovery made by Le Vaillant. He had shot a female Cuckoo, and on opening its mouth in order to stuff it with tow, he found an egg lodged very snugly within the throat.

The peculiar note of the Cuckoo is so well known as to need no particular description, but the public is not quite so familiar with the fact that the note changes according to the time of year. When the bird first begins to sing, the notes are full and clear; but towards the end of the season they become hesitating, hoarse, and broken, like the breaking voice of a young lad. This peculiarity was noticed long ago by observant persons, and many are the country rhymes which bear allusion to the voice and the sojourn of the Cuckoo. For example:—

> " In April
> Come he will.
> In May
> He sings all day.
> In June
> He alters his tune.
> In July
> He prepares to fly.
> In August
> Go he must."

In general appearance the Cuckoo bears some resemblance to a bird of prey, but it has little of the predaceous nature. It is rather curious that small birds have a tendency to treat the Cuckoo much as they treat the hawks and owls, following it wherever it flies in the open country, and attending it through the air.

The colour of the plumage is bluish grey above, with the exception of the wings and tail, which are black, and barred with white on the exterior feathers. The chin, neck, and breast are ashen grey, and the abdomen and under wing-coverts are white, barred with slaty grey.

COLUMBÆ; OR DOVES AND PIGEONS.

THE large order of COLUMBÆ, or the Pigeon tribe, comes now under our notice. It contains many beautiful and interesting birds ; but as its members are so extremely numerous, only a few typical examples can be mentioned.

All the Pigeons may be distinguished from the poultry, and the gallinaceous birds in general, by the form of the bill, which is arched towards the tip, and has a convex swelling at the base, caused by a gristly kind of plate which covers the nasal cavities, and which in some species is very curiously developed.

AMONG the most extraordinary of birds, the PASSENGER PIGEON may take very high rank, not on account of its size or beauty, but on account of the extraordinary multitudes in which it sometimes migrates from one place to another. The scenes which take place during these migrations are so strange, so wonderful, and so entirely unlike any events on this side of the Atlantic, that they could not be believed but for the trustworthy testimony by which they are corroborated.

Wilson, who was fortunate enough to witness some of these migrations, has written a most vivid account of them. After professing his belief that the chief object of the migration is the search after food, and that the birds having devoured all the nutriment in one part of the country take wing in order to feed on the beech-mast of another region, he proceeds to describe a breeding-place seen by himself in Kentucky, which was several miles in breadth, was said to be nearly forty miles in length, and in which every tree was absolutely loaded with nests. All the smaller branches were destroyed by the birds, many of the large limbs were broken off and thrown on the ground, while no few of the grand forest-trees themselves were killed as surely as if the axe had been employed for their destruction. The Pigeons had arrived about the 10th of April, and left it by the end of May.

THE STOCK-DOVE derives its name from its habit of building its nest in the stocks or stumps of trees. It is one of our British Pigeons, and is tolerably common in many parts of England.

It is seldom found far northward, and even when it does visit such localities it is only as a summer resident, making its nest in warmer districts. As has already been mentioned, the nest of this species is made in the stocks or stumps of trees, the birds finding out some convenient hollow, and placing their eggs within. Other localities are, however, selected for the purpose of incubation, among which a deserted rabbit burrow is among the most common. The nest is hardly worthy of the name, being a mere collection of dry fibrous roots, laid about three or four feet within the entrance, just thick enough to keep the eggs from the ground, but not sufficiently woven to constitute a true nest.

The head, neck, and back and wing-coverts are bluish grey, the primary quill feathers of the wing taking a deeper hue, the secondaries being pearl-grey deepening at the tips, and the tertials being blue-grey with two or three spots. The chin is blue-grey, the sides of the neck slaty-grey glossed with green, and the breast purplish red. The specific name of *œnas*, or "wine-

PASSENGER PIGEON.—(*Ectopistes migratorius.*)

STOCK-DOVE.—(*Columba œnas.*)

coloured," is given to the bird on account of the peculiar hue of the throat The whole of the under surface is grey, and the tail-feathers are coloured with grey of several tones, the outside feathers having the basal portion of the outer web white. The beak is deep orange, the eyes scarlet, and the legs and toes red. The total length is about fourteen inches, the female being a little smaller.

THE bird which now comes before our notice is familiar to all residents in the country under the titles of RING-DOVE, WOOD PIGEON, WOOD GUEST, and CUSHAT.

This pretty dove is one of the commonest of our British birds, breeding in almost every little copse or tuft of trees, and inhabiting the forest grounds in great abundance. Towards and during the breeding season, its soft compla-cent cooing is heard in every direction, and with a very slight search its nest may be found. It is a strange nest, and hardly deserving that name, being nothing more than a mere platform of sticks resting upon the fork of a bough, and placed so loosely across each other that when the maternal bird is away the light may sometimes be seen through the interstices of the nest, and the outline of the eggs made out. Generally the Ring-dove chooses a rather lofty branch for its nesting-place, but it occasionally builds at a very low elevation.

The eggs are never more than two in number, and per-fectly white, looking some-thing like hen's eggs on a small scale, save that the ends are more equally rounded.

The food of this Dove con-sists of grain and seeds of various kinds, together with the green blades of newly sprung corn and the leaves of turnips, clover, and other vegetables. Quiet and harmless as it may look, the Ring-dove is a won-derful gormandizer, and can consume great quantities of food. The crop is capacious to suit the appetite, and can con-tain a singular amount of solid food, as indeed seems to be the case with most of the Pigeon tribe, so that when the birds as-

RING-DOVE.—(*Columba palumbus.*)

semble together in the autumn, the flocks will do great damage to the farmer.

The Ring-dove may be easily known by the peculiarity from which it derives its name, the feathers upon the side of the neck being tipped with white so as to form portions of rings set obliquely on the neck.

THE many varieties of size, form, and colour which may be seen in the accompanying illustration afford an excellent example of the wonderful variations of which animals are susceptible under certain circumstances. Different as are the DOMESTIC PIGEONS, which are most ably figured on the next page by a practical pigeon-fancier as well as an accomplished artist, they all are modifications of the common BLUE ROCK PIGEON, and, if permitted to mix freely with each other, display an inveterate tendency to return to the original form, with its simple plumage of black bars across the wing, just as the finest breeds of lop-eared rabbits will now and then produce upright-eared young.

The Rock-dove derives its popular name from its habit of frequenting rocks rather than trees, an idiosyncrasy which is so inherent in its progeny, that even the domestic Pigeons, which have not seen anything except their wooden cotes for a long series of generations, will, if they escape, take to rocks or buildings, and never trouble themselves about trees, though they should be at hand.

This species seems to have a very considerable geographical range, for it is common over most parts of Europe, Northern Africa, the coasts of the Mediterranean, and has even been found in Japan.

From this stock, the varieties that have been reared by careful management are almost innumerable, and are so different in appearance that if they were seen for the first time, almost any systematic naturalist would set them down as belonging not only to different species, but to different genera. Such, for example, as the pouter, the jacobin, the trumpeter, and the fantail, the last-mentioned bird having a greater number of feathers in its tail than any of the others.

GROUP OF PIGEONS.

THE world-famed TURTLE-DOVE is, although a regular visitor of this country, better known by fame and tradition than by actual observation. This bird has, from classic time until the present day, been conventionally accepted as the type of matrimonial perfection, loving but its mate and caring for no other until death steps in to part the wedded couple. Yet it is by no means the only instance of such conjugal affection among the feathered tribes, for there are hundreds of birds which can lay claim to the same excellent qualities, the fierce eagle and the ill-omened raven being among their number.

The Turtle-dove seems to divide its attention pretty equally between Africa and England, pausing for some little time in Southern Italy as a kind of half-way house. It arrives here about the beginning of May, or perhaps a little earlier in case the weather be warm, and after resting for a little while, sets about making its very simple nest and laying its white eggs. The nest of this bird is built lower than is generally the case with the wood-pigeon, and is usually placed on a forked branch of some convenient tree, about ten feet or so from the ground. The eggs are laid rather late in the season, so that there is seldom more than a single brood of two young in the course of the year.

The Turtle-dove may be readily known by the four rows of black feathers, tipped with white, which are found on the sides of the neck.

THE position held by the celebrated DODO among birds was long doubtful, and was only settled in comparatively late years by careful examination of the few relics which are our sole and scanty records of this very remarkable bird.

So plentiful were the Dodos at one time, and so easily were they killed, that the sailors were in the habit of slaying the birds merely for the sake of the stones in their stomachs, these being found very efficacious in sharpening their clasp-knives. The nest of the Dodo was a mere heap of fallen leaves gathered together on the ground, and the bird laid but one large egg. The weight of one full-grown Dodo was said to be between forty and fifty pounds. The colour of the plumage was a greyish brown in the adult males, not unlike that of the

TURTLE DOVE. — (*Turtur auritus.*)

ostrich, while the plumage of the females was of a paler hue.

LEAVING the pigeons, we now come to the large and important order of birds termed scientifically the Gallinæ, and, more popularly, the Poultry. Sometimes they are termed Rasores, or Scrapers, from their habit of scraping up the ground in search of food. To this order belong our domestic poultry, the grouse, partridges, and quails, the turkeys, pheasants, and many other useful and interesting birds.

Our first example of these birds is the CRESTED CURASSOW, the representative of the genus Crax, in which are to be found a number of truly splendid birds. All the Curassows are natives of tropical America, and are found almost wholly in the forests.

The Crested Curassow inhabits the thickly wooded districts of Guiana, Mexico, and Brazil, and is very plentifully found in those countries. It is a really handsome bird, nearly as large as the turkey, and more imposing in form and colour. It is gregarious in its habits, and assembles together in large troops, mostly perched on the branches of trees. It is suscepti-

DODO. — (*Didus ineptus.*)

ble of domestication, and, to all appearances, may be acclimatized to this country as well as the turkey or the pheasant.

B B

In their native country the Curassows build among the trees, making a large and rather clumsy-looking nest of sticks, grass stems, leaves, and grass blades. There are generally six or seven eggs, not unlike those of the fowl, but larger and thicker shelled.

The colour of the Crested Curassow is very dark violet, with a purplish green gloss above and on the breast, and the abdomen is the purest snowy white, contrasting beautifully with the dark velvety plumage of the upper parts. The bright golden yellow of the crest adds in no small degree to the beauty of the bird

CRESTED CURASSOW.—(*Crax Alector.*)

SEVERAL very singular birds are found in Australia and New Guinea, called by the name of Megapodinæ, or Great-footed birds, on account of the very large size of their feet ; a provision of nature which is necessary for their very peculiar mode of laying their eggs and hatching their young.

The first of these birds is the AUSTRALIAN JUNGLE FOWL, which is found in several parts of Australia, but especially about Port Essington. In that country great numbers of high and large mounds of earth exist, which were formerly thought to be the tombs of departed natives, and, indeed, have been more than once figured as such. The natives, however, disclaimed the sepulchral character, saying that they were the artificial ovens in which the eggs of the Jungle Fowl were laid, and which, by the heat that is always disengaged from decaying vegetable substances, preserved sufficient warmth to hatch the eggs.

The size of these tumuli is sometimes quite marvellous ; in one instance, where measurements were taken, it was fifteen feet in perpendicular height, and sixty feet in circumference at its base. The whole of this enormous mound was made by the industrious Jungle Fowl, by gathering up the earth, fallen leaves, and grasses with its feet, and throwing them backwards while it stands on the other leg. If the hand be inserted into the heap, the interior will always be found to be quite hot. In almost every case the mound is placed under the shelter of densely-leaved trees, so as to prevent the sun from shining upon any part of it.

The bird seems to deposit her eggs by digging holes from the top of the mound, laying the egg at the bottom, and then making its way out again, throwing back the earth that it had scooped away. The direction, however, of the holes is by no means uniform, some running towards the centre and others radiating towards the sides. They do not seem to be dug quite per-pendicularly ; so that although the holes in which the eggs are found may be some six or seven feet in depth, the eggs themselves may be only two or three feet from the surface.

The colouring of this bird is simple, but the tints are soft and pleasing. The head is rich ruddy brown, the back of the neck black-ish grey, and the back and wings brownish cinnamon, deepening into dark chestnut on the tail-coverts. The whole under surface

AUSTRALIAN JUNGLE FOWL.—(*Megapodius tumulus.*)

is blackish grey. The legs are orange, and the bill rusty brown.

THE BRUSH TURKEY is principally found in the thick brushwood of New South Wales. Mr. Gould, who first brought it before the public, gives this curious account of their nests :— " The mode in which the materials composing these mounds are accumulated is equally singu-lar, the bird never using its bill, but always grasping a quantity in its foot, throwing it backwards to one common centre, and thus clearing the surface of the ground for a considerable distance so completely that scarcely a leaf or a blade of grass is left. The heap being accumulated, and time allowed for a sufficient heat to be engendered, the eggs are de-posited, not side by side as is ordinarily the case, but planted at the dis-tance of nine or twelve inches from each other, and buried at nearly an arm's depth, perfectly upright, with the large end upwards. They are covered up as they are laid, and allowed to remain until hatched. I am credibly informed, both by natives and settlers living near their haunts, that it is not an unusual event to obtain nearly a bushel of eggs at one time from a single heap ; and as they are delicious eating they are eagerly sought after."

When the Brush Turkey is disturbed, it either runs through the tangled underwood with singular rapidity, or springs upon a low branch of some tree, and reaches the summit by a succession of leaps from branch to branch. This latter peculiarity renders it an easy prey to the sportsman.

THE large family of the PEACOCKS, or Pavonidæ, now claims our attention.

The PEACOCK may safely be termed one of the most magnificent of the feathered tribe, and may even lay a well-founded claim to the chief rank among birds in splendour of plumage and effulgence of colouring. We are so familiar with the Peacock that we think little of its real splendour ; but if one of these birds had been brought to Europe for the first time, it would create a greater sensation than even the hippopotamus or the gorilla.

BRUSH TURKEY.—(*Talegallus Lathami.*)

The Peacock is an Asiatic bird, the ordinary species being found chiefly in India, and the Javanese Peacock in the country from which it derives its name. In some parts of India the Peacock is extremely common, flocking together in bands of thirty or forty in number, covering the trees with their splendid plumage, and filling the air with their horridly dissonant voices. Captain Williamson, in his " Oriental Field Sports," mentions that he has seen at least twelve or fifteen hundred peacocks within sight of the spot where he stood.

They abound chiefly in close-wooded forests, particularly where there is an extent of long grass for them to range in. They are very thirsty birds, and will only remain where they can have access to water. Rhur plantations are their favourite shelter, being close above so as to keep off the solar rays, and open at the bottom sufficiently to admit a free passage for the air. If there be trees near such spots, the Peacocks may be seen mounting into them every evening towards dark to roost ; and in which they generally continue till the sun rises, when they descend to feed, and pass the mid-day in the heavy coverts.

Though Pea-fowls invariably roost in trees, yet they make their nest on the ground, and ordinarily on a bank raised above the common level, where in some sufficient bush they collect leaves, small sticks, &c., and sit very close. I have on several occasions seen them in their nests, but as I refrained from disturbing them, they did not offer to move, though they could not fail to know that they were discovered. They usually sit on about a dozen or fifteen eggs. They are generally hatched about the beginning of November; and from January to the end of March, when the corn is standing, are remarkably juicy and tender. When the dry season comes on, they feed on the seeds of weeds and insects, and their flesh becomes dry and muscular.

The train of the male Peacock, although popularly called its tail, is in reality composed of the upper tail-coverts, which are enormously lengthened, and finished at their extremities with broad rounded webs, or with spear-shaped ends. The shafts of these feathers are almost bare of web for some fourteen or fifteen inches of their length, and then throw out a number of long loose vanes of a light coppery green. These are very brittle, and apt to snap off at different lengths. In the central feathers the extremity is modified into a wide flattened battledore-shaped form, each barbule being coloured with refulgent emerald green, deep violet purple, greenish bronze, gold and blue, in such a manner as to form a distinct "eye," the centre being violet of two shades, surrounded with emerald, and the other tints being arranged concentrically around it. In the feathers that edge the train there is no "eye," the feathers coming to a point at the extremity, and having rather wide but loose emerald-green barbules on its outer web and a few scattered coppery barbules in the place of the inner web. The tail-feathers are only seven or eight inches in length, are of a greyish brown colour, and can be seen when the train is erected, that being their appointed task.

PEACOCK.—(*Pavo cristatus.*)

THE PHEASANTS come next in order, and the grandest and most imposing of this group, although there are many others that surpass its brilliant colouring, is the ARGUS PHEASANT so called in remembrance of the ill-fated Argus of mythology, whose hundred eyes never slept simultaneously until charmed by the magic lyre of Mercury.

This magnificent bird is remarkable for the very great length of its tail-feathers and the extraordinary development of the secondary feathers of the wings. While walking on the ground, or sitting on a bough, the singular length of the feathers is not very striking, but when the bird spreads its wings, as shown in the figure, they come out in all their beauty. As might be supposed from the general arrangement of the plumage, the bird is by no means a good flyer, and when it takes to the air only flies for a short distance. In running its wings are said to be efficient aids.

Although the Argus is hardly larger than an ordinary fowl the plumage is so greatly developed that its total length measures more than five feet. The head and back of the neck are covered with short brown feathers, and the neck and upper part of the breast are warm chestnut-brown, covered with spots of yellow and black, and similar tints are formed on the back. The tail is deep chestnut, covered with white spots, each spot being surrounded with a black ring.

The Argus Pheasant inhabits Sumatra and neighbouring localities.

THE well known PHEASANT affords a triumphant instance of the success with which a bird of a strange country may be acclimatized to this island with some little assistance from its owners.

Originally the Pheasant was an inhabitant of Asia Minor, and has been by degrees introduced into many European countries, where its beauty of form and plumage and the delicacy of its flesh made it a welcome visitor.

The food of this bird is extremely varied. When young it is generally fed on ants' eggs, maggots, grits, and similar food, but when it is fully grown it is possessed of an accommodating appetite, and will eat many kinds of seeds, roots, and leaves. The tubers of the common buttercup form a considerable item in its diet, and the bird will also eat beans, peas, acorns, and berries of various kinds.

ARGUS PHEASANT.—(*Argus giganteus.*)

The Pheasant is a ground-loving bird, running with great speed, and always preferring to trust to its legs rather than its wings. It is a crafty creature, and when alarmed, instead of rising on the wing, it slips quietly out of sight behind a bush or through a hedge, and then runs away with astonishing rapidity, always remaining under cover until it reaches some spot where it deems itself to be safe.

The nest of the Pheasant is a very rude attempt at building, being merely a heap of leaves and grasses collected together upon the ground, and with a very slight depression, caused apparently quite as much by the weight of the eggs as by the art of the bird. The eggs are numerous, generally about eleven or twelve, and their colour is a uniform olive-brown. Their surface is very smooth.

THE BANKIVA JUNGLE FOWL is now supposed to be the original stock of the domesticated poultry.

It is a native of Java, and the male very closely resembles the gamecock of England. It is a splendid creature, with its light scarlet comb and wattles, its drooping hackles, its long arched tail, and its flashing eye. The comb and wattles are of brightest scarlet, the long hackles of the neck and lower part of the back are fine orange-red, the upper part of the back is deep blue-black, and the shoulders are ruddy chestnut. The secondaries and greater coverts are deep steely blue, and the quill feathers of the wing are blackish brown edged with rusty yellow. The long, arched and drooping tail is blue-black glossed with green, and the breast and under parts black, so that in general aspect it is very like the black-breasted red gamecock.

PHEASANT.—(*Phasianus Colchicus.*)

The domesticated bird is of all the feathered tribe the most directl y useful to man, and is the subject of so many valuable treatises that the reader is referred to them for the best mode of breeding, rearing, and general management of poultry. On the accompanying illustration are shown some of the most useful or remarkable of the varieties of this bird.

THE now well-known TURKEY is another example of the success with which foreign birds can be acclimatized in this country.

The Turkey is spread over many parts of America, such as the wooded parts of Arkansas, Louisiana, Alabama, Indiana, &c., but does not seem to extend beyond the Rocky Mountains. It begins

DOMESTIC FOWL.—(*Gallus Bankiva.*)

to mate about the middle of February, and the males then utter those ludicrous gobbling sounds which have caused the bird to be called Gobbler, or Bubbly-Jock by the whites, and Oo-coo-coo by the Cherokees.

The female makes her nest in some secluded spot, and is very guarded in her approaches, seldom employing the same path twice in succession: and, if discovered, using various wiles by which to draw the intruder from the spot. As soon as the young are hatched she takes them under her charge, and the whole family go wandering about to great distances, at first returning to the nest for the night, but afterwards crouching in any suitable spot. Marshy places are avoided by the Turkey, as wet is fatal to the young birds until they have attained their second suit of clothes and wear feathers instead of down. When they are about a fortnight old they are able to get up into trees, and roost in the branches, safe from most of the numerous enemies which beset their path through life.

TURKEY.—(*Meleagris gallopavo.*)

The Turkey is a very migratory bird, passing over great distances and retaining the habit in its tamed state, giving no small amount of trouble to the poultry owner.

THE prettily spotted GUINEA FOWL, or PINTADO, sometimes called GALLINI, is, although now domesticated in England, a native of Africa, and has much of the habits and propensities of the turkey.

Both in the wild and the captive state the Guinea Fowl is wary and suspicious, and particularly careful not to betray the position of its nest, thus often giving great trouble to the farmer. Sometimes when the breeding season approaches, the female Pintado will hide herself and nest so effectually that the only indication of her proceedings is her subsequent appearance with a brood of young round her. The number of eggs is rather large, being seldom below ten, and often double that number. Their colour is yellowish red, covered with very little dark spots, and their size is less than that of the common fowl. Their shells are extremely hard and thick, and when boiled for the table require some little exertion to open properly.

Everyone knows the curious, almost articulate cry of the Guinea Fowl, its " Come-back ! come-back," being continually uttered wherever the bird is kept, and often affording a clue to its presence.

The forehead of the Guinea Fowl is surmounted with a horny casque, and the naked skin round the eyes falls in wattles below the throat. In the male the wattles are purplish red, and in the female they are red without any mixture of blue, and are of smaller size. The legs are without spurs. The pretty spotted plumage of this bird is too well known to need description.

OF the many members of the Perdicine group, we shall only take one example, the well known English PARTRIDGE.

This bird, so dear to British sportsmen, is found spread over the greater part of Europe, always being found most plentifully near cultivated ground. It feeds upon various substances, such as grain and seeds in the autumn, and green leaves and insects in the spring and early summer.

Small slugs are a favourite diet with the Partridge, which has a special faculty for discovering them in the recesses where they hide themselves during the day, and can even hunt successfully after the eggs of these destructive creatures. Caterpillars are also eaten by this bird, and the terrible black grub of the turnip is consumed in great numbers. Even the white cabbage butterfly, whose numerous offspring are so hurtful in the kitchen garden, falls a victim to the quick-eyed Partridge, which leaps into the air and seizes it in its beak as it comes fluttering unsuspectingly over the bird's head.

The Partridge begins to lay about the end of April, gathering together a bundle of dry grasses into some shallow depression in the ground, and depositing therein a clutch of eggs, generally from twelve to twenty in number.

GUINEA FOWL —(*Numida meleagris.*)　　　PARTRIDGE.—(*Perdix cinereus.*)

Sometimes a still greater number have been found, but in these cases it is tolerably evident from many observations that several birds have laid in the same nest.

When the young are hatched they are strong on their legs at once, running about with ease, and mostly leaving the nest on the same day. The mother takes her little new-born brood to their feeding-places, generally ant-hills or caterpillar-haunted spots, and aids them in their search after food by scratching away the soil with her feet.

The nests of the wood ant, which are mostly found in fir plantations or hilly ground, being very full of inhabitants, very easily torn to pieces, and the ants and their larvæ and pupæ being very large, are favourite feeding-places of the Partridge, which in such localities is said to acquire a better flavour than among the lower pasture lands.

The young brood, technically called a " covey," associate together, and have a very strong local tendency, adhering with great pertinacity to the same field or patch of land. When together they are mostly rather wild, and dart off at the least alarm with their well-known whirring flight, just topping a hedge or wall and settling on the other side till again put up ; but when the members of the covey are separated they seem to dread the air, and crouch closely to the ground, so that it is the object of the sportsman to scatter the covey and to pick them up singly.

The plumage of the Partridge is brown of several shades above, mingled with grey. The breast is grey, with a horsehoe-like patch of rich chestnut on its lower portion, and the sides and flanks are barred with chestnut. The total length of the male bird is rather more than a foot; the female is smaller than her mate, and the chestnut bars on the flanks are broader than those of the male.

THE odd, short-legged, round-bodied, quick-footed QUAIL is closely allied to the partridge in form and many of its habits. Of these birds there are many species; but as all are much alike, there is no need of many examples.

QUAIL. —(*Coturnix communis.*)

The common Quail is found spread over the greater part of Europe and portions of Asia and Africa, coming to our island in the summer, though not in very great numbers. In England the bird is not sufficiently plentiful to be of any commercial value; but in Italy and some of the warmer lands which the Quails traverse during their periodical migrations, the inhabitants look forward to the arrival of the Quail with the greatest anxiety. In those countries they are shot, snared, and netted by thousands; and it is chiefly from the foreign markets that our game-shops are supplied with these birds. When fat, the flesh of the Quail is very delicious; and the most approved way of cooking the bird is to envelop it in a very thin slice of bacon, tie it up in a large vine-leaf, and then roast it.

CAPERCAILLIE.—(*Tetrao urogallus.*)

In their migrations the Quails fly by night, a peculiarity which has been noted in the Scriptural record of the Exodus, where it is mentioned, that "at even the Quails came up and covered the camp."

It is rather curious that the males precede the females by several days, and are consequently more persecuted by the professional fowlers.

The male bird does not pair like the partridge, but takes to himself a plurality of wives, and, as is generally the case with such polygamists, has to fight many desperate battles with others of its own sex. Although ill provided with weapons of offence, the Quail is as fiery and courageous a bird as the gamecock; and in Eastern countries is largely kept and trained for the purpose of fighting prize-battles, on the result of which the owners stake large sums. The note of the male is a kind of shrill whistle, which is only heard during the breeding season.

The nest of the Quail is of no better construction than that of the partridge, being merely a few bits of hay and dried herbage gathered into some little depression in the bare ground, and generally entrusted to the protection of corn-stalks, clover, or a tuft of rank grass. The number of eggs is generally about fourteen or fifteen, and their colour is buffy white, marked with patches or speckles of brown.

ALTHOUGH once a common inhabitant of the highland districts of Great Britain, the CAPERCAILLIE has now been almost wholly extinct for some years, a straggling specimen being occasionally seen in Scotland, and shot " for the benefit of science." This bird is also known by the following names :—Cock of the Woods, Mountain Cock, Auerhahn, and Capercailzie.

It is now most frequently found in the northern parts of Europe, Norway and Sweden being very favourite homes. From those countries it is largely imported into England by the game-dealers.

The Capercaillie is celebrated not only for its great size and the excellence of its flesh, but for its singular habit just previous to and during the breeding season.

During this season it holds its " play" or love song, called in Norway the " *lek*." He struts about with drooping wings, spread tail, and ruffled feathers, and utters a peculiar cry. This is a call to the hens, and always attracts them. While the bird is thus engaged, he is so intent upon his " play," that however wary he may be at other times, he can easily be approached and shot.

The nest of the Capercaillie is made upon the ground, and contains eight or ten eggs ; when hatched the young are fed upon insects, more especially ants and their pupæ. The adult birds feed mostly on vegetable substances, such as juniper, cranberry, and bilberries, and the leaves and buds of several trees.

The colour of the adult male bird is chestnut-brown, covered with a number of black lines irregularly dispersed ; the breast is black with a gloss of green, and the abdomen is simply black, as are the lengthened feathers of the throat and tail. The female is easily known by the bars of red and black which traverse the head and neck, and the reddish yellow barred with black of the under surface. In size the Capercaillie is nearly equal to a turkey.

THE well-known BLACK GROUSE, or BLACK COCK, is a native of the more southern countries of Europe, and still survives in many portions of the British Isles, especially those localities where the pine-woods and heaths afford it shelter, and it is not dislodged by the presence of human habitations.

BLACK GROUSE.—(*Tetrao tetrix.*)

Like the two preceding species, the male bird resorts at the beginning of the breeding season to some open spot where he utters his love-calls, and displays his new clothes to the greatest advantage, for the purpose of attracting to his harem as many wives as possible. The note of the Black Cock when thus engaged is loud and resonant, and can be heard at a considerable distance. This crowing sound is accompanied by a harsh, grating, stridulous kind of cry, which has been likened to the noise produced by whetting a scythe.

In the autumn the young males separate themselves from the other sex, and form a number of little bachelor establishments of their own, living together in harmony until the next breeding season, when they all begin to fall in love. The apple of discord is then thrown among them by the charms of the hitherto repudiated sex, and their rivalries lead them into determined and continual battles, which do not cease until the end of the season restores them to peace and sobriety.

The general colour of the adult male bird is black glossed with blue and purple, except a white band across each wing. The under tail-coverts are white.

CURSORES.

WITH the OSTRICH commences a most important group of birds, containing the largest and most powerful members of the feathered tribe, and termed Cursores, or Running Birds, on account of their great speed of foot and total impotence of wing. All the birds belonging to this order have their legs developed to an extraordinary degree, the bones being long, stout, and nearly as solid as those of a horse, and almost devoid of the air-cells which give such lightness to the bones of most birds. The wings are almost wanting externally, their bones, although retaining the same number and form as in ordinary birds, being very small, as if suddenly checked in their growth.

This magnificent creature, the largest of all existing birds, inhabits the hot sandy deserts of Africa, for which mode of life it is wonderfully fitted. In height it measures from six to eight feet, the males being larger than their mates, and of a blacker tint. The food of the Ostrich consists mostly of wild melons, which are so beneficently scattered over the sandy wastes.

The Ostrich is a gregarious bird, associating in flocks, and being frequently found mixed up with the vast herds of quaggas, zebras, giraffes, and antelopes, which inhabit the same desert plains. It is also polygamous, each male bird having from two to seven wives.

OSTRICH.—(*Struthio Camelus.*)

The nest of the Ostrich is a mere shallow hole scooped in the sand, in which are placed a large number of eggs, all set upright, and with a number of supplementary eggs laid round the margin.

The eggs are hatched mostly by the heat of the sun ; but, contrary to the popular belief, the parent birds are very watchful over their nests, and aid in hatching the eggs by sitting upon them during the night. Both parents give their assistance in this task. The eggs which are laid around the margin of the nest are not sat upon, and consequently are not hatched, so that when the eggs within the nest are quite hard, and the young bird is nearly developed, those around are quite fit for food. Their object is supposed to be to give nourishment to the young birds before they are strong enough to follow their parents and forage for themselves. These eggs are put to various useful purposes. Not only are they eaten, but the shell is carefully preserved and chipped into spoons and ladles, or the entire shell employed as a water vessel, the aperture at the top being stuffed with grass.

The feathers are too well known to need description. On an average, each feather is worth about a shilling.

The flesh of the Ostrich is tolerably good, and is said to resemble that of the zebra. It is, however, only the young Ostrich that furnishes a good entertainment, for the flesh of the old bird is rank and tough. The fat is highly valued, and when melted is of a bright orange colour. It is mostly eaten with millet flour, and is also stirred into the egg while roasting, so as to make a rude but well-flavoured omelet.

The voice of the Ostrich is a deep, hollow, rumbling sound, so like the roar of the lion that even practised ears have been deceived by it, and taken the harmless Ostrich for a prowling lion. In its wild state the Ostrich is thought to live from twenty to thirty years.

In the male bird, the lower part of the neck and the body are deep glossy black, with a few white feathers, which are barely visible except when the plumage is ruffled. The plumes of the wings and tail are white. The female is ashen brown, sprinkled with white, and her tail and wing plumes are white, like those of the male. The weight of a fine adult male seems to be between two and three hundred pounds.

THE EMEU inhabits the plains and open forest country of Central Australia, where it was in former days very common, but now seems to be decreasing so

EMEU.—(*Dromaius Novæ Hollandiæ.*)

rapidly in numbers that Dr. Bennett, who has had much personal experience of this fine bird, fears that it will, ere many years, be numbered with the Dodo, the Great Auk, the Nestor, and other extinct species.

The food of the Emeu consists of grass and various fruits. Its voice is a curious, hollow, booming, or drumming kind of note, produced by the peculiar construction of the windpipe. The legs of this bird are shorter and stouter in proportion than those of the ostrich, and the wings are very short, and so small that when they lie closely against the body they can hardly be distinguished from the general plumage.

The nest of the Emeu is made by scooping a shallow hole in the ground in some scrubby spot, and in this depression a variable number of eggs are laid. Dr. Bennett remarks that " there is always an odd number, some nests having been discovered with nine, others with eleven, and others, again, with thirteen." The colour of the eggs is, while fresh, a rich green, of varying quality, but after the shells are emptied and exposed to the light, the beautiful green hue fades into an unwholesome greenish brown. The parent birds sit upon their eggs, as has been related of the ostrich. The Emeu is not polygamous, one male being apportioned to a single female.

AMERICA is not without representatives of this fine group of birds, three distinct species being in the gardens of the Zoological Society.

THE RHEA is a native of South America, and is especially plentiful along the River Plata. It is generally seen in pairs, though it sometimes associates together in flocks of twenty or thirty in number. Like all the members of this group, it is a swift-footed and wary bird, but possesses so little presence of mind that it becomes confused when threatened with danger, runs aimlessly in one direction and then in another, thus giving time for the hunter to come up and shoot it, or bring it to the ground with his " bolas " — a terrible weapon, consisting of a cord with a heavy ball at each end, which is flung at the bird, and winds its coils round its neck and legs, so as to entangle it and bring it to the ground.

The food of the Rhea consists mainly of grasses, roots, and other vegetable substances, but it will occasionally eat animal food, being known to come down to the mud-banks of the river for the purpose of eating the little fish that have been stranded in the shallows.

THE well-known CASSOWARY is found in the Malaccas.

This fine bird is notable for the glossy black hair-like plumage, the helmet-like protuberance upon the head, and the light azure, purple, and scarlet of the upper part of the neck. The " helmet " is a truly remarkable apparatus, being composed of a honeycombed cellular bony substance, made on a principle that much resembles the structure of the elephant's skull, mentioned in an earlier portion of this work.

The plumage of the body is very hair-like, being composed of long and almost naked shafts, two springing from the same tube, and one always being longer than the other. At the roots of the shafts there is a small tuft of delicate down, sufficiently thick to supply a warm and soft inner garment, but yet so small as to be hidden by the long hair-like plumage. Even the tail is furnished with the same curious covering, and the wings are clothed after a similar manner, with the exception of five black, stiff, strong, pointed

CASSOWARY.—(*Casuarius.*)

quills, very like the large quills of the porcupine, and being of different lengths, the largest not exceeding one foot, and generally being much battered about the point. When stripped of its feathers, the whole wing only extends some three inches in length, and is evidently a mere indication of the limb.

The food of this bird in a wild state consists of herbage and various fruits, and in captivity it is fed on bran, apples, carrots, and similar substances, and is said to drink nearly half a gallon of water per diem.

PERHAPS the very strangest and most weird-like of all living birds is the APTERYX, or KIWI-KIWI.

This singular bird is a native of New Zealand, where it was once very common, but, like the dinornis, is in a very fair way of becoming extinct, a fate from which it has probably been hitherto preserved by its nocturnal and retiring habits.

In this bird there is scarcely the slightest trace of wings, a peculiarity which has gained for it the title of Apteryx, or "wingless." The plumage is composed of rather curiously-shaped flat feathers, each being wide and furnished with a soft, shining, silken down, for the basal third of its length, and then narrowing rapidly towards the extremity, which is a single shaft with hair-like webs at each side. The quill portion of the feathers is remarkably small and short, being even overlapped by the down when the feather is removed from the bird.

The skin is very tough, and yet flexible, and the Chiefs set great value upon it for the manufacture of their state mantles, permitting no inferior person to wear them, and being extremely unwilling to part with them even for a valuable consideration. The bird lives mostly among the fern ; and as it always remains concealed during the day in deep recesses of rocks, ground or tree roots, and is remarkably fleet of foot, diving among the heavy fern-leaves with singular adroitness, it is not very easy of capture. It feeds upon insects of various kinds, more especially on worms, which it is said to attract to the surface by jumping and striking on the ground with its powerful feet. The

APTERYX.—(*Apteryx Australis.*)

natives always hunt the Kiwi-kiwi at night, taking with them torches and spears. The speed of this bird is very considerable, and when running it sets its head rather back, raises its neck, and plies its legs with a vigour little inferior to that of the ostrich.

The fine specimen in the Zoological Gardens has already proved a very valuable bird, as she has laid several eggs, thereby setting at rest some disputed questions on the subject, and well illustrates the natural habits of the species.

Upon her box is placed, under a glass shade, the shell of one of her eggs. These eggs are indeed wonderful, for the bird weighs a little more than four pounds, and each egg weighs between fourteen and fifteen ounces, its length being four inches and three-quarters, and its width rather more than two inches, thus being very nearly one-fourth of the weight of the parent bird.

The long curved beak of the Apteryx has the nostrils very narrow, very small, and set on at each side of the tip, so that the bird is enabled to pry out the worms, and other nocturnal creatures on which it feeds, without trusting only to the eyes. The general colour of the Apteryx is chestnut-brown, each feather being tipped with a darker hue, and the under parts are lighter than the upper. The height is about two feet.

Several species of the Apteryx are known.

ALTHOUGH the progress of civilization has conferred many benefits on this country, it has deprived it of many of its aboriginal inhabitants, whether furred or feathered, the GREAT BUSTARD being in the latter category.

This splendid bird, although in former days quite a usual tenant of plains and commons, and having been an ordinary object of chase on Newmarket Heath, is now so rare, that an occasional specimen only makes its appearance at very distant intervals.

The Great Bustard is not fond of flying, its wings having but a slow and deliberate movement ; but on foot it is very swift, and tests the speed of dog and horse before it can be captured.

The nest—if a hole in the ground may be called a nest—of this bird is generally made among corn, rye, &c., although it is sometimes situated in rather unexpected localities. The eggs are two or three in number, and of an olive-brown colour, splashed with light brown in which a green tinge is perceptible. The food of the bird is almost wholly of a vegetable nature, though it it is said to feed occasionally upon

GREAT BUSTARD.—(*Otis tarda.*)

mice, lizards, and other small vertebrates. The flesh of the Bustard is very excellent, but the extreme rarity of these birds prevents it from being often seen upon English tables. When caught young, the Bustard can be readily tamed, and soon becomes quite familiar with those who treat it kindly.

The head and upper part of the neck are greyish white, and upon the side of the neck there is a small patch of slaty blue bare skin, almost concealed by the curious feather tuft which hangs over it. The upper part of the body is pale chestnut barred with black, and the tail is of similar tints with a white tip, and a very broad black band next to the white extremity. The wing-coverts, together with the tertials, are white, and the primaries black. The under surface of the body is white. The total length of an adult male is about forty-five inches.

THE Wading Birds are well furnished with legs and feet formed for walking, and in many species the legs are greatly elongated, so as to enable them to walk in the water while they pick their food out of the waves.

In the British Museum the Plovers head the list of Waders.

THE well-known LAPWING, or PEEWIT, is celebrated for many reasons. Its wheeling flapping flight is so peculiar as to attract the notice of everyone who has visited the localities in which it resides, and its strange, almost articulate, cry is equally familiar. When it fears danger, it rises from its nest, or rather from the eggs, into the air, and continually wheels around the intruder, its black and white plumage flashing out as it inclines itself in its flight, and its mournful cry almost fatiguing the ear with its piercing frequency. " Wee-whit ! wee-e-whit !" fills the air, as the birds endeavour to draw away attention from their home ; and the look and cry are so weird-like that the observer ceases to wonder at the superstitious dread in which these birds were formerly held. The French call the Lapwing " *Dix-huit,*" from its cry.

It is the male bird which thus soars above and around the intruder, the female sitting closely on her eggs until disturbed, when she runs away, tumbling and flapping about as if she had broken her wing, in hopes that the foe may give chase and so miss her eggs. It is certainly very tempting, for she imitates the movements of a wounded bird with marvellous fidelity.

The eggs of the Lapwing are laid in a little depression in the earth, in which a few grass stalks are loosely pressed. The full number of eggs is four, very large at one end and very sharply pointed at the other, and the bird always arranges them with their small end inwards, so that they present a somewhat cross-like shape as they lie in the nest.

LAPWING. —(*Vanellus cristatus.*)

Their colour is olive, blotched and spotted irregularly with dark blackish brown, and they harmonize so well with the ground on which they are laid that they can hardly be discerned from the surrounding earth at a few yards' distance. Under the title of " Plover's eggs " they are in great request for the table, and are sought by persons who make a trade of them, and who attain a wonderful expertness at the business.

The food of the Lapwing consists almost wholly of grubs, slugs, worms, and insects. It is easily tamed, and is often kept in gardens for the purpose of ridding them of these destructive creatures. In the garden next our own a Lapwing was kept, and lived for some years, tripping featly over the grass and thoroughly at home.

THE GOLDEN-BREASTED TRUMPETER is a handsome bird, remarkable for the short velvety feathers of the head and neck, and their beautiful golden green lustre on the breast. The body of this bird is hardly larger than that of a fowl, but its legs and neck are so long as to give it the aspect of being much larger than it really is. Like most birds of similar structure, it trusts more to its legs than its wings, and is able to run with great speed and activity. It is generally found in the forests.

The name of Trumpeter is derived from the strange hollow cry which it utters without seeming to open the beak. This cry is evidently produced by means of the curiously-formed windpipe, which is furnished with two mem-branous expansions, and, during the utterance of the cry, puffs out the neck very forcibly, just as the rhea does when grunting. The nest of the Trumpeter is said to be a hole scratched in the ground at the foot of a tree, and to contain about ten or twelve light green eggs. The head and neck are velvety black, and on the breast the feathers become large, and more scale-like, and their edges beautifully bedecked with rich shining green, with a purplish gloss in some lights and a lustrous golden hue in others. The back is grey, the

feathers being long and silken, and hanging over the wings. The wings, under surface, and tail are black, and the feathers of the tail are soft and short.

ALTHOUGH in former days tolerably common in England, the CRANE has now, with the bustard, almost disappeared from this land, a single specimen being seen at very long and increasing intervals. In some parts of England and Ireland the popular name of the heron is the Crane, so that the occasional reports which sometimes find admission into local newspapers respecting the Crane have often reference, not to that bird, but to the heron.

THE GOLDEN-BREASTED TRUMPETER. —(*Psophia crepitans.*)

The Crane makes its nest mostly on marshy ground, placing it among osiers, reeds, or the heavy vegetation which generally flourishes in such localities. Sometimes, however, it prefers more elevated situations, and will make its nest on the summit of an old deserted ruin. The eggs are two in number, and their colour is light olive, covered with dashes of a deeper hue and brown. The well-known plumes of the Crane are the elongated tertials, with their long drooping loose webs, which, when on the wings of the bird, reach beyond the primaries.

The forehead, top of the head, and neck are. rather dark slaty ash, and a patch of greyish white extends from behind the eyes, partially down the neck on each side. The general surface of the body is soft ashen grey, and the primaries are black. The long plumy tertials form two crest-like ornaments, which can be raised or depressed at will. The eyes are red, and the beak is yellow with a green tinge. The total length of the adult crane is about four feet, but it is rather variable in point of size, and the males are rather larger than the females.

The DEMOISELLE, or NUMIDIAN CRANE is common in many parts of Africa, and has been seen in some portions of Asia, and occasionally in Eastern Europe.

It is a very pretty bird, the soft texture of the flowing plumage, and the delicate greys of the feathers, harmonizing with each other in a very agreeable manner. The general tint of the plumage is blue-grey, taking a more leaden tone on the head and neck, and offering a beautiful contrast to the snowy white ear-tufts, issuing from velvety black, which decorate the head. There is also a tuft of long flowing plumes of a deep black-grey hanging from the breast. Its secondaries are much elongated, and hang over the primaries and tail-feathers. In height the De-moiselle Crane is about three feet six inches.

THE well-known HERON was once one of our com-monest English birds, but on account of the draining of swamps and their conver-sion into fertilized and ha-bitable ground, is now sel-dom to be seen except in certain localities which still

CRANE.—(*Grus cinerea.*)

retain the conditions that render them so acceptable to this bird. There are some places where Herons are yet plentiful, especially those localities where the owner of the land has established or protected the nests, or where a wide expanse of wild uncultivated ground affords them a retreat. Only a few days ago I came suddenly on three of these beautiful birds fishing quietly in the Avon, and permitting my approach within a few yards before they spread their wide wings for flight.

The food of the Heron consists mostly of fish and reptiles, but it will eat small mammalia, such as mice, or even water-rats. In the stomach of one of these birds were found seven small trout, a mouse, and a thrush. Eels also are a favourite food of the Heron, but on account of their lithe bodies and active wrigglings are not so easy to despatch as ordinary fish, and are accordingly taken on shore and banged against the ground until disabled.

Like many other birds, the Heron is able to disgorge the food which it has swallowed, and resorts to this measure when it is chased by birds of prey while going home after a day's fishing.

C C 2

While engaged in its search for food, the Heron stands on the water's edge mostly with its feet or foot immersed, and there remains still, as if carved out of wood, with its neck retracted, and its head resting between the shoulders. In this attitude its sober plumage and total stillness render it very inconspicuous, and as it mostly prefers to stand under the shadow of a tree, bush, or bank, it cannot be seen except by a practised eye, in spite of its large size.

The long beak of the Heron is very sharp and dagger-like, and can be used with terrible force as an offensive weapon. The bird instinctively aims its blow at the eye of its adversary, and if incautiously handled is sure to deliver a stroke quick as lightning at the captor's eye. The beak of a species of Heron set upon a stick is used by some savage tribes as a spear.

DEMOISELLE CRANE.—(*Scops Virgo.*)   HERON.—(*Ardea cinerea.*)

The nest of the Heron is almost invariably built upon some elevated spot, mostly the top of a large tree, but sometimes on rocks near the coast. It is a large and rather clumsy-looking edifice, made of sticks and lined with wool. The eggs are from four to five in number, and their colour is pale green.

THE BITTERN is now seldom seen in this country, partly because it is a rare bird and becoming scarcer almost yearly, and partly because its habits are nocturnal, and it sits all day in the thickest reeds or other aquatic vegetation. The marshy grounds of Essex seem to be the spots most favoured by this bird at the present day, although specimens are annually killed in various parts of the country.

In habits and food, the Bittern resembles the heron, except that it feeds by night instead of by day. Like that bird, it uses its long sharp beak as a weapon of offence, and chooses the eye of its adversary as the point at which to aim. The feet and legs are also powerful weapons, and when disabled from flight, the Bittern will fling itself on its back and fight desperately with foot and bill.

The nest of the Bittern is placed on the ground near water, and concealed among the rank vegetation that is found in such localities. It is made of sticks and reeds, and generally contains about four or five pale brown eggs. The voice of the Bittern varies with the season of the year. Usually it is a sharp harsh cry uttered on rising, but in the breeding season the bird utters a loud booming cry that can be heard at a great distance.

The general colour of this fine bird is rich brownish buff covered with irregular streaks and mottlings of black, dark brown, grey and chestnut. The top of the head is black with a gloss of bronze, the cheeks are buff, and the chin white tinged with buff. Down the front of the neck the feathers are marked with bold longitudinal dashes of blackish and reddish brown, and the feathers of the breast are dark brown broadly edged with buff. The under surface of the body is buff streaked with brown, the beak is greenish yellow, and the feet and legs are green.

BITTERN.—(*Botaurus stellaris.*)

In total length the Bittern measures about thirty inches.

THE well-known SPOONBILL affords an instance of the endless variety of forms assumed by the beak.

It has a very wide range of country, being spread over the greater part of Europe and Asia, and inhabiting a portion of Africa. This species is one of the Waders, frequenting the waters, and obtaining a subsistence from the fish, reptiles, and smaller aquatic inhabitants, which it captures in the broad spoon-like extremity of its beak. It is also fond of frequenting the sea-shore, where it finds a bountiful supply of food along the edge of the waves and in the little pools that are left by the retiring waters, where shrimps, crabs, sand-hoppers, and similar animals are crowded closely together as the water sinks through the sand. The bird also eats some vegetable substances, such as the roots of aquatic herbage, and when in confinement will feed upon almost any kind of animal or vegetable matter, providing it be soft and moist. The beak of an adult Spoonbill is about eight inches in length, very much flattened, and is channelled and

THE WHITE SPOONBILL.
(*Platalea leucorodia.*)

grooved at the base. In some countries the beak is taken from the bird,

scraped very thin, and polished, and is then used as a spoon, and is thought a valuable article, being sometimes set in silver.

The breeding-places of the Spoonbill are usually open trees, the banks of rivers, or in little islands and tufts of aquatic herbage. In the latter cases the nest is rather large, and is made of reeds piled loosely together, and set on a foundation of water-weeds heaped sufficiently high to keep the eggs from the wet. There is no lining to the nest. The eggs are generally four in number, and their colour is greyish white, spotted with rather pale rusty brown.

THE STORK is another of the birds which, in the olden days, were tolerably frequent visitors to the British Islands, but which now seldom make their appearance in such inhospitable regions, where food is scarce and guns are many.

It is sufficiently common in many parts of Europe, whither it migrates yearly from its winter quarters in Africa, makes its nest, and rears its young.

The Stork attaches itself to man and his habitations, building its huge nest on the top of his house, and walking about in his streets as familiarly as if it had built them. It especially parades about the fish-markets, where it finds no lack of subsistence in the offal ; and in Holland, where it is very common, it does good service by destroying the frogs and other reptiles which would be likely to become a public nuisance unless kept down by the powerful aid of this bird.

The Stork is fond of making its nest upon some elevated spot, such as the top of a house, a chimney, or a church spire ; and, in the ruined cities of the East, almost every solitary pillar has its Stork's nest upon the summit. The nest is little more than a heterogeneous bundle of sticks, reeds, and similar substances heaped together and with a slight depression for the eggs. These are usually three or four in number, and their colour is white with a tinge of buff.

STORK.—(*Ciconia alba.*)

The colour of the adult Stork is pure white, with the exception of the quill feathers of the wings, the scapularies, and greater wing-coverts, which are black. The skin round the eye is black, the eyes are brown, and the beak, legs, and toes red. The length of the full-grown bird is about three feet six inches, and when erect, its head is about four feet from the ground.

THERE are several remarkable members of this group, one of which is the well-known ADJUTANT, or ARGALA, of India, the former name being derived from its habit of frequenting the parade-grounds.

This fine bird is notable for the enormous size of the beak, which is capable of seizing and swallowing objects of considerable size, a full-grown cat, a fowl, or a leg of mutton being ingulfed without any apparent difficulty. The Adjutant is a most useful bird in the countries which it inhabits, and is protected with the utmost care, as it thoroughly cleans the streets and public places of the various offal which is flung carelessly in the way, and would be left to putrefy but for the constant services of the Adjutant and creatures of similar habits.

It is easily tamed, and soon attaches itself to a kind owner ; sometimes indeed becoming absolutely troublesome in its familiarity. Mr. Smeathman mentions an instance where one of these birds was domesticated, and was

ADJUTANT.—(*Leptoptilos Argala.*,

accustomed to stand behind its master's chair at dinner-time and take its share of the meal. It was, however, an incorrigible thief, and was always looking for some opportunity of stealing the provisions, so that the servants were forced to keep watch with sticks over the table. In spite of their vigilance it was often too quick for them ; and once it snatched a boiled fowl off the dish and swallowed it on the spot.

The exquisitely fine and flowing plumes, termed " Marabou feathers," are obtained from the Adjutant and a kindred species, the Marabou of Africa (*Leptoptilos Marabou*).

The general colour of the Adjutant is delicate ashen grey above, and white beneath. The great head and proportionately large neck are almost bare of covering, having only a scanty supply of down instead of feathers. From the lower part of the neck hangs a kind of dewlap, which can be inflated at the will of the bird, but generally hangs loose and flabby.

THE SACRED IBIS is one of a rather curious group of birds. With one exception they are not possessed of brilliant colouring, the feathers being mostly white and deep purplish black. The Scarlet Ibis, however, is a most magnificent, though not very large bird, its plumage being of a glowing scarlet, relieved by a few patches of black.

SACRED IBIS.—(*Ibis religiosa.*)   CURLEW.—(*Numenius arquata.*)

The Sacred Ibis is so called because it figures largely in an evidently sacred character on the hieroglyphs of ancient Egypt. It is a migratory bird, arriving in Egypt as soon as the waters of the Nile begin to rise, and remaining in that land until the waters have subsided, and therefore deprived it of its daily supplies of food. The bird probably owes its sacred character to the fact that its appearance denotes the rising of the Nile, an annual phenomenon on which depends the prosperity of the whole country.

By the natives of Egypt it is called the Abou Hannes, *i.e.*, Father John, or Abou Menzel, *i.e.*, Father Sickle-bill, the former name being in use in Upper and the other in Lower Egypt.

The colour of the adult bird is mostly pure silvery white, the feathers being glossy and closely set, with the exception of some of the secondaries, which are elongated and hang gracefully over the wings and tail. These, together with the tips of the primaries, are deep glossy black, and the head and neck are also black, but being devoid of feathers have a slight brownish tinge, like that of an ill-blacked boot, or an old crumpled black kid glove. While young, the head and neck are clothed with a blackish down, but when the bird reaches maturity, even this slender covering is shed, and the whole skin is left bare. The body is little larger than that of a common fowl.

ANOTHER species, the GLOSSY IBIS, is also an inhabitant of Northern Africa, but is sometimes found in this country, where the fishermen know it by the name of Black Curlew. It is probably the Black Ibis mentioned by Herodotus.

THE CURLEW, or WHAUP, is mostly found upon the sea-shore and open moorlands, and partly on account of its wild, shy habits, and partly because its flesh is very delicate and well flavoured, is greatly pursued by sportsmen. These birds are mostly annoying to a gunner who does not understand their ways, having a fashion of keeping just out of gun range, rising from the ground with a wild mournful cry which has the effect of alarming every other bird within hearing, and flying off to a distance, where they alight only to play the same trick again. Moreover, they are strong on the wing and well feathered, so that they require a sharp blow to bring them down, and necessitate the use of large shot.

The breeding-grounds of the Curlew are inland, the locality varying according to the character of the district, wild heath and high hilly grounds being chosen in some places, while marshy and boggy soils are favoured in others. The nest of this bird is very slight, being only a small heap of dry leaves or grasses scraped together under the shelter of a tuft of heather or a bunch of rank grass. There are usually four eggs, placed, as is customary with such birds, with their small ends together, and being much larger at one end than at the other.

Their colour is brownish green, with some blotches and splashes of dark brown and a darker green.

THE AVOCET is one of the most remarkable among English birds, and is easily recognizable by its long, curiously-curved beak, and its boldly pied plumage.

It is not a common bird in England, and is now but seldom seen, though in former days it used to be tolerably plentiful on the sea-coasts and in marshy lands. The long and oddly curved beak is very slender and pointed, and from its peculiar shape has earned

AVOCET.—(*Recurvirostra avoceita.*)

for its owner the name of Cobbler's-Awl Bird. The food of the Avocet consists almost wholly of worms, insects, and little crustaceans ; and while the bird is engaged in the search after these creatures it paddles over the oozy mud with its webbed feet, and traverses the soft surface with much ease and some celerity.

The nest of the Avocet is placed on the ground in some convenient hollow, and the eggs are yellowish brown with black marks.

LIKE many other birds which depend for their existence upon marshy and uncultivated grounds, the RUFF is gradually being turned out of England, and may in time be nothing more than a rare and occasional visitor.

It is one of the migratory species, arriving in this country in April and leaving by the end of September. Formerly it was so common in the fenny districts that six dozen have been taken by one bird catcher in a single day.

The Ruff is a most pugnacious bird, rivalling if not exceeding the game-cock in irritability of temper and reckless courage. The attitude of fighting is not unlike that of the cock, but as they have no spurs, they cannot inflict severe wounds, and after a fierce contest neither party will be much the worse. Prolonged and obstinate combats are waged among the Ruffs for the possession of the females, popularly called Reeves, and as the birds make a great noise about their affairs, and in their eager combat trample down the grass on the little hills where they love to resort, the fowler knows well where to lay his nets.

The Ruff is chiefly remarkable for the peculiarity from which it derives its name, the projecting ruff of long, closely set feathers, which surrounds the neck, and can be raised or lowered at pleasure. This ruff only belongs to the

RUFF.—(*Philomachus pugnax.*)　　　SNIPE.—(*Numenius scolopacinus.*)

adult males, and is assumed by them during the short breeding season, being in greatest perfection about the beginning of June, and falling off by degrees from July to August and September.

THE COMMON SNIPE is too well known to need much description. Its habits, however, are interesting, and deserve some notice.

This bird may be seen all over England wherever damp and swampy places are found. When first flushed it shoots off in a straight line for a few yards, and then begins to twist and turn in a strangely zigzag fashion, and at last darts away, thereby puzzling juvenile sportsmen greatly, and often escaping before its enemy has got his aim.

The male bird has a curious habit of rising to a great height in the air, circling repeatedly over the same ground, and uttering continually a peculiar cry like the words "Chic ! chic ! chic-a, chic-a, chic-a," constantly repeated. Every now and then the bird makes a downward stoop, and then emits a very singular sound, something between the bleating of a goat and the buzzing of a slack harp-string. How this sound is produced has long been a controversy, but I am convinced that it is produced by the wings—at all events that it is not from the mouth.

During a recent stay in the New Forest, I set myself to the elucidation of this problem, and in company with two friends went towards sunset to an excellent cover near a large marsh, in which Snipes were almost as plentiful as sparrows. From this post we could watch the Snipes to great advantage, and the birds would come circling over our heads, piping and drumming vigorously. On several occasions, when a Snipe was passing over us at so low an elevation that his long drooping beak was distinctly visible, he stooped over our heads and uttered his "chic-a ! chic-a !" simultaneously with the "drumming," both sounds being distinctly heard at the same time. The first time that we clearly heard the double sound was on June 27, but we repeatedly heard it on subsequent occasions. The Snipe remains a long time upon the wing while thus engaged, contrary to its usual habit, which is to fly for a short distance and then to pitch again.

The nest of the Snipe is a simple heap of leaves placed under the shelter of a tuft of furze, heath, or grass, and the eggs are four in number, of an olive-white, spotted and dashed with brown of different tones towards and upon the large end. The mother bird has been known to carry away her young when threatened by danger.

THE WOODCOCK is nearly as well known, though not so plentiful as the snipe, to which bird it bears a considerable resemblance in form, plumage, and many habits.

Generally it is only a winter visitor, arriving about October, and leaving England in March or April. Sometimes, however, it will breed within the British Isles, and there remain throughout the summer. During their migration the Woodcocks fly at a great altitude, and descend almost perpendicularly upon the spot where they intend to rest. They fly in companies of varying numbers, and prefer hazy and calm weather for their journey.

The food of the Woodcock consists mostly of worms, which it obtains with extraordinary skill, thrusting its beak as far as the nostrils into the soft moist earth, and hitting upon the hidden worms with unerring skill. A tame Woodcock has been seen to probe large turfs with its bill, and to draw out a worm at every thrust of the long slender beak. It is thought that the sense of smell enables the bird to discover the worms beneath the surface. It moves about chiefly on misty days, and

WOODCOCK.—(*Scolopax rusticola.*)

is said by experienced woodcock shooters to prefer the northern side of a hill to the southern.

It is a very silent bird, seldom uttering its cry except when first starting for its feeding-places, and hardly even crying when flushed. The flight of the Woodcock is wonderfully swift, although the wings do not appear to move very fast ; and the bird has a custom of jerking and dodging about so quickly when it sees the sportsman, that it often escapes his shot. One bird, mentioned by Mr. Thompson, used to baffle an experienced sportsman by always feeding near an archway, and slipping through it before the gun could be brought to bear.

The nest of the Woodcock is made of leaves—those of the fern being favourites—closely laid together, but without any particular skill in arrangement, and without lining. The full number of eggs is four, and their colour is buffy white with rusty brown blotches.

THE JACANAS are found in Asia, Africa, and America. Their light bodies and widely extended claws enable them to walk on the leaves of aquatic plants with equal ease and safety. As their weight is just sufficient to sink the leaf a little below the surface, they have quite the appearance of walking on the water itself. The Common Jacana inhabits the hotter parts of South America, and is abundant in Brazil and Guiana. It possesses large and sharp spurs on the wing. It is not a very large bird, barely exceeding a pigeon in bulk.

CHINESE JACANA.
(*Hydrophasianus Sinensis.*)

WE now come to the large family of the RAILS, a curious group of birds, formed for rapid movement, either on the ground or through the water, but not particularly adapted for long flights. Many species inhabit England.

THE well-known CORNCRAKE, or LANDRAIL, is common in almost every part of the British Islands, its rough grating call being heard wherever the hay-grass is long enough to hide the utterer.

The bird runs with wonderful speed through the tall grass, and its cry may be heard now close at hand, now in the distance, now right, and now left, without any other indication of the bird's whereabouts; for so deftly does it thread the grass-stems that scarcely a shaken blade indicates its presence, and it is so wary that it keeps itself well hidden among the thick herbage. The cry of the Corncrake may be exactly imitated by drawing a quill or a piece of stick smartly over the large teeth of a comb, or by rubbing together two jagged strips of bone. In either case the bird may be decoyed within sight by this simple procedure.

The nest of the Corncrake is placed on the ground, and is made of dry grass arranged in a suitable depression. It generally contains from eight to twelve eggs, of a buffy white covered with rusty brown spots. The shell is rather thick, and the size of the egg large in proportion to the dimensions of the bird.

CORNCAKE.—(*Ortygometra crex.*)

The upper parts of the body are elegantly mottled with dark blackish brown, ashen, and warm chestnut ; the first tint occupying the centre of each feather, the second the edges, and the third the tips. The wing-coverts are rusty red. The throat and abdomen are white, and the breast is greenish ash, warming into reddish rust striped with white on the sides. In total length the Cornrake is not quite ten inches.

OUR most familiar example of the Gallinules is the WATER HEN, sometimes called the MOOR HEN.

This bird may be seen in plenty in every river in England, and mostly on every pond or sheet of water where the reedy or rushy banks offer it a refuge. When startled it often dives on the instant, and, emerging under floating weeds or rubbish, just pokes its bill above the surface, so that the nostrils are uncovered by the water, and remains submerged until the danger is past, holding itself in the proper position by the grasp of its strong toes upon the weeds.

The nesting of this bird is very peculiar. The Water Hen builds a large edifice of sedges, sticks, and leaves, either on the bank close to the water's edge, upon little reedy islands, or on low banks overhanging the water, and generally very conspicuous. The mother bird has a habit of scraping leaves and rushes over her eggs when she leaves the nest—not, as some persons fancy, to keep the eggs warm, but to hide them from the prying eyes of crows and magpies, jays, and other egg-devouring birds.

WATER HEN.—(*Gallinula chloropus.*)

The young are able to swim almost as soon as hatched, and for some time remain close to their parents. I once, to my great regret, shot by mistake several young Moor Hens, still in their first suit of black puffy down, and paddling about among the water-lilies and other aquatic herbage where I could not see them. Pike are rather apt to carry off the little creatures, by coming quickly under the weeds and jerking them under water before they take the alarm.

THE COMMON COOT, or BALD COOT, as it is sometimes called, is another of our familiar British water-birds, being seen chiefly in lakes, large ponds, and the quiet banks of wide rivers.

The habits of the Coot much resemble those of the water hen, and it feeds after a similar fashion upon molluscs, insects, and similar creatures, which it finds either in the water or upon land.

The nest of the Coot is a huge edifice of reeds and rank-water herbage, sometimes placed at the edge of the water, and sometimes on little islands at some distance from shore. I have often had to wade for thirty or forty yards to these nests, which have been founded upon the tops of little hillocks almost covered with water. The whole nest is strongly though rudely made ; and if the water should suddenly rise and set the nest floating, the Coot is very little troubled at the change, but sits quietly on her eggs waiting for the nest to be stranded. The eggs are generally about eight or ten in number, and their colour is olive-white sprinkled profusely with brown. The shell is rather thick in proportion to the size of the egg, so that Coots's eggs can be carried away in a handkerchief without much danger of being broken.

THE well-known FLAMINGO brings us to the large and important order of Anseres, or the Goose tribe.

The common Flamingo is plentiful in many parts of the Old World, and may be seen in great numbers on the sea-shore, or the banks of large and pestilential marshes, the evil atmosphere of which has no effect on these birds, though to many animals it is most injurious, and to man certain death. When feeding, the Flamingo bends its neck, and, placing the upper mandible of the curiously-bent beak on the ground or under the water, separates the nutritive portions with a kind of spattering sound, like that of a duck when feeding. The tongue of the Flamingo is very thick, and of a soft oily consistence, covered with curved spines pointing backwards, and not muscular.

COOT.—(*Fulica atra.*)

A flock of these birds feeding along the sea-shore has a curious appearance, bending their long necks in regular succession as the waves dash upon the shore, and raising them as the ripple passes away along the strand. At each wing is always placed a sentinel bird, which makes no attempt to feed but remains with neck erect and head turning constantly about to detect the least indication of danger. When a flock of Flamingos is passing overhead, they have a wonderfully fine effect, their plumage changing from pure white to flashing rose as they wave their broad wings.

When at rest and lying on the ground, with the legs doubled under the body, the Flamingo is still graceful, bending its neck into snaky coils, and preening every part of its plumage with an ease almost incredible. Its long and apparently clumsy legs are equally under command, for the bird can scratch its cheeks with its toes as easily as a sparrow or a canary.

When flying, the Flamingo still associates itself with its comrades, and the flock form themselves into regular shapes, each band evidently acting under the command of a leader. The eggs are white, their number is two or three, and the young birds are all able to run at an early age. Like many other long-legged birds, the Flamingo has a habit of standing on one leg, the other being drawn up and hidden among the plumage.

The curious beak of this bird is orange-yellow at the base, and black at the extremity ; and the cere is flesh-coloured. When in full plumage the colour is brilliant scarlet, with the exception of the quill feathers, which are jetty black. A full-grown bird will measure from five to six feet in height.

THE BERNICLE GOOSE is found on our shores, and seems to prefer the western to the eastern coasts.

The name of Bernicle Goose is given to this bird because the olden voyagers thought that it was produced from the common barnacle shell, and this notion had taken so strong a hold of their minds that they published several engravings representing the bird in various stages of its transformation.

The Bernicle Goose generally assembles in large flocks and haunts large salt-marshes near the coast, and feeds on grasses and various algæ. It is a very wary bird, and not easily approached. The eggs of this species are large and white. The flesh is considered good. The bill of the Bernicle

FLAMINGO.—(*Phœnicopterus ruver.*)

Goose is black, with a reddish streak on each side. The cheeks and throat are white, a black streak runs from the beak to the eye, the upper parts are bold and marked with black and white, and the lower parts are white. It is rather a small bird, the total length barely exceeding two feet.

THE beautiful SWANS now come before our notice. There are nine or ten species of these fine birds, which are well represented in the British Isles, four species being acknowledged as English birds.

Our most familiar species is the TAME or MUTE SWAN, so called from its silent habits. This elegant and graceful bird has long been partially domesticated throughout England, and enjoys legal protection to a great extent ; heavy penalties being proclaimed against anyone who kills a Swan without a legal right.

BERNICLE GOOSE.—(*Bernicla leucopsis.*)

The food of the Swan consists mostly of vegetable substances, and the bird can be readily fattened on barley, like ordinary poultry. The young birds, called cygnets, ought not to be killed after November, as they then lose their fat, and the flesh becomes dark and tough.

The nest of the Swan is a very large mass of reeds, rushes, and grasses set upon the bank, close to the water, in some sheltered spot. Generally the bird prefers the shore of a little island as a resting-place for its nest. Like other water-birds, the Swan will raise the nest by adding fresh material before the rising of the water near which it is placed. There are generally six or seven eggs, large, and of a dull greenish white. The young are of a light bluish grey colour, and do not assume the beautiful white plumage until maturity.

THE MUTE SWAN.—(*Cygnus olor.*)

The mother is very watchful over her nest and young, and in company with her mate assaults any intruder upon the premises. During the first period of their life the young Swans mount on their mother's back, and are thus carried from one place to another. If in the water, the Swan is able to sink herself so low that the young can scramble upon her back out of the water; and if on land, she helps them up by means of one leg.

THE WHISTLING SWAN.—(*Cygnus ferus.*)

The HOOPER, ELK SWAN, or WHISTLING SWAN, may at once be distinguished from the preceding species by the shape and colour of the beak, which is slender, without the black tubercle, and is black at the tip and yellow at the base, the latter colour stretching as far as the eye.

The nest of the Hooper is like that of the Mute Swan, and the eggs are pale brownish white. The length of the Hooper is about the same as that of the mute species, *i.e.*, five feet.

HOWEVER emblematical of ornithological fiction a BLACK SWAN might have been in ancient times, it is now almost as familiar to English eyes as any of the white species.

This fine bird comes from Australia, where it was first discovered in 1698. It is a striking and handsome bird, the blood-red bill and the white primaries contrasting beautifully with the deep black of the plumage. It is not so elegant in its movements as the white Swan, and holds its neck stiffly, without the easy serpentine grace to which we are so well accustomed in our British Swans.

THERE are very many species of DUCKS, of which we can take but a few examples.

The well-known WIDGEON is very plentiful in this country, arriving about the end of September or the beginning of October, and assembling in large flocks.

These birds, although wary on some occasions, are little afraid of the proximity of man and his habitations, feeding boldly by day, instead of postponing their feeding-time to the night, as is often the case with water-fowl. The food of the Widgeon mostly consists of grass, which it eats after the fashion of the common goose. The nest of the

BLACK SWAN.—(*Cygnus atratus.*)

(WIDGEON.—*Mareca Penelope.*)　(MALLARD.—*Anas boschas.*)

Widgeon is made of decayed reeds and rushes, and is lined with the soft down torn from the parent's body. The eggs are rather small, and of a creamy white colour. The number of eggs is from five to eight. The flesh of this bird is very delicate, and it is largely sold in our markets.

THE common MALLARD, or WILD DUCK, now comes before our notice.

D D

This is by no means one of the least handsome of its tribe ; the rich glossy green of the head and neck, the snowy white collar, and the velvet black of the odd little curly feathers of the tail, giving it a bold and striking appearance, which, but for its familiarity, would receive greater admiration than it at present obtains. It is the stock from which descended our well-known domestic Duck, to which we are so much indebted for its flesh and its eggs.

In its wild state the Mallard arrives in this country about October, assembling in large flocks, and is immediately persecuted in every way that the ingenuity of man can devise.

The nest of the Mallard is made of grass, lined and mixed with down, and is almost always placed on the ground near water, and sheltered by reeds, osiers, or other aquatic plants. Sometimes, however, the nest is placed in a more inland spot, and it now and then happens that a Duck of more than usual eccentricity builds her nest in a tree at some elevation from the ground, so that, when her young are hatched, she is driven to exert all her ingenuity in conveying them safely from their lofty cradle to the ground or the water. Such a nest has been observed in an oak-tree twenty-five feet from the ground, and at Heath Wood, near Chesterfield, one of these birds usurped

TEAL.—(*Querquedula Crecca.*)

possession of a deserted crow's nest in an oak-tree. Many similar instances are on record.

The eggs of the Mallard are numerous, but variable, according to the individual which lays them, some being far more prolific than others. The eggs are rather large, and of a greenish white colour.

THE pretty little TEAL is the smallest and one of the most valuable of the British Ducks, its flesh being peculiarly delicate and its numbers plentiful.

IN the southern parts of England the EIDER DUCK is only a winter visitant, but remains throughout the year in the more northern portions of our island, and in the north of Scotland.

This bird is widely celebrated on account of the exquisitely soft and bright down which the parent plucks from its breast and lays over the eggs during the process of incubation. Taking these nests is with some a regular business, not devoid of risk, on account of the precipitous localities in which the Eider Duck often breeds. The nest is made of fine sea-weeds, and after the mother bird has laid her compliment of eggs, she covers them with the soft down, adding to the heap daily until she completely hides them from view.

The plan usually adopted is to remove both eggs and down, when the female lays another set of eggs and covers them with fresh down. These are again taken, and then the male is obliged to give his help by taking down from his own breast and supplying the place of that which was stolen. The down of the male bird is pale-coloured, and as soon as it is seen in the nest, the eggs and down are left untouched in order to keep up the breed.

EIDER DUCK. — (*Somateria mollissima.*)

WE now come to the family of Colymbidæ, or Divers.

The GREAT NORTHERN DIVER is common on the northern coasts of the British Islands, where it may be seen pursuing its arrowy course through and over the water, occasionally dashing through the air on strong pinions, but very seldom taking to the shore, where it is quite at a disadvantage.

The eggs of the Northern Divers are generally two in number, and of a dark olive-brown, spotted sparingly with brown of another tone. They are laid upon the bare ground, or on a rude nest of flattened herbage near water, and the mother bird does not sit, but lies flat on the eggs. If disturbed, she scrambles into the water and dives away, cautiously keeping herself out of gunshot, and waiting until the danger is past. Should she be driven to fight, her long beak is a dangerous weapon, and is darted at the foe with great force and rapidity.

The head of the adult Northern Diver is black, glossed with green and purple, and the cheeks and back of the neck are black without the green gloss. The back is black, variegated with short white streaks, lengthening towards the breast, and the neck and upper part of the breast are white, spotted with black, and cinctured with two collars of deep black. The breast and abdomen are white. The total length of the bird is not quite three feet. The immature bird is greyish black above, each feather being edged with a lighter hue, and the under parts of the body are dull white. In some places this bird is called the Loon.

THE sub-family of the Grebes is represented in England by several well-known species. All these birds may be readily distinguished by the peculiar form of the foot, in which each toe is furnished with a flattened web, the whole foot looking something like a horse-chestnut leaf with three lobes.

GREAT NORTHERN DIVER.—(*Colymbus glacialis.*)

The best known of the English Grebes is the common DABCHICK, or LITTLE GREBE, the smallest and the commonest of British species. It is a pretty little bird, quick and alert in its movements. When alarmed, it dives so instantaneously that the eye can hardly follow its movements, and if at the moment of its emergence it perceives itself still in danger, it again dives, not having been on the surface for a single second of time. Like many other aquatic birds, it can sink itself in the water slowly, and often does so when uneasy, rising again if relieved from its anxiety, or disappearing as if jerked under the surface from below. I have often seen them in a little pond only a few yards across, thus diving and popping up again with almost ludicrous rapidity.

This bird can fly moderately well, and can rise from the water without difficulty, when it will circle about the spot whence it rose, and keep some five or six feet above the surface, uttering the while its curious rattling crv.

The nest of this bird is made of water-weeds, and is placed among the rank aquatic herbage. It is scarcely raised above the surface, and is mostly wet.

The eggs are five or six in number, and their normal colour is white, though they soon become stained with the decaying vegetable matter on which they rest, and before hatching are of a muddy brown hue.

The food of the Dabchick consists of insects, molluscs, little fish, and the smaller crustaceans.

The CRESTED GREBE is found in some of the fens of the Midland counties of England, and also inhabits parts of Scotland. This bird, together with the other Grebes, builds its nest of a mass of roots and reeds, among sedges. The female, like the water-hen, covers up her eggs when she leaves her nest, which, unlike the nests of most of the aquatic birds, floats upon the surface of the water.

THE sub-family of the Alcinæ, or Auks, has several British representatives, among which the GREAT AUK is the rarest.

This bird, formerly to be found in several parts of Northern Europe, in Labrador, and very rarely in the British Islands, has not been observed for many years, and is as completely extinct as the Dodo. Almost the last living specimens known were seen in the Orkneys, and were quite familiar to the inhabitants under the name of the King and Queen of the Auks.

DABCHICK.—(*Podiceps minor.*)

According to Mr. Lloyd, this bird formerly frequented certain parts of Iceland, a certain locality called the Auk-Skär being celebrated for the number of Auks which nested upon it. The Skär, however, is so difficult of approach on account of the heavy surf which beats upon it, that few persons have the daring to land. In 1813 a number of Auks were taken from the Skär, and, horrible to relate, they were all eaten except one.

The eggs are variable in size, and colour, and markings, some being of a silvery white, and others of a yellowish white ground ; and the spots and streaks are greatly different in colour and form, some being yellowish brown and purple, others purple and black, and others intense blue and green.

The upper surface of this bird is black, except a patch of pure white round and in front of the eye and the ends of the secondaries, which are white. The whole of the under surface is white, and in winter the chin and throat are also white. The total length of the bird is thirty-two inches.

THE odd little PUFFIN, so common on our coasts, is remarkable for the singular shape, enormous size, and light colours of its beak, which really looks as if it had been originally made for some much larger bird. Owing to the dimensions and shape of the beak it is often called the SEA PARROT, or the COULTERNEB.

The Puffin can fly rapidly and walk tolerably, but it dives and swims supremely well, chasing fish in the water, and often bringing out a whole row of sprats at a time ranged along the side of its bill, all the heads being within the mouth and all the tails dangling outside. It breeds upon the rocks and in the rabbit-warrens near the sea, finding the ready-made burrows of the rabbit very convenient for the reception of its eggs, and fighting with the owner for possession of the burrow. Where rabbits do not exist,

GREAT AUK.—(*Alca impennis.*)

the Puffin digs its own burrow, and works hard at its labour. The egg is generally placed several feet within the holes, and the parents defend it vigorously. Even the raven makes little of an attack, for the Puffin grips his foe as he best can, and tries to tumble into the sea, where the raven is soon drowned, and the little champion returns home in triumph. The egg is white, but soon becomes stained by the earth. The food of this bird consists of fish, crustaceans, and insects.

The top of the head, the back, and a ring round the neck are black, and the cheeks and under surfaces are white. The beak is curiously striped with orange upon bluish grey, and the legs and toes are orange. The length of this bird is about one foot.

CRESTED GREBE.—(*Podiceps cristatus.*)

THE PENGUINS form a very remarkable sub-family, all its members having their wings modified into paddles useless for flight, but capable of being employed as fore-legs in terrestrial progression when the bird is in a hurry, and probably as oars or paddles in the water. There are many species of Penguins, but as they are very similar in general habits, we must be content with a single example.

The CAPE PENGUIN is very common at the Cape of Good Hope and the Falkland Islands. From the extraordinary sound it produces while on shore, it is called the Jackass Penguin. Darwin gives the following interesting account of this bird:—" In diving, its little plumeless wings are used as fins, but on the land, *as front legs*. When crawling (it may be said on four legs) through the tussocks, or on the side of a grassy cliff, it moved so very quickly that it might readily have been mistaken for a quadruped. When at sea and fishing, it comes to the surface for the purpose of breathing, with such a spring, and dives again so instantaneously, that I defy anyone at first

PUFFIN.—(*Fratercula arctica.*)

sight to be sure that it is not a fish leaping for sport."

These birds feed their young in a very singular manner. The parent bird gets on a hillock, and apparently delivers a very impassioned speech for a few minutes, at the end of which, it lowers its head and opens its beak. The young one, who has been a patient auditor, thrusts its head into the open beak of the mother, and seems to suck its subsistence from the throat of the parent bird. Another speech is immediately made, and the same process repeated, until the young is satisfied.

This Penguin is very courageous, but utterly destitute of the better part of courage — discretion ; for it will boldly charge at a man just as Don Quixote charged the windmills, and with the same success, as a few blows from a stick is sufficient to lay a dozen birds prostrate.

THE common GUILLEMOT is an example of the next sub-family.

This bird is found plentifully on our coasts throughout the year, and may be seen swimming and diving with a skill little inferior to that of the divers. It can, however, use its legs and wings tolerably well, and is

(CAPE PENGUIN.—*Spheniscus demersus.*)

said to convey its young from the rocks on which it is hatched, by taking it on the back and flying down to the water.

The Guillemot lays one egg, singularly variable in colour. I possess several eggs, all unlike, and Mr. Champley has five hundred, no two of which

are similar, the ground colouring being of every shade, from pure white to intense red, and from pale stone-colour to light and dark green.

THE curious family of the PETRELS now comes before us. A well-known British example is the STORMY PETREL, known to sailors as MOTHER CAREY'S CHICKEN, and hated by them after a most illogical manner because it foretells an approaching storm.

This bird has long been celebrated for the manner in which it passes over the waves, pattering with its webbed feet and flapping its wings so as to keep itself just above the surface. It thus traverses the ocean with wonderful ease, the billows rolling beneath its feet and passing away under the bird without in the least disturbing it. It is mostly on the move in windy weather, because the marine creatures are flung to the surface by the chopping waves and can be easily picked up as the bird pursues its course. It feeds on the little fish, crustaceans, and molluscs, which are found in abundance on the surface of the sea, especially on the floating masses of algæ, and will for days keep pace with a ship for the sake of picking up the refuse food thrown overboard. Indeed to throw the garbage of fish into the sea is a tolerably certain method of attracting these birds, who are sharp-sighted, and seldom fail to perceive anything eatable. The name of Petrel is given to the bird on account of its powers of walking on the water, as is related of St. Peter.

This Petrel breeds on our northern coasts, laying a white egg in some convenient recess, a rabbit burrow being often employed for the purpose.

THE well-known WANDERING ALBATROSS is the largest of all the species.

This fine bird is possessed of wondrous powers of wing, sailing along for days together without requiring rest, and hardly ever flapping its wings, merely

GUILLEMOT.—(*Uria Troile.*)

swaying itself easily from side to side with extended pinions. It is found in the Southern Seas, and is very familiar to all those who have voyaged through that portion of the ocean. Like the petrel, it follows the ships for the sake of obtaining food, and so voracious is the bird that it has been observed to dash at a piece of blubber weighing between three and four pounds, and gulp it down entire.

The Albatross makes its home on the lofty precipices of Tristan d'Acunha, the Crozettes, the Marion Islands, and other similar localities.

The FULMAR PETREL is rather a large bird, being about nineteen inches long, and stoutly built. It is very plentiful at St. Kilda, and is used for various purposes, furnishing down and oil, besides being itself eaten. Like several other Petrels, the Fulmar is able, when alarmed, to eject from the mouth the oil with which they are so liberally supplied. The egg—for there is never more than one—of the Fulmar Petrel is laid upon a narrow ledge of cliff, and always at a considerable distance from the summit and the bottom of the rock.

THE large family of the GULLS is here represented by two species, both of which are among our British birds.

The GREAT BLACK-BACKED GULL is a very fine bird, not very plentiful on our coasts, but spread over the greater part of the British shores.

WANDERING ALBATROSS.—(*Diomedea exulans.*)

This bird prefers low-lying and marshy lands, and is found on the flat shores of Kent and Essex at the mouth of the Thames, where it is popularly known under the name of the Cob. It is very plentiful on the shores of Sweden and Norway, and on some of the islands of Shetland and Orkney it breeds in abundance, the eggs being highly valued on account of their rich flavour and their large size.

It is a fierce bird, and when wounded will fight vigorously for its liberty. The nest of this species is of grass, and generally contains three eggs of greenish dun flecked with grey and brown. In the summer plumage the head and neck of the Great Black-backed Gull are white ; the upper surface of the body is dark leaden grey, with some white upon the quill feathers of the wings ; the whole of the under surface is pure white, and the legs and feet are pinkish. The length of this bird is about thirty inches.

THE common TERN, or SEA SWALLOW, is very plentiful on our coasts, and may be seen flying along on rapid wing, its long forked tail giving it so decidedly

FULMAR PETREL.—(*Procellaria glacialis.*)

a swallow-like air, that its popular name of Sea Swallow is well applied.

The Tern breeds on low-lying lands, and makes a very rude nest, being indeed nothing more than a shallow depression in the earth, into which are scraped a few sticks, stones, and dry grasses. The Tern reaches this country about May and departs in September. An adult bird in summer plumage has the tip of the head and nape of the neck jet-black, the upper part of the body ashen grey, the under surface white, and the legs, feet, and bill coral-red, the bill deepening into black at the tip. The length of the Tern rather exceeds fourteen inches; much of it is due to the long forked feathers of the tail.

TERN.—(*Sterna hirundo.*)

We now arrive at the last family of birds, the PELICANS, a group which includes many species, all remarkable for some peculiarity, and many of them really fine and handsome birds.

As its name implies, the TROPIC BIRD is seldom to be seen outside the tropics unless driven by storms. It is wonderfully powerful on the wing, being able to soar for a considerable period, and passing whole days in the air without needing to settle.

As a general fact they do not fly to very great distances from land, three hundred miles being about the usual limit; but Dr. Bennett observed them on one occasion when the nearest land was about one thousand miles distant. The long tail-shafts of the Tropic Bird are much valued in many lands, the natives wearing them as ornaments, or weaving them into various implements.

The Tropic Bird breeds in the Mauritius. The total length of this bird is about two feet six inches, of which the tail-feathers occupy about fifteen inches.

TROPIC BIRD.—(*Phaëton æthereus.*)

THE GANNET, SOLAN GOOSE, or SPECTACLED GOOSE, is a well-known resident on our coasts, its chief home being the Bass Rock in the Frith of Forth, on which it congregates in vast numbers.

The Gannet is a large bird, nearly three feet long; and being powerful on the wing, and possessed of a large appetite, it makes great havoc among the fish which it devours. Herrings, pilchards, sprats, and similar fish are the favourite food of the Gannet, and as soon as the shoals of herrings approach the coast the Gannets assemble in flocks and indicate to the fishermen the presence and position of the fish.

The nest of the Gannet is a heap of grass, seaweed, and similar substances, on which is laid one very pale blue egg, which, however, does not long retain its purity. The young are clothed with white puffy down, which after a while changes to nearly black feathers, the white plumage not being assumed until the bird has reached full age. The head and neck of the full-grown bird are buff, the primaries black, and the rest of the plumage white. The yearling bird is almost wholly black covered with streaks and triangular marks of greyish white. The total length of this bird is about thirty-four inches.

THE common CORMORANT is well known for its voracious habits, its capacities of digestion having long become proverbial.

This bird is common on all our rocky coasts, where it may be seen sitting on some projecting ledge, or diving and swimming with great agility, and ever and anon returning to its resting-place on the rock. It is an admirable swimmer and a good diver, and chases fish with equal perseverance and success, both qualities being needful to satisfy the wants of its ever-craving maw.

GANNET OR SOLAN GOOSE.—(*Sula Bassanea.*)

The Cormorant can easily be tamed, and in China, where everything living or dead is utilized, the bird is employed for the purpose of catching fish. The Cormorants are regularly trained to the task, and go out with their master in a boat, where they sit quietly on the edge until they receive his orders. They then dash into the water, seize the fish in their beaks, and bring them to their owner. Should one of these birds pounce upon a fish too large for it to carry alone, one of its companions will come to its assistance, and the two together will take the fish and bring it to the boat. Sometimes a Cormorant takes an idle fit, and swims playfully about instead of attending to its business, when it is recalled to a sense of duty by its master, who strikes the water with his oar and shouts at the bird, who accepts the rebuke at once and dives after its prey. When the task is completed, the birds are allowed their share of fish. A detailed and interesting account of these birds may be found in Mr. Fortune's work on China.

CORMORANT.—(*Graculus Carbo.*)

The nest of the Cormorant is made of a large mass of sticks, seaweed, and grass, and the eggs are from four to six in number, rather small in proportion to the dimensions of the parent bird, and of a curious chalky texture externally, varied with a pale greenish blue.

ANOTHER well-known British species of this genus is the CRESTED COR-MORANT, GREEN CORMORANT, or SHAG, a bird which can at once be distinguished from the preceding species by the green colour of the plumage and the difference in size, the length of an adult male being only twenty-seven inches. In habits this species resembles the common Cormorant.

WE now arrive at the well-known PELICAN, which is universally accepted as the type of the family. This bird is found spread over many portions of Africa and Asia, and is also found in some parts of Southern Europe.

The pouch of the Pelican is enormously large, capable of containing two gallons of water, and is employed by the bird as a basket wherein to carry the fish which it has caught. The Pelican is a good fisherman, hovering above the water watching for a shoal of fish near the surface. Down sweeps the bird, scoops up a number of fish in its capacious pouch, and then generally goes off homeward.

The nest of the Pelican is placed on the ground in some retired spot, usually an island in the sea, or the borders of some inland lake or a river. It is made of grasses, and contains two or three white eggs. The female sits on the eggs, and her mate goes off to fish for her; and when the young are hatched they are fed by the parents, who turn the fish out of their pouches into the mouths of the young.

PELICAN.—(*Pelecanus onocrotalus.*)

The colour of the Pelican is white, with a delicate roseate tinge like that of a blush rose. On the breast the feathers are elongated and of a golden yellow. The quill feathers are black, and the bill is yellow tipped with red. The length of the bird is almost six feet, and the expanse of wing about twelve feet.

THE last bird on our list is the well-known FRIGATE BIRD, SEA HAWK, or MAN-OF-WAR BIRD, an inhabitant of the tropical seas. It derives its name of Man-of-war Bird from its habit of watching the gannets when they fish, and then swooping upon them and robbing them of their prey.

The long black feathers of the tail are in great request among the Society Islanders, being woven as ornaments into the head-dresses of the chiefs. The nest of the Frigate Bird is sometimes built upon trees and bushes where the low shores afford no cliffs, but its usual locality for breeding is on the summit of some rocky cliff. On the rock there is no nest, but when the bird breeds among

FRIGATE BIRD.—(*Atagen Aquila*).

trees, it makes a rude scaffolding of sticks like the nest of the wood pigeon. There is only one egg, of a peculiar chalky whiteness, and while sitting, the bird is very bold and will not stir even if pushed with a stick, snapping and biting at the obnoxious implement. The voice of this bird is rough and harsh, and is likened to the sound produced by turning a winch.

The colour of the adult Frigate Bird is shining black glossed with green, the female being dull black above, and white streaked with cinnamon upon the head, breast, and under parts. The pouch on the throat is scarlet, and when distended has a very curious effect against the dark black of the throat and neck. Including the long tail, the male measures three feet in length, but the body is extremely small. The expanse of the wings is about eight feet.

REPTILES.

REPTILES.

THE remarkable beings which are classed together under the general title of REPTILES, or creeping animals, are spread over those portions of the globe where the climate is tolerably warm, and are found in the greatest profusion under the hotter latitudes.

Some reptiles inhabit the dry and burning deserts, but the generality of these creatures are semi-aquatic in their habits, are fitted by their structure for progression on land or in water, and are able to pass a considerable time below the surface without requiring to breathe. This capacity is mostly the result of the manner in which the circulation and aëration of their blood is effected.

In all Mammalia and Birds, the heart is divided into a double set of compartments, each having a direct communication with the other. In the Reptiles, however, this structure is considerably modified, so that the blood is never so perfectly aërated as in the higher animals. The blood is consequently much colder than in the creatures where the oxygen obtains a freer access to its particles.

In consequence of this organization the whole character of the Reptiles is widely different from that of the higher animals. Dull sluggishness seems to be the general character of a Reptile, for though there are some species which whisk about with lightning speed, and others, especially the larger lizards, can be lashed into a state of terrific frenzy by love, rage, or hunger, their ordinary movements are inert, their gestures express no feeling, and their eyes, though bright, are stony, cold, and passionless.

The young of Reptiles are produced from eggs, mostly being hatched after they have been laid, but in some cases the young escape from the eggs before they make their appearance in the world. As a general fact, however, the eggs of Reptiles are placed in some convenient spot where they are hatched by the heat of the sun.

TORTOISES.

THE very curious reptiles which are known by the general name of TOR-TOISES are remarkable for affording the first example of a skeleton brought to the exterior of the body, a formation which is frequent enough in the lower orders, the crustaceans and insects being familiar examples thereof. In these reptiles the bones of the chest are developed into a curious kind of box, more or less perfect, which contains within itself all the muscles and the viscera, and in most cases can receive into its cavity the head, neck, and limbs ; in one genus so effectually, that when the animal has withdrawn its limbs and head, it is contained in a tightly-closed case without any apparent opening.

In the true Tortoises the feet are club-shaped and the claws blunt, and the neck can be wholly withdrawn within the shell.

Perhaps the best known species of these creatures is the COMMON LAND TORTOISE, so frequently exposed for sale in our markets, and so favourite an inhabitant of gardens.

This appears to be the only species that inhabits Europe, and even in that continent it is by no means widely spread, being confined to those countries which border the Mediterranean.

It is one of the vegetable feeders, eating various plants, and being very fond of lettuce leaves, which it crops in a very curious manner, biting them off sharply when fresh and crisp, but dragging them asunder when stringy, by putting the fore-feet upon them, and pulling with the jaws. This Tortoise will drink milk, and does so by opening its mouth, scooping up the milk in its lower jaw as if with a spoon, and then raising its head to let the liquid run down its throat.

One of these animals, which I kept for some time, displayed a remarkable capacity for climbing, and was very fond of mounting upon various articles of furniture, stools being its favourite resort. It revelled in warmth, and could not be kept away from the hearthrug, especially delighting to climb upon a footstool that generally lay beside the fender.

This Tortoise had a curious kind of voice, not unlike the mewing of a little kitten. The Common Tortoise is known to live to a great age.

COMMON LAND TORTOISE.—(*Testudo Græca.*)

Another specimen, a very large one, has been in my possession for several years. At the end of autumn it burrows under a heap of leaf-mould, and waits there until the warm days of spring. It feeds mostly on grass, and eats its way in a line, leaving a groove of cut grass to mark its track. With the exception of strawberry eating, it does no harm in the garden. It has a most inexplicable objection to rain, of which not one drop can penetrate its shell ; and whenever a shower comes, it makes its way to an earth-bank, forces itself partly into the loose soil, and remains there with retracted head and limbs until the rain has ceased.

WE now come to a group of Tortoises called TERRAPINS.

These creatures are inhabitants of the water, and are mostly found in rivers. They are carnivorous in their diet, and take their food while in the water. They may be known by their flattened heads, covered with skin, sometimes hard, but often of a soft consistency, and their broad feet with the toes webbed as far the claws.

The CHICKEN TORTOISE is found in North America.

It is very common in ponds, lakes, or marshy grounds, and though very plentiful, and by no means quick in its movements, is not easily caught, owing to its extreme wariness.

The Chicken Tortoise swims well, but not rapidly, and as it passes along with its head and neck elevated above the surface, it looks so like the dark water-snake of the same country, that at a little distance it might readily be mistaken for that reptile.

I have kept several of these reptiles, and found no difficulty in preserving them in health. They lived in a tank in which were several large stones that projected above the surface of the water. On the top of these stones the Chicken Tortoises loved to sit, and so exactly did their bodies harmonize with the stones, that it was not easy to decide at a hasty glance whether the stones were bare or covered with the little Tortoises. At first, the least movement or sound would send them tumbling into the water, but after a while they became used to captivity, and would even feed out of the hand

CHICKEN TORTOISE.—(*Emys reticularia.*)

Their diet consisted of meat, either raw or cooked. They used to seize it in their mouths, and then, placing a foot on its side, push away the meat, so as to cut a piece completely out with their sharp-edged jaws. They will even seize fish and serve them in like manner, and indeed it is not safe to place them in tanks wherein are any other living creatures.

It is rather a small species, seldom exceeding ten inches in length. Its flesh is remarkably excellent, very tender and delicately flavoured, something like that of a young chicken, so that this Tortoise is in great request as an article of food, and is largely sold in the markets, though not so plentifully as the common salt-water terrapin. Its colour is dark brown above, and the plates are scribbled with yellow lines, and wrinkled longitudinally. The neck is long in proportion to the size of the animal, so long indeed that the head and neck together are almost as long as the shell. The lower jaw is hooked in front.

THE well-known CARET, or HAWKSBILL TURTLE, so called from the formation of the mouth, is a native of the warm American and Indian seas, and is common in many of the islands of those oceans.

The Hawksbill Turtle is the animal which furnishes the valuable "tortoiseshell" of commerce, and is therefore a creature of great importance. The scales of the back are thirteen in number, and as they overlap each other for about one-third of their length, they are larger than in any other species where the edges only meet. In this species, too, the scales are thicker stronger, and more beautifully clouded than in any other Turtle.

The uses to which this costly and beautiful substance are put are innumerable. The most familiar form in which the tortoiseshell is presented to us is the comb, but it is also employed for knife-handles, boxes, and many other articles of ornament or use.

THE best known of all the Turtles is the celebrated GREEN TURTLE, so called from the green colour of its fat.

This useful animal is found in the seas and on the shores of both continents, and is most plentiful about the Island of Ascension and the Antilles, where it is subject to incessant persecution for the sake of its flesh. The shell of this reptile is of very little use, and of small value, but the flesh is remarkably rich and well-flavoured, and the green fat has long enjoyed a world-wide and fully deserved reputation.

GREEN TURTLE.—(*Chelonia viridis.*)

The eggs of the Turtle are thought as great delicacies as its flesh. It is while the female Turtle is visiting the shore for the purpose of depositing her eggs that she is usually captured, as these sea-loving reptiles care little for the shore except for this purpose.

CROCODILES AND ALLIGATORS.

ACCORDING to the arrangement of the national collection in the British Museum, the link next to the tortoise tribe is formed of an important group of reptiles, containing the largest of the reptilian order, larger indeed than most present inhabitants of the earth.

These great reptiles are divided, or rather fall naturally, into two families, namely, the Crocodiles and the Alligators. All the members of these families can be easily distinguished by the shape of their jaws and teeth, the lower canine teeth of the Crocodiles fitting into a *notch* in the edge of the upper jaw, and those of the Alligators fitting into a *pit* in the upper jaw. This peculiarity causes an obvious difference in the outline of the head, the muzzle of the Crocodiles being narrowed behind the nostrils, while that of the Alligators forms an unbroken line to the extremity. A glance therefore at the head will suffice to settle the family to which any species belongs. In the Crocodiles, moreover, the hind-legs are fringed behind with a series of compressed scales.

THE most peculiar of these reptiles is the long-celebrated CROCODILE of Northern Africa.

This terrible creature is found chiefly in the Nile, where it absolutely swarms, and though a most destructive and greatly dreaded animal, is without doubt as valuable in the water as the hyæna and vulture upon the land. Living exclusively on animal food, and rather preferring tainted or even putrefying to fresh meat, it is of great service in devouring the dead animals that would otherwise pollute the waters and surrounding atmosphere.

Human beings have a great dread of this voracious reptile. Many instances are known where men have been surprised near the water's edge, or captured when they have fallen into the river. There is, it is said,

EGYPTIAN CROCODILE.—(*Crocodilus vulgaris.*)

only one way of escape from the jaws of the Crocodile, and that is to turn boldly upon the scaly foe, and press the thumbs into his eyes, so as to force him to relax his hold or relinquish the pursuit.

The eggs of the Crocodile are about as large as those of the goose, and many in number, so that these terrible reptiles would overrun the country were they not persecuted in the earliest stages by many creatures, who discover and eat the eggs almost as soon as they are laid. It is curious that the Crocodile is attended by a bird which warns it of danger, just as the rhinoceros has its winged attendant, and the shark its pilot fish. The Crocodile bird is popularly called the Ziczac, from its peculiar cry.

ALLIGATOR.—(*Alligator Mississipensis.*)

WE now come to the ALLIGATORS, the second family of those huge reptiles, which may be known, as has already been mentioned, by the lower canine teeth fitting into pits in the upper jaw.

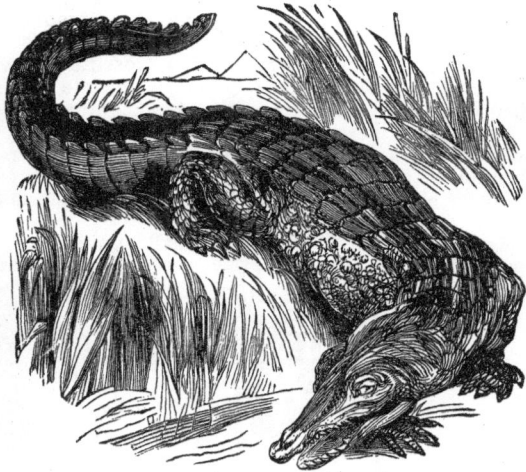

The COMMON ALLIGATOR inhabits Northern America, and is plentifully found in the Mississippi, the lakes and rivers of Louisiana and Carolina, and similar localities. It is a fierce and dangerous reptile, in many of its habits bearing a close resemblance to the crocodiles and the other members of the family.

Unlike the crocodile, however, it avoids the salt water, and is but seldom seen even near the mouths of rivers, where the tide gives a brackish taste to their waters. It is mostly a fish-eater, haunting those portions of the rivers where its prey most abounds, and catching them by diving under a passing shoal, snapping up one or two victims as it passes through them, tossing them in the air for the purpose of ejecting the water which has necessarily filled its mouth, catching them adroitly as they fall, and then swallowing them.

The eggs of the Alligator are small and numerous. The parent deposits them in the sand of the river side, scratching a hole with her paws, and placing

SCALY LIZARD.—(*Zootoca vivipara*.)

the eggs in a regular layer therein. She then scrapes some sand, dry leaves, grass, and mud over them, smoothes it, and deposits a second layer upon them. These eggs are then covered in a similar manner, and another layer deposited until the mother reptile has laid from fifty to sixty eggs. Although they are hatched by the heat of the sun and the decaying vegetable matter, the mother does not desert her young, but leads them to the water and takes care of them until their limbs are sufficiently strong and their scales sufficiently firm to permit them to roam the waters without assistance.

During the winter months the Alligator buries itself in the mud, but a very little warmth is sufficient to make it quit its retreat and come into the open air again. While lively, especially at night, it is a most noisy animal, bellowing in so loud a tone and in so singular a cadence that even the nightly concert of jaguars and monkeys is hardly heard when the Alligators are roaring.

It sometimes attains to a great size, and is then formidable to man. Mr. Waterton mentions a case where one of these creatures was seen to rush out of the water, seize a man, and carry him away in spite of his cries and struggles. The beast plunged into the river with his prey, and neither Alligator nor man was afterwards seen.

THE true LIZARDS have four limbs, generally visible, but in a few instances hidden under the skin. Their body is long and rounded, and the tail is tapering and mostly covered with scales set in regular circles or " whorls."

ENGLAND possesses at least two examples of the true Lizards, one of which, the SCALY LIZARD, is extremely common. This pretty little reptile is extremely plentiful upon heaths, banks, and commons, where it may be seen darting about in its own quick, lively manner, flitting among the grass stalks with a series of sharp, twisting springs, snapping up the unsuspecting flies as they rest on the grass-blades, and ever and anon slipping under shelter of a gorse bush, or heather tuft, only to emerge, in another moment, brisk and lively as ever.

This is one of the reptiles that produces living young, the eggs being hatched just before the young Lizards are born. With reptiles, the general plan is to place the eggs in some spot where they are exposed to the heat of the sunbeams ; but this Lizard, together with the viper, is in the habit of lying on a sunny bank before her young ones are born, apparently for the purpose of gaining sufficient heat to hatch the eggs. This process is aided by the thinness of the membrane covering the eggs.

UNTIL comparatively recent years, the SAND LIZARD was confounded with the Scaly Lizard, which has just been described.

SAND LIZARD.—(*Lacerta agiles.*)

Though quick and lively in its movements, it is not so dashingly active as the scaly lizard, having a touch of deliberation as it runs from one spot to another, while the scaly lizard seems almost to be acted upon by hidden springs.

Unlike the scaly lizard, this species lays its eggs in a convenient spot, and then leaves them to be hatched by the warm sunbeams. Sandy banks with a southern aspect are the favoured resorts of this reptile, which scoops out certain shallow pits in the sand, deposits her eggs, covers them up, and then leaves them to their fate. Mr. Bell, who has paid great attention to this subject, has remarked that the eggs are probably laid for a considerable period before the young are hatched from them.

A SECOND tribe of Lizards now comes before our notice. These are the GEISSOSAURI, a title derived from two Greek words, the former signifying " the eaves of a house," and the latter " a lizard." As in this tribe there are many families, and more than eighty genera, it will be impossible to give more than a very slight account of these reptiles, or even to mention more than a small number selected as types of the large or small groups which they represent.

THE large and important family of the SKINKS contains between forty and fifty genera, nearly each of which possesses one or more species, concerning which there is something worthy of notice.

This family finds a familiar representative in the common BLINDWORM, or SLOW-WORM, of England, which, from its snake-like form and extreme fragility, might well deserve the title of the English Glass Snake. In this reptile there is no external trace of limbs, the body being uniformly smooth

BLINDWORM.—(*Anguis fragilis.*)

as that of a serpent, and even more so than in some of the snakes, where the presence of the hinder pair of limbs is indicated by a couple of little hook-like appendages. Under the skin, however, the traces of limbs may be discovered, but the bones of the shoulders, the breast, and the pelvis are very small and quite rudimentary.

This elegant little reptile is very common throughout England, and is spread over the greater part of Europe and portions of Asia, not, however, being found in the extreme north of Europe. In this country it is plentiful along hedge-rows, heaths, forest lands, and similar situations, where it can find immediate shelter from its few enemies, and be abundantly supplied with food. It may often be seen crawling leisurely over a beaten footpath, and I have once captured it while crossing a wide turnpike road near Oxford.

Why the name of the Blindworm should have been given to this creature I cannot even conjecture, for it has a pair of conspicuous though not very large eyes, which shine as brightly as those of any animal, and are capable of good service. Indeed, all animals which prey upon insects and similar moving things must of necessity possess well-developed eyes, unless they are gifted with the means of attracting their prey within reach, as is the case with some well-known fishes, or chase it by the senses of hearing and touch, as is done by the mole. Moreover, the chief food of the Blindworm consists of slugs, which glide so noiselessly that the creature needs the use of its eyes to detect the soft mollusc as it slides over the ground on it slimy course. Speed is not needful for such a chase, and the Blindworm accordingly is slow and delibe-rate in all its movements, except when very young, when it twists and wriggles about in a singular fashion as often as it is touched.

The great fragility of the Blindworm is well known. By a rather curious structure of the muscles and bones of the spine, the reptile is able to stiffen itself to such a degree, that on a slight pressure or trifling blow, or even by the voluntary contraction of the body, the tail is snapped away from the body, and, on account of its proportionate length, looks just as if the creature had been broken in half. The object of this curious property seems to be to ensure the safety of the animal. The severed tail retains, or rather acquires, an extraordinary amount of irritability, and for several minutes after its amputation leaps and twists about with such violence that the attention of the foe is drawn to its singular vagaries, and the Blindworm itself creeps quietly away to some place of shelter.

When the tail of the Blindworm is thus snapped off, the scales of the body project all round the fractured portion, forming a kind of hollow into which the broken end of the tail can be slipped.

According to popular notions, the Blindworm is a terribly poisonous creature, and by many persons is thought to be even more venomous than the viper, whereas it is perfectly harmless, having neither the will nor the ability to bite, its temper being as quiet as its movements, and its teeth as innocuous as its jaws are weak. I fancy that the origin of this opinion may be found in the habit of constantly thrusting out its broad, black, flat tongue with its slightly forked tip ; for the popular mind considers the tongue to be the sting, imagining it to be both the source of the venom and the weapon by which it is injected into the body, and so logically classes all creatures with forked tongues under the common denomination of poisonous animals.

It is said that this reptile will bite when handled, but that its minute teeth and feeble jaws can make no impression upon the skin ; and also that when it has thus fastened on the hand of its captor, it will not release its hold unless its jaws be forced open. For my own part, and I have handled very many of these reptiles, I never knew them attempt to bite, or even to assume a threatening attitude. They will suddenly curl themselves up tightly, and snap off their tails, but to use their jaws in self-defence is an idea that seldom appears to occur to them.

The specimen whose portrait is given in the illustration was in my posses-sion for a considerable time, and was fed on the white garden slug.

In its wild state the Blindworm feeds mostly on slugs, but will also eat worms and various insects. Some persons assert that it devours mice and reptiles ; but that it should do so is a physical impossibility, owing to the very small dimensions of the mouth and the structure of the jaw, the bones of which are firmly knitted together, and cannot be separated while the prey is being swallowed, as is the case with the snakes.

In captivity it seems to reject almost any food except slugs, but these molluscs it will eat quite freely.

The Blindworm generally retires to its winter quarters towards the end of August, or even sooner, should the weather be chilly. The localities which it chooses for this purpose are generally dry and warm spots, where the dried leaves and dead twigs of decayed branches have congregated into heaps, so as to afford it a safe refuge. Sometimes it bores its way into masses of rotten wood ; and on heathery soils, where the ground slopes considerably, it selects a spot where it will be well sheltered from the winter's rains and snows, and burrows deeply into the dry loose soil.

Like the snakes, the Blindworm casts its skin at regular intervals, seeming to effect its object in various modes, sometimes pulling it off in pieces, but usually stripping it away, like the snakes, by turning it inside out, just as an eel is skinned.

A NEW group now comes before our notice, the members of which are distinguished by the formation of their tongues, which, instead of being flat and comparatively slender, as in the preceding Lizards, are thick, convex, and have a slight nick at the end. On account of this structure, the species of this sub order are termed PACHYGLOSSÆ, or Thick-tongued Lizards.

These reptiles are divided into sundry groups, the first of which is termed the NYCTISAURA, or Nocturnal Lizards. These creatures have eyes formed for seeing in the dusk ; circular eyelids, which, however, cannot meet over the eyeball, and in almost every case the pupil is a long narrow slit like that of the cat. The body is always flattened. The limbs are four in number, tolerably powerful, and are used in progression.

OF these Lizards, the first family is the GECKOTIDÆ, or GECKOS, a very curious group of reptiles, common in many hot countries, and looked upon with dread or adoration by the natives, sometimes with both, where the genius of the nation leads them to reverence the object of their fears, and to form no other conception of supreme power than the capability of doing harm.

GECKO.—(*Gecko verus.*)

The COMMON GECKO, or RINGED GECKO, is an Asiatic species, being as common in India as the preceding species in North Africa. It may be easily known from allied Fan-foot by the large tubercles upon the back.

This reptile has much the same habits as the fan-foot, and possesses equally the ability to run over a perpendicular wall. During the day-time it conceals itself in some chink or dark crevice, but in the evening it leaves its retreat, moving rapidly and with such perfectly silent tread, that the ignorant natives may well be excused for classing it among supernatural beings. The Gecko occasionally utters a curious cry, which has been compared to that peculiar clucking sound employed by riders to stimulate their horses, and in some species the cry is very distinct and said to resemble the word " Geck-o," the last syllable being given smartly and sharply. On account of this cry, the Geckos are variously called Spitters, Postilions, and Claquers.

During the cold months of the year the Geckos retire to winter quarters, and are thought to retain their condition during this foodless season by means of two fatty masses at the base of the abdomen, which are supposed to nourish them as the camel is nourished by the hump. The male is smaller than the female, and the eggs are very spherical, and covered with a brittle chalky shell. The colour of the Gecko is reddish grey with white spots.

The scales of the back are flat and smooth, and there is also a series of rather large tubercular projections arranged in twelve distinct rows.

WE now arrive at an important tribe of Lizards, called by the name of STROBILOSAURA, a title derived from two Greek words, one signifying a "fir-cone" and the other "a lizard," and given to these creatures because the scales that cover their tails are set in regular whorls, and bear some resemblance to the projecting scales of the fir-cone. In all these reptiles the tongue is thick, short, and very slightly nicked at the tip. The eyes have circular pupils, and are formed for day use.

The first family of these Lizards consists of those creatures which are grouped together under the general title of IGUANA. Our illustration depicts the COMMON IGUANA.

This conspicuous, and, in spite of its rather repulsive shape, really handsome Lizard, is a native of Brazil, Cayenne, the Bahamas, and neighbouring localities, and was at one time very common in Jamaica, from which, however, it seems to be in process of gradual extirpation.

In common with those members of the family which have their body rather compressed and covered with squared scales, the Iguana is a percher on trees, living almost wholly among the branches, to which it clings with its powerful feet, and on which it finds the greater part of its food. It is almost always to be found on the trees that are in the vicinity of water, and especially favours those that

IGUANA.—(*Iguana tuberculata.*)

grow upon the banks of a river where the branches overhang the stream.

Though not one of the aquatic Lizards, the Iguana is quite at home in the water, and if alarmed will often plunge into the stream, and either dive or swim rapidly away. While swimming, it lays its fore-legs against the sides, so as to afford the smallest possible resistance to the water, stretches out the hinder legs, and, by a rapid serpentine movement of its long and flexible tail, passes swiftly through the waves. It has considerable power of enduring immersion, as indeed is the case with nearly all reptiles, and has been known to remain under water for an entire hour, and at the end of that time to emerge in perfect vigour.

From the aspect of this long-tailed, dewlapped, scaly, spiny Lizard, most persons would rather recoil than feel attracted, and the idea of eating the flesh of so repulsive a creature would not be likely to occur to them. Yet, in truth, the flesh of the Iguana is justly reckoned among one of the delicacies of the country where it resides, being tender, and of a peculiarly delicate flavour, not unlike the breast of a spring chicken. There are various modes of cooking the Iguana, roasting and boiling being the most common. Making it into a fricassee, however, is the mode which has met with the largest general approval, and a dish of Iguana cutlets, when properly dressed, takes a very high place among the delicacies of a well-spread table.

The eggs, too, of which the female Iguana lays from four to six dozen, are very well flavoured and in high repute. It is rather curious that they contain very little albumen, the yellow filling almost the entire shell. As is the case with the eggs of the turtle, they never harden by boiling, and only assume a little thicker consistence. Some persons of peculiar constitutions cannot eat either the flesh or the eggs of the Iguana, and it is said that this diet is very injurious in some diseases. The eggs are hid by the female Iguana in sandy soil, near rivers, lakes, or the sea-coast, and after covering them with sand, she leaves them to be hatched by the heat of the sun.

THE family which comes next in order is that in which are included the AGAMAS, a group of Lizards which have been appropriately termed the Iguanas of the Old World. In the members of this family the teeth are set upon the edge of the jaws, and not upon their inner side, as in the true Iguanas of the New World. Between thirty and forty genera are contained in this family, and some of the species are interesting as well as peculiar beings.

Perhaps the most curious of all this family, if not, indeed, the most curious of all the reptiles, is the little Lizard which is well known under the title of the FLYING DRAGON.

This singular reptile is a native of Java, Borneo, the Philippines, and neighbouring islands, and is tolerably common. The most conspicuous characteristic of this reptile is the singularly developed membranous lobes on either side, which are strengthened by certain slender processes from the first six false ribs, and serve to support the animal during its bold leaps from branch to branch. Many of the previously mentioned Lizards are admirable leapers, but they are all outdone by the Dragon, which is able, by means of the membranous parachute with which it is furnished, to sweep through distances of thirty paces, the so-called flight being almost identical with that of the flying squirrels and flying fish.

FLYING DRAGON.—(*Draco volans.*)

When the Dragon is at rest, or even when traversing the branches of trees, the parachute lies in folds along the sides, but when it prepares to leap from one bough to another, it spreads its winged sides, launches boldly into the air, and sails easily, with a slight fluttering of the wings, towards the point on which it had fixed, looking almost like a stray leaf blown by the breeze. As if in order to make itself still more buoyant, it inflates the three membranous sacs that depend from its throat, suffering them to collapse again when it has settled upon the branch. It is a perfectly harmless creature, and can be handled with impunity. The food of the Flying Dragon consists of insects.

THE last tribe of the Lizards contains but one genus and very few species. From their habit of constantly living on trees, these creatures are called DENDROSAURA, or TREE LIZARDS. In these, the scales of the whole body are small and granular, and arranged in circular bands. The tongue is very curious, being cylindrical and greatly extensile, reminding the observer of a common earthworm, and swollen at the tip. The eyes are as peculiar as the tongue, being very large, globular, and projecting, and the ball is closely covered with a circular lid, through which a little round hole is pierced, much like the wooden snow-spectacles of the Esquimaux. The body is rather compressed, the ears are concealed under the skin, and the toes are separated into two opposable groups, so that the creature can hold very firmly upon the boughs. The tail is very long and prehensile, and is almost invariably seen coiled round the bough on which the reptile is standing.

THE most familiar example of the Dendrosaura is the common CHA-MELEON, a reptile which is found both in Africa and Asia.

This singular reptile has long been famous for its power of changing colour ; a property, however, which has been greatly exaggerated, as will be presently seen. Nearly all the Lizards are constitutionally torpid, though some of them are gifted with great rapidity of movement during certain seasons of the year. The Chameleon, however, carries this sluggishness to an extreme, its only change being from total immobility to the slightest imaginable degree of activity.

When it moves along the branch upon which it is clinging, the reptile first raises one foot very slowly indeed, and will sometimes remain foot in air for a considerable time, as if it had gone to sleep in the interim. It then puts the foot as slowly forward, and takes a good grasp of the branch. Having satisfied itself that it is firmly secured, it leisurely unwinds its tail, which has been tightly twisted round the branch, shifts it a little forward, coils it round again, and then rests for a while. With the same elaborate precaution, each foot is successively lifted and advanced, so that the forward movements seem but little faster than the hour-hand of a watch.

If placed on level ground, it is perforce obliged to walk, but it does so very awkwardly, though it gets over the ground faster than would be imagined from its movements on a tree.

The food of the Chameleon consists of insects, mostly flies, but, like many other reptiles, the Chameleon is able to live for some months without taking food at all. This capacity for fasting, together with the singular manner in which the reptile takes its prey, gave rise to the absurd fable that the Chameleon lived only upon air. To judge by external appearance, there never was an animal less fitted than the Chameleon for capturing the winged and active flies ; but when we come to examine its structure, we find that it is even better fitted for this purpose than many of the more active insect-eating Lizards.

The tongue is the instrument by which the fly is captured, being first deliberately aimed, like a billiard-player aiming a stroke with his cue, and then darted out with singular velocity. This member is very muscular, and is furnished at the tip with a kind of viscid secretion which causes the fly to adhere to it. Its mouth is well furnished with teeth, which are set firmly into its jaw, and enable it to bruise the insects after getting them into its mouth by means of the tongue.

The eyes have a most singular appearance, and are worked quite independently of each other, one rolling backwards, while the other is directed forwards or upwards. There is not the least spark of expression in the eye of the Chameleon, which looks about as intellectual as a green-pea with a dot of ink upon it.

A few words on the change of colour will not be out of place.

I kept a Chameleon for a long time, and carefully watched its changes of colour. Its primary hue was grey-black, but other colours were constantly passing over its body. Sometimes it would be striped like a zebra with light yellow, or covered with circular yellow spots. Sometimes it was all chestnut and black like a leopard, and sometimes it was brilliant green. Sometimes it would be grey, covered with black spots ; and once, when it was sitting on a branch, it took the hue of the autumnal leaves so exactly that it could scarcely be distinguished from them. A detailed account of this specimen is given in my " Glimpses into Petland," published by Messrs. Bell and Daldy.

The young of the Chameleon are produced from eggs, which are very spherical, white in colour, and covered with a chalky and very porous shell. They are placed on the ground under leaves, and there left to hatch by the heat of the sun and the warmth produced by the decomposition of the leaves. The two sexes can be distinguished from each other by the shape of the tail, which in the male is thick and swollen at the base.

CHAMELEON.—(*Chameleo vulgaris.*)

THE large and important order at which we now arrive consists of reptiles which are popularly known as SNAKES, or more scientifically as Ophidia, and to which all the true Serpents are to be referred.

The movements of the Serpent tribe are performed without the aid of limbs, and are, as a general rule, achieved by means of the ribs and the large curved scales that cover the lower surface. Each of these scales overlaps its successor, leaving a bold horny ridge whenever it is partially erected by the action of the muscles. The reader will easily see that a reptile so constructed can move with some rapidity by successively thrusting each scale a little forward, hitching the projecting edge on any rough substance, and drawing itself forward until it can repeat the process with the next scale. The movements are consequently very quiet and gliding, and the creature is able to pursue its way under circumstances of considerable difficulty.

The tongue of the Snake is long, black, and deeply forked at its extremity, and when at rest is drawn into a sheath in the lower jaw. In these days it is perhaps hardly necessary to state that the tongue is perfectly harmless, even in a poisonous serpent, and that the popular idea of the " sting " is entirely erroneous. The Snakes all seem to employ the tongue largely as a feeler, and may be seen to touch gently with the forked extremities the objects over which they are about to crawl, or which they desire to examine. The external organs of hearing are absent.

The vertebral column is most wonderfully formed, and is constructed with a special view to the peculiar movements of the serpent tribe. Each vertebra is rather elongated, and is furnished at one end with a ball and at the other with a corresponding socket, into which the ball of the succeeding vetebra exactly fits, thus enabling the creature to writhe and twine in all directions without danger of dislocating its spine. This ball-and-socket principle extends even to the ribs, which are jointed to certain rounded projections of the vertebræ in a manner almost identical with the articulation of the vertebræ upon each other, and, as they are moved by very powerful muscles, perform most important functions in the economy of the creature to which they belong.

The bones of the jaws are very loosely constructed, their different portions being separable, and giving way while the creature exerts its wonderful powers of swallowing. The great python Snakes are well known to swallow animals of great proportionate size, and anyone may witness the singular process by taking a common field snake, keeping it without food for a month or so, and then giving it a large frog. As it seizes its prey, the idea of getting so stout an animal down that slender neck and through those little jaws appears too absurd to be entertained for a moment, and even the leg which it has grasped appears to be several times too large to pass through the throat. But by slow degrees the frog disappears, the mouth of the snake gradually widening until the bones separate from each other to some distance, and are only held by the ligaments, and the whole jaw becoming dislocated, until the head and neck of the Snake look as if the skin had been stripped from the reptile, spread thin and flat, and drawn like a glove over the frog.

The Serpents, in common with other reptiles, have their bodies covered by a delicate epidermis, popularly called the skin, which lies over the scales, and is renewed at tolerably regular intervals. Towards the time of changing its skin, the Snake becomes dull and sluggish, the eyes look white and blind, owing to the thickening of the epidermis that covers them, and the bright colours become dim and ill-defined. Presently, however, the skin splits upon the back, mostly near the head, and the Snake contrives to wriggle itself out of the whole integument, usually turning it inside out in the process. This shed skin is transparent, having the shape of each scale impressed upon it, being fine and delicate as goldbeater's skin, and being applicable to many of the same uses, such as shielding a small wound from the external air.

THE first sub-order of Snakes consists of those serpents which are classed under the name of VIPERINA.

All these reptiles are devoid of teeth in the upper jaw, except two long poison-bearing fangs, set one at each side and near the muzzle. The lower jaw is well furnished with teeth, and both jaws are feeble. The scales of the abdomen are bold, broad, and arranged like overlapping bands. The head is large in proportion to the neck, and very wide behind, so that the head of these Snakes has been well compared to an ace of spades. The hinder limbs are not seen.

In the first family of the Viperine Snakes, called the Crotalidæ, the face is marked with a large pit or depression on each side, between the eye and the nostril. The celebrated and dreaded RATTLESNAKE belongs to this family.

This dreaded reptile is a native of North America, and is remarkable for the singular termination to the tail, from which it derives its popular name. At the extremity of the tail are a number of curious loose horny structures, formed of the same substance as the scales, and varying greatly in number according to the size of the individual. It is now generally considered that the number of joints on the "rattle" is an indication of the reptile's age, a fresh joint being gained each year immediately after it changes its skin and before it goes into winter quarters.

The joints of this remarkable apparatus are arranged in a very curious manner, each being of a somewhat pyramidal shape, but rounded at the edges, and being slipped within its predecessor as far as a protuberant ring which runs round the edge. In fact, a very good idea of the structure of the rattle may be formed by slipping a number of thimbles loosely into each other. The last joint is smaller than the rest, and rounded. As was lately mentioned, the number of these joints is variable, but the average number is from five or six to fourteen or fifteen. There are occasional specimens found that possess more than twenty joints in the rattle, but such examples are very rare.

When in repose, the Rattlesnake usually lies coiled in some suitable spot, with its head lying flat, and the tip of its tail elevated in the middle of the coil. Should it be irritated by a passenger, or feel annoyed or alarmed, it instantly communicates a quivering movement to the tail, which causes the joints of the rattle to shake against each other, with a peculiar skirring ruffle very much like the sound of the escaping steam of a railway engine.

Fortunately for the human inhabitants of the same land, the Rattlesnake is slow and torpid in its movements, and seldom attempts to bite unless it is provoked, even suffering itself to be handled without avenging itself. Mr. Waterton tells me, in connection with these reptiles : " I never feared the bite of a snake, relying entirely on my own movements. Thus, in the presence of several professional gentlemen, I once transferred twenty-seven Rattlesnakes from one apartment to another with my hand alone. They hissed and rattled when I meddled with them, but they did not offer to bite me." When about to inflict the fatal blow, the reptile seems to swell with anger, its throat dilating, and its whole body rising and sinking as if inflated by bellows. The tail is agitated with increasing vehemence, the rattle sounds its threatening war-note, with sharper ruffle, the head becomes flattened as

RATTLESNAKE.—(*Uropsophus durissus.*)

it is drawn back ready for the stroke, and the whole creature seems a very incarnation of deadly rage. Yet, even in such moments, if the intruder withdraw, the reptile will gradually lay aside its angry aspect, the coils settle down in their place, the flashing eyes lose their lustre, the rattle become stationary, and the serpent sink back into its previous state of lethargy.

The general colour of the Rattlesnake is pale brown. A dark streak runs along the temples from the back of the eye, and expands at the corner of the mouth into a large spot. A series of irregular dark brown bands are drawn across the back, a number of round spots of the same hue are scattered along the sides, upon the nape of the neck, and back of the head.

WE now come to the second great family of poisonous serpents, namely, the VIPERS, or Viperidæ. All the members of this family may be distinguished by the absence of the pit between the eyes and the nostrils. There are no teeth in the upper jaw, except the two poison-fangs.

The terrible Puff Adder belongs to this family.

This reptile is a native of Southern Africa, and is one of the commonest, as well as one of the most deadly, of poisonous snakes. It is slow and apparently torpid in all its movements, except when it is going to strike, and the colonists say that it is able to leap backwards so as to bite a person who is standing by its tail.

There is in nature no more fearful object than a full-grown Puff Adder. It grovels on the sand, winding its body so as to bury itself almost wholly in the tawny soil, and just leaving its flat, cruel-looking head, lying on the ground and free from sand. The steady, malignant, stony glare of those eyes is absolutely freezing as the creature lies motionless, confident in its deadly

PUFF ADDER.—(*Clotho arietans.*)

powers, and when roused by the approach of a passenger, merely exhibiting its annoyance by raising its head an inch or two, and uttering a sharp angry hiss. Even horses have been bitten by this reptile, and died within a few hours after the injury was inflicted. The peculiar attitude which is exhibited in the illustration is taken from life, one of the Puff Adders in the collection of the Zoological Society having been purposely irritated, so as to force it to raise itself from the sand, and assume the attitude of offence.

The Bushmen are in the habit of procuring from the teeth of this serpent the poison with which they arm their tiny but most fearful arrows. In the capture of the Puff Adder they display very great courage and address. Taking advantage of the reptile's sluggish habits, they plant their bare feet upon its neck before it has quite made up its reptilian mind to action, and, holding it firmly down, cut off its head and extract the poision at their

F F

leisure. In order to make it adhesive to the arrow-point, it is mixed with the glutinous juice of the amaryllis.

The colòur of the Puff Adder is brown, chequered with dark brown and white, and with a reddish band between the eyes. The under parts are paler than the upper.

THE true CERASTES, or HORNED VIPER, is a native of Northern Africa, and divides with the cobra of the same country the questionable honour of being the "worm of Nile" to whose venomous tooth Cleopatra's death was due.

The bite of this most ungainly-looking serpent is extremely dangerous, though perhaps not quite so deadly as that of the cobra, and the creature is therefore not quite so much dreaded as might be imagined. The Cerastes has a most curious appearance, owing to a rather large horn-like scale which projects over each eye, and which, according to the natives, is possessed of wonderful virtues.

The Cerastes has, according to Bruce, an awkward habit of crawling until it is alongside of the creature whom it is about to attack, and then making a side-long leap at its victim. He relates an instance where he saw a Cerastes perform a certainly curious feat :—" I saw one of them at Cairo crawl up the side of a box in which there were many, and there lie still as if hiding himself, till one of the people who brought them to us came near him, and though in a very disadvantageous position, sticking, as it were, perpendicularly to the side of the box, he leaped near the distance of three feet, and fastened between the man's fore-finger and thumb, so as to bring the blood."

The Cerastes usually lives in the driest and hottest parts of Northern Africa, and lies half buried in the sand until its prey should come within reach. Like

CERASTES, OR HORNED VIPER.—(*Cerastes Hasselquistii.*)

many serpents, it can endure a very prolonged frost without appearing to suffer any inconvenience ; those kept by Bruce lived for two years in a glass jar without partaking of food, and seemed perfectly brisk and lively, casting their skins as usual, and not even becoming torpid during the winter.

The colour cf the Cerastes is pale brownish white, covered irregularly with brown spots. Its length is about two feet.

THE common VIPER, or ADDER, is very well known in many parts of England, but in some localities is very plentiful, while in others it is never seen from one year's end to another.

Many persons mistake the common grass snake for the Viper, and dread it accordingly. They may, however, always distinguish the poisonous reptile from the innocuous, by the chain of dark spots that runs along the spine, and forms an unfailing guide to its identification. Fortunately for ourselves, it is the only poisonous reptile inhabiting England, the variously coloured specimens being nothing more than varieties of the same species.

Like most reptiles, whether poisonous or not, the Viper is a very timid creature, always preferring to glide away from a foe rather than to attack, and only biting when driven to do so under great provocation.

The head of the Viper affords a very good example of the venomous apparatus of the poisonous serpents, and is well worthy of dissection, which is better accomplished under water than in air. The poison-fangs lie on the sides of the upper jaw, folded back and almost undistinguishable until lifted with a needle. They are singularly fine and delicate, hardly larger than a lady's needle, and are covered almost to their tips with a muscular envelope through which the points just peep. The poison-secreting glands and the reservoir in which the venom is stored are found at the back and sides of the head, and give to the venomous serpents that peculiar width of head which is so unfailing a characteristic. The colour of the poison is a very pale yellow, and its consistence is very like that of salad oil, which, indeed, it much resembles, both in look and taste. There is but little in each individual, and it is possible that the superior power of the

VIPER, OR ADDER.—(*Pelias Berus.*)

large venomous snakes of other lands, especially those under the tropics, may be due as much to its quantity as its absolute intensity. In a full-grown rattlesnake, for example, there are six or eight drops of this poison, whereas the Viper has hardly a twentieth part of that amount.

On examining carefully the poison-fangs of a Viper, the structure by which the venom is injected into the wound will be easily understood. On removing the lower jaw, the two fangs are seen in the upper jaw, folded down in a kind of groove between the teeth of the palate and the skin of the head, so as to allow any food to slide over them without being pierced by their points. The end of the teeth reach about halfway from the nose to the angle of the jaw, just behind the corner of the eye.

Only the tips of the fangs are seen, and they glisten bright, smooth, and translucent, as if they were curved needles made from isinglass, and almost as fine as a bee's sting. On raising them with a needle or the point of the forceps, a large mass of muscular tissue comes into view, enveloping the tooth for the greater part of its length, and being, in fact, the means by which the fang is elevated or depressed. When the creature draws back its head and opens its mouth to strike, the depressing muscles are relaxed, the opposite series are contracted, and the two deadly fangs spring up with their points ready for action. It is needful while dissecting the head to be exceedingly careful, as the fangs are so sharp that they penetrate the skin with a very slight touch, and their poisonous distilment does not lose its potency even after the lapse of time.

There are generally several of the fangs in each jaw, lying one below the other in regular succession. From the specimen which has just been described, I removed four teeth on each side, varying in length from half to one-eighth the dimensions of the poison-fangs.

The ordinary food of the Viper is much the same as that of the common snake, and consists of mice, birds, frogs, and similar creatures. It is, however, less partial to frogs than the common snake, and seems to prefer the smaller mammalia to any other prey.

WE now arrive at a very important family of serpents, including the largest species found in the order. These snakes are known by the popular title of BOAS, and scientifically as Boidæ, and are all remarkable, not only for their great size and curious mode of taking their prey, but for the partial development of their hinder limbs, which are externally visible as a pair of horny spurs, set one on each side of the base of the tail, and moderately well developed under the skin, consisting of several bones jointed together.

BOA CONSTRICTOR.—(*Boa Constrictor.*)

THE BOA CONSTRICTOR is a native of Southern and Tropical America, and is one of those serpents that were formerly held sacred and worshipped with divine honours. It attains a very large size, often exceeding twenty feet in length, and being said to reach thirty feet in some cases. It is worthy of mention, that before swallowing their prey the Boas do not cover it with saliva, as has been asserted. Indeed, the very narrow and slender forked tongue of the serpent is about the worst possible implement for such a purpose. A very large amount of this substance is certainly secreted by the reptile while in the act of swallowing, and is of great use in lubricating the prey so as to aid it in its passage down the throat and into the body, but it is only poured upon the victim during the act of swallowing, and is not prepared and applied beforehand.

The dilating powers of the Boa are wonderful. The skin stretches to a degree which seems absolutely impossible, and the comparison between the diameter of the prey and that of the mouth through which it has to pass, and the throat down which it has to glide, is almost ludicrous in its apparent impracticability, and, unless proved by frequent experience, would seem more like the prelude to a juggler's trick than an event of every-day occurrence. To such an extent is the body dilatable that the shape of the animal swallowed can often be traced through the skin, and the very fur is visible through the translucent eyes, as the dead victim passes through the jaws and down the throat.

AN equally celebrated snake, the ANACONDO, is a native of Tropical America, where it is known under several names, La Culebra de Agua, or "water serpent," and El Traga Venado, or "deer swallower," being the most familiar.

Sir R. Ker Porter has some curious remarks on the Anaconda :—" This serpent is not venomous, nor known to injure men (at least not in this part of the New World) ; however, the natives stand in great fear of it, never bathing in waters where it is known to exist. Its common haunt, or rather domicile, is invariably near lakes, swamps, and rivers ; likewise close to wet ravines produced by inundations of the periodical rains ; hence, from its aquatic habits, its first appellation (*i.e.*, Water Serpent). Fish, and those animals which repair there to drink, are the objects of its prey. The creature lurks watchfully under cover of the water, and while the unsuspecting animal is drinking, suddenly makes a dart at the nose, and, with a grip of its back-reclining double range of teeth, never fails to secure the terrified beast beyond the power of escape."

Compression is the only method employed by the Anacondo for killing its prey, and the pestilent breath which has been attributed to this reptile is wholly fabulous. Indeed, it is doubtful whether any snake whatever possesses a fetid breath, and Mr. Waterton, who has handled snakes, both poisonous and inoffensive, as much as most living persons, utterly denies the existence of any perceptible odour in the snake's breath. It is very possible that the pestilent and most horrible odour which can be emitted by many snakes, when they are irritated, may have been mistaken for the scent of the breath. This evil odour, however, is produced from a substance secreted in the glands near the tail, and has no connection with the breath.

WE now come to another section of the serpents, termed the COLUBRINÆ, the members of which are known by the broad band-like plates of the abdomen, the shielded head, the conical tail, and the teeth of both jaws. Some of them are harmless and unfurnished with fangs, whereas some are extremely venomous and are furnished with poison-fangs in the upper jaw. These, however, do not fold down like those of the viper and rattlesnake, but remain perfectly erect.

OUR common GRASS SNAKE, or RINGED SNAKE, is a good example of these reptiles.

It is extremely plentiful throughout England, being found in almost every wood, copse, or hedgerow, where it may be seen during the warm months of the year sunning itself on the banks, or gently gliding along in search of prey, always, however, betraying itself to the initiated ear by a peculiar rustling among the herbage. Sometimes it may be detected while in the act of creeping up a perpendicular trunk or stem, a feat which it accomplishes, not by a spiral movement, as is generally represented by artists, but by pressing itself firmly against the object, so as to render its body flatter and wider, and crawling up by the movement of the large banded scales of the belly, the body being straight and rigid as a stick, and ascending in a manner that seems almost inexplicable.

The Ringed Snake is perfectly harmless, having no venomous fangs, and all its teeth being of so small a size that, even if the creature were to snap at the hand, the skin would not be injured.

The food of the Ringed Snake consists mostly of insects and reptiles, frogs being the favourite prey. I have known snakes to eat the common newt, and in such cases the victim was invariably swallowed head first, whereas the frog is eaten in just the opposite direction. Usually the frog, when pursued by the serpent, seems to lose all its energy, and instead of jumping away, as it would do if chased by a human being, crawls slowly like a toad, dragging itself painfully along as if paralysed. The snake, on coming up with its prey, stretches out its neck and quietly grasps one hind foot of the frog, which thenceforward delivers itself up to its destroyer an unresisting victim.

The whole process of swallowing a frog is very curious, as the creature is greatly wider than the mouth of the snake, and in many cases, when the frog is very large and the snake rather small, the neck of the serpent is hardly as wide as a single hind-leg of the frog, while the body is so utterly disproportioned that its reception seems wholly impossible. Moreover, the snake generally swallows one leg first, the other leg kicking freely in the air. However, the serpent contrives to catch either the knee or the foot in its mouth during these convulsive struggles, and by slow degrees swallows both legs. The limbs seems to act as a kind of wedge, making the body follow easily, and in half an hour or so the frog has disappeared from sight, but its exact position in the body of the snake is accurately defined by the swollen abdomen. Should the frog be small, it is snapped up by the side and swallowed without more ado.

The Ringed Snake is fond of water, and is a good swimmer, sometimes diving with great ease and remaining below the surface for a considerable length of time, and sometimes swimming boldly for a distance that seems very great for a terrestrial creature to

RINGED SNAKE, OR GRASS SNAKE.—
(*Tropidonotus natrix.*)

undertake. This reptile will even take to the sea, and has been noticed swimming between Wales and Anglesea.

During winter the snake retires to some sheltered spot, where it remains until the warm days of spring call it again to action. The localities which it chooses for its winter quarters are always in some well-sheltered spot, generally under the gnarled roots of ancient trees, under heaps of dry brushwood, or deep crevices. In these places the snakes will congregate in great numbers, more than a hundred having been taken from one hollow. A few years ago I saw a hole from which a great number of Ringed Snakes had been taken ; it was situated in a bank at some depth. The colour of the Ringed Snake is greyish green above and blue-black below, often mottled with deep black. Behind the head is a collar of golden yellow, often broken in the middle so as to look like two patches of yellow. Behind the yellow collar is another of black, sometimes broken in the middle also. Along the back run two rows of small dark spots, and a row of large oblong spots is arranged down each side. Both the colour and the shape of the spots are very variable.

ONE group of snakes is composed of the TREE SERPENTS, or Dendro-phidæ, so called from the habit of residing among the branches of trees.

Our first example of this family is the well-known BOOMSLANGE of South-ern Africa. In pronouncing this word, which is of Dutch or German origin, and signifies Tree Snake, the reader must remember that it is a word of

BOOMSLANGE. — (*Bucephalus Capensis.*)

three syllables. The Boomslange is a native of Southern Africa, and is among the most variable of serpents in colouring, being green, olive, or brown, of such different colours, that it has often been separated into several distinct species.

Dr. A. Smith has given the following valuable description of the Boom-slange and its habits;—

" The natives of South Africa regard the Boomslange as poisonous. but in their opinion we cannot concur, as we have not been able to discover the existence of any gland manifestly organized for the secretion of poison. The fangs are inclosed in a soft pulpy sheath, the inner surface of which is commonly coated with a thin glairy secretion. This secretion possibly may have something acrid and irritating in its quality, which may, when it enters a wound, occasion pain and swelling, but nothing of great importance.

" The Boomslange is generally found on trees, to which it resorts for the purpose of catching birds, upon which it delights to feed. The presence of a specimen in a tree is generally soon discovered by the birds of the neighbourhood, who collect around it and fly to and fro, uttering the most piercing cries, until some one, more terror-struck than the rest, actually scans its lips, and almost without resistance becomes a meal for its enemy. During such a proceeding the snake is generally observed with its head raised about ten or twelve inches above the branch round which its body and tail are entwined, with its mouth open and its neck inflated, as if anxiously endeavouring to increase the terror which it would almost appear it was aware would sooner or later bring within its grasp some one of the feathered group.

" Whatever may be said in ridicule of fascination, it is nevertheless true that birds, and even quadrupeds also, are, under certain circumstances, unable to retire from the presence of certain of their enemies ; and, what is even more extraordinary, unable to resist the propensity to advance from a situation of actual safety into one of the most imminent danger. This I have often seen exemplified in the case of birds and snakes ; and I have heard of instances equally curious, in which antelopes and other quadrupeds have been so bewildered by the sudden appearance of crocodiles, and by the grimaces and contortions they practised, as to be unable to fly or even move from the spot towards which they were approaching to seize them."

WE now come to one of the most deadly of the serpent tribe, the well-known COBRA DI CAPELLO, or HOODED COBRA, of India.

This celebrated serpent has long been famous, not only for the deadly power of its venom, but for the singular performances in which it takes part. The Cobra inhabits many parts of Asia, and in almost every place where it is found certain daring men take upon themselves the profession of serpent charmers, and handle these fearful reptiles with impunity, cause them to move in time to certain musical sounds, and assert that they bear a life charmed against the bite of their reptilian playmates. One of these men will take a Cobra in his bare hands, toss it about with perfect nonchalance, allow it to twine about his naked breast, tie it round his neck, and treat it with as little ceremony as if it were an earthworm. He will then take the same serpent—or apparently the same—make it bite a fowl, which soon dies from the poison, and will then renew his performances.

Some persons say that the whole affair is but an exhibition of that jugglery in which the Indians are such wondrous adepts ; that the serpents with which the man plays are harmless having been deprived of their fangs, and that a really venomous specimen is adroitly substituted for the purpose of killing the fowl. It is moreover said, and truly, that a snake, thought to have been rendered innocuous by the deprivation of its fangs, has bitten one of its masters and killed him, thus proving the imposture.

Still, neither of these explanations will entirely disprove the mastery of man over a venomous serpent. In the first instance, it is surely as perilous an action to substitute a venomous serpent as to play with it. Where was it hidden, why did it not bite the man instead of the fowl, and how did the juggler prevent it from using its teeth while he was conveying it away ? And, in the second instance, the detection of an impostor is by no means a proof that all who pretend to the same powers are likewise impostors. The following narrative of Mr. H. E. Reyne, quoted by Sir J. E. Tennent in his " Natural History of Ceylon," seems to be a sufficient proof that the man did possess sufficient power to induce a truly poisonous serpent to leave its hole and to perform certain antics at his command:—" A snake-charmer

COBRA DI CAPELLO.—(*Naja tripudians.*)

came to my bungalow in 1854, requesting me to allow him to show me his snakes dancing. As I had frequently seen them, I told him I would give him a rupee if he would accompany me to the jungle and catch a Cobra that I knew frequented the place.

"He was willing, and as I was anxious to test the truth of the charm, I counted his tame snakes, and put a watch over them until I returned with him. Before going, I examined the man, and satisfied myself he had no snake about his person. When we arrived at the spot, he played upon a small pipe, and after persevering for some time, out came a large Cobra from an ant-hill which I knew it occupied. On seeing the man it tried to escape, but he caught it by the tail and kept swinging it round until we reached the bungalow. He then made it dance, but before long it bit him above the knee. He immediately bandaged the leg above the bite, and applied a snake-stone to the wound to extract the poison. He was in great pain for a few minutes, but after that it gradually went away, the stone falling off just before he was relieved.

" When he recovered, he held up a cloth, at which the snake flew, and caught its fangs in it. While in that position, the man passed his hand up its back, and, having seized it by the throat, he extracted the fangs in my presence and gave them to me. He then squeezed out the poison on to a leaf. It was a clear oily substance, and, when rubbed on the hand, produced a fine lather. I carefully watched the whole operation, which was also witnessed by my clerk and two or three other persons."

One notable peculiarity in the Cobra is the expansion of the neck, popularly called the hood. This phenomenon is attributable not only to the skin and muscles, but to the skeleton. About twenty pairs of the ribs of the neck and fore-part of the back are flat instead of curved, and increase gradually from the head to the eleventh or twelfth pair, from which they decrease until they are merged into the ordinary curved ribs of the body.

When the snake is excited, it brings these ribs forward so as to spread the skin, and then displays the oval hood to the best advantage. In this species the back of the hood is ornamented with two large eye-like spots, united by a curved black stripe, so formed that the whole mark bears a singular resemblance to a pair of spectacles.

It is rather curious that many persons fancy that the Cobra loses a joint of its tail every time that it sheds its poison, this belief being exactly opposite to the popular notion that the rattlesnake gains a new joint to its rattle for every being which it has killed.

THE BATRACHIANS are separated from the true reptiles on account of their peculiar development, which gives them a strong likeness to the fishes, and affords a good ground for considering these animals to form a distinct order. On their extrusion from the egg they bear no resemblance to their parents, but are in a kind of intermediate existence, closely analogous to the caterpillar or larval state of insects, and called by the same name. Like the fish, they exist wholly in the water, and breathe through gills instead of lungs, obtaining the needful oxygen from the water which washes the delicate gill-membranes. At this early period they have no external limbs, moving by the rapid vibration of the flat and fan-like tail with which they are supplied. While in this state they are popularly called tadpoles, those of the frog sometimes bearing the provincial name of pollywogs. The skin of the Batrachians is not scaly, and in most instances is smooth and soft. Further peculiarities will be mentioned in connection with the different species.

These creatures fall naturally into two sub-orders, the Leaping or Tail-less Batrachians, and the Crawling Batrachians. The Leaping Batrachians, comprising the Frogs and Toads, are familiar in almost all lands, and in England are well known on account of their British representatives.

The most familiar of all the Batrachians, is the COMMON FROG of Europe. The general form and appearance of this creature are too well known to need much description. It is found plentifully in all parts of England, wandering to considerable distances from water, and sometimes getting into pits, cellars, and similar localities, where it lives for years without ever seeing water. The food of the adult Frog is wholly of an animal character, and consists of slugs, possibly worms, and insects of nearly every kind, the wire-worm being a favourite article of diet. A little colony of Frogs is most useful in a garden, as they will do more to keep down the various insect vermin that injure the garden than can be achieved by the constant labour of a human being.

The chief interest of the Frog lies in the curious changes which it under-goes before it attains its perfect condition. Every-one is familiar with the huge masses of transpa-rent jelly-like substance, profusely and regularly dotted with black spots, which lie in the shallows of a river or the ordinary ditches that intersect the fields. Each of these little black spots is the egg of a Frog, and is surrounded with a glo-bular gelatinous envelope about a quarter of an inch in diameter.

In process of time, certain various changes take place in the egg, and at the proper period the form of the young Frog begins to become appa-rent. In this state it is

COMMON FROG.—(*Rana temporaria.*)

a black grub-like creature, with a large head and a flattened tail. By degrees it gains strength, and at last fairly breaks its way through the egg and is launched upon a world of dangers, under the various names of tadpole, pollywog, toe-biter, or horsenail.

As it is intended for the present to lead an aquatic life, its breathing apparatus is formed on the same principle as the gills of a fish, but is visible externally, and when fully developed con-sists of a double tuft of finger-like appendages on each side of the head. The tadpole, with the fully-developed branchiæ, is shown at Fig. *a* on the accompanying illustration. No sooner, how-ever, have these organs attained their size than they begin again to diminish, the shape of the body and head being at the same time much altered, as is seen in Fig. *b*. In a short time they entirely disappear, being drawn into the cavity of the chest and guarded externally by a kind of gill-cover.

TADPOLES.

Other changes are taking place meanwhile. Just behind the head two little projections appear through the skin, which soon develop into legs, which, however, are not at all employed for progression, as the tadpole wriggles its way

through the water with that quick undulation of the flat tail which is so familiar to us all. The creature then bears the appearance represented in Fig. *c*. Presently another pair of legs make their appearance in front, the tail is gradually absorbed into the body—not falling off, according to the popular belief—the branchiæ vanish, and the lungs are developed. Fig. *d* represents a young Frog just before the tail is fully absorbed.

The internal changes are as marvellous as the external. When first hatched, the young tadpole is to all intents and purposes a fish, has fish-like bones, fish-like gills, and a heart composed of only two chambers, one auricle and one ventricle. But, in proportion to its age, these organs receive corresponding modifications, a third chamber for the heart being formed by the expansion of one of the large arteries, the vessels of the branchiæ

TOAD.—(*Bufo vulgaris.*)

becoming gradually suppressed and their place supplied by beautifully cellular lungs, formed by a development of certain membranous sacs that appear to be analogous to the air-bladders of the fishes.

THE celebrated EDIBLE FROG, or GREEN FROG of Europe (*Rana esculenta*), also belongs to this large genus. This handsome species is common in all the warmer parts of the Continent, but in the vicinity of large cities is seldom seen, except in the ponds where it is preserved, and whence issues a horrid nocturnal concert in the breeding-time. The proprietors of these froggeries supply the market regularly, and draw out the Frogs with large wooden rakes as they are wanted. In Paris these creatures are sold at a rather high price for the table, and as only the hind-legs are eaten, a dish of Frogs is rather an expensive article of diet.

WE now arrive at another section of Batrachians, including those creatures which are known under the title of Toads, and of which the COMMON TOAD of Europe is so familiar an example. The members of this section may be known by the absence of teeth in the jaws and the well-developed ears.

The general aspect and habits of this creature are too well known to require more than a cursory notice. Few creatures, perhaps, have been more reviled and maligned than the Toad, and none with less reason. In the olden days, the Toad was held to be the very compendium of poison, and to have so deadly an effect upon human beings, that two persons were related to have died from eating a leaf of a sage-bush under which a Toad had burrowed.

In France this poor creature is shamefully persecuted, the idea of its venomous and spiteful nature being widely disseminated and deeply rooted. The popular notion is that the Toad is poisonous throughout its life, but that after the age of fifty years it acquires venomous fangs like those of the serpents.

In point of fact, the Toad is a most useful animal, devouring all kinds of insect vermin, and making its rounds by night when the slugs, caterpillars,

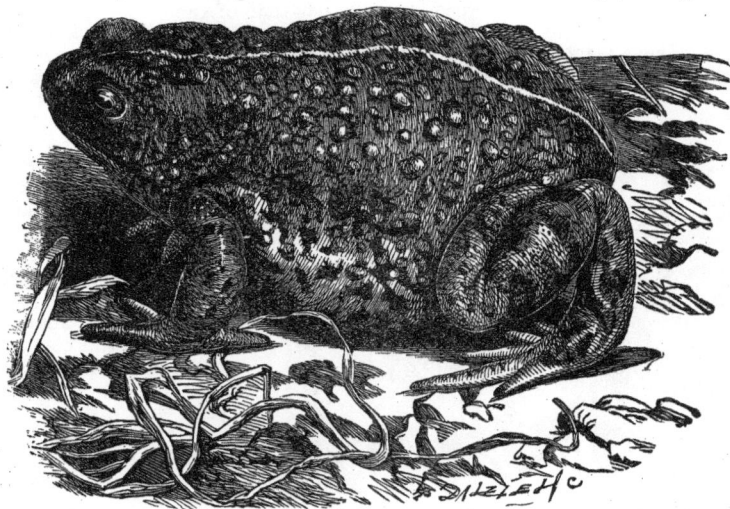

NATTERJACK.—(*Bufo calamita.*)

earwigs, and other creatures are abroad on their destructive mission. Many of the market gardeners are so well aware of the extreme value of the Toad's services, that they purchase Toads at a certain sum per dozen, and turn them out in their grounds.

Last year, my children had several large Toads which were quite tame. They used to carry the Toads in their hands round the garden, and then hold them up to flowers on which insects had settled. The Toads were quite accustomed to this mode of feeding, and always caught the insects.

Entomologists sometimes make a curious use of the Toad. Going into the fields soon after daybreak, they catch all the Toads they can find, kill them, and turn the contents of their stomachs into water. On examining the mass of insects that are found in the stomach, and which are floated apart in the water, there are almost always some specimens of valuable insects, generally beetles, which, from their nocturnal habits, small dimensions, and sober colouring, cannot readily be detected by human eyes.

The Toad will also eat worms, and in swallowing them it finds its forefeet of great use. The worm is seized by the middle, and writhes itself frantically into such contortions that the Toad would not be able to swallow it but by the aid of its fore-feet, which it uses as if they were hands. Sitting quietly down with the worm in its mouth, the Toad pushes it further between the jaws, first with one paw and then with another, until it succeeds by alternate gulps and pushes in forcing the worm fairly down its throat.

This animal is extremely tenacious of life, and is said to possess the power of retaining life for an unlimited period if shut up in a completely air-tight cell. Many accounts are in existence of Toads which have been discovered in blocks of stone when split open, and the inference has been drawn that they were enclosed in the stone while it was still in the liquid state, some hundreds of thousands of years ago, according to the particular geological period, and had remained without food or air until the stroke of the pick brought them once more to the light of day.

The development of the Toad is much like that of the Frog, except that the eggs are not laid in masses, but in long strings, containing a double series of eggs placed alternately. These chains are about three or four feet in length, and one-eighth of an inch in diameter. They are deposited rather later than those of the Frog, and the reptiles, which are smaller and blacker than the Frog larvæ, do not assume their perfect form until August or September.

ANOTHER species of Toad, the NATTERJACK, is found in many parts of England. It may be known from the common species by the short hind-legs, the more prominent eyes, the less webbed feet, the yellow line along the middle of the back, and the black bands on the legs. It is not so aquatic as the common Toad, haunting dry places, and seldom approaching water except during the breeding season. Its ordinary length is about three inches.

This is really a pretty creature, its colour being green, diversified with a line of bright yellow along the back. I kept several of these Toads for a long time, feeding them with various insects. No matter how large or active the insect might be which was put into the box, sooner or later the Natterjacks were sure to catch it.

GREEN TREE-FROG.—(*Hyla arborea.*)

WE now come to the TREE-FROGS, or TREE-TOADS, so called from their habits of climbing trees and attaching themselves to the branches or leaves by means of certain discs on the toes, like those of the geckos.

The best known species is the common GREEN TREE-FROG of Europe, now so familiar from its frequent introduction into fern cases and terrestrial vivaria.

This pretty creature is mostly found upon trees, clinging either to their branches or leaves, and being generally in the habit of attaching itself to the under side of the leaves, which it resembles so strongly in colour, that it is almost invisible even when its situation is pointed out. When kept in a fern-case, it is fond of ascending the perpendicular glass sides, and there sticking firmly and motionless, its legs drawn closely to the body, and its abdomen flattened against the glass.

The food of the Tree-Frog consists almost entirely of insects, worms, and similar creatures, which are captured as they pass near the leaf whereto their green foe is adhering. It is seldom seen on the ground except during the breeding season, when it seeks the water, and there deposits its eggs much in the same manner as the common Frog. The tadpole is hatched rather late in the season, and does not attain its perfect form until two full months have elapsed. Like the Toad, the Tree-Frog swallows its skin after the change. The common Tree-Frog is wonderfully tenacious of life, suffering the severest wounds without seeming to be much distressed, and having even been frozen quite stiff in a mass of ice without perishing.

The colour of this species is green above, sometimes spotted with olive, and a greyish yellow streak runs through each eye towards the sides, where it becomes gradually fainter, and is at last lost in the green colour of the skin. In some specimens there is a greyish spot on the loins. Below, it is of a

POUCHED FROG.—(*Nototrema marsupiatum.*)

paler hue, and a black streak runs along the side, dividing the vivid green of the back from the white hue of the abdomen.

IN the POUCHED FROG we find a most singular example of structure, the female being furnished with a pouch on her back, in which the eggs are placed when hatched, and carried about for a considerable period.

This pouch is clearly analogous to the living cradle of the marsupial animals. It is not merely developed when wanted, but is permanent, and lined with skin like that of the back. The pouch does not attain its full development until the creature is of mature age, and the male does not possess it at all. When filled with eggs, the pouch is much dilated, and extends over the whole back nearly as far as the back of the head. The opening is not easily seen without careful examination, being very narrow, and hidden in folds of the skin.

THE RHINOPHRYNE is remarkable as being the only known example among the Frogs where the tongue has its free end pointing forward, instead of being directed towards the throat.

This curious species inhabits Mexico, and can easily be recognized by the peculiar form of its head, which is rounded, merged into the body, and has the muzzle abruptly truncated, so as to form a small circular disc in front. The gape is extremely small, and the head would, if separated, be hardly recognizable as having belonged to a Frog. The legs are very short and thick, and the feet are half-webbed. Each hind-foot is furnished with a flat,

RHINOPHRYNE.—(*Rhinophrynus dorsalis.*)

oval, horny spur formed by the development of one of the bones. There are no teeth in the jaws, and the ear is imperfect. The colour of the Rhinophryne is slate-grey, with yellow spots on the sides and a row of similar spots along the back. Sometimes these latter spots unite so as to form a jagged line down the back.

WE now arrive at the Crawling Batrachians, technically called Amphibia Gradientia. All these creatures have a much-elongated body, a tail which is never thrown off as in the frogs and toads, and limbs nearly equal in development, but never very powerful. Like the preceding sub-order, the young are hatched from eggs, pass through the preliminary or tadpole state, and, except in very few instances, the gills are lost when the animal attains its perfect form. Both jaws are furnished with teeth, and the palate is toothed in some species. The skin is without scales, and either smooth or covered with wart-like excrescences. There is no true breast-bone, but some species have ribs.

THE celebrated SALAMANDER, the subject of so many strange fables, is a species found in many parts of the continent of Europe.

This creature was formerly thought to be able to withstand the action of fire, and to quench even the most glowing furnace with its icy body. It is singular how such ideas should have been so long promulgated, for although Aristotle repeated the tale on hearsay, Pliny tried the experiment by putting a Salamander into the fire, and remarks, with evident surprise, that it was burned to a powder. A piece of cloth dipped in the blood of a Salamander was said to be unhurt by fire, and certain persons had in their

SALAMANDER.—(*Salamandra maculosa.*)

possession a fireproof fabric, made, as they stated, of Salamander's wool, but which proved to be asbestos.

The Salamander is a terrestrial species, only frequenting the water for the purpose of depositing its young, which leave the egg before they enter into independent existence. It is a slow and timid animal, generally hiding itself in some convenient crevice during the day, and seldom venturing out except at night or in rainy weather. It feeds on slugs, insects, and similar creatures. During the cold months it retires into winter quarters, generally the hollow of some decaying tree, or beneath mossy stones, and does not reappear until the spring.

The ground colour of this species is black, and the

COMMON NEWT.—(*Triton cristatus.*)

spots are light yellow. Along the sides are scattered numerous small tubercles.

THE common NEWT, ASKER, EFFET, EFT, or EVAT, as it is indifferently termed, is well known throughout England. At least two species of Newt

G G

inhabit England, and some authors consider that the number of species is still greater. We shall, however, according to the system employed in this work, follow the arrangement of the British Museum, which accepts only two species, the others being merely noted as varieties.

The CRESTED NEWT derives its popular name from the membranous crest which appears on the back and upper edge of the tail during the breeding season, and which adds so much to the beauty of the adult male.

This creature is found plentifully in ponds and ditches during the warm months of the year, and may be captured without difficulty. It is tolerably hardy in confinement, being easily reared even from a very tender age, so that its habits can be carefully noted.

At Oxford we had some of these animals in a large slate tank through which water was constantly running, and which was paved with pebbles, and furnished with vallisneria and other aquatic plants, for the purpose of imitating as nearly as possible the natural condition of the water from which the creatures had been taken. Here they lived for some time, and here the eggs were hatched and the young developed.

It was a very curious sight to watch the clever manner in which the female Newts secured their eggs; for which purpose they used chiefly to employ the vallisneria, its long slender blades being exactly the leaves best suited for that purpose. They deposited an egg on one of the leaves, and then, by dexterous management of the feet, twisted the leaf round the egg, so as to conceal it, and contrived to fasten it so firmly that the twist always retained its form. The apparent shape of the egg is oval and semi-transparent, but on looking

SMOOTH NEWT.—(*Lophinus punctatus.*)

more closely it is seen to be nearly spherical, of a very pale yellow-brown, and enclosed within an oval envelope of gelatinous substance.

When the young Newt is hatched it much resembles the common tadpole, but is of a lighter colour, and its gills are more developed. It rapidly increases in size until it attains a length of nearly two inches, the fore-legs being then tolerably strong, and the hinder pair very small and weak.

Towards the breeding season the male changes sensibly in appearance; his colours are brighter, and his movements more brisk. The beautiful waving crest now begins to show itself, and grows with great rapidity, until it assumes an appearance not unlike that of a very thin cock's-comb, extending from the head to the insertion of the hinder limbs, and being deeply toothed at the edge. The tail is also furnished with a crest, but with smooth edges. When the animal leaves the water, this crest is hardly visible, because it is so delicate that it folds upon the body and is confounded with the skin; but when supported by the water, it waves with every movement of its owner, and has a most graceful aspect.

After the breeding season, the crest diminishes as rapidly as it arose, and in a short time is almost wholly absorbed. Some remnants of it, however, always remain, so that the male may be known even in winter by the line of irregular excrescences along the back. The use of this crest is not known, but it evidently bears a close analogy to the gorgeous nuptial plumage of many birds, which at other times are dressed in quite sober garments.

The Newt feeds upon small worms, insects, and similar creatures, and may be captured by the simple process of tying a worm on a thread by the middle, so as to allow both ends to hang down, and then angling as if for fish. The Newt is a ravenous creature, and when it catches a worm, closes its mouth so firmly that it may be neatly landed before it loosens its hold. Some writers recommend a hook, but I can assert, from much practical experience, that the hook is quite needless, and that the Newt may be captured by the simple worm and thread, not even a rod being required.

It is curious to see the Newt eat a worm. It seizes it by the middle with a sudden snap, as if the jaws were moved by springs, and remains quiet for a few seconds, when it makes another snap, which causes the worm to pass farther into its mouth. Six or seven such bites are usually required before the worm finally disappears.

The skin or epidermis of the Newt is very delicate, and is frequently changed, coming off in the water in flakes. I found that my own specimens always changed their skin as often as I changed the water, and it was very curious to see them swimming about with the flakes of transparent membrane clinging to their sides. The skin of the paws is drawn off just like a glove, every finger being perfect, and even the little wrinkles in the palms being marked. These gloves look very pretty as they float in the water, but if removed they collapse into a shapeless lump.

PROTEUS.—(*Proteus anguinus.*)

The food of the Newt consists of worms, insects, and even the young of aquatic reptiles. I have seen a large male Crested Newt make a savage dart at a younger individual of the same species, but it did not succeed in eating the intended victim.

THE next order of Crawling Batrachians is called by the name of MEANTIA, and contains a very few but very remarkable species. In all these creatures the body is long and smooth, without scales, and the gills are very conspicuous, retaining their position throughout the life of the animal. There are always two or four limbs furnished with toes, but these members are very weak, and indeed rudimentary, and both the palate and the lower jaw are toothed.

The first example of this order is the celebrated PROTEUS, discovered by the Baron de Zois, in the extraordinary locality in which it dwells.

At Adelsberg, in the Duchy of Carniola, is a most wonderful cavern, called the Grotto of the Maddalena, extending many hundred feet below the surface of the earth, and consequently buried in the profoundest darkness. In this cavern exists a little lake, roofed with stalactites, surrounded with masses of rock, and floored with a bed of soft mud, upon which the Proteus may be seen crawling uneasily, as if endeavouring to avoid the unwelcome light by which its presence is known. These creatures are not always to be found in the lake, though after heavy rains they are tolerably abundant, and the road by which they gain admission is at present a mystery.

The theory of Sir H. Davy is, "that their natural residence is a deep subterraneous lake, from which in great floods they are sometimes forced through the crevices of the rocks into the places where they are found ; and it does not appear to me impossible, when the peculiar nature of the country is considered, that the same great cavity may furnish the individuals which have been found at Adelsberg and at Sittich."

Whatever may be the solution of the problem, the discovery of this animal is extremely valuable, not only as an aid to the science of comparative anatomy, but as affording another instance of the strange and wondrous forms of animal life which still survive in hidden and unsuspected nooks of the earth.

Many of these animals have been brought in a living state to this country, and have survived for a considerable time when their owners have taken pains to accommodate their condition as nearly as possible to that of their native waters. I have had many opportunities of seeing some fine specimens, brought by Dr. Lionel Beale from the cave at Adelsberg. They could hardly be said to have any habits, and their only custom seemed to be the systematic avoidance of light.

The gills of the Proteus are very apparent, and of a reddish colour, on account of the blood that circulates through them. I have often witnessed this phenomenon by means of the ingenious arrangement invented by Dr. Beale, by which the creature was held firmly in its place while a stream of water was kept constantly flowing through the tube in which it was confined. The blood-discs of this animal are of extraordinary size; so large, indeed, that they can be distinguished with a common pocket magnifier, even while passing through the vessels. Some of the blood-corpuscles of the specimen described above are now in my possession, and, together with those of the lepidosiren, form a singular contrast to the blood-corpuscles of man, the former exceeding the latter in dimensions as an ostrich egg exceeds that of a pigeon.

The colour of the Proteus is pale faded flesh tint, with a wash of grey. The eyes are quite useless, and are hidden beneath the skin, those organs being needless in the dark recesses where the Proteus lives. Its length is about a foot.

FISHES.

FISHES.

IN the FISHES, the last class of vertebrated animals, the chief and most obvious distinction lies in their adaptation to a sub-aqueous existence, and their unfitness for life upon dry land.

There are many vertebrate animals which pass the whole of their lives in the water, and would die if transferred to the land, such as the whales and the whole of the cetacean tribe, an account of which may be found in page 127. But these creatures are generally incapable of passing their life beneath the waters, as their lungs are formed like those of the mammalia, and they are forced to breathe atmospheric air at the surface of the waves. And though they would die if left upon land, their death would occur from inability to move about in search of food, and in almost every case a submersion of two continuous hours would drown the longest breathed whale that swims the seas.

The Fishes, on the contrary, are expressly formed for aquatic existence; and the beautiful respiratory organs, which we know by the popular term of "gills," are so constructed that they can supply sufficient oxygen for the aëration of the blood.

The reason that Fishes die when removed from the water is not because the air is poisonous to them, as some seem to fancy, but because the delicate gill-membranes become dry and collapse against each other, so that the circulation of the blood is stopped, and the oxygen of the atmosphere can no longer act upon it. It necessarily follows, that those Fishes whose gills can longest retain moisture will live longest on dry land, and that those whose gills dry most rapidly will die the soonest. The herring, for example, where the delicate membranes are not sufficiently guarded from the effects of heat and evaporation, dies almost immediately it is taken out of the water; whereas the carp, a fish whose gill-covers can retain much moisture, will survive for an astonishingly long time upon dry land; and the anabas, or climbing perch, is actually able to travel from one pool to another, ascending the banks, and even traversing hot and dusty roads.

The power by which the Fishes propel themselves through the water is almost entirely obtained by the lateral movement of the tail. The fins are scarcely employed at all in progression, but are usually used as balancers, and occasionally to check an onward movement.

Before proceeding further, I may mention that all the fins of a Fish are distinguished by appropriate names. As they are extremely important in determining the species and even the genus of the individual, and as these members will be repeatedly mentioned in the following pages, I will briefly describe them.

Beginning at the head and following the line of the back, we come upon a fin, called from its position the "dorsal" fin. In very many species there are two such fins, called, from their relative positions, the first and the second dorsal fins. The extremity of the body is furnished with another fin, popularly called the tail, but more correctly the "caudal" fin. The fins which are set on that part of the body which corresponds to the shoulders are termed the "pectoral" fins; that which is found on the under surface and in front of the vent is called the "abdominal" fin, and that which is also on the lower surface, and between the vent and tail, is known by the name of the "anal" fin. All these fins vary extremely in shape, size, and even in position.

The gill-cover, or operculum as it is technically called, is separated into four portions, and is so extensively used in determining the genus and species, that a brief description must be given. The front portion, which starts immediately below the eye, is called the " præoperculum," and immediately behind it comes the " operculum." Below the latter is another piece, termed from its position the " sub-operculum, " and the lowest piece, which touches all the three above it, is called the " inter-operculum." Below the chin and reaching to the sub-operculum, are the slender bones, termed the " branchi-ostegous rays," which differ in shape and number according to the kind of Fish.

The scales with which most of the Fish are covered are very beautiful in structure, and are formed by successive laminæ, increasing therefore in size according to the age of the Fish. They are attached to the skin by one edge, and they overlap each other in such a manner as to allow the creature to pass through the water with the least possible resistance. The precise mode of overlapping varies materially in different genera. Along each side of the Fish runs a series of pores, through which passes a mucous secretion formed in some glands beneath. In order to permit this secretion to reach the outer surface of the body, each scale upon the row which comes upon the pores is pierced with a little tubular aperture, which is very perceptible on the exterior, and constitutes the " lateral line." The shape and position of this line are also used in determining the precise position held by any species. In comparing the scales taken from different Fishes, it is always better to take those from the lateral line.

The heart of the Fish is very simple, consisting of two chambers only, one auricle and one ventricle. The blood is in consequence cold.

The hearing of Fishes appears in most cases to be dull, and some persons have asserted that they are totally destitute of this faculty. It is now, how-ever, known that many species have been proved capable of hearing sounds, and that carp and other fish can be taught to come for their food at the sound of a bell or whistle. The internal structure of the ear is moderately developed, and there are some curious little bones found within the cavity, technically called the otoliths.

THE fishes comprised in the first order are called by the rather harshly sounding title of Chondropterygii, a term derived from two Greek words, the former signifying " cartilage" and the latter " a fin," and given to these crea-tures because their bones contain a very large amount of cartilaginous sub-stance, and are consequently soft and flexible. The bones of the head are rather harder than those of the body and fins.

The first family, of which the common STURGEON is a good and familiar example, is at once known by the cartilaginous or bony shields with which the head and body are at intervals covered.

In this remarkable fish the mouth is placed well under the head, and in fact seems to be set almost in the throat, the long snout appearing to be entirely a superfluous ornament. The mouth projects downwards like a short and wide tube, much wider than long, and on looking into this tube no teeth are to be seen. Between the mouth and the extremity of the snout is a row of fleshy finger-like appendages, four in number, and apparently organs of touch.

One or two species of Sturgeon are important in commerce, as two valuable articles, namely, isinglass and caviare, are made from them. The former substance is too well known to need a description, and the mode of preparing it for use is briefly as follows. The air-bladder is removed from the fish, washed carefully in fresh water, and then hung up in the air for a day or two, so as to stiffen. The outer coat or membrane is then peeled off, and the remainder is cut up into strips of greater or lesser length, technically called staples, the long staples being the most valuable. This substance affords so large a quantity of gelatinous matter, that one part of isinglass dissolved in a hundred parts of boiling water will form a stiff jelly when cold.

Caviare is made from the roe of this fish, and as nearly three millions of eggs have been taken from a single fish, the amount of caviare that one Sturgeon can afford is rather large.

The body of the Sturgeon is elongated, and slightly five-sided from the head to the tail. Along the body run five rows of flattened bony plates, each plate being marked with slight grooves in a radiating fashion, and having a pointed and partly conical spine on each plate, the points being directed towards the tail. The plates along the summit of the back are the largest.

STURGEON.—(*Acipenser attilus.*)

THE fishes belonging to the next group have their gills fixed by their outer edge to the divisions in the gill-openings at the side of the neck. This group includes the Sharks and the Rays, many representatives of which creatures are found on the British coasts.

The first family of this large and important group is known by the name of Scyllidæ, and its members can be recognized by several distinguishing characteristics. They have spout-holes on the head, and the gill-openings are five in number on each side. Sometimes there only seem to be four openings, but on closer examination the fourth and fifth are found set closely together, the opening of the fifth appearing within that of the fourth. The teeth are sharp and pointed, and the tail is long, notched on the outer side, and is not furnished with a fin.

One of the commonest British species is the LITTLE DOG-FISH, called by several other names, as is usual with a familiar species that is found in many localities. Among such names are SMALL SPOTTED DOG-FISH, LESSER SPOTTED SHARK, MORGAY, and ROBIN HUSS.

This fish is plentiful on our coasts, especially in the southern extremity of England, and is often thought a great nuisance by fishermen, whose bait it takes instead of the more valuable fish for which the hook was set.

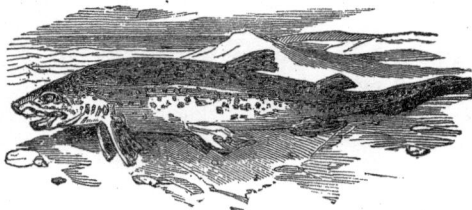

LITTLE DOG FISH.—(*Scyllium caniculum.*)

It generally remains near the bottom of the water, and is a voracious creature, feeding upon crustaceans and small fish. It often follows the shoals of migrating fish, and on account of that custom is called the Dog-fish. Generally its flesh is neglected, but when properly dressed it is by no means unpalatable, and is said to be sometimes trimmed and dressed in fraudulent imitation of more valuable fish.

The skin of this and other similar species is rough and file-like, and is employed for many purposes. The handles of swords, where a firm hold is required, are sometimes bound with this substance ; and joiners use it in polishing the surface of fine woods so as to bring out the grain. It is also employed instead of sand-paper upon match-boxes.

The egg of this species is very curious in form and structure, and is often found on the sea-shore, flung up by the waves, especially after a storm. These objects are familiar to all observant wanderers by the sea-shore, under the name of mermaid's purses, sailor's purses, or sea purses. Their form is oblong, with curved sides, and at each angle there is a long tendril-like appendage, having a strong curl, and in form not unlike the tendrils of the vine. The use of these appendages is to enable the egg to cling to the growing seaweed at the bottom of the ocean, and to prevent it from being washed away by the tide.

For the escape of the young shark, when strong enough to make its own way in the wider world of waters, an outlet is provided in the opened end of the envelope, which opens when pushed from within, and permits the little creature to make its way out, though it effectually bars the entrance against any external foe.

The head of the Little Dog-fish is rather flat upon the top, there is a little spiracle or blow-hole behind each eye, and the shape of the mouth is somewhat like a horse-shoe. The general colour of the body is pale reddish on the upper parts, covered with many little spots of dark reddish brown; below it is yellowish white. The length of this species is about eighteen inches. The colour is beautiful slate-blue above, and white below.

THE remarkable fish depicted in the illustration affords a striking instance of the wild and wondrous modifications of form assumed by certain creatures, without any ascertained purpose being gained thereby. We know by analogous reasoning that some wise purpose is served by this astonishing variation in form ; but as far as is yet known there is nothing in the habits of this species that accounts for the necessity of this strange shape.

The shape of the body is not unlike that of the generality of sharks, but it is upon the head that the attention is at once riveted. As may be seen from the figure the head is expanded laterally in a most singular manner, bearing, indeed, no small resemblance to the head of a hammer. The eyes are placed at either end of the projecting extremities, and the mouth is set quite below, its corners just coinciding with a line drawn through the two projecting lobes of the head.

This species attains to a considerable size, seven or eight feet being a common measurement, and specimens of eleven or twelve feet having been known. Its flesh is said to be almost uneatable, being hard, coarse, and ill flavoured. The HAMMER-HEADED SHARK produces living young, and from the

Egg of Dog-fish.

interior of a very fine specimen captured near Tenby in 1839, and measuring more than ten feet in length, were taken no less than thirty-nine young, all perfectly formed, and averaging nineteen inches in length. The general colour of this species is greyish brown above, and greyish white below.

THE dreaded WHITE SHARK, the finny pirate of the ocean, is happily almost a stranger to our shores, though a stray specimen may now and then visit the British Islands, there to find but scant hospitality.

This is one of the large species that range the ocean, and in some seas they are so numerous that they are the terror of sailors and natives. One individual, whose jaws are still preserved, was said to have measured thirty-seven feet in length; and when we take into consideration the many instances where the leg of a man has been bitten off through flesh and bone as easily as if

HAMMER-HEADED SHARK.—(*Sphyrnias zygæna.*)

it had been a carrot, and even the body of a boy or woman severed at a single bite, this great length will not seem to be exaggerated.

Many portions of this fish are used in commerce. The sailors are fond of cleaning and preparing the skull, which, when brought to England, is sure of a ready sale, either for a public museum, or to private individuals who are struck with its strange form and terrible armature. The spine, too, is frequently taken from this fish, and when dried it passes into the hands of walking-stick makers, who polish it neatly, fit it with a gold handle, and sell it at a very high price. One of these sticks will sometimes fetch six or seven pounds. There is also a large amount of oil in the shark, which is thought rather valuable, so that in Ceylon and other places a regular trade in this commodity is carried on.

The fins are very rich in gelatine, and in China are, as is said, employed largely in the

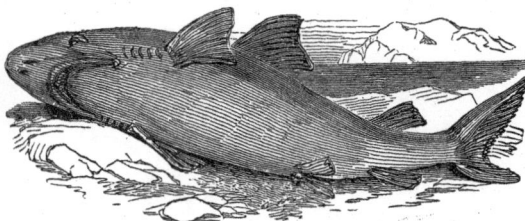

WHITE SHARK, OR LAMIA.—(*Carcharodon Rondeletii.*)

manufacture of that gelatinous soup in which the soul of a Chinese epicure delights, and of which the turtle soup of our metropolis is thought by Chinese judges to be a faint penumbra or distant imitation. The flesh is eaten by the natives of many Pacific islands ; and in some places the liver is looked upon as a royal luxury, being hung on boards in the sun until all the contained oil has drained away, and then carefully wrapped up in leaves and reserved as a delicacy.

The colour of the White Shark is ashen brown above, and white below.

WE now arrive at the RAYS. The first family of these fishes is evidently intermediate between the sharks and the skates, and is in many respects a very interesting and remarkable group of fishes. The common SAW-FISH, so well known from the singular development of the snout, is a good example of this family.

It has a very wide range of locality, being found in almost all the warmer seas, and even in the cold regions near the pole.

The snout of this fish is greatly prolonged, and flattened like a sword-blade. On either edge it bears a row of tooth-like projections, firmly imbedded in the bone, few, short, and wide apart at the base of the beak, but becoming larger and set closer together towards the point. The form of the sockets into which the teeth are received, and their rather enlarged termination, are conspicuously indicated on the surface of the saw-blade. The tip of the saw is covered with hard granular scales. The number of teeth is not the same in every individual; in a specimen in my possession

SAW-FISH.—(*Pristis antiquorum.*)

there are twenty-eight on each side of the saw.

It is said that, like the sword-fish, this creature will attack the whale, thrusting its armed beak into the soft blubber-covered body of the huge cetacean, and avoiding, by its superior agility, the strokes of the tortured animal's tail, any blow of which, if it succeeded in its aim, would crush the assailant to death. The Saw-fish does certainly use this weapon for the destruction of fish. Colonel Drayson has informed me that when lying becalmed off the Cape, he has more than once seen a Saw-fish come charging among a shoal of fishes, striking right and left with the serrated edges of the saw, and killing or disabling numbers of the fish by this process.

In all the Saw-fishes the skin is covered with minute rounded or hexagonal scales, arranged like the stones of a mosaic. The temporal orifices are very large, and are set some distance behind the eyes. The mouth is on the under surface of the head, and is furnished with a crushing apparatus, made exactly on the principle of the stone-crushing machines of the present day.

IN the true Rays, or Raidæ, the fore-part of the body is flattened and formed into a disc-like shape, by the conjunction of the breast-fins with the snout.

Our first example of the Rays is the TORPEDO, a fish long celebrated for its power of

EYED TORPEDO.—(*Torpedo oculata.*)

emitting at will electrical shocks of considerable intensity. In consequence of this property, it is sometimes called the CRAMP-FISH, CRAMP RAY, ELECTRIC RAY, or NUMB-FISH.

The object of this strange power seems to be twofold, namely, to defend itself from the attacks of foes, and to benumb the swift and active fish on which it feeds, and which its slow movements would not permit it to catch in fair chase. It does not always deliver the electric shock when touched, though it is generally rather prodigal of exercising its potent though invisible arms, but will allow itself to be touched, and even handled, without inflicting a shock. But if the creature be continually annoyed, the shock is sure to come at last, and in such cases with double violence. It has been observed, moreover, that the fish depresses its eyes just before giving its shock.

That the stroke of the Torpedo is veritable electricity is a fact which was once much disputed, but is now conclusively proved by a host of experiments. Needles have been magnetized by it just as if the shock had been that of a galvanic battery, the electrometer showed decided proofs of the nature of the fluid that had been sent through it, and even the electric spark has been obtained from the Torpedo—very small, it is true, but still recognizably apparent. It is rather curious, that in the course of the experiments it was discovered that the upper surface of the Torpedo corresponded with the copper plate of a battery, and the lower surface with the zinc plate.

The structure of the electrical organ is far too complex to be fully described in this work, as it would require at least forty or fifty pages, and a large number of illustrations. Any of my readers who would like to examine it in detail will find ample information in an article on the subject by Dr. Coldstream, in the "Cyclopædia of Anatomy and Physiology," and from a valuable series of wax models in the museum of the College of Surgeons.

This fish is found in the Mediterranean and the Indian and Pacific Oceans, and occasionally off the Cape, and has now and then been captured on our coasts. Happily, the Torpedo does not attain a very great size, one of the largest specimens being about four feet long, and weighing sixty or seventy pounds.

THE Rays are well represented in England by several large and curious species. One of the commonest examples is the THORNBACK SKATE, or RAY, so called from the large number of thorny projections which are scattered over its back, and especially along the spine.

The Thornback is one of our common Rays,

THORNBACK SKATE.—(*Raia clavata.*)

and is taken plentifully on the shores of England, Scotland, and Ireland. As is the case with many of the same genus, the flesh is considered rather good, and is eaten both when fresh and when salted for consumption during stormy weather. Autumn and winter are the best seasons for procuring this fish, as the flesh is then firm and white, while during the rest of the year it is rather liable to become flabby. Thornbacks taken in November are thought to be the best.

This species, like the rest of the Rays, feeds on crustacea, flat-fish, and molluscs ; and as many of these creatures possess very hard shells, the Rays are furnished with a crushing-mill of teeth, which roll on each other in such a way that even the stony shell of a crab is broken up under the pressure.

. The young of this and other Skates are produced from eggs, whose form is familiar to every visitor to the sea-shore, where they go by the popular name of Skate-barrows. Their colour is black, their texture leathery, thin, and tough, and their form wonderfully like a common hand-barrow, the body of the barrow being represented by the middle of the egg, and the handles by the four projections at the angles. The empty cases are continually thrown on the beach, but it is seldom that the young are found enclosed, except after a violent storm, or when obtained by means of the dredge.

This species is notable for certain thorny appendages to the skin, which are profusely sown over the back and whole upper surface, and among which stand out conspicuously a few very large tubercular spines, with broad, oval, bony bases, and curved, sharp-pointed projections. Fifteen or sixteen of these bony thorns are found on the back. Along the spine runs a single row of similar spines, and at the commencement of the tail it is accompanied by another row on either side, making that member a very formidable instrument of offence. In point of fact, the tail is as formidable a weapon as can be met with, and the manner in which this living quarter-staff is wielded adds in no slight degree to its power. When angered, the Skate bends its body into a bow-like form, so that the tail nearly touches the snout, and then, with a sudden fling, lashes out with the tail in the direction of the offender, never failing to inflict a most painful stroke if the blow should happen to take effect.

: The colour of the Thornback Skate is brown, diversified with many spots of brownish grey, and the under parts are pure white.

THE COMMON SKATE, sometimes called the TINKER, is so well known that only a very short description is needed.

This fish is found on all our coasts in great plenty, and sometimes attains to a really large size, a fine specimen having been known to weigh' two hundred pounds. The fishermen have a custom of calling the female Skate a Maid, and the male, in consequence of the two elongated appendages at the base of the tail, is called the Three-tailed Skate. It is a very voracious creature, eating various kinds of fish, crustaceans, and other inhabitants of the deep.

WE now arrive at the vast order of the Spine-finned Fishes, known scientifically as the ACANTHOPTERYGII. In all these fishes the skeleton is entirely bony, and part of the rays of the dorsal, anal, and ventral fins are formed into spines, in some species very short, and in others of extraordinary length.

The first family is well represented in England by many pretty and interesting species, of which the creature figured in the engraving is a familiar example.

THE THREE-SPINED STICKLEBACK is one of our commonest British fishes, and is known in different parts of England under the names of TITTLEBAT, PRICKLE-FISH, and SHARPLIN. It is a most bold and lively little fish, hardly knowing fear, pugnacious to an absurd degree, and remarkably interesting in its habits. Even more voracious than the perch, it renders great service to mankind by keeping within due bounds the many aquatic and terrestrial insects which, although performing their indispensable duties in the world, are so extremely prolific that they would render the country uninhabitable were they allowed to increase without some check.

Should the reader be disposed to place specimens in an aquarium, he must make up his mind that they will fight desperately at first, and until they have satisfactorily settled the championship of the tank their intercourse will be of the most aggressive character. Never were such creatures to fight as the

Sticklebacks, for they will even go out of their way to attack anything which they think may possibly offend them, and they have no more hesitation in charging at a human being than at one of their own species. I have known one of these belligerent fish make repeated dashes at my walking-stick, knocking his nose so hard against his inanimate antagonist that he inflicted a perceptible jar upon it, and, in spite of the blows which his nose must have suffered, returning to the combat time after time with undiminished spirit.

These combats are, however, most common about the breeding season, when every adult Stickleback challenges every other of his own sex, and they do little but fight from morning to evening. They are as jealous as they are courageous, and will not allow another fish to pass within a certain distance of their home without darting out and offering battle.

Anyone may see these spirited little combats by quietly watching the inhabitants of a clear streamlet on a summer day. The two antagonists dart at each other with spears in rest, snap at each other's gills or head, and retain their grasp with the tenacity of a bull-dog. They whirl round and round in the water, they drop, feint, attack, and retreat, with astonishing quickness, until one confesses itself beaten and makes off for shelter, the conqueror snapping at its tail and inflicting a parting bite.

Then is the time to see the triumphant little creature in all the glory of his radiant apparel ; for with his conquest he assumes the victor's crown ; his back glows with shining green, his sides and head are glorious with gold and scarlet, and his belly is silvery white. It is a little creature, certainly, but even among the brilliant inhabitants of the southern seas a more gorgeously-coloured fish can hardly be found. If the conqueror Stickleback could only be enlarged to the size of a full-grown perch or roach, it would excite the greatest admiration. It is curious, that the vanquished antagonist loses in brillance as much

THREE-SPINED STICKLEBACK.—(*Gasterosteus aculeatus.*)

as the conqueror has gained ; he sneaks off ignominiously after his defeat, and hides himself, dull and sombre, until the time comes when he too may conquer in fight, and proudly wear the gold and scarlet insignia of victory.

These struggles are not only for mastery, but are in so far praiseworthy that they are waged in defence of home and family.

The Stickleback is one of the very few fish who build houses for their young, as a defence against the many foes which are ever lying in wait for the destruction of the eggs or the newly-hatched young. These nests are built of various vegetable substances, fastened together with a kind of slime that exudes from the body of the male. The Fifteen-spined Stickleback, a marine species, also makes a nest, though hardly of so careful a construction.

The Three-spined Stickleback is very fond of inhabiting the mouths of rivers where they empty themselves into the sea, the brackish water appearing to suit its constitution. It can therefore be easily acclimatized to new conditions, and a specimen that has been taken from an inland stream can soon be brought to inhabit the water of a marine aquarium, though such water is usually, in consequence of evaporation, more salt than that of the sea.

WE now come to the large and important family of the PERCHES, which comprises many of the handsomest and most valuable fishes. The members of this family are found in all parts of the globe.

THE COMMON PERCH is well known as one of our handsomest river fish, and, on account of its boldness and the voracious manner in which it takes the bait, and the active strength with which it struggles against its captor, is a great favourite with many anglers. Moreover, when captured and placed in an aquarium, it very soon learns to distinguish the hand that feeds it, and will come to the surface and take food from the fingers. It has a fashion of seizing its food with a rather sharp jerk, and then snatches it away with such violence, that when it takes the hook it will drag a stout cork-float several inches below the surface, and by the force of its own stroke will mostly hook itself without any exertion on the part of the angler. Bold-biting, however, as is its reputation, there are some seasons of the year when it is almost impossible to catch a Perch, and even the shy and gently nibbling roach is an easier prey.

The Perch is a truly voracious fish, feeding upon all kinds of aquatic worms, insects, and fishes, preferring the latter diet as it becomes older and larger. The smaller fish, such as minnows, young roach, dace, and gudgeons, are terribly persecuted by the Perch, and a bait formed of either of these fish, or a good imitation of them, will generally allure the finest Perches to the hook. Although generally inhabiting mid or deep water, it will sometimes come to the surface to snap up a casual fly that has fallen into the water, and on several occasions has been captured by anglers when fishing with a fly for trout.

Practical fishermen say that the Perch is almost the only fish which the pike does not venture to attack, and that if a pike should make one of its rushing onslaughts on a Perch, the intended prey boldly faces the enemy, erects the dorsal fin with its array of formidable spines, and thus baffles the ever-hungry aggressor. Still it is an article of faith with some anglers, that a young Perch from which the dorsal fin has been removed is one of the surest baits for pike. Perhaps they think that the pike is so delighted to find a Perch unarmed, that it seizes the opportunity to feed upon a luxury which it can seldom obtain.

PERCH.—(*Perca fluviatilis.*)

The Perch is not a large fish, from two to three pounds being considered rather a heavy weight. Individuals, however, of much greater dimensions have been, though rarely, captured. One of the finest Perches ever taken in England was captured in the river Avon, in Wiltshire, by a night-line baited with a roach; its weight was eight pounds. Specimens of five or six pounds are occasionally taken, but are thought so valuable that the captor generally sends the account of his success to some journal.

The colour of the Perch is rich greenish-brown above, passing gradually into golden white below. Upon the sides is a row of dark transverse bands, generally from five to seven in number. The first dorsal fin is brown, with a little black between two or three of the first and last rays; the second dorsal and the pectoral are pale brown, and the tail and other fins are bright red.

PASSING by many large genera, which cannot be noticed for lack of space, we come to a very odd-looking fish, called perforce, for want of a popular title, the OREOSOMA, a name formed from two Greek words, and literally signifying hilly-bodied. As the reader may see by reference to the engraving, the name is very appropriate. The upper figure shows its aspect from below.

This remarkable little fish was captured in the Atlantic by Peron, and has ever been esteemed as one of the curiosities of the animal kingdom. Upon the body there are no true scales, but their place is supplied by a number of bony or horny protuberances, of a conical shape, and serving no ascertained purpose. These cones may be divided into two distinct sets, the larger set being arranged into two ranks, four on the back and ten on the abdomen, and among them are placed the smaller set. The body of this fish is very deep in proportion to its length, and the operculum has two ridges, terminating in flattened angles. There are two dorsal fins, the first armed with five spines.

WE now arrive at a large family, containing a series of fishes remarkable for their extraordinary shape, their bold and eccentric colouring, and their curious habits. In Dr. Günther's elaborate arrangement of the Acanthopterygiian fishes, this family is called by the name of Squa-

OREOSOMA.—(*Oreosoma Atlanticum.*)

mipennes, or scaly-finned fishes, because " the vertical fins are more or less densely covered with small scales;" the spinous portions sometimes not scaly. They are nearly all carnivorous fishes, and for the most part are exclusively inhabitants of the tropical seas or rivers. Their bodies are very much compressed and extremely deep in proportion to their length, and the mouth is usually small and placed in front of the snout.

The first group of this family, or sub-family as it might be called, is termed Chætodontina, from the large typical genus of the group. Their mouths are small, and furnished with several rows of very tiny, slender, and bristle-like teeth, a peculiarity of structure that has gained for them their scientific name Chætodontina, a term composed of two Greek words, the former signifying " hair," and the latter " a tooth." The colours of the species belonging to this group are brilliant in tint, and are generally arranged in bold stripes or spots. Black and yellow are the prevailing hues, but blue and green are found in some species.

The figure given on page 466 represents a most remarkable species, called, from the form of its mouth, the BEAKED CHÆTODON.

H H

The curiously elongated muzzle is employed by this fish in a rather unexpected manner, being used as a gun or bow, a drop of water taking the place of the arrow or bullet.   Perhaps the closest analogy is with the cele- brated, "sumpitan," or blow-gun, of the Macoushi Indians, a tube through which an arrow is driven by the force of the breath.   The Beaked Chætodon feeds largely on flies and other insects, but is not forced to depend, as is the case with nearly every other fish, on the accidental fall of its prey into the water. If it sees a fly or other insect resting on a twig or grass-blade that ovei- hangs the water, the Chætodon approaches very quietly, the greater part of its body submerged, and its nose just showing itself above the surface, the point directed towards the victim.   Suddenly, it shoots a drop of water at the fly,

BEAKED CHÆTODON.—(*Chelmo rostratus.*)

with such accuracy of aim that the unsuspecting insect is knocked off its perch, and is snapped up by the fish as soon as it touches the surface of the water.

This habit it continues even in captivity, and is in consequence in great estimation as a household pet by the Japanese.   They keep the fish in a large bowl of water, and amuse themselves by holding towards it a fly upon the end of a slender rod, and seeing the finny archer strike its prey into the water.   Another fish, called the "Archer," inhabiting the East Indian and Polynesian seas, possesses the same faculty, but is not so remarkable for its eccentric form and the bold beauty of its tints.

The Beaked Chætodon inhabits the Indian and Polynesian seas, and has been taken off the west coast of Australia, where it is usually found in or near the mouths of rivers.   Over the head and body of this species are drawn five brownish cross-bands edged with darker brown and white, and in the middle of the soft dorsal fin there is a rather large circular black spot edged with white.

THE large and important family of the Triglidæ, or GURNARDS, is represented by several British fishes. This family contains a great number of species, many of which are most remarkable, not only for their beautiful colours, which alone are sufficient to attract attention, but also for the strange and weird shape and large development of the fins. They are carnivorous fish, mostly inhabiting the seas, a very few species being able to exist in fresh water. They are not swift or strong swimmers, and therefore remain for the most part in deep water. Some, however, are able by means of their largely-developed pectoral fins to raise themselves into the air, and for a brief space to sustain themselves in the thinner element. The mouth is mostly large, and in some cases the gape is so wide and the head and jaws so strangely shaped, that the general aspect is most repulsive.

WE now come to a very familiar and not very prepossessing fish ; the well-

BULL-HEAD.—(*Cottus gobio.*)

known BULL-HEAD, or MILLER'S THUMB, sometimes called by the name of TOMMY LOGGE.

This large-headed and odd-looking fish is very common in our brooks and streams, where it is generally found under loose stones, and affords great sport to the juvenile fisherman. In my younger days, the chase of the Bull-head was rather an exciting one, and was carried out without hook or line, or indeed any aid but the hands. This fish has a habit of hiding itself under loose stones, and on account of its flat, though wide head, is enabled to push itself into crevices which are apparently much too small to contain it.

The name of Miller's Thumb is derived from the peculiarly wide and flattened head, which is thought to bear some resemblance to the object whence its name is taken. A miller judges of the quality of the meal by rubbing it with his thumb over his fingers as it is shot from the spout, and by the continual use of this custom the thumb becomes gradually widened and flattened at its extremity. The name of Bull-head also alludes to the same width and flatness of the skull.

The Bull-head is a voracious little fish, feeding on various water insects, worms, larvæ, and the young fry of other fish. It is a representative of a rather large genus, comprising about twenty-six or twenty-seven known

H H 2

species, which are spread over all the northern and temperate parts of the world. In Russia the Bull-head is believed by the general public to possess the same quality as is attributed to the kingfisher by our own rustic population, and to indicate the direction of the wind by always keeping its head turned to windward when it is dried and suspended horizontally by a thread.

The mouth of this little fish is very wide, and contains numerous minute teeth. There is one spine on the præoperculum, and the operculum ends in a flattened point. The general colour of the smooth skin is very dark brown on the back, white on the abdomen, and greyish white on the flanks. The rays of the fins are spotted with dark blackish brown and white, rather variable in different individuals, and the fins are marked with dark brown dots. The eyes are yellow, and the pupil very dark blue. It is but a small fish, averaging four, and seldom exceeding five, inches in length.

WE now come to the typical genus of this family, which is represented by several well-known British species.

THE RED GURNARD, or CUCKOO GURNARD, as it is sometimes called, from the sound it utters when taken out of the water, is very common on the English coast. It is rather a small fish, rarely exceeding fourteen inches in length. The colours of its body when living are very beautiful, the upper part being bright red, and the under parts silvery white.

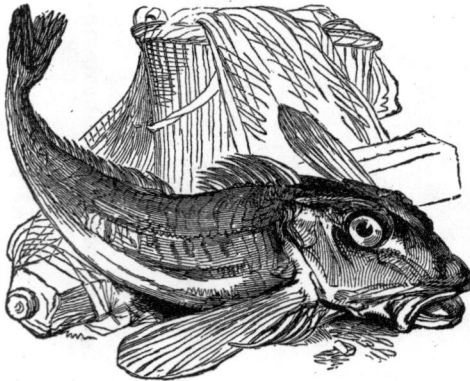

GURNARD.—(*Trigla cuculus.*)

There are nine species of Gurnard known to frequent the coasts of England, some, as the SAPPHIRINE and the MAILED GURNARDS, being most extraordinary in form.

THE FLYING GURNARD is common in the Indian seas. Its pectoral fins are so much enlarged, that when it springs out of the water, when pursued by the dolphin or bonito, the wide quivering fins are able to sustain it in the air for a limited period.

This fish has often been confounded by voyagers with the true flying fish (*Exocœtus*), which belongs to an entirely different order.

THE important, though not very large family of the Scomberidæ, contains many species that are almost invaluable as food, and others that are beautiful in form and interesting in habits.

OUR first example of these fishes is the MACKEREL, so well known for the exceeding beauty of its colours and the peculiar flavour of its flesh. This is one of the species that are forced by the irresistible impulse of instinct to migrate in vast shoals at certain times of the year, directing their course towards the shores, and as a general rule frequenting the same or neighbouring localities from year to year.

This fish is taken both by nets and lines, the nets being of two kinds, one called the drift-net, and the other the seine. The drift-net is, as its name implies, allowed to be drifted out by the tide, and is suspended along a cord called the drift-rope. The whole length of one of these nets when shot is sometimes a mile and a half, these enormous dimensions being attained by attaching a number of nets together at the ends. Each of these nets is one hundred and twenty feet long and twenty feet deep, and along the upper edge are fastened a series of cork floats. When the net is to be shot, a large buoy is attached to the end of the drift-rope, the buoy is thrown overboard, and the sails set. As the boat dashes away from the spot, the nets, which have already been attached to the drift-rope, are thrown successively overboard, until all the nets are paid out and hang in the water like a net wall. The strain of the buoy at one end of the drift rope and the boat at the other keeps the rope straight and the net upright.

As the Mackerel come swimming along, they are arrested by the net, which they cannot see, on account of the thin twine of which it is made, and the large meshes, which are about two-and-a-half inches in diameter. The head slips through the meshes, but the middle of the body is too large and cannot pass. When the fish attempts to recede, the open gill-covers become hitched in the meshes, and so retain it in that uncomfortable position until the net is hauled in.

This is a delicate and difficult operation, especially when the take of fish is heavy. Mr. Yarrell mentions that in June 1808 the nets were so heavily loaded that the fishermen could not haul them in, or even keep them afloat, so that they were forced to cut the drift-ropes and let the nets sink and be lost. The nets on this occasion were worth nearly sixty pounds, not including the value of the fish.

In the seine-net, the fish are taken by surrounding the shoal with the net, which is made with very small meshes, and either gently hauled to the surface, so that the enclosed fish can be dipped out, or even drawn ashore and then emptied.

Fishing for Mackerel with a line is also a profitable mode of taking these fish, although they cannot be caught in such multitudes as with the net. The Mackerel is a very voracious fish, and will bite at almost any glittering substance drawn quickly through the water, a strip of scarlet

MACKEREL. —(*Scomber scomber.*)

cloth being a very favourite bait. A tapering strip of flesh cut from the side of a Mackerel is found to be the most successful of any bait, and the method of angling is simply to pass the hook through the thicker end of the strip —technically called a "lask"—and to throw it overboard a boat in full sail, so that it is towed along without trouble. The hook is kept below the surface of the water by means of a leaden plummet fixed to the line a short distance above the hook, and the Mackerel on seizing the flying bait is immediately caught. On a favourable day, when the sky is not too bright and the wind is tolerably brisk, two or three men can take the fish as fast as they can bait and throw overboard.

The colour of the Mackerel is rich green upon the back, variegated with deep blue and traversed with cross bands of black, straight in the males, but undulating in the females. The abdomen and sides are silvery white, with golden reflections. These colours are most brilliant during the life of the fish, and as they fade soon after it has left the water, their brilliancy affords a good test of its freshness.

THE celebrated TUNNY belongs to this family, and is closely allied to the Mackerel.

This magnificent and most important fish does not visit our coasts in sufficient numbers to be of any commercial importance; but on the shores of the Mediterranean, where it is found in very great abundance, it forms one of the chief sources of wealth of the sea-side population.

In May and June, the Tunnies move in vast shoals along the shores, seeking for suitable spots wherein to deposit their spawn. As soon as they are seen on the move, notice is given by a sentinel who is constantly watching from some lofty eminence, and the whole population is at once astir, preparing nets for the capture, and salt and tubs for the curing of the expected fish. There are two modes of catching the Tunny, one by the seine-net and the other by the "madrague." The mode of using the seine is identical with that which has already been described when treating of the Mac-

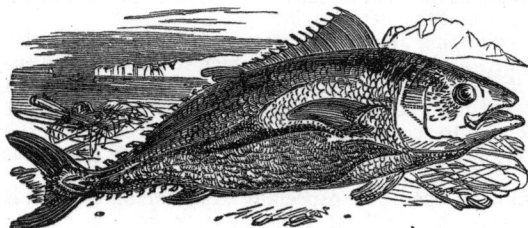

TUNNY.—(*Thynnus Thynnus.*)

kerel, but the madrague is much more complicated in its structure and management. The principle of the madrague is precisely the same as of the "corral" by which elephants are entrapped in Ceylon.

A vast enclosure of united nets, nearly a mile in length, and divided into several chambers, is so arranged that as the Tunnies pass along the coast they are intercepted by a barrier, and, on endeavouring to retreat, are forced to enter one of the chambers. When a number of Tunnies have fairly entered the net, they are driven from one chamber to another, until they are forced into the last and smallest, called significantly the chamber of death. This chamber is furnished with a floor of net, to which are attached a series of ropes, so that by hauling in the ropes the floor of the net is drawn up and the fish brought to the surface. The large and powerful fish struggle fiercely for liberty, but are speedily stunned by blows from long poles, and lifted into the boats. The flesh of the Tunny is eaten both fresh and salted. It is most extensively used, being pickled in various ways, boiled down into excellent soup, and is also made into pies, which are thought to be very excellent, and possess the valuable property of remaining good for nearly two months. The different parts of the fish are called by appropriate names, and are said to resemble beef, veal, and pork. The food of the Tunny consists mostly of smaller fish, such as herrings and pilchards, and the cuttle-fish also forms some portion of its diet.

SUCKING FISH.—(*Echeneis remora.*)

In general shape the Tunny is not very unlike the mackerel, but in size it is vastly superior, generally averaging four feet in length, and sometimes attaining the dimensions of six or seven feet. The colour of the upper part of the body is very dark blue, and the abdomen is white decorated with spots of a silvery lustre. The sides of the head are white.

EVERYONE has heard of the SUCKING FISH, and there are few who are not acquainted with the wild and fabulous tales narrated of its powers.

This little fish was reported to adhere to the bottom of ships, and to arrest their progress as suddenly and firmly as if they had struck upon a rock. The winds might blow, the sails might fill, and the masts creak, but the unseen fish below could hold the vessel by its single force, and confine her to the same spot as if at anchor. It is wonderful how fully this fable was received, and how many years were needed to root the belief out of prejudiced minds. Both scientific names refer to this so-called property, *echeneis* signifying "ship-holder," and *remora* meaning "delay."

That the Sucking Fish is able to adhere strongly to smooth surfaces is a well-known fact, the process being accomplished by means of the curious shield or disc upon the upper surface of the head and shoulders, the general shape of which can be understood by reference to the engraving. This disc is composed of a number of flat bony laminæ, arranged parallei to each other in a manner resembling the common wooden window-blind, and capable of being raised or depressed at will. It is found by anatomical investigation that these laminæ are formed by modifications of the spinous dorsal fin, the number of laminæ corresponding to that of the spines. They are moved by a series of muscles set obliquely, and when the fish presses the soft edge of the disc against any smooth object and then depresses the laminæ, a vacuum is formed, causing the fish to adhere tightly to the spot upon which the disc is placed.

When the creature has once fixed itself it cannot be detached without much difficulty, and the only method of removing it, without tearing the body or disc, is to slide it forwards in a direction corresponding with

JOHN DORY.—(*Zeus faber.*)

the set of the laminæ. In the opposite direction it cannot be moved, and the fish, therefore, when adhering to a moving body, takes care to fix itself in such a manner that it cannot be washed off by the water through which it is drawn. Even after death, or when the disc is separated from the body, this curious organ can be applied to any smooth object, and will hold with tolerable firmness. In order to accommodate the disc, the upper part of the skull is flattened and rather widened.

It is rather a voracious fish, and takes the hook eagerly if baited with a piece of raw flesh. When hooked, however, it is by no means secured, for as soon as it feels the prick of the sharp point and the pull of the line, it darts to the side of the vessel, dives deeply, and affixes itself so strongly to the bottom that the hook may be torn out of the mouth before the fish will relax its hold. It is therefore necessary to draw the Sucking Fish smartly out of the water as soon as it is fairly hooked, and in this manner great numbers can be caught. The flesh is thought to be very good, and is said to resemble that of the eel, but without its richness. The colour of this species is dusky brown, darker on the back than on the abdomen. The fins are darker than the body, and are of a dense leathery consistence. The length of this fish seldom exceeds eight inches.

THE well-known JOHN DORY, so dear to epicures, is found in the British seas, and is frequently seen in the fishmongers' shops, where its peculiar shape seldom fails of attracting attention even from those who are not likely to purchase it, or even to have seen it on the table.

The name of John Dory is thought to be a corruption of the French name *jaune dorée*, a title given to the fish on account of the gilded yellow which decorates its body. It was called Zeus by the ancients, because they considered it to be the king of eatable fish ; and the name of Faber, or "blacksmith," has probably been earned by the smoky tints which cloud its back. The dark and conspicuous spots on the side are thought in many places to be imprinted upon the fish as a memorial of the honour conferred upon its ancestor in times past, when St. Peter took the tribute money from the mouth of the Dory, and left the print of his finger and thumb as a perpetual remembrance of the event. Some persons, however, contend that the marks are due, not to St. Peter, but St. Christopher; and the Greeks, who hold to the latter tradition, call the fish Christophoron.

The flesh of the Dory is remarkably excellent and, as it is rather improved by the lapse of twenty-four hours after the fish has been taken from the sea, it is peculiarly valuable to those who live far inland and cannot hope for the more delicate fishes, which must be eaten almost as soon as caught. Although a common fish, it always commands a high price, and as, when cooked, the head occupies so large a space, it never affords an economical dish.

SWORD-FISH.—(*Xiphias gladius.*)

The shape of the Dory is very peculiar. The body is very deep, and greatly compressed. The head is oddly shaped, and the mouth can be protruded to a surprising extent. The spines of the first dorsal fin are much prolonged, and behind each ray is given off a very long waving filament, three times as long as the ray in front of it. Along the base of the dorsal and anal fins are arranged two rows of spiny scales, their points being directed backward, and one row being set at each side of the fin.

THE well-known SWORD-FISH derives its popular name from the curious development of the snout, which projects forward, and is greatly prolonged into a shape somewhat resembling a sword-blade. The sword is formed by the extension of certain bones belonging to the upper part of the head. This fine fish is found in the Mediterranean Sea, and also in the Atlantic Ocean, and in the former locality is often very plentiful. The Sicilian fishermen are accustomed to pursue the Sword-fish in boats, and mostly employ the harpoon in its capture. The weapon is not very heavy, and by a strong and practised hand can be hurled to some distance. The flesh of the Sword-fish is always eatable and nourishing, and in all specimens is white and well flavoured.

The use of the "sword" is not clearly ascertained. In all probability, the fish employs this curious weapon in gaining its subsistence, but the precise mode of so doing is not known. It is an ascertained fact that the Sword-fish will sometimes attack whales, and stab them deeply with its sharp beak ; and it is also known that this fish has several times driven its beak so deeply

SWORD-FISH.   Page 472.

The Popular Natural History

into a ship that the weapon has been broken off by the shock. In such cases, the blow is so severe, that the sailors have fancied that their vessel has struck upon a rock. Several museums possess examples of pierced planks and beams, but it is possible that the fish may have struck them by accident, and not in a deliberate charge. The Sword-fish generally go in pairs.

The food of this creature is rather varied, consisting of cuttle-fish, especially the squid, and of small fishes, neither of which animals would in any way fall victims to the sword. It certainly has been said that the weapon is used for transfixing the flat fish as they lie on the bed of the sea, but this assertion does not appear to be worthy of credit.

The young and adult specimens are very different from each other. In the young, the body is covered with projecting tubercles, which gradually disappear as it increases in size, and when it has attained the length of three feet they are seldom to be seen. Those on the abdomen remain longer than the others. The dorsal fin extends in the young specimens from the back of the head to the root of the tail, but the membranes and spines of its centre are so extremely delicate that they are soon rubbed away, and the adult specimen then appears to have two dorsal fins.

The colour of the Sword-fish is bluish black above, and silvery white below. The whole body is rough, and the lateral line is almost invisible. The usual length of the Sword-fish is from ten to twelve feet, but specimens have been seen which much exceed those dimensions. A few examples of the Sword-fish have been captured in British waters ; one, that measured seven feet in length, was taken off Margate.

WE now arrive at the large family of the GOBIES, which include many curious fish, and of which the British coasts present many representatives,

THE BLACK GOBY, sometimes known as the ROCK-FISH, is a moderately common example of the enormous genus to which it belongs, and which contains more than a hundred and fifty authenticated species. The members of this genus may easily be recognized by the peculiar form of the ventral fins, which are united together so as to form a hollow disc, by which they can attach themselves to rocks or stones at pleasure. In fact, this disc, although differing in shape, acts on exactly the same principle as that of the sucking fish.

The Black Goby prefers the rocky to the sandy coasts, and may be found in the pools left by the retreating tide. Some naturalists deny that the disc is used for adhesion, but I have caught and kept many Gobies, and have frequently seen them sticking to the sides of the vessel in which they were confined. The adhesion was achieved with astonishing rapidity, and the little fish contrived to hold itself with great tenacity. The surface of the Black Goby is very slippery, owing to the abundant mucous secretion which is poured from the appropriate glands ; but after it has been in spirits for some time, the edges of the scales begin to project through the mucus, and are exceedingly rough to the touch.

In some places along the sea-coast the Gobies are known by the popular appellation of Bull-routs, and are rather feared on account of the sharp bite which their strong jaws and pointed teeth can inflict upon the bare hand.

ANOTHER small family now comes before us, called the Batrachidæ, or Frog-fishes, from the frogish aspect of the body, and especially of the head.

THE FISHING FROG, ANGLER-FISH, or WIDE GAB, is not unfrequent on the British coasts, and has long been famous for the habit from which it has derived its popular name.

The first dorsal fin is almost wholly wanting, its place being occupied merely by three spines, moveable by means of certain muscles. The manner in which these spines are connected with the body is truly marvellous. The first, which is furnished at its tip with a loose shining slip of membrane, is developed at its base into a ring, through which passes a staple of bone that proceeds from the head. The reader may obtain a very perfect idea of this beautiful piece of mechanism by taking a common iron skewer, slipping a staple through its ring, and driving the staple into a board. It will then be seen that the skewer is capable of free motion in every direction. The second spine is arranged after a somewhat similar fashion, but is only capable of being moved backwards and forwards. The use of these spines is no less remarkable than their form.

The Fishing Frog is not a rapid swimmer, and would have but little success if it were to chase the swift and active fishes on which it feeds. It therefore buries itself in the muddy sand, and continually waves the long filaments with their glittering tips. The neighbouring fish, following the instincts of their inquisitive nature, come to examine the curious object, and are suddenly snapped up in the wide jaws of their hidden foe. Many fishes can be attracted by any glittering object moved gently in the water, and it is well known by anglers how deadly a bait is formed of a spoon-shaped piece of polished metal, furnished with hooks and quickly drawn through the water.

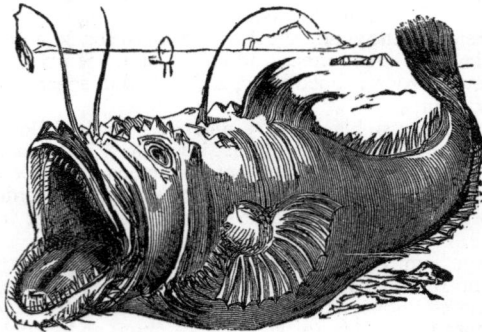

FISHING FROG.—(*Lophius piscatorius.*)

It is impossible to mistake this fish for any other inhabitant of the ocean, its huge head—wide, flattened, and toad-like—its enormous and gaping mouth, with the rows of sharply-pointed teeth, its eyes set on the top of the head, and the three long spines, being signs which cannot be misunderstood. The general colour of this fish is brown above and white below; the ventral and pectoral fins are nearly white, and that of the tail almost black. The throat, just within the jaws, is composed of loose skin, which forms a kind of bag. The average length of the adult Fishing Frog is about a yard.

The family in which this fish is placed may be distinguished by the peculiar structure of the pectoral fins, which are mounted on a sort of arm produced by an elongation of the carpal bones. From this peculiarity, the family is termed Pediculati, or foot-bearing fishes, as the prolonged fins enable them to walk along wet ground almost like quadrupeds.

THE very odd-looking creature, called the WALKING FISH, which is shown in the illustration on page 475, is one of the strange and weird forms that sometimes occur in nature, and which are so entirely opposed to all preconceived ideas, that they appear rather to be the composition of human ingenuity than beings actually existing. The traveller who first discovered this remarkable fish would certainly have been disbelieved if he had contented himself with making a drawing of it, and had not satisfied the rigid scrutiny of scientific men by bringing home a preserved specimen.

In the fishes of this genus, the carpal bones, *i.e.* those bones which represent the wrist in man, are very greatly lengthened, more so than in the preceding genus, and at their extremity are placed the pectoral fins, which are short, stiff ,and powerful, the pointed rays resembling claws rather than fins. In all the fishes of this genus, the body is much compressed and decidedly elevated ; but in the present species, these peculiarities are carried to an almost exaggerated extent. The first dorsal spine, with its membranous appendages, is placed as usual just above the snout, and the second ray is set immediately behind it. The third, however, is placed at a very great distance from the second, and forms part of the soft dorsal fin.

Dr. Günther remarks upon the fishes of this genus, that they are so extremely variable in form, colour, and the greater or less development of the dorsal spines, that hardly two specimens are found sufficiently alike to

WALKING FISH.—(*Antennarius hispidus.*)

enable the systematic naturalist to decide upon their precise situation in the zoological scale. Moreover, their geographical range is exceedingly wide, some species ranging over the Atlantic and Indian Oceans ; and the learned ichthyologist above mentioned is of opinion that many specimens which he has at present been compelled to admit into the list of separate species will be ultimately found to be mere casual varieties.

The colour of this species is yellow, diversified with many spots and streaks of brown, some of the streaks radiating from the eye, and others extending over the dorsal fin. It is a native of the Indian seas.

THE important family of the BLENNIES comes next in order. They are all carnivorous fishes, many being extremely voracious, and are spread over the shores of every sea on the globe. They mostly reside on or near the bottom.

THE SEA WOLF, SEA CAT, or SWINE-FISH, is one of the fiercest and most formidable of the finny tribes that are found on our coast, and has well earned the popular names by which it is known.

The Sea Wolf possesses a terrible armature of teeth, not only in the jaws, but arranged in a double band on the palate, and by means of these powerful weapons it can crush with ease the hard shelled molluscs and crustaceans on which it feeds. As may be imagined, the aspect of the Sea Wolf is far from prepossessing, its fierce head with the armed jaws, strong and cruel as those of the tiger or hyæna, and the smooth, slime-covered skin, giving it a most repulsive aspect.

The Sea Wolf is sometimes taken with the hook, but is mostly found entangled in the nets together with other fish, and in either case it struggles violently as soon as it perceives the loss of its liberty. It will tear the nets to pieces with its teeth, and when hauled out of the water it still flounces about with such vigour, and bites at every object with such ferocity, that the boatmen usually stun it by a blow on the head before lifting it into the vessel, a very heavy stroke being required for the purpose.

The general colour of the Sea Wolf is brownish grey, with a series of brown vertical stripes and spots over the upper parts ; the under parts are white. On our shores it attains a length of six or seven feet, but in the northern seas, where it thrives best, it greatly exceeds those dimensions. There is an American variety where the vertical streaks are modified into round spots of blackish brown.

THE typical genus of this family is represented by several British specimens, of which the EYED BLENNY is one of the most conspicuous. This pretty fish is not very common, but has been taken on the southern coasts of England. From the elevated dorsal fin, and the bold dark brown spot that decorates it, this Blenny has sometimes been called the Butterfly Fish. In the Mediterranean it is tolerably common, and lives mostly among the seaweed, where it finds abundance of the smaller crustacea and molluscs.

THE extraordinary fish, called, from its habits, the CLIMBING PERCH, is a native of Asia, and is remarkable for its apparent disregard of certain natural laws.

This singular creature has long been celebrated for its powers of voluntarily leaving the failing streams, ascending the banks, and proceeding over dry land towards some spot where its unerring instinct warns it that water is yet to be found.

Several species, of which the ANABAS SCANDENS has been chosen as the best example, possess this singular property of walking over dry land, so that the old proverb of a fish out of water is in these cases quite inapplicable. Several instances of this remarkable propensity have been collected by Sir J. Emerson Tennent, and have been inserted in his valuable work on the Natural History of Ceylon. The following account is written by Mr. Morris, the Government agent in Trincomalee :—

" I was lately on duty inspecting the bund of a large tank at Nade-cadua, which being out of repair, the remaining water was confined in a small hollow in the otherwise dry bed. Whilst there, heavy rains came on, and as we stood on the high ground we observed a pelican on the margin of the shallow pool gorging himself : our people went towards him, and raised a cry of 'Fish ! fish !' We hurried down, and found numbers of fish struggling upward through the grass, in the rills formed by the trickling of the rain. There was scarcely water to cover them, but nevertheless they made rapid progress up the bank, on which our followers collected about two baskets of them at a distance of about forty yards from the tank.. They were forcing their way up the knoll, and had they not been interrupted first by the pelican and afterwards by ourselves, they would in a few minutes have gained the highest point, and descended on the other side into a pool which formed another portion of the tank.

" . . . As the tanks dry up, the fish congregate in the little pools, till at last you find them in thousands in the moistest parts of the beds, rolling in the blue mud, which is at that time about the consistence of thick gruel.

"As the moisture further evaporates, the surface fish are left uncovered, and they crawl away in search of fresh pools. In one place I saw hundreds diverging in every direction from the tank they had just abandoned, to a distance of fifty or sixty yards, and still travelling onwards. In going this distance, however, they must have used muscular exertion enough to have taken them half a mile on level ground, for at these places all the cattle and wild animals of the neighbourhood had latterly come to drink, so that the surface was everywhere indented with footmarks in addition to the cracks in the surrounding baked mud, into which the fish tumbled in their progress.

CLIMBING PERCH.—(*Anabas scandens.*)

In those holes which were deep, and the sides perpendicular, they remained to die, and were carried off by kites and crows.

"My impression is, that this migration must take place at night or before sunrise, for it was only early in the morning that I have seen them progressing, and I found that those I brought away with me in the chatties appeared quiet by day, but a large proportion managed to get out of the chatties by night—some escaped altogether, others were trodden on and killed.

"One peculiarity is the large size of the vertebral column, quite disproportioned to the bulk of the fish. I particularly noticed that all in the act of migrating had their gills expanded."

It is known of the Climbing Perch that the fishermen of the Ganges, who subsist largely on these fishes, are accustomed to put them into an earthen pan or chatty as soon as caught ; and although no water is supplied to them, they exist very well without it, and live this strange life for five or six days.

On opening the head of this fish, the curious structure which enables it to perform such marvellous feats is clearly seen. Just within the sides of the head, the pharyngeal bones, *i.e.* the bones that support the orifice between the mouth and gullet, are much enlarged, and modified into a series of labyrinthine cells and duplications, so that they retain a large amount of water in the interstices, and prevent the gill membranes from becoming dry. Some writers say that this fish is capable of climbing up the rough stems of palmtrees, in search of the water that lodges between the bases of the dead leaves and the stem, or perhaps in search of food. In the Tamoule language it is called *Paneiri*, or tree-climber.

THE FLAT FISHES, as they are popularly called, or the Pleuronectidæ, as they are named scientifically, are among the most remarkable of the finny tribe. The latter name is of Greek origin, and signifies " side-swimmer," in allusion to the mode of progression usually adopted by these fishes.

The popular but erroneous idea of these creatures is, that their bodies are flattened so that the abdomen rests on the ground and the back remains uppermost ; but a brief examination suffices to show that the form of these fishes is really compressed, so that when a turbot or a sole is placed on the ground, it lies upon one side or the other. Though varying in colour, it is found that the upper side is always of a dark tint, the under side being quite if not wholly white. This arrangement is most useful in guarding them against the attacks of enemies, their flat dark upper surface bearing so great a resemblance to the sand on which they love to creep, that they can scarcely be distinguished, even when the eye is directed towards them.

While at their ease, they slide themselves over the bed of the sea in a kind of creeping fashion and have an odd custom of lying with the head raised in a manner that irresistibly reminds the observer of the cobra. If alarmed, they start at once into the vertical position usually assumed by fishes, and dash off with astonishing speed. As they swim, the flat fishes undulate through the water in a most graceful manner, and it is very interesting to watch one of the common flat fishes, such as the plaice or the sole, swim with serpentine ease and elegance, and then suffer itself to sink slowly to the bottom, where it sits, with raised head and watchful eyes.

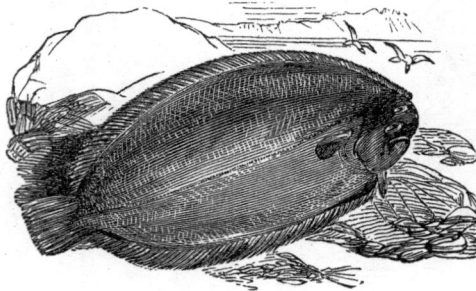

SOLE.—(*Solea vulgaris.*)

It is evident that if the eyes of the flat fishes were placed in the manner customary among the finny race, one eye would be rendered useless as long as the fish was lying upon its side. This difficulty is therefore met by a most extraordinary modification of the bones of the head, by which means both the eyes are brought to that side which remains uppermost, and are thereby enabled to command a wide view around. There have been one or two instances where the eyes have been placed one on each side, but these may be considered as simple variations from the ordinary rule.

THE COMMON SOLE is one of the most familiar of our British flat fishes, and is found on all our coasts, those of the southern shores being the most plentiful, and attaining the largest dimensions.

The Sole can be taken by the line, but the fishermen always use the trawl-net, a kind of huge dredge, with a mouth that often exceeds thirty feet in width. As these nets are drawn along the bed of the sea, the great beam which edges the mouth scrapes the mud and sand, and alarms the fishes to such an extent that they dash wildly about, and mostly dart into the net, whence they never escape. Vast numbers of Soles are taken by this method of fishing, and as the trawls bring to the surface enormous quantities of crustaceans, molluscs, zoophytes, and other marine inhabitants, the energetic naturalist cannot employ his time better than in taking a sail in one of these boats, and enduring a few hours' inconvenience for the sake of the rich harvest which he is sure to reap. Some of the rarest and most valuable British animals have been taken in the trawl-nets. The finest Sole that I ever saw I took in a trawl-net in one of the creeks of the Medway. As to length and breadth, it was not very remarkable, but it was almost half as thick again as any Sole that I have seen.

The Sole is in condition throughout the greater part of the year, the only time when it is not worth eating being from the end of February to the last week in March, when the fishes are full of roe, and the flesh is rather soft and watery. It is a hardy fish, and can soon be acclimatized to live in fresh water ; and it is said that under such circumstances the fish can be readily fattened, and become nearly twice as thick as when bred in the sea. Sometimes the Soles venture into the mouths of rivers, passing about four or five miles into the fresh water, and depositing their multitudinous eggs in such localities.

The colour of the Sole is almost always brown on the right side and white on the left, but examples of reversed Soles are not uncommon, where the left side is brown and the other is white. The scales are small, and give a rough, rasp-like sensation to the hand. The dimensions of this fish are very variable, an average specimen weighing about a pound or eighteen ounces. Much larger examples, however, occur occasionally, and Mr. Yarrell mentions one instance where a Sole measured twenty-six inches in length, eleven and a half in width, and weighed nine pounds.

THE well-known TURBOT, so widely and so worthily celebrated for the firm delicacy of its flesh, inhabits many of the European coasts, and is found in tolerable abundance off our own shores. Like all flat fishes, it mostly haunts the sandy bed of the sea, but will sometimes swim boldly

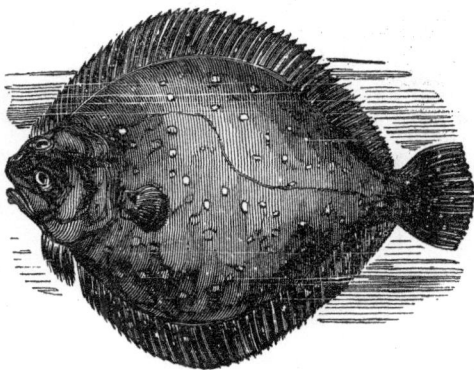

TURBOT.—(*Pleuronectes maximus.*)

to the surface of the water. It is a restless and wandering fish, traversing considerable distances as it feeds, and generally moving in small companies.

The Turbot is known in Scotland by the title of BANNOCK FLEUK, or SPAWN FLEUK, the former name being given to it on account of its flat shape, which resembles a bannock or oatcake, and the latter because it is thought to be at the best while in roe. After spawning, *i.e.* about August, its flesh loses its peculiar firmness, but in a very short time the fish regains its condition.

The colour of the Turbot is brown of different shades on one side, usually the left, and the whole of that side is spotted with little round bony tubercles, which may be found in the skin after boiling. The size of this fish is extremely variable. The average weight is six or seven pounds, but Turbots are often taken of far greater dimensions. The largest specimen of which an authentic notice is preserved was taken near Plymouth in the year 1730, and weighed seventy pounds.

THE PLAICE is well known by the bright red spots which are scattered over its dark side. I have caught numbers of Plaice, some measuring six or seven inches in length, by merely wading into the muddy sand, holding them down with the feet, and picking them out with the hands. Their terrified wriggle is easily felt by the bare feet, as the fishes find themselves pressed into the sand whither they had fled for refuge, and by a little dexterous management they may be captured by inserting the fingers under the foot, seizing them firmly across the body.

The colour of the Plaice is light brown, variegated with a number of bright red spots upon the body and the dorsal and anal fins. When young, the Plaice has often a dark spot in the centre of each red mark.

THE FLOUNDER, MAYOCK FLEUK, or BUTT, is quite as common as the plaice, and is found in salt, brackish, or fresh water, sometimes living in the sea, sometimes inhabiting the mouths of rivers, and sometimes passing up the stream for many miles.

In former days the Flounder has been known to ascend the Thames as high as Hampton Court, and has there been observed actively chasing the minnows and driving them into shallow water. I have often taken small Flounders in the Thames just above Erith.

THE well known COD-FISH is a native of many seas, and in some localities is found in countless legions.

This most useful fish is captured in vast numbers at certain seasons of the year, and is always taken with the hook and line. The lines are of two descriptions, namely, the long lines to which a great number of short lines are attached, and the simple hand-lines which are held by the fishermen. The long lines sometimes run to an extraordinary length, and shorter lines, technically called snoods, are affixed to the long line at definite distances.

COD.—(*Gades morrhua.*)

To the end of each snood is attached a baited hook, and as the sharp teeth of the fish might sever a single line, the portion of the snood which is near the hook is composed of a number of separate threads fastened loosely together, so as to permit the teeth to pass between the strands. At each end of the long line is fastened a float or buoy, and when the hooks have been baited with sand launce, limpets, whelks, and similar substances, the line is ready for action.

The boat, in which the line is ready coiled, makes for the fishing place, lowers a grapnel or small anchor, to which is attached the buoy at one end of the line, and the vessel then sails off, paying out the line as it proceeds, and always "shooting" the line across the tide, so as to prevent the hooks from being washed against each other or twisted round the line, which is usually shot in the interval between the ebb and flow of the tide, and hauled in at the end of about six hours.

As soon as the long line has been fairly shot, and both ends firmly affixed to the grapnels, the fishermen improve the next six hours by angling with short lines, one of which is held in each hand. They thus capture not only Cod-fish, but haddock, whiting, hake, pollack, and various kinds of flat fishes. On favourable occasions the quantity of fish captured by a single boat is very great, one man having taken more than four hundred Cod alone in ten hours.

The Cod is sometimes sent away in a fresh state, but is often split and salted on the spot; packed in flats on board, and afterwards washed and dried on the rocks. In this state it is called Klip-fish or Rock-fish. The liver produces a most valuable oil, which is now in great favour for the purpose of affording strength to persons afflicted with delicate lungs or who show symptoms of decline. The best oil is that which drains naturally from the livers as they are thrown into a vessel which is placed in a pan filled with boiling water. The oil is then carefully strained through flannel, and is ready for sale.

The roe of the Cod is useful for bait, the sardine in particular being very partial to that substance. Much of the roe is stupidly wasted by the fisherman, who carelessly flings into the sea a commodity of which he can sell any amount, and for which he can obtain ten or eleven shillings per cwt. In Norway the dried heads of the Cod are used as fodder for cows, and, strange to say, the graminivorous quadrupeds are very fond of this aliment.

Like several other marine fish, the Cod can be kept in a pond, provided the water be salt; and if the pond should communicate with the sea, these fishes can be readily fattened for the table. Several such ponds are in existence, and it is the custom to transfer to them the liveliest specimens that have been caught during the day's fishery, the dead or dying being either sold or cut up as food for their imprisoned relatives. These fishes are extremely voracious and will eat not only the flesh of their kinsmen, but that of whelks and other molluscs, which are abundantly thrown to them. It is found that under this treatment the Cod is firmer, thicker, and heavier in proportion to its length than if it had been suffered to roam at large in the sea.

In the large and important group of fishes to which our attention is now drawn, and popularly known as EELS, the ventral fins are wholly wanting, the body is long, snake-like, smooth and slimy on the exterior, and in many cases covered with very little scales hidden in the thick soft skin.

THE two EELS represented in the engravings are examples of some very common and useful British fish.

THE SHARP-NOSED EEL derives its name from the shape of its head, and by that structure may be distinguished from the second species. In their habits the Eels are so similar that the present species will be taken as an example of the whole genus.

Of the general habits of the Eel, the Hon. Grantley F. Berkeley has given the following short and interesting account. During hot, still sunny weather, day and night, in the month of June, the Eels are chiefly on the top of the water. Wherever masses of weeds lie, and what is called the cow-weed grows the longest, there Eels do congregate, to bask in the sun by day, to enjoy by night the warmth left in the weeds by the sun, and there, while thus luxuriating, to snap at and catch the myriads of gnats, moths, flies, and other insects that seek the weeds for food or rest, and by damping their wings become an easy prey to their ambushed assailants. In waiting for the otter, or watching the river, I have often sat in my boat embayed in weeds, and seen and heard the Eels thus occupied; and near and within these weeds, in the particular weather alluded to, the wire-traps, nets, and snig-pots take best. The haunts of Eels are quite as variable as the weather. In warm, still weather, seek them on the rapids and near weeds either waving on the surface of the water or in floating masses of detached weeds that the eddies of the stream have wound and kept in one place. In blowing, cooler, or rainy weather, then look for them in the still, deep ditches. If a flush of water comes, and a little shallow stream running from or into the main river becomes fuller than usual, then let all the capturing gear be set to take them on, to them, this delicious change of ground, for against this stream they will work as long as it is freshened. In one night, in a little stream of this sort, I took thirty pounds weight of Eels.

Like several fishes which have already been mentioned, Eels are very tenacious of life, and are able to live for a long time when taken out of water, owing to a simple but beautiful modification of structure, which retains a sufficient amount of moisture to keep the gills in a damp state and able to perform their natural functions. These fishes have been seen crawling over considerable distances, evidently either in search of water, their own dwelling-place being nearly dried, or in quest of some running stream in whose waters they might descend to the sea after the manner of their race.

SHARP-NOSED EEL.—(*Anguilla acutirostris.*)

Towards the latter end of summer the Eels migrate towards the sea, and it is found that these fishes can live either in fresh or salt water with equal ease, the mouths of rivers being favoured localities. It sometimes happens that even in our seaport towns and marine watering-places, the common river Eel is caught by those who are angling in the sea for marine fish. This quality is peculiarly valuable in the Eel, as it enables the Dutch fishermen, who annually supply our markets with vast numbers of these fish, to bring them across the sea in vessels that are fitted with "wells" pierced for the transmission of the sea-water through which the vessel is sailing.

The tenacity of life possessed by this fish is really remarkable; and it is worthy of notice that the best mode of killing Eels is to grasp them by the neck and slap their tails smartly against a stone or post. The muscular irritability of the body is wonderfully enduring, and after the creature has been cut up into lengths each separate piece moves about as if alive, while at the touch of a pin's point it will curve itself as if it felt the injury. When all such irritability has ceased, the portions will flounce about vigorously if placed in boiling water; and even after they have remained quiet under its influence, the addition of salt will make them jump about as vigorously as ever. Of course there can be no real sensation, the spinal cord having been severed.

The reproduction of the Eel has long been a subject of discussion, some persons thinking that the young are produced in a living state, and others holding that they are hatched from eggs. This question has, however, been set at rest by that universal revealer, the achromatic microscope, which has shown that the masses of oily-looking substance generally called fat are really the aggregated clusters of eggs, and that these objects, minute though they may be, not so large as the dot over the letter *i*, are quite perfect, and under the microscope are seen to be genuine eggs.

THE well-known CONGER EEL is a marine species, very common in our seas, and being most usually found on the rocky portion of the coast.

CONGER.—(*Conger vulgaris.*)

This useful fish has, of late years, come into more general use than formerly, and its good qualities are more appreciated. The flesh, though not very palatable if dressed unskilfully, is now held in some estimation, and for the manufacture of soup is thought to be almost unrivalled. The fishermen can now always obtain a ready sale for the Congers ; and those which are not purchased for the table are mostly bought up and made into isinglass.

The colour of the Conger is pale brown above and greyish white below. It often attains to a very great size, measuring ten feet in length and weighing more than 100 lbs.

The ELECTRIC EEL is even more remarkable for its capability of delivering powerful electric shocks than the torpedo, but as it is never found in the British seas it is not so well known as that fish.

THE Electric Eel is a native of Southern America, and inhabits the rivers of that warm and verdant country. The organs which enable it to produce such wonderful effects are double, and lie along the body, the one upon the other.

In the native country of these fishes they are captured by an ingenious but somewhat cruel process.

ELECTRIC EEL.—(*Gymnotus electricus.*)

A number of wild horses are driven to the spot and urged into the water. The alarmed Gymnoti, finding their domains thus invaded, call forth all the terrors of their invisible artillery to repel the intruders, and discharge their pent-up lightnings with fearful rapidity and force. Gliding under the bellies of the frightened horses, they press themselves against their bodies, as if to economize all the electrical fluid, and by shock after shock generally succeed in drowning several of the poor quadrupeds.

Horses, however, are but of slight value in that country, hardly, indeed, so much valued as pigeons in England, and as fast as they emerge from the water in frantic terror, are driven back among their dread enemies. Presently the shocks become less powerful, for the Gymnotus soon exhausts its store of electricity, and when the fishes are thoroughly fatigued they are captured with impunity by the native hunters. A most interesting account of this process is given by Humboldt, but is too long to be inserted in these pages.

Several of these wonderful fish have been brought to England in a living state ; and many of my readers may remember the fine Gymnotus that lived in the Polytechnic Institution. Numbers of experimenters were accustomed daily to test its powers ; and the fatal, or at all events the numbing, power of the stroke was evident when the creature was supplied with the fish on which it fed. Though blind, it was accustomed to turn its head towards the spot designated by the splashing of the attendant's finger, and as soon as a fish was allowed to fall into the water the Gymnotus would curve itself slightly, seemed to stiffen its muscles, and the victim turned over on its back, struck as if dead by the violence of the shock.

When full grown, the Electric Eel will attain a length of five or six feet, and is then a truly formidable creature. The body is rounded, and the scales small and barely visible. According to Marcgrave, the native name for this fish is Carapo.

WE now come to that most valuable family of fishes, the HERRING tribe called technically Clupeidæ, from the Latin word *clupea,* " a herring."

THE well-known ANCHOVY is properly a native of the Mediterranean Sea, though it often occurs on our coasts, and has once or twice been captured in our rivers. Indeed, one practical writer on British fishes thinks that the capture of the Anchovy off our shores is a task that would be highly remunerative if properly undertaken, and that, with proper pains, the British markets might be fully supplied with Anchovies from our own seas.

ANCHOVY.—(*Engraulis encrasicholus.*)

This little fish has long been famous for the powerful and unique flavour of its flesh, and is in consequence captured in vast quantities for the purpose of being made into anchovy sauce, anchovy paste, and other articles of diet in which the heart of an epicure delights. Unfortunately, however, the little fish is so valuable, that in the preparations made from its flesh the dishonest dealers too often adulterate their goods largely, and palm off sprats and other comparatively worthless fish for the real Anchovy. As the head is always removed before the process of potting is commenced, the deception is not easily detected—the long head with its projecting upper jaw and deeply-cleft gape affording so clear an evidence of the identity of the fish, that no one would venture to pass off one fish for the other if the heads were permitted to remain in their natural places. The flavour of the veritable Anchovy is rudely imitated by various admixtures, and its full rich colour is simulated by bole ammoniac and other abominations.

The very long generic title of this fish was given to it in ancient times, and is still retained, as being at once appropriate and sanctioned by the verdict of antiquity. Its literal signification is " gall-tinctured," and the name has been given to it on account of the peculiar bitter taste of the head, in which part the ancients supposed the gall to be placed. The colour of the Anchovy is bluish green on the back and upper part of the head, and the remainder of the body silvery white ; the fins have a tinge of green, and are beautifully transparent. The scales are large, and fall off almost at a touch. The length of the Anchovy varies from five to seven inches.

THE HERRING is undoubtedly the most valuable of our British fishes, and the one which could least be spared. It is at once the luxury of the rich and the nourishment of the poor, capable of preservation throughout a long period, easily packed, quickly and simply dressed, and equally good, whether eaten fresh or salted, smoked or potted.

During the greater part of the year the Herring lives in deep water, where its habits are entirely unknown. About July or August the Herring is urged, by the irresistible force of instinct, to approach the shores for the purpose of depositing its spawn in the shallow waters, where the warm rays of the sun may pour their vivifying influence upon the tiny eggs that will hereafter produce creatures of so disproportionate a size, and where the ever-moving tides may fill the water with free oxygen as the waves dash on the shores and fall back in whitened spray, thus giving to the water that sparkling freshness so needful for the development of the future fish.

HERRING. — (*Clupea harengus.*)

The Herrings, when they once begin to move, arise in vast shoals, and direct their course towards some part of the shore. In their choice of locality they are most capricious fish, sometimes frequenting one spot for many successive years, then deserting it for a length of time, and again returning to it without any apparent reason for either course of proceeding. They are essentially gregarious while on the move; and each shoal is so closely compacted, and its limits so well defined, that while one net will be filled almost to bursting with Herrings, another net, only a yard or two distant, will be left as empty as when it was shot.

The Herring is one of the fish that cannot endure absence from water, dying almost immediately after it is taken out of the sea, and thus giving rise to the familiar saying " as dead as a herring."

The food of the Herring is extremely varied, even in the comparatively shallow waters, and its subsistence during the time it is submerged in the deep is necessarily unknown. In the stomach of the Herring have been found crustacea of various kinds, molluscs, the spawn, and fry of other fish, and even the young of its own kind. It can be taken with a hook, and has been known to seize a limpet that was used as a bait. The colour of the Herring is blue above with greenish reflections, and the rest of the body is silvery white. After the fish has been dead for some hours, the cheeks and gill-covers become red, as if from injected blood.

PILCHARD.—(*Clupea pilchardus.*)

THE value of the HERRING family to man is almost incalculable. The PILCHARD and the herring are very similar in appearance, but may be easily known by the position of the dorsal fin, which in the Pilchard is so exactly in the centre of the body, that if the fish is held by it, the body exactly balances; while in the herring the dorsal fin is placed rather backwards, so that when suspended, the fish hangs with its head downwards.

Unlike the herring, which visits every part of our coast, the Pilchard is only found on the shores of Devonshire and Cornwall. Here, however, the enormous shoals that annually make their appearance fully compensate for the limited space occupied by them. Occasionally a few shoals are seen on the southern coast of Ireland. The coasts of France and Spain are tolerably frequent resorts of this fish.

The fish are usually taken in an enormous building of nets, called "sean nets." The nets used in the sea fishery are two, a large net called the "stop sean," about a quarter of a mile in length, and a hundred feet in depth; and a smaller net, called the "tuck sean," about a furlong in length, and a hundred and twenty feet in depth, the average value of the two nets being 500*l*.

When the fishermen see a shoal of pilchards approaching, they immediately set out in two fishing boats, one of which carries the tuck sean and the other the stop sean. Guided by signs from the master-seamen, they silently surround the shoal with the nets, the larger of which is used to enclose a large number of fish, and the smaller to pass within the other net, to bring the mass of fish into a small compass, and finally to prevent them from escaping until the fishermen have leisure to remove them to the boats.

When landed, the Pilchards are taken to the storehouses, salted, and after remaining in heaps for five or six days, are pressed into casks by powerful levers. During the pressure, which lasts about a fortnight, fresh layers of fish being added as the former are pressed close, an abundance of excellent oil escapes from holes made in the cask for the purpose. The entire refuse of the fish, consisting of the superabundant salt, the scales and other rejected portions, is sold to the farmers as a valuable manure. The refuse of each pilchard is calculated to manure one square foot of land.

FLYING-FISH.—(*Exocœtus volitans.*)

THE far-famed FLYING FISH exists in many of the warmer seas, and derives its popular name from its wonderful powers of sustaining itself in the air.

The passage of this fish through the atmosphere can lay no just claim to the title of flight, for the creature does not flap the wing-like pectoral fins on which it is upborne, and is not believed even to possess the power of changing its course.

In allusion to the habits of this remarkable fish, Mr. F. D. Bennett, in his "Narrative of a Whaling Voyage," has the following valuable remarks :—

"The principal external agents employed in this mode of locomotion are the large lobe of the tail fin and the broad transparent pectoral fins, which, on this occasion, serve at least as a parachute, and which, being situated close o the back, place the centre of suspension higher than the centre of gravity. It is also curious to notice how well the specific gravity of the fish can be regulated, in correspondence with the element through which it may move. The swim-bladder, when perfectly distended, occupies nearly the entire cavity of the abdomen, and contains a large quantity of air ; and in addition to this, there is a membrane in the mouth which can be inflated through the gills ; these two reservoirs of air affording good substitutes for the air-cells so freely distributed within the bones of birds, and having the additional advantage of being voluntary in their function.

"The pectoral fins, though so large when expanded, can be folded into an exceedingly slender, neat, and compact form ; but whether they are employed in swimming in the closed or expanded state, I have been unable to determine."

The ancients were well acquainted with the Flying-fish, and in their narratives even improved upon its powers, as was customary with the voyagers of those days, and asserted that, as soon as night came on, this fish left the ocean, flew ashore, and slept until morning safe from the attacks of its marine enemies. The generic name of *exocœtus*, literally "a sleeper-out," refers to this supposed habit. About thirty species of Flying-fish are known, mostly belonging to the Mediterranean Sea, but others occur in the North Sea and the Atlantic and Pacific Oceans.

THE fierce and voracious PIKE has well earned its titles of Fresh-water Shark and River Pirate, for though perhaps not one whit more destructive to animal life than the roach, gudgeon, and other harmless fish, the prey which it devours are of larger size, and its means of destruction are so conspicuous and powerful, that its name has long been a byword for pitiless rapacity.

The Pike is found in almost every English river, and, although supposed to have been artificially introduced into our country, has multiplied as rapidly as

PIKE.—(*Esox lucius.*)

any indigenous fish. The Pike is the master of the waters in which it resides, destroying without mercy every other fish that happens to come near its residence, none seeming able to escape except the perch, whose array of sharp spines daunts even the voracious Pike from attempting its capture. As if to show that the Pike really desires to eat the perch, and is only withheld from so doing by a wholesome dread of its weapons, there is no better bait for a Pike than a young perch from which the dorsal fin has been removed. It will even feed upon its own kind, and a young Pike, or Jack as it is then called, of three or four inches in length, has little chance of life if it should come across one of its larger kindred.

After hatching, the growth of the young Jack is extremely rapid, and, according to Bloch, it will attain a length of ten inches in the first year of its life. If well fed, the growth of this fish continues at a tolerably uniform rate of about four pounds per year, and this increase will be maintained for six or seven successive years.

The voracity of the Pike is too well known to need much comment. A tiny Jack of five inches in length has been known to capture and try to eat a gudgeon of its own size, and to swim about quite unconcernedly with the tail of its victim protruding from its mouth.   Had it been suffered to live, it would probably have finished the gudgeon in course of time, as the head was found to have been partially digested.   Three water-rats have been found in the stomach of one Pike, accompanied by the remains of a bird too far decomposed to be recognizable, but supposed to be the remnants of a duck.   So universal is the appetite of this fish, that it has even been known to seize the paste bait which had been used for other and less voracious inhabitants of the waters.

When the Pike attains a tolerable size, it takes possession of some particular spot in the bank, usually a kind of hole or cave which is sheltered by overhanging soil or roots, and affords a lair where it can lurk in readiness to pounce upon its passing prey.

The Pike seems to have no limit to its size, for it is a very long-lived fish, and seems always to increase in dimensions provided it be well supplied with food.   A fish of ten or twelve pounds' weight is considered to be a fine specimen, though there have been examples where the Pike has attained more than five times the latter weight.   These huge fishes of sixty or seventy pounds are, however, of little value for the table.

The colour of the Pike is olive-brown on the back, taking a lighter hue on the sides, and being variegated with green and yellow.   The abdomen is silvery white.

THE SALMON is undoubtedly the king of British river-fish ; not so much for its dimensions, which are exceeded by one or two giant members of the finny tribe, but for the silvery sheen of its glittering scales, its wonderful dash and activity, affording magnificent sport to the angler, the interesting nature of its life from the egg to full maturity, and last, but not least, for the exquisite flavour and nutritive character of its flesh.

SALMON.—(*Salmo Salar.*)

In former days, before civilization had substituted man and his dwellings for the broad meadows and their furred and feathered inmates, the Salmon was found in many an English river.   Now, however, there are but few streams where this splendid fish can be seen, for in the greater number of British rivers the water has been so defiled by human agency, that the fastidious Salmon will not suffer itself to be poisoned by such hateful mixture of evil odours and polluted waters ; and in the few streams where the water is still sufficiently pure for the Salmon to venture into them, the array of nets, weirs, and all kinds of Salmon traps is so tremendous, that not one tithe of the normal number are now found in them.

The Salmon is a migratory fish, annually leaving the sea, its proper residence, and proceeding for many miles up rivers for the purpose of depositing its spawn.   This duty having been accomplished, it returns to the sea in the spring.   The perseverance of this fish in working its way up the stream is perfectly wonderful.   No stream is rapid enough to daunt it, nor is it even checked by falls. These it surmounts by springing out of the water, fairly passing over the fall. Heights of fourteen or fifteen feet are constantly leaped by this powerful fish, and when it has arrived at the higher and shallower parts of the river, it scoops furrows in the gravelly bottom, and there deposits its spawn. The young, called "fry," are hatched about March, and immediately commence their retreat to the sea. By the end of May the young Salmon, now called "smolts," have almost entirely deserted the rivers, and in June not one is to be found in fresh water. Small Salmon, weighing less than two pounds, are termed "salmon peel," all above that weight are called "grilse."

The havoc wrought among Salmon by foes of every description is so enormous, that notwithstanding the great fecundity of the fish, it is a matter of surprise that so many escape destruction; for although the fish are preserved from their human foes by many stringent regulations, yet other foes, such as otters, who devour the large fish, and other fish who devour the spawn, have but little respect for laws and regulations.

While in the rivers, multitudes of Salmon are annually caught, usually by stake nets, which are capable of confining an immense number of fish at one time. Salmon-spearing is a favourite amusement. This animated and exciting sport is usually carried on by torch-light. The torches, when held close to the surface of the water, illumine the depths of the river, and render every fish within its influence perfectly visible. The watchful spearman, guided by slight indications bearing no meaning to an unpractised eye, darts his unerring spear, and brings up in triumph the glittering captive, writhing in vain among the barbed points. In the northern rivers this destructive pursuit is carried on to a great extent, more than a hundred salmon being frequently taken in an evening. Anglers also find considerable sport in using the fly for this beautiful and active fish, whose strength makes it no mean antagonist.

NEXT to the salmon, the bright-scaled, carmine-speckled, active TROUT is perhaps the greatest favourite of anglers, and fully deserves the eulogies of all lovers of the rod; its peculiarly delicate flesh,

TROUT.—(*Salmo fario.*)

its fastidious voracity, and the mixture of strength, agility, and spirited courage with which it endeavours to free itself from the hook, forming a combination of excellences rarely met with in any individual fish.

The Trout is found in rapid and clear-running streams, but cares not for the open and shallow parts of the river, preferring the shelter of some stone or hole in the bank, whence it may watch for prey. Like the pike, it haunts some especial hiding-place, and in a similar manner is sure to take possession of a favourite haunt that has been rendered vacant by the demise of its predecessor or its promotion to superior quarters. Various baits are used in fishing for trout, such as the worm, the minnow, and the fly, both natural and artificial, the latter being certainly the neatest and most artistic method. The arcana of angling are not within the province of this work; and for information on that subject, the reader is referred to the many valuable works which have been written by accomplished masters of the art.

THOUGH not so brightly spotted as the trout, nor so desperately active when hooked, and very inferior in flesh, the CARP is yet in much favour with anglers, on account of its extreme cunning, which has earned for the fish the name of Fox of the waters. As the number of British fish is so great, and our space so small, it will be needful to compress the descriptions as much as possible, and to omit everything that does not bear directly on the subject.

The Carp is found both in rivers and lakes, and in some places, among which the royal palaces of France may be mentioned, will often grow to an enormous size and become absurdly tame, crowding to the bank on the least encourage-

CARP.—(*Cyprinus carpio.*)

ment, and poking their great snouts out of the water in anxious expectation of the desired food. It is most curious to watch these great creatures swimming lazily along, and to see how completely they have lost the inherent dread of man by the exercise of their reasoning powers, which tell them that the once feared biped on the bank will do them no harm, but, in all probability, will be the means of indulging their appetite with favourite food.

The Carp is one of the fish that retains its life for a lengthened period even when removed from the water, and if carefully packed in wet moss so as to allow a free circulation of air, will survive even for weeks. Anglers never seem sure of the Carp—taking plenty on one day and none at all for a week afterwards, the fish having been aroused to a sense of their danger, and declining to meddle with anything that looks as if it might hide a hook. Even the net, that is so effectual with most fish, is often useless against the ready wiles of the Carp, which will sometimes bury itself in the mud as the ground line approaches so as to allow the net to pass over it; or, if the ground be too hard for such a manœuvre, will shoot boldly from the bottom of the water, leap over the upper edge of the net and so escape into the water beyond.

BARBEL.—(*Cyprinus barbus.*)
GOLD-FISH.—(*Cyprinus auratus.*)

THE beautiful GOLD-FISH (*Cyprinus auratus*), so familiar as a pet and so elegant as it moves round the glass globe in which it is usually kept, is another member of this large and important genus. It seems to have been brought to this country from China, and has almost acclimatized itself to the cold seasons of England. Its habits and splendid clothing are too well-known to need description.

ANOTHER well-known member of the same genus is the BARBEL, a fine but not brilliant fish which is common in many of the English rivers.

This fish may easily be known, from the four fleshy appendages, called beards or barbules, which hang from the head, two being placed on the nose and the other two at each angle of the mouth. It is one of the mud-loving fish, grubbing with its nose in the soft banks for the purpose of unearthing the aquatic larvæ of various insects which make their home in such places, and being, in all probability, aided by its barbules in its search after food.

The Barbel is sometimes so sluggish in its movements, and so deeply occupied in rooting about the bank, that an accomplished swimmer will occasionally dive to the bed of the river, feel for the Barbel along the banks, and bring them to the surface in his bare hand. From this habit of grubbing in the mud, the Barbel has earned the name of Fresh-water Pig.

The colour of the Barbel is brown above with a green wash, and yellowish green on the sides. All the scales have a metallic lustre, and the cheeks and gill-covers have also a polished look as if covered with very thin bronze. The abdomen is white. The Barbel is somewhat long in proportion to its weight, which is extremely variable, seldom, however, exceeding eleven or twelve pounds.

THE TENCH is hardly so common as the other two species, preferring the slowest and muddiest rivers, and thriving well in ponds and lakes, or even in clay-pits. No water, indeed, seems to be too thick, muddy, or even fetid, for the Tench to inhabit, and it is rather curious that in such cases, even where the fishermen could scarcely endure the stench of the mud adhering to their nets, the fish were larger sized, and of remarkably sweet flavour.

In the winter months the Tench is said to bury itself in the mud, and there to remain, in a semi-torpid condition, until the succeeding spring calls it again to life and action. The colour of the Tench is greenish olive, darker above than below, and with a fine golden wash.

THE ease with which the GUDGEON is taken has passed into a proverb. This pretty little fish is usually found in shallow parts of rivers, where the bottom is gravelly. If the gravel is stirred up, the Gudgeons immediately flock to the place, and a worm suspended amid the turbid water is eagerly snapped at by them. The fishermen usually take them in nets, and keep them alive in well-boats. They are largely purchased as baits for trolling.

GUDGEON. —(*Gcbio fluviatilis.*)
BREAM.—(*Abramis brama.*)

The flesh of the Gudgeon is particularly delicate, and although its length rarely exceeds seven inches, yet from the ease with which numbers can be obtained, it forms a dish by no means to be despised. The BREAM is mostly found in large lakes or in slowly running rivers, the lakes of Cumberland being favourite resorts of this fish. Although the flesh of the Bream is not held in any great estimation, being poorly flavoured and very full of bones, so that, in spite of the great depth of its body, there is scarcely sufficient flesh to repay the trouble of cooking, still, the fish was formerly in much repute as a delicacy ; so that either the fish seems to have deteriorated, or the present generation to have become more fastidious.   Spring and autumn furnish the best Bream, and the flesh can be dried something like that of the cod-fish.

The colour of the Bream is yellowish white, except the cheeks and gill-covers, which have a silvery lustre without any tinge of yellow.   Sometimes the Bream attains a considerable size, reaching a weight of twelve or fourteen pounds.

THE last of the three is the ROACH, a fish especially dear to scientific anglers on account of its capricious habits and the delicate skill required to form a successful roach-fisher.

An angler accomplished in this art will catch Roach where no one without special experience would have a chance of a bite, and will succeed in his beloved sport through almost every season of the year, the winter months being the favourites.   So capricious are these fish and so sensitive to the least change of weather that a single hour will suffice to put them off their feed, and the angler may be suddenly checked in the midst of his sport by an adverse breeze or change in the temperature.

The Roach is a gregarious fish, swimming in shoals and keeping tolerably close to each other.   It is not a large species, all over a pound being considered as fine specimens, and any that weigh more than two pounds are thought rare.   It is a pretty fish, the upper parts of the head and body being greyish green glossed with blue, the abdomen silvery white, and the sides passing gradually into white from the darker colours of the back. The pectoral, ventral, and anal fins are bright red, the former having a tinge of yellow ; and the dorsal and tail fins are brownish red.

ROACH.—(*Leuciscus rutilus.*)
DACE.—(*Leuciscus vulgaris.*)

CLOSELY allied to the roach is the DACE (*Leuciscus vulgaris*), a common and small species that inhabits most of our streams.   The well-known CHUB (*Leuciscus cephalus*) also belongs to this genus, as does the BLEAK (*Leuciscus alburnus*), in many countries called the TAILOR BLAY by the ignorant, from the idea that whenever any other fish, especially the pike, wounds its skin, it immediately seeks the aid of the Bleak, which, by rubbing its body against the wound, causes the torn skin to close.   The beautifully white crystalline deposit beneath the scales was much used in the manufacture of artificial pearls, hollow glass beads being washed in the interior with a thin layer of this substance, and then filled with white wax.   The scales of the whitebait were also used for the same purpose.   The MINNOW (*Leuciscus phoxinus*) is another member of this large genus, and is too well known to need description.

A VERY curious order of fishes now comes before our notice.   These creatures are called Pectognathi, or "fixed jaws," because their jaws are fused together and cannot be opened and shut.

Our example of this curious order is the well-known SUN-FISH, which looks just as if the head and shoulders of some very large fish had been abruptly cut off, and a fin supplied to the severed extremity.

Several specimens of this odd-looking fish have been captured in British waters, and in almost every case the creature was swimming, or rather floating, in so lazy a fashion, that it permitted itself to be taken without attempting to escape. In the seas where this fish is generally found, the harpoon is usually employed for its capture, not so much on account of its strength, though a large specimen will sometimes struggle with amazing force and fury; but on account of its great weight, which renders its conveyance into a boat a matter of some little difficulty, and the leverage obtained by the harpoon quite necessary.

The flesh of the Sun-fish is white and well-flavoured, and is in much request among sailors, who always luxuriate in fresh meat after the monotony of salted provisions. In flavour and aspect it somewhat resembles that of the Skate. Its liver is rather large, and yields a very considerable amount of oil, which is prized by the sailors as an infallible remedy against sprains, bruises, and rheumatic affections.

SUN-FISH.—(*Orthragoriscus mola.*)

One of the most curious peculiarities of this fish is the structure of the eye, which is bedded in a mass of very soft and flexible folds belonging to the outer membranous coat, while it rests behind on a sac filled with a gelatinous fluid. When the creature is alarmed, it draws the eye back against the sac of fluid, which is thus forced into the folds of skin, and distends them so largely as nearly to conceal the entire organ behind them.

While swimming quietly along, and suffered to be undisturbed, it generally remains so near the surface that its elevated dorsal fin projects above the water. Only in warm, calm weather is it seen in this attitude, and during a stormy season it remains near the bed of the sea, and contents itself with feeding on the seaweeds which grow so luxuriantly at the bottom of the shallower ocean waters.

The colour of the Sun-fish is greyish brown, darker upon the back than on the sides of the abdomen, and the skin is hard and rough. It often attains a very great size, one that was harpooned on the equator measuring six feet in length. Several species of Sun-fish are known.

The family of the Syngnathidæ is represented by several British species.

The SEA HORSE is common in many European seas, and is sometimes captured on the British coasts. In all these fishes there is only one dorsal fin, set far back, and capable of being moved in a marvellous fashion, that reminds the observer of a screw propeller, and evidently answers a similar purpose. The tail of the Sea Horse, stiff as it appears to be in dried

specimens, is, during the life of the creature, almost as flexible as an elephant's proboscis, and is employed as a prehensile organ, whereby its owner may be attached to any fixed object. The specimen represented in the engraving is shown in the attitude which the creatures are fond of assuming. The head of the Sea Horse is wonderfully like that of the quadruped from which it takes its name, and the resemblance is increased by two apparent ears

SEA HORSE.—(*Hippocampus brevirostris.*)

that project partly from the sides of the neck. These organs are, however, fins, and when the fish is in an active mood, are moved with considerable rapidity. It is rather a remarkable fact that the Sea Horse, like the chameleon, possesses the power of moving either eye at will, quite independently of the other, and therefore must be gifted with some curious modification in the sense of sight which enables it to direct its gaze to different objects without confusing its vision.

The colour of this interesting little fish is light ashen brown, relieved with slight dashes of blue on different parts of the body, and in certain lights gleaming with beautiful iridescent hues that play over its body with a changeful lustre. About twenty species of Sea Horses are known, several of which have been exhibited alive in the aquarium at the Crystal Palace. The Cyclostomi, or Circular-mouthed fishes, are represented by several British examples.

THE well-known LAMPREY and its kin are remarkable for the wonderful resemblance which their mouths bear to that of a leech.

They are all long-bodied snake-like fish, and possess a singular apparatus

LAMPREY.—(*Petromyson marinus.*)

of adhesion, which acts on the same principle as the disc of the sucking fish, or the ventral fins of the goby, though it is set on a different part of the body. Several fishes are popularly known by the name of Lamprey, but the only one to which the title ought properly to be given is that shown in the engraving.

The Lamprey is a sea-going fish, passing most of its time in the ocean, but ascending the rivers for the purpose of spawning.

The flesh of the Lamprey is peculiarly excellent, though practically unknown to the great bulk of our population, and the juvenile student in history is always familiar with the fatal predilection of British royalty for this fish. Though it spends so much of its time in the sea, it is seldom captured except during its visit to the rivers, and even in that case is only in good condition during part of its sojourn. Practically, therefore, the Lamprey is less persecuted than most of the finny tribe who are unfortunate enough to possess well-flavoured flesh, and whose excellences are publicly known.

When the Lamprey deposits its spawn, it is obliged to form a hollow in the bed of the river, in which it can leave the eggs in tolerable safety and performs this operation with great speed and no small skill. The fish is not gifted with any great power of fin, and cannot make much head against a sharp current, needing to rest at intervals, and for that purpose fastening on to some large stone over which the stream has no control.

But when it sets to work upon its nursery, it takes advantage of the current to help it in its labours, and, by the mingled force of the stream and its own muscular action, soon contrives to carry away the pebbles that would interfere with the well-being of its future young.

The process is simple enough. When the Lamprey has fixed on the convenient spot to which it is urged by its unfailing instinct, it surveys the locality for a short time, and then sets vigorously to work. Fastening itself to one of the obnoxious pebbles, and disposing its body so as to gain the strongest hold upon the rushing stream, it "backs water" with wonderful energy, and fish and stone are soon seen tumbling together down the current.

In this way, the Lamprey will remove stones of such a magnitude that a fish of three times its dimensions would appear unable even to stir them. As soon as the stone has been moved a yard or two away, the Lamprey wriggles its way back again, and takes possession of another stone. By a repetition of this process the hollow is soon made, and the industrious fish is able to deposit its eggs therein.

The colour of the Lamprey is olive-brown, spotted and mottled with dark brown and deep greenish olive. Its ordinary length is from sixteen to twenty inches.

THE LAMPERN is plentiful in many of the English rivers, and, if the generality of residents near the water were only aware of its excellence for the table, would soon be thinned in numbers. The prejudice that exists against the eel and the lamprey is absolutely mild when compared with the horror with which the Lampern is contemplated in many parts of England. Not only do the ignorant people refuse to eat it, but they believe it to

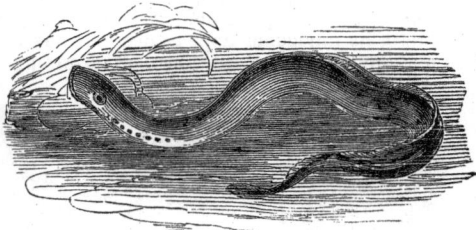

LAMPERN.—(*Lampetra fluviatilis.*)

be actually poisonous, and would sooner handle an angry viper than a poor harmless Lampern. It is fortunate for the fish that its evil reputation is so widely and firmly established, for, under shelter of its name, it passes scathless through many a stream, from which it would be nearly extirpated if its right character and good qualities were better known.

Granted the bad reputation, the creature certainly behaves in a manner well calculated to strengthen any unfavourable reports; for, as soon as grasped, it writhes about in a viperine, not to say venomous fashion, and is sure to fix its sucker of a mouth on the imprisoning hand. Few uninitiated captors can endure to any further extent, and when they feel the cold lips pressed to the skin, and the quick suck by which the fish attaches itself, they generally utter a scream of terror, and fling the Lampern away as far as their arm can jerk it. Yet the creature has no idea of using its mouth as a weapon of offence, and when it fixes itself to the hand, is only seeking for a point of support as a fulcrum for its struggles.

Certainly, it has teeth, and under proper circumstances can use them in the task for which teeth were made, but it seems either to be unable or unwilling to employ them as weapons. I have caught thousands of these fish with the bare fingers, and had six or seven fixed on my hand at the same time ; but they never did the least harm, and though I am afflicted with a peculiarly delicate skin, they did not even leave the least mark of their presence.

Like the sea lamprey, it scoops hollows in the pebbly bed of some stream for the purpose of depositing its eggs, and removes the stones in like manner. Sometimes a pair of Lamperns settle upon one spot, and, by dint of tugging and hauling, make a cradle for their special benefit. But it often happens that a great number of these fish, fifty or sixty, for example, will settle themselves in the same locality, and make a hollow as large as the rim of an ordinary pail.

The flesh of the Lampern is remarkably excellent, and in many places, remote from its habitation, is in great repute, and is indeed admired by many who have not the least idea of the fish they are eating. A large part of the "eel" pies so famous in the metropolis is composed of Lampern flesh, and in the opinion of competent judges the substitute is better than the reality. It can be dressed in a variety of ways, stewing and potting being the favourites. Yet, as a general rule, the poorer portion of the community refuse to eat the fish, and suffer the pangs of cruel hunger rather than avail themselves of the rich banquet at their very doors.

THE MYXINE, or GLUTINOUS HAG-FISH, is so remarkably worm-like in its form and general appearance, that it was classed with the annelids by several authors, and was only placed in its proper position among the fishes after careful dissection.

The Myxine is seldom taken when at large in the sea, but is captured while engaged in devouring the bodies of other fish, to which it is a fearful enemy in spite of its innocuous appearance. It has a custom of getting inside the cod and similar fishes, and entirely consuming the interior, leaving only the skin and the skeleton remaining. The fishermen have good reason to detest the Myxine, for it takes advantage of the helpless state in which the cod-fish hangs on the hook, makes its way into the interior, and if the fish should

HAG-FISH, OR MYXINE.—(*Myxine glutinosa.*)

happen to be caught at the beginning of the tide, will leave but little flesh on the bones. The cod thus hollowed are technically called "robbed" fish. Six Myxines have been found within the body of a single haddock.

The name of Glutinous Hag-fish is derived from the enormous amount of mucous secretion which the Myxine has the power of pouring, from a double row of apertures, set along the whole of the under surface, from the head to the tail.

Around the lips of the Myxine are eight delicate barbules, which are evidently intended as organs of touch ; the mouth is furnished with a single hooked tooth upon the palate, serving apparently as an organ of prehension, and the tongue is supplied with a double row of smaller, but powerful teeth on each side, acting on the principle of a rasp. The Myxine can scarcely be said to possess any bones, the only indication of a skeleton being the vertebral column, which is nothing more than a cartilaginous tube, through which a probe can be passed in either direction.

The colour of the Hag-fish is dark brown above, taking a paler tint on the sides, and greyish yellow below. Its length is generally about a foot or fifteen inches.

THE last of the fishes is a creature so unfishlike that its real position in the scale of nature was long undecided, and the strange little being has been banded about between the vertebrate and invertebrate classes. Between these two great armies the LANCELET evidently occupies the neutral ground, its structure partaking with such apparent equality of the characteristics of each class, that it could not be finally referred to its proper rank until it had been submitted to the most careful dissections. In fact, it holds just such a position between the vertebrates as does the lepidosiren between the reptiles and the fishes.

It has no definite brain, at all events it is scarcely better defined than in

LANCELET.—(*Amphioxus lanceolatus.*)

many of the insect tribe, and only marked by a r ther increased and blunted end of the spinal cord. It has no true heart, the place of that organ being taken by pulsating vessels, and the blood being quite pale. It has no bones, the muscles being merely attached to soft cartilage, and even the spinal cord is not protected by a bony or even horny covering. The body is very transparent, and is covered by a soft delicate skin without any scales. There are no eyes, and no apparent ears, and the mouth is a mere longitudinal fissure under that part of the body which we are compelled, for want of a better term, to call the head, and its orifice is crossed by numerous cirrhi, averaging from twelve to fifteen on each side. Altogether, it really seems to be a less perfect and less developed animal than many of the higher molluscs.

The general aspect of the Lancelet is not unlike that of another fish called the leptocephalus, the delicate transparent body and the diagonal arrangement of the muscles causing a considerable resemblance between the two. But the leptocephalus is at once distinguished by its head, which, although very small in proportion to the body, is yet perfect, possessing well-developed eyes, gill-covers, jaws, and teeth; whereas the Lancelet has no particular head, and neither eyes, gill-covers, jaws, nor teeth.

INVERTEBRATE ANIMALS.

INVERTEBRATE ANIMALS.

MOLLUSCS.

WE now come to the second great division into which all animated beings have been distinguished. All the creatures which we have hitherto examined, however different in form they may be, the ape and the eel being good examples of this external dissimilarity, yet agree in one point, namely, that they possess a spinal cord, protected by vertebræ, and are therefore termed Vertebrated animals.

But with the fishes ends the division of vertebrates, and we now enter upon another vast division in which there is no true brain and no vertebra. These creatures are classed together under the name of Invertebrate animals; a somewhat insufficient title, as it is based upon a negative and not on a positive principle. Whatever may be its defects, it has been too long received, and is too generally accepted, to be disturbed by a new phraseology, and though it be founded on the absence and not the presence of certain structures, it is concise and intelligible.

THE first order of Invertebrate animals is called MOLLUSCA, a name given to these creatures on account of the soft envelope which surrounds their bodies.

THE highest of the molluscs are those beings which are classed together under the title of CEPHALOPODA. This term is derived from two Greek words, the former signifying " a head," and the latter " a foot," and is applied to these creatures because the feet, or arms as they might also be called, are arranged in a circular manner round the mouth.

They are all animals of prey, and are furnished with a tremendous apparatus for seizure and destruction. Their long arms are furnished with round hollow discs, set in rows, each disc being a powerful sucker, and, when applied to any object, retaining its hold with wonderful tenacity. The mode by which the needful vacuum is made is simple in the extreme. The centre of the disc is filled with a soft, fleshy protuberance, which can be withdrawn at the pleasure of the owner. When therefore the edges of the disc are applied to an object, and the piston-like centre withdrawn, a partial vacuum is formed, and the disc adheres like a cupping-glass or a boy's leather sucker.

These discs are all under the command of the owner, which can seize any object with an instantaneous grasp, and relax its hold with equal celerity. The arms are as movable and as useful to the cuttle-fish as the proboscis to the elephant, for besides answering the purposes which have been mentioned, they are also used as legs, and enable the creature to crawl on the ground, the shell being then uppermost.

OUR first example is the celebrated ARGONAUT, or PAPER NAUTILUS, the latter title being given on account of the extreme thinness and fragility of the shell, which crumbles under a heedless grasp like the shell of an egg, and the former in allusion to the pretty fable which was formerly narrated of its sailing powers. It is rather remarkable, by the way, that the shell of the Argonaut is, during the life of its owner, elastic and yielding, almost as if it were made of thin horn.

Two of the arms of the Argonaut are greatly dilated at their extremities; and it was formerly asserted, and generally believed, that the creature was accustomed to employ these arms as sails, raising them high above the shell, and allowing itself to be driven over the surface by the breeze, while it directed its course by the remaining arms, which were suffered to hang over the edge of the shell into the water and acted like so many oars. In consequence of this belief, the creature was named the Argonaut, in allusion to the old classical fable of the ship *Argo* and her golden freight.

Certainly, the *Argo* herself could not have carried a more splendid cargo than is borne by the shell of the Argonaut when its inhabitant is living and in its full enjoyment of life and health. The animal, or "poulp" as it is technically called, is indeed a most lovely creature, despite of its unattractive form. "It appeared," writes Mr. Rang, when describing one of these creatures which had been captured alive, " little more than a shapeless mass, but it was a mass of silver, with a cloud of spots of the most beautiful rose-colour, and a fine dotting of the same, which heightened its beauty. A long semicircular band of ultramarine blue, which melted away insensibly, was very decidedly marked at one of its extremities, that is of the keel. A large mem

ARGONAUT OR PAPER NAUTILUS.—(*Argonauta Argo.*)

brane covered all, and this membrane was the expanded velation of the arms, which so peculiarly characterises the poulp of the Argonaut.

"The animal was so entirely shut up in its abode, that the head and base of the arms only were a little raised above the edges of the opening of the shell. On each side of the head a small space was left free, allowing the eyes of the mollusc some scope of vision around, and their sharp and fixed gaze appeared to announce that the animal was watching attentively all that passed around it. The slender arms were folded back from their base, and inserted very deeply round the body of the poulp, in such a manner as to fill in part the empty spaces which the head must naturally leave in the much larger opening of the shell."

M. Rang then proceeds to show the real use of the expanded arms, which is to cover the shell on its exterior, and, as has since been definitely proved, to build up its delicate texture and to repair damages, the substance of the shell being secreted by these arms, and by their broad expansions moulded into shape.

The modes of progression employed by the Argonaut are to the full as wondrous as its fabled habits of sailing. Its progression by crawling has already been casually mentioned. While thus engaged, the creature turns itself so as to rest on its head, withdraws its body as far as possible into its shell, and, using its arms like legs, creeps slowly but securely along the ground, sometimes affixing its discs to stones or projecting points of rocks for the purpose of hauling itself along.

When, however, it wishes to attain greater speed, and to pass through the wide waters, it makes use of a totally different principle.

As has already been mentioned, the respiration is achieved by the passage of water over the double gills or branchiæ ; the water, after it has completed its purpose, being ejected through a moderately long tube, technically called the siphon. The orifice of the siphon is directed towards the head of the animal, and it is by means of this simple apparatus that the act of progression is effected. When the creature desires to dart rapidly through the water, it

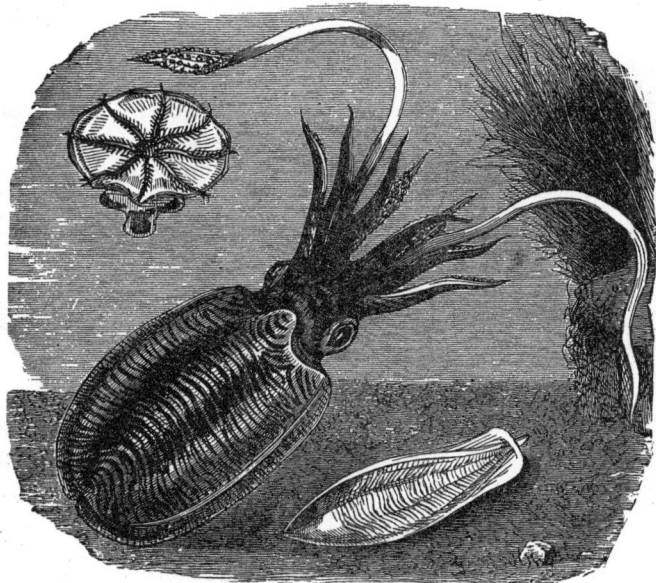

WEBBED SEPIA.—(*Cirrhoteuthis Mulleri.*)
COMMON SEPIA.—(*Sepia officinalis.*)

gathers its six arms in a straight line, so as to afford the slightest possible resistance to the water through which it passes, keeps its velated arms stretched tightly over the shell, and then by violently ejecting water from the siphon, drives itself, by the reaction, in the opposite direction.

As the various cephalopods are so numerous as to preclude all possibility of figuring and describing each species, we must therefore content ourselves with a typical form of each family and a general account of its members.

The lower figure in the illustration represents the common SEPIA of our own seas. It is chiefly remarkable for the chalky internal skeleton, commonly called Cuttle-bone, and much used for the manufacture of tooth-powder. This year, 1875, I found eight of these bones on the sands at Margate, and all within a space of a few yards square. The bone is seen lying on the right of the animal.

The upper figure is that of the WEBBED SEPIA, an inhabitant of Greenland, and very rare. Its colour is violet.

THE species belonging to the family of the Octopodidæ, or Eight-armed Cuttles, possess no external shell like that of the nautilus, its place being taken by two short styles or "pens" in the substance of the mantle. There are eight arms, unequal in length, and furnished with double or single rows of the suckers which have already been described.

They are solitary beings, voracious to a degree, and so active that they find little difficulty in capturing their prey or in escaping from the attacks of their enemies. Even when pursued into the narrow precincts of a rock-pool, the creature is not easily caught. When threatened, or if apprehensive of danger, the Polypus, as the animal was formerly called, darts with arrowy swiftness from one side of the pool to the other.

THE common OCTOPUS is now familiar to all those who have visited the

LITTLE SQUID, OR SEPIOLA. —(*Sepiola Atlantica.*)   OCTOPUS.—(*Octopus vulgaris.*)

great aquaria at Brighton and the Crystal Palace, where its extraordinary movements and great power of the arms are well shown.

THE family of the Teuthidæ, popularly known as Calamaries, or Squids, are distinguished by their elongated bodies, their short and broad fins, and the horny shell or pen which is found in their interior. All the squids are very active, and some species called, FLYING SQUIDS by sailors, and *Ommastrephes* by systematic naturalists, are able to dash out of the sea and dart to considerable distances.

OUR present example of this family is the LITTLE SQUID, or SEPIOLA, of which genus six species are known, inhabiting most parts of the world, and living on our own shores.

The celebrated "ink" of these creatures, from which the valuable colour called "sepia" was formerly obtained, deserves a brief notice.

This substance is liquid, and is secreted in a sac popularly termed, from its office, the "ink-bag." The sac is filled with a spongy kind of matter, in which the ink lies, and from which it can be forcibly expelled at the will of the animal. The ink-bag is not always in the same position, but some species have it in the liver, others near the siphon, and others among the viscera. There is a communication between the ink-bag and the siphon, so that when the ink is ejected it is forcibly thrown out together with the water. Thus the very effort for escape serves the double purpose of urging the creature away from danger and discolouring the water in which it swims.

The animal can eject the ink with such force that it has been known to dedecorate a naval officer's white duck trousers with its liquid missile, the aggrieved individual always asserting that it took a deliberate aim for that purpose.

Generally, the animal throws out its ink on the least alarm, a circumstance of some importance in geology. It was discovered by Dr. Buckland that in many specimens of fossil cephalopods, called scientifically Geoteuthis, *i e.*, Earth Squid, the ink-bag remained in the animal untouched by its long sojourn within the earth, and even retained its quality of rapid mixture with water. A drawing was actually made by Sir F. Chantrey, with a portion of " sepia " taken from a fossil species, and the substance proved to be of such excellent quality, that an artist to whom the sketch was shown was desirous of learning the name of the colourman who prepared the tint.

The curious skeleton of the Sepia, popularly called " Cuttle-bone," is composed of many tiers of tiny chalk pillars, which can only be seen by the aid of the microscope.

ANOTHER order of cephalopods is called by the name of Tetrabranchiata or Four-gilled animals, because the organs of respiration are composed of four branchiæ. These creatures possess a very strong external shell, which is divided into a series of gradually-increasing compartments connected together by a central tube called the siphuncle. As the animal grows, it continues to enlarge its home, so that its age can be inferred from the number of chambers comprising its shell.

In former days these creatures were very abundant, but in our day the only known living representative is the CHAMBERED or PEARLY NAUTILUS.

While the animal still lives, the short tubes that pass through the walls of the chambers are connected by membranous pipes, and even in a specimen that has been long dead, these connecting links hold their places, provided that the shell has not been subjected to severe shocks. In one of these shells now before me, which I have very cautiously opened, the whole series of membranous tubes can be seen in their places, black and shrivelled externally, but perfect tubes nevertheless.

The colour of the shell is very beautiful. The ground is white, over which are drawn, as with single dashes of a painter's brush, sundry bold streaks of reddish chestnut, mostly coalescing above, and reaching nearly to the centre of the spiral. This porcelain-like material is, however, only an outer coat laid on the real pearly substance of the shell, which is seen on looking into the hollow or into any of the chambers. The Chinese avail themselves of this double coating, and, with the untiring perseverance of their laborious nature, take the greatest trouble to spoil the finest shells by covering them with their grotesquely unperspective carvings of figures and landscapes, cut so as to relieve the deep colour of the raised figures by the white pearly background. Unlike the shell of the argonaut, which is almost as fragile as if made of sugar, that of the Nautilus is firm and strong, and will bear a considerable amount of rough handling before betraying any signs of injury.

THE order which now comes before our notice is composed of animals which crawl upon a broad muscular organ, termed, from its use, the foot. It is an enormously large order, containing all the snails, whether terrestrial, aquatic, or marine, the whelks, limpets, and similar animals not so familiarly known. Many species are much used as food, while others are of great service in the arts, furnishing employment to many hundreds of workmen. As the shell of these creatures consists of one piece or valve only, they are sometimes termed Univalves, in contradistinction to the oysters, mussels, scallops, and similar shells, which are termed Bivalves in allusion to their double shell.

There is a structure belonging to these animals which must be described before proceeding further, inasmuch as its shape and comparative dimensions often afford valuable indications by which a species, or even a genus may be distinguished. This structure is called "operculum," and its use,

CHAMBERED NAUTILUS.—(*Nautilus Pompilius.*)

when fully developed, is to close the aperture of the shell when the animal has withdrawn itself into the recesses of its home.

The operculum can be well seen in the water-snails, where it attains its full size, and exactly fits the opening which it is intended to protect. The material of which the operculum is essentially composed is a horny substance, but in some species the horn is strengthened by layers of the same nacreous matter which lines the shell, and becomes so thick and heavy, that when found separate from its owner it is often mistaken for some species of shell. The operculum is very variable both in its form and comparative dimensions, and even in its presence or absence. Sometimes it is circular, like a flat plate, and composed of concentric circles, while in some species it assumes a regularly spiral form like a flattened watch-spring.

THE shells that are included in the family of the Muricidæ may readily be distinguished by the straight beak or canal in front, and the absence of any such canal behind. All the animals belonging to this family are not only carnivorous, but rapacious, preying on other molluscs, and destroying them with the terrible armature called the tooth-ribbon, which, when examined with the microscope, proves to be a set of adamantine teeth, sharp-edged and pointed as those of the shark, and cutting their way through the hard shells of their victims as the well-known cordon saw passes through thick blocks of hard wood.

About one hundred and eighty species are known to belong to the typical genus, and there is hardly a portion of the world where a Murex of some kind may not be found.

THORNY WOODCOCK.—(*Murex tenuispinis.*)

The illustration represents the shell which is popularly known under the name of THORNY WOODCOCK, the latter title being given to it, in common with several of its congeners, on account of its long beak, which is thought to bear some resemblance to that of the woodcock, and the former in allusion to the vast number of lengthened spines or thorns which are arranged regularly over its surface. It has also received the equally appropriate and more poetical name of VENUS'S COMB.

This shell is found in the Indian Ocean, and varies greatly in dimensions, four or five inches being about the average length. It is evident that as nothing is ever made in vain, or to be wasted, the wonderful array of external spines must play some important part in nature, if not in the economy of the particular species. But what that part may be, and what may be the object of these beautiful structures, is a problem which seems almost insoluble, at all events with our present means of discovery.

The colour of the shell is very pale brown, each ridge being slightly tuberculated and edged with white. The spines are uniform drab or very pale brown, with an almost horny translucence.

WE now arrive at another and rather larger family, of which the common WHELK is a familiar example.

This is one of the most carnivorous of our molluscs, and among the creatures of its own class is as destructive as the lion among the herds of antelopes. Its long tongue, armed with row upon row of curved and sharp-edged teeth, harder than the notches of a file, and keen as the edge of a lancet, is a most irresistible instrument when rightly applied, drilling a circular hole through the thickest shells as easily as a carpenter's centre-bit works its way through a deal board.

WHELK.—(*Buccinum undatum.*)

The front of the tongue often has its teeth sadly broken, or even wanting altogether, but their place is soon supplied by others, which make their way gradually forward, and are brought successively into use as wanted.

As a general rule, there are about a hundred rows of teeth in the Whelk's tongue ; each row contains three teeth, and each tooth is deeply cleft into several notches, which practically gives the creature so many additional teeth.

Vast quantities of Whelks are taken annually for the markets, and are consumed almost wholly by the poorer classes, who consider them in the light of a delicacy. They are, however, decidedly tough and stringy in texture, and, like the periwinkle, which is also largely eaten, are not particularly digestible. The mode of taking these molluscs is very simple. Large wicker baskets are baited with the refuse portions of fish and lowered to the bottom of the sea by ropes. The ever-hungry Whelks instinctively discover the feast, crowd into the basket by thousands, and are

IMPERIAL HARP-SHELL.—(*Harpa imperialis.*)

taken by merely raising the laden basket to the surface and emptying it into a tub. Sometimes the Whelk is captured by the dredge, but the baited basket is the quickest and surest method. Besides its use as an article of human consumption, it is sometimes employed by the fishermen as bait for their hooks.

The reader will doubtless have observed on the sea-shore considerable masses of little yellowish capsules, mostly empty, and so light as to be drifted on the surface of the sea like so many masses of corks. These are the empty egg-cases of the Whelk. At the proper season of the year, when the unhatched egg-clusters are flung on the shores by the gales, the little Whelks can be discovered within the capsules, several shells being found in each case. Later in the season, the egg-capsules will be seen to be split open at one end, so as to allow the young to escape.

When hatched, the young escape into the sea through a round hole in the capsule.

The sweeping curves, broad swelling lip, and regular ridges of the next genus of shells have earned for them the popular title by which they are known.

About nine or ten species belong to this pretty genus, some of which are rare and costly. The IMPERIAL HARP-SHELL, which is represented in the engraving on page 508, is still a valuable shell; but in former days, when the facilities of commerce were far less than at present, it could only be purchased at a most extravagant rate.

The Harp-shells are only found in the hottest seas, and are taken mostly on the shores of the Mauritius, Ceylon, and the Philippine Islands. They frequent the softer and more muddy parts of the coast, and prefer deep to shallow water. None of the harp-shells possess the operculum.

The colour of the Imperial-harp is pale chestnut and white, with a dash of yellow, arranged in tolerably regular and slightly spiral bands.

ONE of the strangest, though not the most beautiful, of shells is the MAGILUS, a native of the Red Sea and the Mauritius.

On reference to the illustration, the reader will see two figures, one repre-

MAGILUS.—(*Magilus antiquus.*)

senting a group of madrepores, in which a small and delicate shell is lying, and the other a long, crumpled, and partly spiral tube, with a shell at one end and an opening at the other. Strange as the assertion may seem, these two figures represent the same animal in two stages of its development.

For the purpose, apparently, of carrying out some mysterious object, the Magilus resides wholly in the masses of madrepore, and in its early youth is a thin delicate shell without anything remarkable about it. As it advances in age, it enlarges in size, as is the case with most creatures; but its growth is confined to one direction, and instead of enlarging in diameter, it merely increases in length. The cause of the continual addition made to its length is probably to be found in the growth of the madrepore in which it is sheltered, and which would soon inclose the Magilus within it stony walls, did not the mollusc provide against such a fate by lengthening its shell and taking up its residence in the mouth.

The most curious point, however, in the economy of the Magilus is, that as fast as it adds a new shell in front, it fills up the cavity behind with a solid concretion of shelly matter, very hard, and of an almost crystalline structure, so as to leave about the same amount of space as in the original shell. The animal is always to be found in the very front of the shelly tube, and closes the aperture with a strong operculum that effectually shields it against all foes.

WE now pass to the CONE-SHELLS, or Conidæ, a family so called on account of their form. All the Cones have a similar external outline : the aperture is long and narrow, the head of the living animal is more or less lengthened, the foot is splay and abruptly cut off in front, the tentacles are rather widely separate, and the eyes are placed upon these organs.

The right-hand figure represents the TEXTILE CONE-SHELL, brought from the Mauritius. This handsome species is about four or five inches in length and its markings are curiously disposed, so that it is impossible to say which

ADMIRAL CONE.—(*Conus ammiralis.*)   TEXTILE CONE.—(*Conus textilis.*)

is the ground colour. The dark, narrow, angular lines are dark brown, accompanied by white, and variegated by dashes of yellow umber. The bold triangular spots are pure white, and the inside of the shell is of the same colour.

The empty shell lying on the ground, at the left of the illustration, is the ADMIRAL CONE, and is placed so as to exhibit the peculiarities of the long and narrow aperture. This species, in common with the other members of the genus, haunts the fissures and holes in rocks, and the warmer pools in coral reefs. They all take a moderate range of depth, varying from one to forty fathoms.

WE now come to the family of the Cowries, or Cypræidæ, two representatives of which family are given on page 511.

All the Cowries are lovers of the shallow waters near shore, and are carnivorous in their habits, feeding mostly upon the numerous zoophytes that inhabit the same coasts. These shells change their forms in a truly remarkable manner. When young, the shell is very like that of a volute having a prominent spire and a rather wide-spreading lip ; but in process of time the lobes of the mantle expand over it on either side, and by degrees deposit so thick a layer of smooth, shining substance, that the spire is entirely hidden. The pale streak which generally exists along the back of the Cowries indicates the line where the edges of the mantle nearly meet.

This little Cowry is so well known as to need no description.

The celebrated MONEY COWRY (*Cypræa moneta*) belongs to this genus. These little white shells are well known as being the medium of barter in many parts of Western Africa; and vast multitudes are gathered from their home in the Pacific and Eastern seas, and imported into this country for the purpose of immediate exportation to the African coast. Sixty tons' weight of Money Cowries have been freighted at a single British port in one year.

MONEY COWRY.
(*Aric a moneta.*)

The grooved or wrinkled edges of the lip are well known to everyone who has handled a Cowry, and these ridges assume a remarkable development in the DEEP-TOOTHED COWRY, a figure of which is here given, the empty shell being laid so as to exhibit the opening and the lips. The colour of this shell is extremely variable, but is mostly a mottled wood-brown, sometimes diversified with bands, and dark inside. It is not a very large species.

WE now arrive at a vast army of shells called the SEA SNAILS, and distinguished by having the edges of the aperture without notches, the shell spiral or limpet-shaped, and the operculum either horny or covered with hard, smooth, shelly matter.

One of the most curious of these shells is the SPINED NERITINA. The operculum is shelly, with a flexible border, and has some small teeth on its strait edge. All the Neritinæ are globular in their general shape, darkly

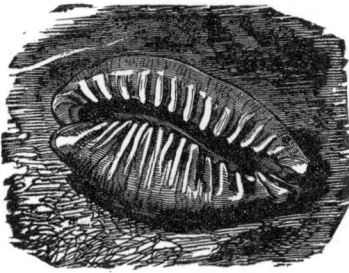

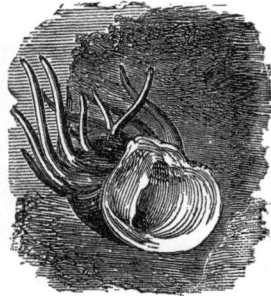

DEEP-TOOTHED COWRY.—(*Cypræa caurica.*) SPINED NERITINA.—(*Neritina spinosa*).

spotted or banded with black and purple, and covered with a polished bone-like epidermis. The colour of the Spined Neritina is deep green black on the exterior and blackish white within. The shell is thick and solid at the aperture, but becomes thinner towards the interior.

IN the family of the Turritellidæ, the shell is either tubular or spiral; the aperture is not waved, notched, or formed into canals; the foot is very small, the muzzle is short, and the eyes sunk rather deeply into the base of the tentacles.

THE lower figure in the accompanying engraving represents the empty shell of the STAIR-CASE or PRECIOUS WENTLETRAP, in former days one of the scarcest and most costly of the specimens of which a conchologist's cabinet could boast. There was hardly any sum which a wealthy connoisseur or virtuoso, as the fashion was then to call those who were fond of natural history, would not give for an especially large and perfect example of this really pretty shell. Now, however, its glory has departed, for a tolerably good specimen may be procured for a few shillings, and a Wentletrap which would a few years ago have been sold for fifty pounds can now be purchased for fifteen shillings.

Putting aside, however, the question of rarity or cost, this shell is a very interesting one, both for its beauty and the mode of its construction. It is purely white, and partly transparent, the elevated ridges being of a more snowy white than the body of the shell, on account of their superior thickness, which does not permit the light to pass through them as in the case of the thinner body. The whorls of this shell are separate from each other and, apparently bound together only by the projecting ridges, so that the general appearance is as if a long conical tube had been loosely coiled, and each whorl kept in its place by a succession of shelly elevations. This beautiful shell is found in the Indian and Chinese seas.

The smaller figure, showing the shell attached to the animal, represents

COMMON WENTLETRAP.—(*Scalaria communis.*)
STAIRCASE WENTLETRAP.—(*Scalaria pretiosa.*)

the COMMON or FALSE WENTLETRAP, a species tolerably common upon our coasts.

In this shell, the whorls are united together and furnished with a number of circular elevations, which, however, are not nearly so bold as those of the preceding species, but thick in proportion to their height, set obliquely on the shell, and smooth.

WE now arrive at another family, termed the Litorinidæ, or Shore Molluscs, because the greater number of them frequent the coasts, and feed upon the various algæ. The shell is always spiral and never pearly, by which latter characteristic it may be distinguished from certain shells belonging to another family, but somewhat similar in external appearance. The aperture is rounded. The animal has its eye set at the outer base of the tentacles, and the foot is remarkable for a longitudinal groove along the sole, so that in the act of walking each side advances in its turn. The tongue is rather long, and is armed with a formidable series of sharp teeth that serve admirably for the purpose of scraping away the vegetable matter on which the animal feeds. The operculum is horny, and rather spiral. The common PERIWINKLE (*Litorina litorea*) is the most familiar example of this family, and is too well known to need any detailed description.

The Periwinkle is found upon our rocks in great profusion, occupying the zone between high and low water and always being found near the edge of the tide.

IN former days, the PHEASANT-SHELLS were articles of great price and rarity, some specimens almost rivalling the precious wentletrap in the enormous sums asked and obtained for them. Now, however, that their habitations have been discovered and more frequent voyages are made, they have become comparatively plentiful, although, from the fragility of their structure, a perfect specimen is not at all common, and will still bring a good price in the conchological market.

PERIWINKLE.—(*Litorina litorea.*)

The Pheasant shells are now found in great numbers on the sandy beaches of several shores, being especially plentiful on the coast of Port Western in Bass's Straits. The high tide sweeps them towards the shore, where they are left by the receding waters, and seek for shelter beneath the masses of sea-weed that are always flung on the beach by the tide. On lifting these sheltering weeds, the Pheasant-shells may be found crowded together under

COMMON TOP.—(*Trochus zizyphinus.*)
AUSTRALIAN PHEASANT SHELL.—(*Phasianella Australis.*)

their wet fronds. They can move with some speed, the duplicate nature of the foot aiding them greatly in progression.

The little pointed shell on the left side of the illustration is the well-known TOP of our own shores.

This little shell, which is here represented of the natural size, is one of the

L L

most plentiful species of the British coasts, and may be found by hundreds either crawling among the sea weeds at low water, or flung upon the sands by the tide. The shell of this creature is beautifully pearly, and when the outer coating is removed the iridescent nacre below has a very lovely appearance. Jewellers and lapidaries employ these shells largely in their art, polishing them carefully and then stringing them together so as to form bracelets and necklaces, or affixing them as ornaments to various head-dresses.

LIMPET..—(*Patella vulgaris.*)

THE well-known univalves, so familiar under the name of LIMPETS, are divided into several families, on account of certain variations in the structure of the shell. The first family is termed Fissurellidæ, on account of the fissure which appears either at the apex or in the front edge of the shell.

All the Limpets are strongly adhesive to rocks, as is well known by everyone who has tried to remove one of these molluscs from the stony surface to which they clung. The means by which the animal is able to attach itself with such firmness is analogous to the mode in which the suckers of the cuttle-fish adhere to the objects which they seize, the formation of a vacuum, and the consequent pressure of the atmoshere, being the means employed. The foot of the Limpet is rounded, broad, thick, and powerful ; and when the animal wishes to cling tightly to any substance, it presses the foot firmly upon the surface, and retracts its centre, while its edges remain affixed to the rock. A partial vacuum is therefore formed, and the creature becomes as firmly attached to the rock as a boy's leathern sucker to the stone on which he has pressed it.

LIMPET.
(Showing underside.)

WE now come to the curious family of molluscs called appropriately Chitonidæ, or Mailshells, because their shells are jointed together like the pieces of plate armour. When separated from each other, the plates bear a strong resemblance to the joint of a steel gauntlet, and overlap each other in a similar fashion, a thick and strong mantle taking the place of the leather. There are eight of these plates, and all of them have a somewhat saddle-like shape. A similar arrangement may be observed in the lower abdominal plates of many beetles. Each of these plates is fixed to the mantle by certain rounded processes from their front edge, and when the plates are examined separately the processes will be plainly seen, white and pearly, as the interior of the shell.

MARBLED CHITON.—(*Chiton marmoreus.*)

The Chitons are able to roll themselves up in a partial kind of manner, and present a curious resemblance to the well-known armadillo, or pill woodlouse.

The illustration represents the MARBLED CHITON, a rather prettily-coloured shell, its exterior being rusty red mixed with brown and yellow, and edged with brown.

PASSING from the sea to the land, we come to those Gasteropods which breathe atmospheric air, and are furnished with respiratory organs suited to the element in which they live.

The first family is that of the SNAILS, or Helicidæ, containing a vast number of species. Most of the Snails have a shell large enough to permit the animal to withdraw itself wholly into the protecting domicile.

The genus Helix, which is universally accepted as the type of this family, is of enormous extent, both in numbers and in range of locality, containing more than fourteen hundred species, and spread nearly over the whole earth.

Our present example is the common SNAIL, which is even now largely consumed in many parts of the world, and is regularly fed and fattened for that purpose.

It is thought a delicacy by those who are sufficiently strong-minded to eat it ; and it is quite common to see, even in Paris, the poorer orders dressing their dinner of Snails on an iron plate, heated over burning charcoal.

An allied species, the Edible Snail, (*Helix pomatia*) was introduced into England by the Romans, and still exists in many places.

Towards the end of autumn, the Snail ceases to feed, withdraws itself to some sheltered spot, mostly under grass, moss, or dried leaves, and then sets about making its winter habitation. This process is very curious, and is thus described by Mr. Bell :—

" A large quantity of very viscid mucus is secreted on the under surface of the foot, to which a layer of earth or dead leaves adheres ; this is turned on one side, and, a fresh secretion being thrown out, the layer of earth mixed with mucus is left. The animal then takes another layer of earth on the bottom of the foot, turns it also to the part where he intends to form the wall of his habitation and leaves it in the same manner, repeating the process until the cavity is sufficiently large, and thus making the sides smooth, even, and compact. In forming the dome or arch of the chamber, a similar method is used ; the foot

COMMON SNAIL.—(*Helix aspersa.*)

collecting on its under surface a quantity of earth, and the animal, turning it upwards, leaves it by throwing out fresh mucus ; and this is repeated until a perfect roof is formed.

" As I have very often watched this curious process, I am certain of the facts. On removing very carefully the portion of the roof soon after its completion, I was enabled to see the formation of the operculum. In about an hour, or even less, the whole surface of the collar of the mantle instantaneously pours out the calcareous secretion in considerable quantity.

" This is at first a fluid or thick cream, but very soon acquires exactly the consistence of birdlime, being excessively adhesive and tenacious ; and in about an hour after it is poured out it is perfectly solid."

WE now arrive at the great family of Limacidæ, or SLUGS, a race of beings which many a gardener doubtless wishes extinct.

In these creatures the foot and body are indistinguishable from each other ; the head is retractile ; and the whole creature can be gathered into a short rounded mass, looking so like a pebble that it would escape a casual glance. At the first view, the Slugs appear to be destitute of shell, but on a closer examination the shell is found upon the fore part of the body, and either entirely or partly buried beneath the integuments.

The GREAT GREY SLUG is the largest of the British species, and when furnished with abundant food on which it can fatten itself during the night, and a secure hiding-place, whither it can retreat during the day, often attains an enormous size.

THE Water Snails are represented by the common POND-SNAIL, or LIMNÆA, shown in the act of climbing up the stem of a water-plant. In all the members of this family the shell is thin, and sufficiently capacious to contain the entire animal when it desires to withdraw itself into its home. The aperture is simply rounded, without notches or ridges, and the lip is sharp.

It may be found plentifully in nearly all streams where the water is not polluted, and the current not very swift. I have generally found that the back eddies of "lashers" are favourite haunts of various Water Snails.

DORIS.—(*Doris Johnstoni.*)

WATER SNAIL.—(*Limnæa stagnalis.*)

EOLIS.—(*Eolis coronata.*)

WE now arrive at a very remarkable series of molluscs which have been separated by systematic naturalists into a distinct section, appropriately called Nudibranchidæ, or Naked-gilled Molluscs, because their gills are always external and placed on the back or sides of the animals.

The slug-like animal which is represented crawling on the frond of a laminaria, is the common DORIS of our own shores. All the members of his family to which this creature belongs may be known by the plume-like gills set in a circle on the middle of the back, like the feathery coronet with

which the Blackfoot Indian adorns the head of his horse, and the two
tentacles placed more towards the front. In the skin are imbedded a vast
number of little spiculæ.

The beautiful EOLIS is common on our own coasts, and may be seen
moving over the plants and stones with tolerable activity and always keeping
the tentacles and papillæ in motion, sometimes contracting and sometimes
extending them, while the movement of the water causes them to wave in a
very graceful manner. These papillæ possess the property of discharging
a milky kind of fluid when the animal is irritated. The fluid, however, is
quite harmless, at all events to the human skin. As in the previous case, the
papillæ are liable to fall off at a touch.

A SMALL, but important, group of molluscs now comes before us. These
are the Pteropoda, or Wing-footed Molluscs, so called from the fin-like lobes
that project from the sides, and are evidently analogous to the similar organs
in some of the sea-snails. These appendages are used almost like wings,
the creature flapping its way vigorously through the water, just as a butterfly

HYALEA.—(*Hyalea tridentata.*)   CLEODORA.—(*Cleodora pyramidata.*)
(Empty shell below.)

urges its devious course through the air. They are found in the hotter seas,
swimming boldly in vast multitudes amid the wide waters, and one species
(*Clioborealis*) has long been celebrated as furnishing the huge Greenland
whale with the greater part of its subsistence.

The curious figure on the left hand of the illustration is the HYALEA,
remarkable not only for the two wide fins which are found in all the family
to which it belongs, but for the long appendages which pass through certain
apertures in the shell, and trail behind as the creature proceeds on its course.
It will be also seen that the wings are united by a nearly semicircular lobe.
The empty shell is placed below in order to show its structure.

Just on the right of the Hyalea is a smaller creature, with an odd-looking
three-pointed shell, hanging as it were from the wings. This is the
CLEODORA, a very beautiful and interesting animal, of which Mr. F. D.

Bennett writes as follows :—" On that part of the body which is lodged in the apex of the shell, there is a small, globular, pellucid body, resembling a vesicle, and which at night emits a luminous gleam, sufficiently vivid to be visible even when it is opposed to the strong light of a lamp. It is the only example of a luminous shell-fish I have ever met with ; nor would the luminosity of this species be of any avail, did not the shell possess a structure so vitreous and transparent. Examples were chiefly captured at night or in the evening.

THE next great group of molluscs is that which is known by the technical term of Conchifera. In them each valve corresponds with the right or left side.

In the first family, of which the common OYSTER is a very familiar instance, the two valves are unequal in size, and the animal inhabits the sea. The Oyster is too well known to need description ; but it may be mentioned, that practical naturalists have for some years been carefully studying its habits, for the purpose of breeding the valuable molluscs artificially, and so of securing a constant supply throughout the four months of the year during which the creature is out of condition. In this country the system is being gradually carried out, but in France it is developed to a very large extent, and with great success.

THE next family are termed wing-shells, or Avicularidæ because the apices, or "umbones" as they are called, are flattened and spread on either

OYSTER.—(*Ostrea edulis.*)   SCALLOP.—(*Pecten jacobeus.*)

side something like the wing of a bird. The interior of the valves is pearly and the exterior layer is composed of a kind of mosaic work of five or six sided particles. This structure is easily to be seen by means of a moderately powerful simple lens, merely by holding up a scallop or other shell before the window, so as to allow the light to pass through it.

THE common SCALLOP is found along our southern coasts, and in the seas of Europe. This shell was formerly used as the badge of a Pilgrim to the Holy Land.

> "—— His pilgrim's staff he bore,
> And fix'd the Scallop in his hat before."

It is a singular fact, that in the stomach of a common Scallop is found an earthy deposit, which, when boiled in nitric acid in order to dissolve the animal and other portions, exhibits under a powerful microscope animalcules

precisely similar to those which, in a fossil state, form the earth on which
the town of Richmond in America is built. THE well-known PEARL
OYSTER is one of the most valuable of the shell-bearing molluscs, furnishing
the greater part of the pearls that are set by jewellers and worn by ladies.

The pearls are secreted by the animal in precisely the same manner as the
nacre of the shell, and are, indeed, the same substance, formed into a
globular shape, and disposed in concentric layers so as to give that peculiar
translucency which is quite indescribable, but is known among jewellers by
the name of "water."

The Pearl Oyster does not produce its costly harvest under six or seven
years of age, and it is therefore a matter of importance that the bed should
be so managed that the young Oysters may be suffered to remain in peace
until they have attained an age which renders them capable of repaying the
expense of procuring them, and that no part of the bed should be harried where
the Oysters are too small to produce pearls.

The Oysters are now obtained by means of men who are trained to the
business, and who can remain under water for a considerable time without
being drowned. Each diver takes with him a net-bag for the purpose of

PEARL OYSTER.—(*Meleagrina Margaritifera.*)   MUSSEL.—(*Mytilus Edulis.*)

holding the Oysters, puts his foot into a stirrup, to which hangs a stone
weighing about thirty pounds, and after taking a long breath is swiftly carried
to the bottom. He then flings himself on his face, fills his bag as fast as
he can, and when his breath begins to fail, shakes his rope as a signal, and
is drawn up together with the bag.

WE now come to the large, useful, and even beautiful family of the
MUSSELS.

THE EDIBLE MUSSEL, so common in the fishmonger's shop and the costermonger's barrow, is found in vast profusion on our coasts, where it may be
seen moored to rocks, stones, and fibres, alternately covered with water or
left dry according to the flowing and ebbing of the tide. At some periods
of the year the Mussel is extremely injurious as an article of food, though the
effects seem to depend greatly on the constitution of the partaker. Attempts
have been successfully made to propagate the breed of Mussels; and the

vast plantations, as they may be called, of these creatures have increased to such an extent, that they threaten to obliterate several useful bays for all maritime purposes.

COCKLE.—(*Cardium edule.*)

THE family of the COCKLES, or Cardiadæ, so called from their heart-like shape, is well represented by the common COCKLE (*Cardium edule*) of our British shores. Generally, the Cockle is a marine animal ; but it sometimes prefers brackish water to the salt waves of the ocean.

This mollusc frequents sandy bays, and remains about low-water mark, burying itself in the sand by means of the powerful foot, which enables it to leap to a surprising height.

WE now come to the well-known Solenidæ or RAZOR-SHELLS, so called on account of their shape.

These curious molluscs always live buried in the sand in an upright position, leaving only an opening shaped like a key-hole, which corresponds with the two siphon tubes.

COMMON RAZOR-SHELL.—(*Solen vagina.*)

These creatures are generally found at a depth of one or two feet, and when they make their burrows, as they are often in the habit of doing among the rocks, not even the hooked iron can draw them from their retreat.

PIDDOCK.—(*Pholas dactylus.*)

WE next come to the Pholas, the best example of which is the common species popularly called PIDDOCK, and found in profusion along the sea-coast.

The common Piddock may be found in vast numbers in every sea-covered chalk rock, into which it has the gift of penetrating so as to protect itself from almost every foe.

Mr. Woodward remarks, very justly, that the " condition of the Pholades is always related to the nature of the material in which they are found burrowing ; in soft-sea beds they attain the largest size and greatest perfection, whilst in hard and especially gritty rock they are dwarfed in size, and all prominent points and ridges appear worn by friction. No notice is taken of the hypothesis which ascribes the perforation of rocks, &c., to ciliary action, because in fact there is no current between the shell, or siphon, and the wall of the tube." As soon as the animal has completely buried itself it ceases to burrow, and only projects the ends of the siphon from the aperture of the tunnel.

ALLIED to the preceding molluscs is the SHIPWORM, so called from its depredations on the bottoms of ships and all submerged wooden structures ; it is found in most seas, and on our own coasts works fearful damage by

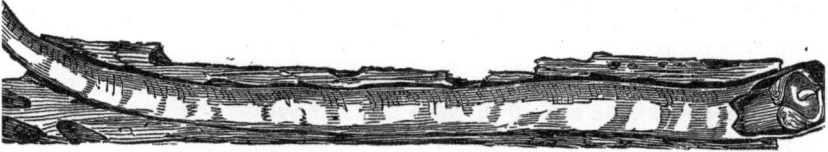

SHIP-WORM.—(*Teredo navalis.*)

eating into piles, planks, or even loose wood that lies tossing about in the ocean.

When removed from the tube the Shipworm is seen to be a long greyish white animal, about one foot in length and half an inch in thickness. At one end there is a rounded head, and at the other a forked tail. The curious three-lobed valves are seen on the right hand as they appear before being separated from the animal. The burrow which the creature forms is either wholly or partially lined with shell, and it is worthy of notice that the Shipworm and its mode of burrowing was the object that gave Sir I. Brunel the idea of the Thames Tunnel.

POLYZOA.

THE very remarkable beings which now come before our notice are appropriately termed POLYZOA, from two Greek words, signifying "many animals,"

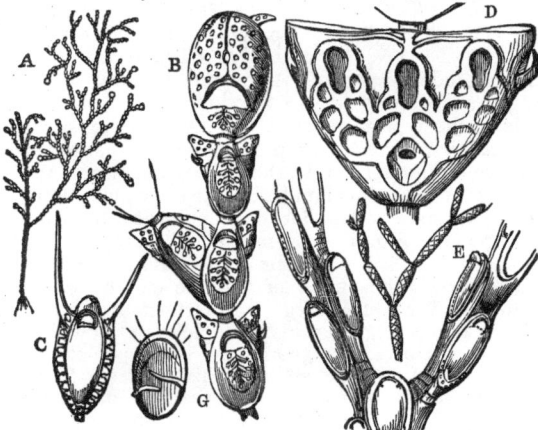

A. *Catenicella lorica.*　　　**B.** *Catenicella hastata.*　　　**C.** *Catenicella cornuta.*
D. *Calpidium ornatum.*　　　**E.** *Salicornaria jarciminoides.*　　　**F.** *Cellularia Peachii.*
G. *Menipea Fuguensis.* (Mouth of a cell.)

because a large number of individuals are massed together in groups of various forms and textures.

The true animal nature of these and many other beings, which had been formerly classed among the vegetables, was at length fairly proved by the researches of two eminent men, Trembley and Ellis, the latter of whom may lay claim to the honour of having produced the best and most comprehensive work of his time.

Fig. A is an example of one of these beings, the LITTLE CHAIN, or BREAST-PLATE, one of those beings that are so plentiful in the sea, and are popularly called zoophytes. This figure is of the natural size ; but in order to show the peculiarities of structure, two examples of species belonging to the same genus are given as they appear when considerably magnified. Fig. B is the *Catenicella hastata*, wherein is seen the shape of the cells, the form of their mouths, the method in which they give out their branches, and the peculiar organs called technically "avicularia" and "vibracula;" the former being processes that in many species bear an almost absurdly close resemblance to the heads of birds; and the latter, curious hair-like projections, which move regularly backward and forward as if impelled by machinery.

The members of the present genus are found most commonly in the Australian seas, seldom in the southern hemisphere, while in the northern hemisphere they are almost entirely unknown. Many specimens have been taken from Bass's Straits, at a depth of forty-five fathoms. As a general rule, however, the Polyzoa prefer the shallower waters, and are most commonly found a little below low-water mark.

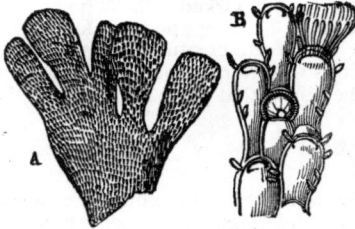

Fig. C is another species belonging to the same genus, and is remarkable for the long pointed spines that project from the margin, like a pair of cow's horns. In allusion to this peculiarity it is called *Catenicella cornuta*.

A. *Flustra foliacea* (Sea Mat.)
B. *Flustra foliacea*-(*magnified.*)

At Fig. D is shown another curious polyzoon, termed *Calpidium ornatum*, also found in Bass's Straits, at the same depth as the preceding species. A magnified figure is given in order to show the singular method of its construction.

An example of the typical genus of this family is given at Fig. E, where the *Salicornaria farciminoides* is represented of the natural size.

WE now arrive at another family, the Cellularidæ, where the general shape resembles that of the preceding family, but the cells, instead of being arranged round an imaginary axis, and so forming cylindrical branches, are arranged on the same plane. Fig. F is a magnified example of this family, the *Cellularia Peachii* so called in honour of the eminent naturalist, Mr. Peach.

At Fig. G is shown the mouth of a single cell, belonging to the genus Menipea, found in Terra del Fuego, and termed from its habitat, *Menipea Fuguensis*. The object of giving this example is to show the curious "operculum" which closes or rather guards the mouth of the cell, and in this genus is in the form of a simple spike. This species is found at low-water.

EVERYONE who has walked along the seashore must have observed the pretty leaf-like Sea Mats strewn on the beach, and admired the wonderful regularity of their structure, perceptible to the naked eye; but when magnified, even by a pocket lens, their beauty increases in proportion to the power employed, and the marvellous arrangement of the cells and the orderly system in which they are placed are almost beyond belief. Beautiful, however, as they are in this state, they are but the dead and lifeless habitations of the creatures who built the wondrous cells, and the only method of showing the Sea Mat in its full glory is to take a living specimen from the stone or shell to which it is affixed, and watch it under the microscope while the creatures are still in full activity. In the illustration, Fig. A, is shown a portion of the common Sea Mat, sometimes called the Hornwrack, of its natural size ; and Fig. B represents a few cells of the same species rather highly magnified.

On the upper-right hand cell may be seen the funnel-shaped group of tentacles belonging to one of the animals, and in the centre is a curiously shaped cell, which is analogous to the birds' heads which we have so lately examined, the place of the lower jaw being supplied by a kind of lid.

INSECTS.

THE INSECTS, to which we must devote a few pages, afford the first examples of the Articulata, *i.e.*, the jointed animals without vertebræ. Their bodies are composed of a series of rings, and they are separated into at least two and mostly three portions ; the head being distinct from the body. They pass through a series of changes before attaining the perfect form ; and when they have reached adult age they always possess six jointed legs, neither more nor less, and two antennæ, popularly called horns or feelers.

In most instances their preliminary forms, technically called the larva and pupa, are extremely unlike the perfect Insect ; but there are some in which, at all events externally, they retain the same shape throughout their entire life. The whole of the growth takes place in the preliminary stages, so that the perfect insect never grows, and the popular idea that a little insect is necessarily a young one is quite incorrect.

Insects breathe in a very curious manner. They have no lungs nor gills, but their whole body is permeated with a network of tubes, through which, the air is conveyed, and by means of which the blood is brought into contact with the vivifying influence of the atmosphere. These breathing tubes, technically called tracheæ, ramify to every portion of the creature, and penetrate to the extremities of the limbs, the antennæ, and even the wings, when those organs exist. Their external orifices are called spiracles, and are set along the sides.

They have very little internal skeleton, the hard materials which protect the soft vital organs being placed on the exterior, and forming a beautiful coat of mail, so constructed as to defend the tender portions within, and yet to permit perfectly free motion on the part of the owner.

There are many other interesting points in the structure of the Insects, such as the eyes, the wings, the tracheæ, &c., which will be described in the course of the following pages.

The first order is called the Coleoptera, a word of Greek origin, signifying sheathed winged animals, and includes all those insects which are more popularly known under the title of Beetles. In these insects the front pair of wings are modified into stout horny or leathery cases, under which the second pair of wings are folded when not in use. The hinder pair of wings are transparent and membranous in their structure, and when not employed are arranged under the upper pair, technically called the elytra, by folds in two directions, one being longitudinal and the other transverse. The mouth is furnished with jaws, often of considerable power, which move horizontally.

PASSING over the details of classification, we come to the first family of Insects, scientifically called the Cicindelidæ, and popularly known by the name of TIGER BEETLES, or Sparklers, both names being very appropriate ; the former on account of their exceeding voracity, their ferocious habits, and the wonderful activity of their movements ; and the latter in allusion to the brilliancy of their colours as they flash along in the sunshine. These Beetles are represented by several British species, among which the common TIGER BEETLE (*Cicindela campestris*) is the most common, and perhaps the most beautiful. Well does this little creature deserve its popular name ; for what the dragon-fly is to the air, what the shark is to the sea, the Tiger Beetle is to the earth ; running with such rapidity that the eye can hardly follow its course ; armed with jaws like two reapers' sickles crossing each other at the points ; furnished with eyes that project from the sides of the head and permit the creature to see in every direction without turning itself; and, lastly, gifted with agile wings that enable it to rise in the air as readily as a fly or a wasp. Moreover, it is covered with a suit of mail, gold embossed, gem studded, and burnished with more than steely brightness, light yet strong,

TIGER BEETLE.

EIGHT-SPOT TIGER BEETLE.—(*Cicind. la octonotata.*)

and though freely yielding to every movement, yet so marvellously jointed as to leave no vulnerable points even when in full action, and, in fine, such a suit of armour as no monarch ever possessed and no artist ever conceived.

Even in its larval state the Tiger Beetle is a terror to other insects, snapping them up as they pass by its burrows and dragging them into the dark recesses of the earth to be devoured. Several American species inhabit trees, and are quite as destructive among the branches as their congeners upon the earth. The typical species which is represented in the illustration is the EIGHT-SPOT TIGER BEETLE of India.

The British Tiger Beetle is remarkable for exuding a powerful scent, much resembling the odour produced by a crushed verbena leaf.

A VERY large and important family of Beetles, the Carabidæ, now comes before us, which is represented in England by very many species, the common GROUND BEETLES being familiar examples.

Of the typical genus of this family we take the Violet Ground Beetle (*Carabus violaceus*) as an example.

This fine Beetle is plentiful in this country, and may be found in gardens, gravel-pits, and similar localities. It is said to be especially common in the midland counties.

The elytra are rather convex, and narrowed at the shoulder, and are finely granulated, *i.e.*, covered with minute rounded projections. They are black; but the margins are edged with a band of coppery or golden violet, sometimes warming into purple. The body is black beneath. The disc of the thorax is black, and the margins are violet; and the head is black. The length of the beetle is about an inch.

The members of this genus are almost wholly inhabitants of temperate climates; and it has been stated that scarcely any species are to be found within thirty degrees from the equator on either side.

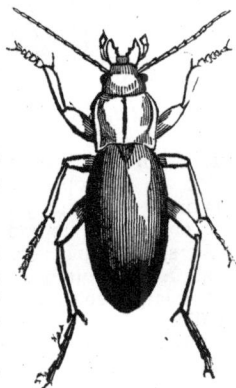

VIOLET GROUND BEETLE.
(*Carabus violaceus.*)

WE now come to the large group of WATER BEETLES, which are divided into several families.

In order to enable them to perform the various movements which are necessary for their aquatic existence, their hind legs are developed into oars with flattened blades and stiff hairy fringes, and the mode of respiration is slightly altered in order to accommodate itself to the surrounding conditions. It has been already mentioned that in all insects the respiration is conducted through a series of apertures set along the sides, and technically called spiracles. In the Water Beetles, the spiracles are set rather high, so as to be covered by the hollowed elytra, and to be capable of breathing the air under those organs. When, therefore, the beetle dives it is in no ways distressed for want of air, as it carries a tolerable supply beneath the elytra. When, however, that supply is exhausted, the beetle rises to the surface, just pushes the ends of the elytra out of the water, takes in a fresh supply of air, and again seeks its subaquatic haunts.

The male of the Great Water Beetle, in common with other species, is

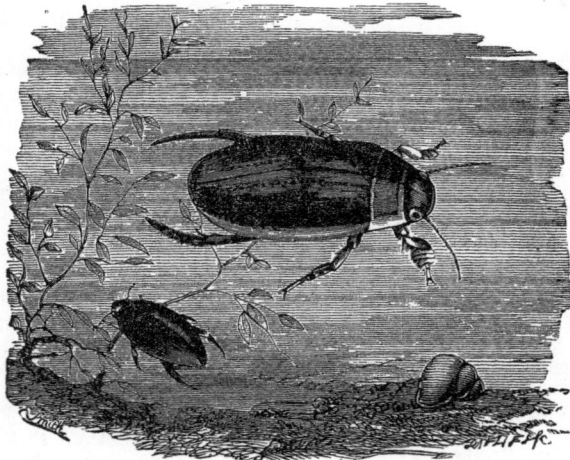

(*Ilybius ater.*)   WATER BEETLES.   (*Dyticus dimidiatus.*)

specially notable for the singular development of the fore-legs, the tarsi of which are developed in a most extraordinary apparatus caused by the dilatation of the three first joints, which are flattened so as to form a nearly circular disc, covered on its under surface with a multitude of wonderfully-constructed suckers, one being very large, another about half its size, and the others very small, and set on pear-shaped footstalks.

The larger specimen in the engraving is the *Dyticus dimidiatus*, and the smaller is the *Ilybius ater*, both British species.

PASSING by several large and interesting families, we come to the curious creatures which will at once be recognised by reference to the illustration. These beetles are popularly known by the name of ROVE BEETLES, or COCKTAILS, the latter name being given to them on account of their habit of curling up the abdomen when they are alarmed or irritated. The common BLACK COCKTAIL has, when it assumes this attitude, standing its ground defiantly with open jaws and elevated tail, so diabolical an aspect that the rustics generally call it the Devil's Coach-horse. It has, moreover, the power of throwing out a most disgusting odour, which is penetrating and persistent to a degree, refusing to be driven off even with many washings.

These beetles are termed Staphylinidæ, or Brachelytra, the latter term signifying short elytra, and being a very apposite name, as the elytra are short, square, and not more than one-fourth the length of the abdomen. If we watch one of these beetles settling after its flight, we shall see the object of its flexible tail. The wings are so large, and the elytra so small, that the process of folding the delicate membranes could not be completed without some external aid. When the insect alights, it suddenly furls its wings into loose folds, and then, by means of its tail, it pushes the wings under the elytra, which are then shut down. This process, although rather elaborate, is effected in a very rapid manner.

(*Ocypus olens.*)   ROVE BEETLES.   (*Creophilus maxillosus.*)

The two species which are represented in the illustration are common in England. The lower figure represents the *Creophilus maxillosus*, which is plentiful in and about drains or dead animal matter, and may be known by the grey hairy look of the elytra.

The upper figure represents the Devil's Coach horse, shown of the natural size.

NEXT to the Staphylinidæ are placed some insects that have become quite famous for their curious and valuable habits. These are the Necrophaga, popularly and appropriately termed Burying Beetles.

It is owing to the exertions of these little scavengers that the carcases of birds, small mammals, and reptiles are seldom seen to cumber the ground, being buried at a depth of several inches, where they serve to increase the fertility of the earth instead of tainting the purity of the atmosphere. These beetles may easily be captured by laying a dead mouse, mole, bird, frog, or even a piece of meat on the ground, and marking the spot so as to be able to find the place where it had been laid. It will hardly have remained there for a couple of hours before some Burying Beetle will find it out, and straightway set to work at its interment. The plan adopted is by burrowing underneath the corpse and scratching away the earth so as to form a hollow, into which the body sinks. When the beetles have worked for some time they are quite hidden, and the dead animal seems to subside into the ground as if by magic.

The strength and perseverance of these beetles are so great that a very short time suffices to bury the creature completely below the ground, and the earth being scraped over it, the process is complete. The object of burying dead animals is to gain a proper spot wherein to deposit their eggs,

BURYING BEETLES.
(*Hister cadaverinus.*)   (*Necrophorus vestigator.*)   (*Silpha opaca.*)

as the larvæ when hatched feed wholly on decaying animal substance.

WE now come to the Lamellicorn Beetles, so called from the beautiful plates, or lamellæ, which decorate the antennæ. This family includes a vast number of species, many of which, as, for example, the Common Cockchafer,

COCKCHAFER.—(*Melolontha vulgaris*)

are extremely hurtful to vegetation both in the larval and adult form. In this family are found the most gigantic specimens of the Coleoptera, some of which look more like crabs than beetles, so huge are they and so bizarre are their shapes. In all these creatures the lamellæ are larger and more beautiful in the female than in the male insect.

The COMMON COCKCHAFER is too familiar to need any description of its personal appearance, but the history of its life is not so widely known as its aspect. The mother beetle commences operations by depositing the eggs in the ground, where in good time the young are hatched. The grubs are unsightly-looking objects, having the end of the body so curved that the creatures cannot crawl in the ordinary fashion, but are obliged to lie on their sides. They are furnished with two terribly trenchant jaws like curved shears, and immediately set to work at their destructive labours.

They feed mostly upon the roots of grasses and other plants, and when in great numbers have been known to ruin an entire harvest. To turf they are especially destructive, shearing away the roots with their scissor-like jaws and killing the vegetation effectually. For three years the future insect continues in its larval state ; and after a brief sojourn in the pupai condition. changes its skin for the last time, and emerges from the ground a perfect Cockchafer. Even in its perfect state it is a terribly destructive insect, working sad havoc among the foliage of trees.

The STAG BEETLE is the largest of our British Coleoptera, and when it has attained its full dimensions is an extremely powerful and rather formidable insect, its enormous mandibles being able to inflict a very painful bite, not

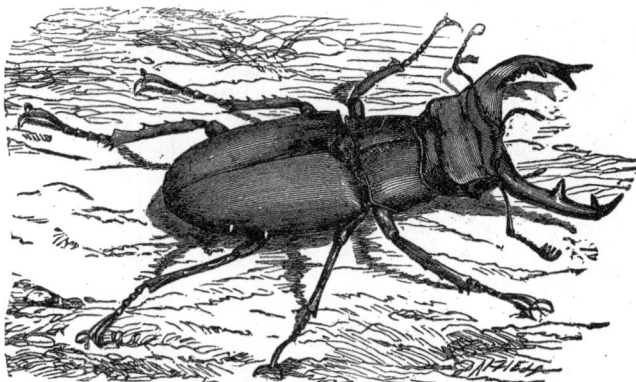

STAG BEETLE.—(*Lucanus cervus.*)

only on account of the powerful muscles by which they are moved, but in consequence of the antler-like projections with which their tips are armed These horn-like jaws only belong to the male, those of the female being simply sharp and curved mandibles, in no way conspicuous.

The larvæ of the Stag Beetle reside in trees, into which they burrow with marvellous facility, and as after they have emerged from their holes they appear to cling to the familiar neighbourhood, they may be found upon or near the trees in which they have been bred.

From the formidable shape of the mandibles it might be supposed that the Stag Beetle was one of the predaceous species. This, however, is not the case, the food of this fine insect consisting mostly, if not wholly, of the juices of vegetables, which it wounds with the jaws so as to cause the sap to flow. It is true that specimens have been detected in the act of assaulting other insects, but they never seem to have been observed in the act of feeding upon their victim. Whether the food be of animal or vegetable nature, it is always liquid, and is lapped, or swept up, by a kind of brush which forms part of the mouth, and looks like a double pencil of shining orange-coloured hairs.

PASSING by one or two families of more or less importance, we arrive at the Buprestidæ, a family of beetles remarkable for the extraordinary gorgeousness of their tints, almost every imaginable hue being found upon these brilliant insects.

They are found in many portions of the globe, but, as is generally the case with insects, their colours take the greatest intensity within the tropics. They fly well, and seem to exult in the hottest sunshine, where the bright beams cause their burnished raiment to flash forth its most dazzling hues. They are, however, slow of foot, and, when alarmed, have a habit of falling to the ground with folded limbs, as if they were dead.

The species that is given in the illustration is one of the finest of this splendid family. The sides of the thorax are covered with little round pits, something like the depressions on the head of a thimble, and are of a fiery copper hue. The head and middle of the thorax are light burnished blue, like that of a well-tempered watch-spring, and the elytra are warm cream-coloured, diversified with a patch of deep purple-blue at each side, and another at the tip. The CHRYSOCHROA is a native of India.

THE celebrated GLOW-WORM belongs to the typical genus of its family.

Contrary to the usual rule among insects, where the male absorbs the whole of the beauty, and the female is comparatively dull and sombre in colour and form, the female carries off the palm for beauty, at all events after dusk, the male regaining the natural ascendency by the light of day. Either through books, or by actual observation, almost everyone is familiar with the Glow-worm, and would recognize its pale green blue light on a summer's evening. Many, however, if they came across the insect by day, would fail to detect the brilliant star of the night in the dull, brown, grub-like insect crawling slowly among the leaves, and still fewer would be able to distinguish the male, so unlike are the two sexes.

It has often been said that the female alone is luminous. This, however, is an error, as I have caught numbers of these beetles of both sexes, and always found that the males were gifted with the power of producing the peculiar phosphorescent light, though in a much smaller degree than their mates, the light looking like two small pins' heads of phosphorus upon the end of the tail.

Seen by day, the male is a much handsomer looking insect than the female, being soft brown in colour, long-bodied, and wide-winged, altogether beetle-like ; while the female is more like a grub than a perfect insect, has no wings at all, and only the slightest indications of elytra.

The larva of the Glow-worm feeds upon molluscs, especially upon the

CHRYSOCHROA.—(*Chrysochroa Bugnetti.*)
GLOW-WORM.—(*Lampyris noctiluca.*)

smaller snails, which it is able to devour even when retracted within the walls of the shell.

THE two insects represented in the accompanying illustration are found in England, and are here given as examples of the family Cantharidæ, of which the BLISTER FLY, sometimes called SPANISH FLY, is the typical species.

OIL BEETLE.—(*Meloe violaceus.*)
BLISTER, OR SPANISH FLY.—(*Cantharis vesicatoria.*)

The Blister Fly is by no means a common species in England, though it has occasionally appeared in considerable numbers. In such cases, however, it is extremely local, and does not appear to be disseminated through the country. Spain is famous for the multitudes of Blister Flies which are found within its limits, and the whole of South-western Europe is prolific in this remarkable beetle.

The Spanish Fly is a handsome insect, nearly an inch in length, and of a rich silken green, with a gold gloss in certain lights.

The second figure represents an insect belonging to the same family, and very common in England. It popularly goes by the appropriate name of OIL BEETLE, because, when handled, it has the property of pouring a yellowish oily fluid from the joints of its legs.

As may be seen by reference to the illustration, the abdomen is extremely large in proportion to the rest of the body, and the short diverging elytra descend but a very little way below the thorax. The oily matter that is poured from the joints is considered in some countries to be a specific for rheumatism, and is expressed from the insect for medicinal purposes. The Oil Beetle is represented of the natural size, and its colour is dull indigo blue.

WE now arrive at a vast group of beetles, embracing several thousand species, which are popularly classed under the name of WEEVILS, and may all be known by the peculiar shape and very elongated snouts. Many of these creatures have their elytra covered with minute but most brilliant scales, arranged in rows, and presenting, when placed under the microscope, a spectacle almost unapproached in splendour. They are mostly slow in their movements, not quick of foot, and many being wholly wingless.

The most brilliant of the Weevils are to be found in the typical family Curculionidæ, to which belong the well-known Diamond Beetles, in such request as objects for the microscope.

The maggots that are so frequently found in nuts, and which leave so black and bitter a deposit behind them that the person who has unfortunately tasted a maggot-eaten nut is forcibly reminded of the Dead Sea apple, with its inviting exterior and bitter dusty contents, also belong to the Weevils, and are the larvæ of the NUT WEEVIL (*Balaninus nucum*). All the members of the genus are remarkable for the extraordinary length of the snout, at the extremity of which are placed the small but powerful jaws.

WE now come to the Longicorn Beetles, so called on account of the extraordinary length of the antennæ in many of the species. These insects are

NUT WEEVIL.
(*Magnified.*)

BLOODY-NOSE
BEETLE.

MUSKBEETLE. -- (*Cerambyx moschatus.*)

well represented in England by many species, the best known being the common MUSK BEETLE.

The beautiful beetles of which the common Musk Beetle is an excellent example, vary considerably in size ; some being several inches in length, while some are hardly one-quarter of an inch long. The extreme length of their antennæ is the most conspicuous property, and by that peculiarity they are at once recognized.

A small moth, Adela de Geerella, possesses the same peculiarity. The length of the moth is about a quarter of an inch, and the length of the antennæ more than an inch and a half. The antennæ wave about with every breath of air, as if the insect had become entangled in a spider's web, and escaped with some of the loose threads floating about it.

The Musk Beetle is a large insect, common in most parts of England. It is extremely common at Oxford, and is found in old willow-trees, with which Oxford is surrounded. Its peculiar scent, something resembling that of roses, often betrays its presence, when its green colour would have kept it concealed. When touched, it emits a curious sound, not unlike that of the bat, but more resembling the faint scratching of a perpendicularly-held slate pencil. Its larva bores deep holes in the trees, which are often quite honeycombed by them.

As in the preceding family, the Longicorn Beetles pass their larval state in wood, sometimes boring to a considerable depth, and sometimes restricting themselves to the space between the bark and the wood. The grubs practically possess no limbs, the minute scaly legs being entirely useless for locomotion, and the movements of the grub being performed by alternate contraction and extension of its ringed body. In order to aid in locomotion the segments are furnished with projecting tubercles, which are pressed against the sides of the burrow.

PASSING by several families, we come to the Chrysomelidæ, which are round-bodied, and in most cases very brilliantly coloured with shining green, purple, blue, and gold, of a peculiar but indescribable lustre. They are slow walkers, but grasp the leaves with a wonderfully firm hold. The British species of Chrysomela are very numerous. One of the genera belonging to this family contains the largest British specimen of these beetles, commonly known by the name of the BLOODY-NOSE BEETLE (*Timarcha tenebricosa*), on account of the bright red fluid which it ejects from its mouth and the joints of its legs when it is alarmed. This fluid is held by many persons to be a specific in case of toothache. It is applied by means of permitting the insect to emit the fluid on the finger and then rubbing it on the gum, and the effects are said to endure for several days. The larva of this beetle is a fat-bodied, shining, dark green grub, which may be found clinging to grass, moss, or hedgerows in the early summer. It is so like the perfect insect that its identity cannot be doubted.

THE family of the Coccinellidæ, or LADYBIRDS, is allied to the Chrysomelidæ, and is well known on account of the pretty little spotted insects with which we have been familiar from our childhood. Though the LADYBIRD is too well known to need description, it may be mentioned that it is an extremely useful insect, feeding while in the larval state on the aphides that swarm on so many of our favourite plants and shrubs. The mother Ladybird always takes care to deposit the eggs in spots where the aphides most swarm, and so secure an abundant supply of food for the future offspring.

EARWIGS.

TAKING leave of the beetles, we now proceed to a fresh order, distinguished by several simple characteristics, among which may be mentioned the soft and leathery elytra or forewings, the wide and membranous hind-wings, and the forceps with which the tail is armed. The insects belonging to this order are popularly known by the name of EARWIGS, and are represented in this country by several species of different dimensions.

The membranous wings of the Earwig are truly beautiful. They are thin and delicate to a degree, very large and rounded, and during the day-time packed in the most admirable manner under the little square elytra. The process of packing is very beautiful, being greatly assisted by the forceps on the tail, which are directed by the creature with wonderful precision, and used as deftly as if they were fingers directed by eyes. The Earwigs seldom fly except by night, and it is not very easy to see them pack up their wings. Some of the smaller species, however, are day-flyers, and in spite of their tiny dimensions may be watched without much difficulty. There are about seven or eight British species, some of them being of very small size.

GIANT EARWIG.

The largest British species is that which is given here. It is of very rare occurrence, and seldom seen, as it only inhabits the seashore, and never shows itself until dusk. I have a fine specimen that was caught on the sands near Folkestone, in the month of July.

ORTHOPTERA.

A LARGE and important order succeeds the Earwigs, containing some of the finest, and at the same time the most grotesquely formed members of the insect tribe. In this order we include the grasshoppers, locusts, crickets, cockroaches, and leaf and stick insects ; and its members are known by the thick parchment-like upper wings, with their stout veinings and their overlapping tips.

THE first family of Orthopteræ is the Blattidæ, a group of insects familiar under the title of Cockroaches.

In these insects the body is flattened, the antennæ are long and threadlike, and the perfect wings are only to be found in the adult male. The Common COCKROACH, so plentiful in our kitchens, and so well known under

Male.   Female.

COCKROACH.—(*Blatta orientalis.*)

the erroneous name of black beetle—its colour being dirty red, and its rank not that of a beetle—is supposed to have been brought originally from India.

The eggs of the Cockroach are not laid separately, but inclosed in a hard membranous case, exactly resembling an apple puff, and containing about sixteen eggs. Plenty of these cases may be found under planks or behind the skirting boards where these insects love to conceal themselves. Along one of the edges of the capsule there is a slit which corresponds with the opening of the puff, and which is strengthened like that part of the pastry by a thickened margin. The edges of the slit are toothed, and it is said that each tooth corresponds with an egg. When the young are hatched, they pour out a fluid which has the effect of dissolving the cement which holds the edges together, the newly-hatched Cockroaches push themselves through the aperture, which opens like a valve, and closes again after their exit, so that the empty capsule appears to be perfectly entire.

A GOOD example of the Cricket is found in the FIELD CRICKET, a noisy creature, inhabiting the sides of hedges and old walls, and making country lanes vocal with its curious cry, if such a word can be applied to a sound produced by friction. The Field Cricket lives in burrows, made at the foot of hedges or walls, and sits at their mouth to sing. It is, however, a very timid creature, and on hearing, or perchance feeling, an approaching foot-step, it immediately retreats to the deepest recesses of the burrow, where it waits until it imagines the danger to have gone by.

Despite of its timidity, however, it seems to be combative in no slight degree, and if a blade of grass or straw be pushed into its hole, it will seize the intruding substance so firmly that it can be drawn out of the burrow before it will loosen its hold. The males are especially warlike, and if two specimens be confined in the same box, they will fight until one is killed. The vanquished foe is then eaten by the victor. In White's "Natural His-tory of Selborne" there is a careful and interesting description of the Field FIELD CRICKET.—(*Gryllus campestris.*) Cricket and its habits.

ONE of the oddest-looking of the British insects is the MOLE CRICKET, so called on account of its burrowing habits and altogether Mole-like aspect. This insect is represented of the natural size, and, as may be seen, attains considerable dimensions.

Like those of the mole, the fore limbs of the Mole Cricket are of enormous comparative size, and turned outwards at just the same angle from the body. All the legs are strong, but the middle and hinder pair appear quite weak and insignificant when compared with the gigantic developments of the front pair. This insect is rather local, but is found in many parts of England,

MOLE CRICKET.—(*Gryllotalpa vulgaris.*)

where it is known by sundry popular titles, Croaker being the name most in vogue near Oxford, where it is found in tolerable plenty.

The colour of the Mole Cricket is brown of different tints, darker upon the thorax than on the wing-covers, both of which organs are covered with a very fine and short down.

As might be surmised from the extraordinary muscular power of the fore-legs, the Mole Cricket can burrow with great rapidity. The excavation is of a rather complicated form, consisting of a moderately large chamber with neatly smoothed walls, and many winding passages communicating with this central apartment. In the chamber are placed from one to four hundred eggs of a dusky yellow colour; and the roof of the apartment is so near the surface of the ground that the warmth of the sunbeams penetrates through the shallow layer of earth, and causes the eggs to be hatched.

The food of the Mole Cricket is mostly of a vegetable nature, but it has been known to feed upon raw meat, upon other insects, and even to exhibit a strong cannibalistic propensity when shut up in company and deprived of the normal food.

MIGRATORY LOCUST.—(*Locusta migratoria.*)

The MIGRATORY LOCUST is a well-known instance of a very large family of insects represented in our own land by many examples. All the Locusts and Grasshoppers are vegetable feeders; and in many cases their voracity is so insatiable, their jaws so powerful, and their numbers so countless, that they destroy every vestige of vegetation wherever they may pass, and devastate the country as if a fire had swept over it.

Such is the case with the Migratory Locust, so called from its habit of congregating in vast armies, which fly like winged clouds over the earth, and wherever they alight, strip every living plant of its verdure. So assiduously do they ply their busy jaws, that the peculiar sound produced by the champing of the leaves, twigs, and grass-blades can be heard at a considerable distance. When they take to flight, the rushing of their wings is like the roaring of the sea; and as their armies pass through the air, the sky is darkened as if by black thunder-clouds.

The warm sunbeams appear to be absolutely necessary for the flight of the Locust, for no sooner does the sun set than the Locusts alight and furl their wings. Woe to the ill-fated spot where they settle, for they consume everything that their jaws can sever, and are not content with eating the green herbage, but devour even linen, blankets, or tobacco. At the approach of the aërial hosts everyone is in fear except the Bushman, who welcomes the Locust with all his heart; for he has no crops to lose, no clothing to be destroyed, and only sees in the swarming insects his greatest luxury, namely, an abund-

LEAF INSECT.—(*Phyllium scythe.*)

ant supply of food without any trouble in obtaining it. In the path of the Locust he kindles large fires, and the insects, being stifled with the smoke, and having their wings scorched by the flames, fall in thousands, and are gathered into heaps, roasted, and eaten. Those that remain, after the Bushman has eaten his fill are then ground between two stones into a kind of meal, which is dried in the sun, and can be kept for a long time without becoming putrid. This substance does not seem very palatable to Europeans, but its distastefulness is probably owing to the careless way in which the insects are scorched over the fire, as Dr. Livingstone speaks highly of the Locust as an article of food, thinking it superior to shrimps. Honey is always eaten together with the Locusts, whenever that sweet condiment can be obtained, as it serves to render the insects more digestible. Our common English grasshoppers belong to the true Locusts.

IN the accompanying illustration is represented a LEAF INSECT, one of the singular species which have such a wonderful resemblance to fallen leaves. The peculiar leaf-like elytra may be seen on reference to the engraving as also the singular manner in which the limbs are furnished with wide flattened appendages, in order to carry out the leafy aspect. Only the females possess the wide, veined wing-covers, those of the male being comparatively short. The wings, however are entirely absent in the female, while in the opposite sex they are very wide and reach to the extremity of the body.

THYSANOPTERA.

THE next order, according to Mr. Westwood's arrangement, that called the Thysanoptera, or Fringe-winged Insects, on account of the manner in which the wings are edged with long and delicate cilia. They are all little insects, seldom exceeding the tenth or twelfth of an inch in length, but, although small, are capable of doing considerable damage. They are mostly to be found on plants and flowers, especially those blossoms where the petals are wide and deep and afford a good shelter. The convolvulus is always a great favourite with them. Greenhouses are sadly liable to their inroads, and owing to their numbers they are very injurious to melons, cucumbers, and similar plants, covering their leaves with a profusion of decayed patches that look as if some powerful acid had been sprinkled over them. Only one family of these insects is acknowleded by entomologists.

NEUROPTERA.

WE now come to an order of insects containing some of the most beautiful and a few of the most interesting members of the class. They are known by the possession of four equal-sized membranous wings divided into a great number of little cells technically called areolets. The mouth is furnished with transversely movable jaws, and the females do not possess a sting or valved ovipositor. In this order are comprised the ant-lions, the dragon-flies, the termites, the lace-wings, and the May-flies.

THE first family in Mr. Westwood's arrangement is that of the Termites, popularly known by the name of WHITE ANTS, because they live in vast colonies, and in many of their abits display a resemblance to the insect from which they take their name. All the Termites are miners, and many of them erect edifices of vast dimensions when compared with the size of their architect. For example, the buildings erected by the Common White Ant (*Termes bellicosus*) will often reach the astonishing height of sixteen or seventeen feet, which in proportion to the size of the insect would be equivalent to an edifice a mile in height if built by man. The dwelling is made of clay, worked in some marvellous manner by the jaws of the insect-architects ; and is of such astonishing hardness, that, although hollow and pierced by numerous galleries and chambers, they will sustain the weight of cattle, which are in the habit of ascending these wonderful monuments of insect labour for the purpose of keeping a watch on the surrounding country,

To give a complete history of the Termites would be a task demanding so much time and space, that it cannot be attempted in these pages ; and we must therefore content ourselves with a slight sketch of their general history, premising that many parts of their economy, and especially those which relate to their development, are still buried in mystery.

The most recent investigations give the following results :—

Each Termite colony is founded by a fruitful pair, called the king and queen, who are placed in a chamber devoted to their sole use, and from which they never stir when once inclosed.   These insects produce a vast quantity of eggs, from which are hatched the remaining members of the colony, consisting of neuters of both sexes, the females being termed workers and the males soldiers, the latter being distinguished by their enormous heads and powerful jaws; of larvæ of two forms, some of which will be fully developed, and others pass all their lives in the worker or soldier

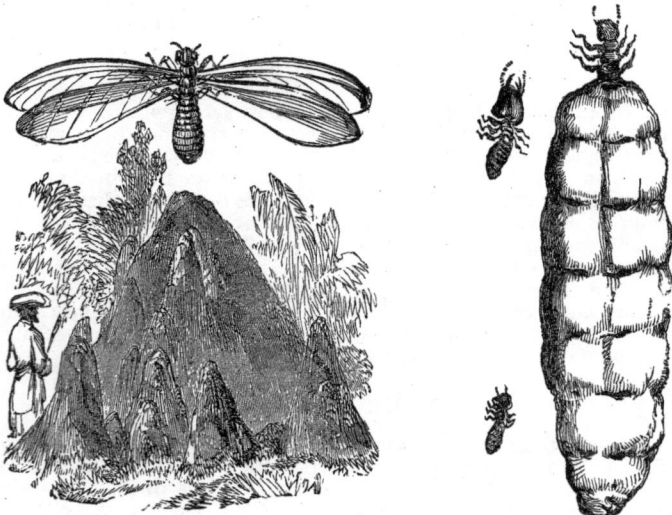

WHITE ANT.—(*Termes bellicosus.*)   Soldier—Worker—Female.

condition ; of pupæ of two forms ; and, lastly, of male and female perfect insects, which are destined to found fresh colonies.   The neuters of either sex are without wings.

In founding a colony, the order is as follows.   The parent pair are taken possession of by the workers, who inclose them in a chamber which is intended as the nucleus of the infant establishment.   The walls of this chamber are pierced by holes which will suffer the workers to pass, but are far too small to afford exit for the king or queen.   Shortly after they have been fairly installed, a wondrous change takes place in the female.   Though her head, thorax, and legs retain their normal dimensions, her abdomen begins to swell in the most preposterous manner, until it is as long as a man's finger and about twice its thickness, thus precluding its owner from advancing a single step.

The queen, thus developed and for ever fixed in her home, is truly the mother of her subjects, producing nearly eighty thousand eggs in each twenty-four hours. The eggsare carried off by the workers as soon as laid, and conveyed to suitable places in the nest, where they are guarded until they are hatched, and are then fed and watched until they have passed through their preliminary stages of existence.

The great bulk of a Termite establishment is composed of workers, who outnumber the soldiers in the proportion of a hundred to one. By the mysterious instinct which is implanted in these insects, the soldiers and workers confine themselves to their respective occupation, the former doing nothing but fight and the latter nothing but labour.

There are many species of Termite, and all are fearfully destructive, being, indeed, the greatest pest of the country wherein they reside. Nothing, unless cased in metal, can resist their jaws; and they have been known to destroy the whole woodwork of a house in a single season. They always work in darkness, and at all expenditure of labour keep themselves under cover, so that their destructive labours are often completed before the least intimation has been given. For example, the Termites will bore through the boards of a floor, drive their tunnels up the legs of the tables or chairs, consume everything but a mere shell no thicker than paper, and yet leave everything apparently in a perfect condition . Many a person has only learned the real state of his furniture by finding a chair crumble into dust as he sat upon it, or a whole staircase fall to pieces as soon as a foot was set upon it. In some cases the Termite lines its galleries with clay, which soon becomes as hard as stone, and thereby produces very remarkable architectural changes. For example, it has been found that a row of wooden columns in front of a house have been converted into stone pillars by these insects.

PASSING by several families of the Neuroptera, we come to the Libellulidæ, or DRAGON-FLIES. These insects are very familiar to us by means of the numerous Dragon-flies which haunt our river sides, and which are known to the rustics by the very inappropriate name of Horse-stingers, they possessing no sting and never meddling with horses. The name of DRAGON-FLY, on the contrary, is perfectly appropriate, as these insects are indeed the dragons of the air, far more voracious and active than even the fabled dragons of antiquity.

Even in their preliminary stages the Dragon-flies preserve their predatory habits, and for that purpose are armed in a most remarkable manner. During the larval and pupal states the Dragon-fly is an inhabitant of the water, and may be found in most of our streams, usually haunting the muddy banks, and propelling itself along by an apparatus as efficacious as it is simple, and exactly analogous to the mode by which the nautilus forces itself through the water. The respiration is carried on by means of the oxygen which is extracted from the water; and the needful supply of liquid is allowed to pass into and out of the body through a large aperture at the end of the tail.

Such are its means of locomotion; those of attack are not less remarkable or less efficacious.

The lower lip, instead of being a simple cover to the mouth, is developed into a strange-jointed organ, which can be shot out to the distance of nearly an inch; or, when at rest, can be folded flat over the face, much as a carpenter's rule can be shut up so as to fit into his pocket, and can be rapidly protruded or withdrawn very like the instrument called a "lazy-tongs." Like that instrument, it is furnished at its extremity with a pair of forceps, and is able to grasp at passing objects with the swiftness and certainty of a serpent's stroke.

The creature remains for some ten or eleven months in the preliminary stages of existence, and when the insect is about to make its final change, the undeveloped wings become visible on the back. When its time has come, the pupa leaves the water and crawls up the stem of some aquatic plant until it has reached a suitable elevation ; it clings firmly with its claws, and remains apparently quiet. On approaching it, however, a violent internal agitation is perceptible, and presently the skin of the back splits along the middle, and the Dragon-fly protrudes its head and part of the thorax. By degrees it withdraws itself from the empty skin, and sits for a few hours drying itself, and shaking out the innumerable folds into which the wide gauzy wings have been gathered. After a series of deep respirations of the unwonted air, and much waving of the wings, the glittering membranes gain strength and elasticity, and the enfranchised insect launches forth into the air in search of prey and a mate.

DEMOISELLE.—(*Calepteryx splendens.*)

There are very many species of Dragon-flies, all very similar in their habits, being fiercely predaceous, strong of wing, and gifted with glittering colours. Unfortunately, the rich azure, deep green, soft carnation, or fiery scarlet of these insects fade with their life, and in a few hours after death the most brilliant Dragon-fly will have faded to a blackish brown. The only mode of preserving the colours is to remove all the interior of the body and to introduce paint of the proper colours. This, however, is but an empirical and unsatisfactory sort of proceeding, and, no matter how skilfully it may be achieved, will never be worth the time bestowed upon it. In many species the sexes are of different colours, as, for example, in the beautiful DEMOISELLE DRAGON-FLIES, where the male is deep purple, with dark spots on the wings, and the female rich green, with the wings uncoloured.

ANT-LION.—(*Myrmeleog transyatus.*)

THE far-famed ANT-LION is one of the insects that are more celebrated in their preliminary than in the perfect stage of existence. As may be seen by reference to the illustration, their perfect form is very light and elegant, and closely resembling that of the dragonflies, save that the wings are lighter, softer, and broader. In their larval condition, however, they are by no means attractive-looking creatures, somewhat resembling flattened maggots with rather long legs and very large jaws, the legs being apparently useless as organs of progression, all movements being made by means of the abdomen. Slow of movement as is this creature, and yet predaceous, feeding wholly on living insects, the mode of obtaining its food seems to be rather a problem. The solution, however, is simple enough, the creature digging a pitfall, and lying ensconced therein while the expected prey approaches.

In order to enable the Ant-Lion to extract the juices of the insects on which it feeds, the inner curve on each mandible is deeply grooved, and another portion of the jaws, technically called the maxilla, plays within the groove. The larva, half-sunk in its pit-fall, is shown in the left-hand lower corner of the illustration.

The MAY-FLY has long been celebrated for its short space of life, a single day sometimes witnessing its entrance into the perfect state and its final departure from the world. The popular idea concerning these insects is, that the whole of their life is restricted to a single day. This, however, is an error, as they have already passed at least two years in their preliminary stages of existence. In the larval and pupal states they are inhabitants of the water, and are fond of hiding themselves under stones, or burrowing into the muddy banks. Under the latter circumstances they make a very curious tunnel, something like a double-barrelled gun.

The May-fly is peculiarly notable for a stage of development which seems to be quite unique among insects. When it has passed through its larval and pupal state, it leaves the water, creeps out of its pupa case, and takes to its wings. After a period, varying from one to twenty hours, it flies to some object, such as the trunk of a tree or the stems of water-plants, and casts off a thin membranous pellicle, which has enveloped the body and wing, the dry pellicle remaining in the same spot, and looking at first like a dead insect.

MAY-FLY.—(*Ephemera vulgata.*) CADDIS-FLY.—(*Phryganea grandis.*)

After this operation the wings become brighter, and the three filaments of the tail increase to twice their length. Some authors call the state between leaving the water and casting the pellicle the " pseudimago" state.

Some of these insects are well known to fishermen under the names of Green and Grey Drake, the former being the pseudimago, and the latter the perfect form of the insect, which is represented in the illustration. Sometimes these insects occur in countless myriads, looking like a heavy fall of snow as they are blown by the breeze, and having on some occasions been so plentiful that they have been gathered into heaps and carted off to the fields for manure.

THE order called Trichoptera, or Hairy-winged insects, is represented by the common CADDIS FLY.

This fly is well known to every angler both in its larval and in its perfect state. The larva is a soft white worm, of which fishes are exceedingly fond, and it therefore requires some means of defence. It accordingly actually makes for itself a movable house of sand, small stones, straws, bits of shells, or even small living shells, in which it lives in perfect security, and crawls about in search of food, dragging its house after it. When it is about to become a pupa, it spins a strong silk grating over the entrance of its case, so that the water necessary for its respiration can pass through, but at the same time all enemies are kept out. When the time for its change has arrived, the pupa bites through the grating, rises to the surface, and crawls out of the reach of the water, which would soon be fatal to it. The skin then splits down the back, and the perfect insect emerges.

The order is called Trichoptera, because the wings, instead of being covered with scales as are those of butterflies, are clothed with hairs.—There are many species of Caddis-flies.

WE now come to the vast order of insects technically called the HYMENOPTERA. In these insects the wings are four in number, transparent, membranous, the veins comparatively few, and the hinder pair smaller than the others. Their mouth is furnished with powerful horny jaws, and with a tongue guarded by the modified maxillæ. The females are armed with a many-valved sting or ovipositor. In this enormous order are included all the bees, wasps, and their kin, the great family of saw-flies, the ichneumons, the gall-flies, and the ants. We will proceed at once to the family of the Tenthredinidæ, or SAW-FLIES.

Rhyssa persuasoria.

In this and the next family, the females are furnished with a peculiar ovipositor, composed of several pieces.

The true Saw-flies are known by the curious piece of animal mechanism from which they derive their name. The females of this family are supplied

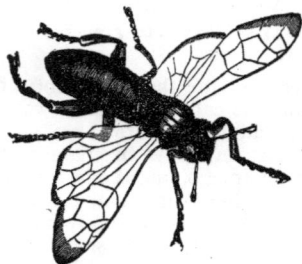

Cimbex femorata.

Urocerus gigas.

with a pair of horny saws, placed side by side on the lower extremity of the abdomen.

These saws are of various forms, according to the particular species to which they belong, and may be seen even in the dried specimens, the top of their sheath slightly projecting, and their shapes plainly visible after the removal of a portion of the abdomen. When taken from the insect and placed under the microscope, they present a very pretty appearance, owing to the gently-curved ribs with which their sides are strengthened and decorated. The saws act alternately, one being pushed forward as the other is being retracted. Their object is to form a groove in some plant, in which the eggs of the mother insect can be deposited, and wherein they shall find a supply of nourishment in order to enable them to complete their development ; for it is a most remarkable fact, that after the egg is deposited in the groove it rapidly increases in size, obtaining twice its former dimensions.

In the genus Cimbex, of which an example is given in the illustration, the larvæ possess twenty-two feet, and have the power of discharging a translucent greenish fluid from certain pores placed on the sides of body just above the spiracles. This feat they can repeat six or seven times in succession.

Crabro cribrarius.   *Philanthus triangulum.*

When they have eaten their way to the next stage of existence, they spin a cocoon of a brownish colour and of a stringy, tough consistency, and either suspend it to the branches of the tree on which they have been feeding, or hide it under fallen leaves. In this cocoon they remain for a comparatively short time, and then emerge as perfect insects.

The fine insect on the same illustration, which is known by the name of the GIANT ICHNEUMON, is an example of the next family, in which the ovipositor is converted into a gimlet instead of a double saw. With this powerful instrument the female is enabled to drill holes into living timber for the purpose of depositing the eggs. When they are hatched, the young grubs immediately begin to gnaw their way through the wood, boring it in every direction, and making burrows of no mean size. Those of the present species prefer fir and pine, and I have had specimens of the wood sent to me which have been riddled by the grubs until they looked as if they had harboured a colony of the ship-worm.

THE next group of the Terebrantia is called Entomophaga, or Insect-eaters, because the greater number of them are parasitic upon other insects, just as the saw-flies are parasitic upon vegetables. In these insects the ovipositor is furnished with two delicate spiculæ, and the last segments of the abdomen are not formed into a telescope-like tube.

The first family is that of the Cynipidæ, or Gall Insects, the creatures by whose means are produced the well-known galls upon various trees, the so-called oak-apple being perhaps the best known, and the ink-gall (also found on the oak) the most valuable. These Galls are formed by the deposition of an egg in the leaf, branch, stem, twig, or even root of the plant, and its consequent growth.

The true Ichneumons, of which a specimen is given in the illustration, form a vast group of insects, the British Ichneumonidæ alone numbering many more than a thousand described and acknowledged species. In them the ovipositor is straight, and is employed in inserting the eggs into the bodies of other insects, mostly in their larval state. In some cases, this slender and apparently feeble instrument is able to pierce through solid wood, and is insinuated by a movement exactly like that which is employed by a carpenter when using a brad-awl. When not engaged in this work, the ovipositor is protected by two slender sheaths that inclose it on either side.

Were it not for the Ichneumons, our fields and gardens would be hopelessly ravaged by caterpillars and grubs of all kinds, for practical entomologists always find that when they attempt to rear insects from the egg or the larval state, they must count upon losing a very large percentage by the Ichneumons.

IN the next great division of Hymenopterous insects, the ovipositor of the female is changed into a sharply-pointed weapon, popularly called a sting, and connected with a gland in which is secreted a poison closely analogous to that which envenoms a serpent's tooth.

First come those curious and interesting insects known popularly by the names of SAND WASPS and WOOD WASPS. These creatures are in the habit of making burrows into the ground or in posts, and placing therein their eggs, together with the bodies of other insects which are destined to serve as food for the future progeny. Spiders are sometimes captured and immured for this purpose. In many instances the captured insects are stung to death before they are placed in the burrow, but it is often found that they only receive a wound sufficient to paralyse them, so that they lead a semi-torpid life until they are killed and eaten by the young grub. Two of these Sand Wasps are tolerably common in England. One of them (*rabro cribarius*), the woodborer, drilling its burrow into posts, palings, and similar substances. It feeds its young with the larvæ of one of the leaf-rolling caterpillars that ives in the oak, and is scientifically known by the name of *Tortrix chlorana.* It also employs for this purpose several two-winged insects. One species of these burrowing wasps prefers the well-known cuckoo-spit insect for this purpose (*Aphrophora spumaria*), pulling it out of its frothy bed by means of its long legs.

Another of these insects, called *Philanthus triangulum*, is in the habit of provisioning its burrow with the hive-bee, which it contrives to master in spite of the formidable weapon possessed by its victim, and then murders or paralyses by means of its sting. M. Latreille mentioned that he saw from fifty to sixty of these insects busily engaged in burrowing into a sandbank not more than forty yards long ; and as each female lays five or six eggs, and deposits a bee with each egg, the havoc made among the hives is by no means inconsiderable.

THE true Ants, as is well known, associate in great numbers, and, as is peculiarly the case with the bees, the great bulk of their numbers is composed of workers, or neuters, which are destined to perform the constant labours needful to regulate so large a community. The perfect insects of either sex take no part in the daily tasks, their sole object being to keep up the numbers of the establishment. In the Ants, moreover, the neuters are without wings, and even the perfect insects only retain these organs for a brief period of their existence.

Everyone has heard of the objects called "ants' eggs," which are so strongly recommended as food for the nightingale and other birds ; and many persons, though they have seen them, have believed them really to be the objects which their popular name would infer. In truth, however, they are simply the cocoons, in which the insects are passing their pupal state before emerging in their winged condition. It has been already mentioned that only the perfect males and females possess wings.

As soon as they gain sufficient strength, they fly upward into the air, where they seek their mates, and soon descend to earth. The males, having now

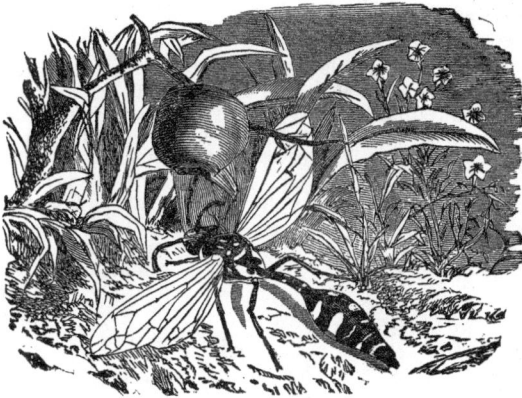

Eumenes arcuatus.

nothing to do, speedily die, as they ought, but the females begin to make provision for their future households. Their first proceeding is a rather startling one, being the rejection of the wings which had so lately borne them through the air. This object is achieved by pressing the ends of the wings against the ground, and then forcing them suddenly downwards. The wing then snaps off at the joint, and the creature, thus reduced to the wingless state of a worker, is seized upon and conveyed to a suitable spot, where she begins to supply a vast quantity of eggs. These are carefully conveyed away and nurtured until they burst forth into the three states of male, female, and neuter, the precise method by which the development is arrested so as to produce the neuter condition not being very accurately known.

The illustration represents an Australian example of the SOLITARY WASPS, many of which are found in England. The curious nest of this insect is shown immediately above, suspended to a branch. The creature makes a separate nest for each egg, the material being clay well worked, and the shape as represented in the engraving. The nest is stocked with larvæ of moths or butterflies.

N N

THE true Wasps, or Vespidæ, are gregarious in their habits, building nests in which a large but uncertain number of young are reared. The common Wasp makes its nest within the ground, sometimes taking advantage of the deserted hole of a rat or mouse, and sometimes working for itself. The substance of which the nest is made is a paper-like materal, obtained by nibbling woody fibres from decayed trees or bark, and kneading it to a paste between the jaws. The general shape of the nest is globular, and the walls are of considerable thickness, in order to guard the cells from falling earth, a circular aperture being left, through which the inhabitants can enter or leave their home.

The cells are hexagonal, and laid tier above tier, each story being supported by little pillars, made of the same substance as the cells, and all the open ends being downwards, instead of laid horizontally, as is the case with the bees. It will thus be seen that, on account of this arrangement, the nurse-wasps are enabled to get at the grubs as they lie, or rather hang, in their cells, with their heads downwards.

THE WASP.—(*Vespa vulgaris.*)

The grubs are fat, white, black-headed creatures, very well known to fishermen, who find them excellent bait after they have been baked so as to render them sufficiently hard to remain on the hook. When they are about to enter the pupal state, they close the mouths of their cells with a silken cover, through which the black eyes are plainly visible, and there wait until they emerge in the perfect state. The grubs are fed with other insects, fruit, sugar, meat, or honey, the mingled mass being disgorged from the stomachs of the nurses and thus given to their charge.

There are separate cells for males, females, and neuters, the two former classes only being produced towards the end of autumn, so as to keep up a supply for the succeeding year.

THERE are, perhaps, few insects so important to mankind as those which procure the sweet substance so well known by the name of honey.

THE HIVE BEE.—(*Apis mellifica.*)

Nearly all the honey-making Hymenoptera are furnished with stings, and in many species the poison is fearfully intense. Some of these insects, such as the HIVE BEE, make waxen cells of mathematical accuracy, the larvæ being placed in separate cells, and fed by the neuters.

This useful little creature is so well known that a lengthened description of it would be useless. A merely general sketch will be quite sufficient.

The cells of the Bee are, as is well known, made of wax. This wax is secreted in the form of scales under six little flaps situated on the under side of the insect. It is then pulled out by the Bee, and moulded with other scales until a tenacious piece of wax is formed. The yellow substance on the legs of the Bees is the pollen of flowers. This is kneaded up by the bees, and is called bee-bread.

The cells are six-sided, a form which gives the greatest space and strength with the least amount of material, but the method employed by the Bees to give the cells that shape is not known. The cells in which the Drone or male Bees are hatched, are much larger than those of the ordinary or worker Bee. The edges of the cells are strengthened with a substance called propolis, which is a gummy material procured from the buds of various trees. This propolis is also used to stop up crevices and to mix with wax when the comb has to be strengthened.

The royal cells are much larger than any others, and are of an oval shape. When a worker larva is placed in a royal cell, and fed in a royal manner, it imbibes the principles of royalty, and becomes a queen accordingly. This practice is adopted if the Queen Bee should die, and there be no other queen to take her place.

The Queen Bee is lady paramount in her own hive, and suffers no other queen to divide rule with her. Should a strange queen gain admittance, there is a battle at once, which ceases not until one has been destroyed.

At the swarming time, the old queen is sadly put out by the encroachments of various young queens, who each wish for the throne, and at last is so agitated that she rushes out of the hive, attended by a large body of subjects, and thus the first swarm is formed. In seven or eight days, the queen next in age also departs, taking with her another supply of subjects. When all the swarms have left the original hive, the remaining queens fight until one gains the throne.

The old method of destroying Bees for the sake of the honey was not only cruel but wasteful, as by burning some dry "puff-ball" the Bees are stupefied, and shortly return to consciousness. The employment of a "cap" on the hive is an excellent plan, as the Bees deposit honey alone in these caps, without any admixture of grubs or bee-bread. Extra hives at the side, with a communication from the original hive, are also useful.

The Queen Bee lays about eighteen thousand eggs. Of these about eight hundred are males or drones, and four or five queens, the remainder being workers.

In some cases, such as the common HUMBLE BEE, the cells are egg-shaped, each cell being either occupied by a larva or filled with honey ; while in some species the eggs are placed parasitically in

THE HUMBLE BEE.
(*Bombus terrestris.*)

the nests of other Bees, so that the larvæ feed either upon the stores of food gathered for the involuntary host, or upon the body of the deluded insect itself.

In gathering honey, the Bees lick the sweet juices from flowers, swallow them, and store them for the time in a membranous sac, popularly called the honey-bag. When this sac is filled, the Bee returns to the hive, and discharges the honey into cells, closing its mouth with wax when it is filled. The structure of the bee-cell, its marvellous adaptation to the several purposes for which it is intended, its mathematical accuracy of construction, whereby the best amount of material is found to afford the greatest amount of space and strength, are subjects too complicated to be here described, but may be found in many works which have been written upon the Hive Bee.

For want of space, we are compelled to pass by many interesting Hymenoptera, such as the Leaf-cutter Bees, the Woodborers, and the Mason Bees, each of which creatures would demand more space than can be given to the whole of the insects.

LEPIDOPTERA.

WE now come to an order in which are included the most beautiful of all insects, namely, the Butterflies and Moths. On account of the feather-like scales with which their wings are covered, and to which the exquisite colouring is due, they are technically called Lepidoptera, or scale-winged insects.

The wings are four in number, and it is occasionally found that the two pairs are connected together by a strong bristle in one and a hook-like appendage in the other, so that the two wings of each side practically become one mem-

SWALLOW-TAILED BUTTERFLY.—(*Papilio machaon.*)

ber, in a manner similar to the formation of many hymenopterous insects. Those species which take any nourishment subsist entirely upon liquid food, which is drawn into the system by suction, and not by means of a brush, as is the case with the liquid-feeding beetles and bees. The wings are strengthened by nervures, which are of great use in determining the position of the insects.

The scales with which the membranous wings are at once protected and adorned are of various shapes, sometimes broad, flat, and overlapping each other like the tiles of a house-roof.

The series of changes undergone by the Lepidoptera are, perhaps, better known than those of any other order, on account of the large dimensions and conspicuous habits of the insects.

Having given this general glance at the order, we will now proceed to our examples.

IN the system which is adopted in this work, the Lepidoptera are divided into two sections, the Butterflies and Moths, technically call Rhopalocera and Heterocera, which may generally be distinguished from each other by the form of the antennæ, those of the Butterflies having knobs at their tips, whilst those of the Moths are pointed. The first family is that of the Papilionidiæ, in which are included the largest and most magnificent specimens of this order.

The beautiful insect represented on p. 548 is not very uncommon in some parts of England, especially in the fenny parts of Cambridgeshire.

It flies with exceeding rapidity, nearly in a straight line, and is very difficult to capture.

The colour of the wings is black, variegated most beautifully with yellow markings, and near the extremity of each hinder wing is a circular red spot, surmounted by a crescent of blue, and the whole surrounded by a black ring.

PEACOCK BUTTERFLY.—(*Vanessa Io.*)

WE now come to another family, called the Pieridæ, which may be known at once by the manner in which the inner edges of the hinder wings are folded, so as to form a kind of gutter in which the abdomen rests. In all these insects the colours are comparatively sober, the upper surface being generally white and black, and the under surface sparingly coloured with red and yellow. To this family belong our common white butterflies, together with the well-known Brimstone Butterfly (the harbinger of spring), all the Marbled Butterflies, the Orange-tip, and the now scarce Veined-white.

THE large and important family of the Nymphalidæ contains a vast number of species, most of which are notable for their brilliant colouring, and many of which are well-known natives of England.

To this family belongs the brightly coloured genus Vanessa, of which the common PEACOCK BUTTERFLY is a familiar British example. This insect, which is one of the finest of our British butterflies, is very common in our own country, and may be seen very plentifully in fields, roads, or woods, when the beauty of its colouring never fails to attract admiration.

The caterpillar of the Peacock Butterfly feeds upon the stinging-nettle, in common with others of the same genus, and therefore the insect is worthy of our protection. Its general shape and appearance may be gathered from a reference to the illustration ; its common colour is black, studded with tiny white points. The chrysalis is one of those which hang suspended during the time of their nonage, and is frequently found to be infested with the ichneumon-fly.

The beautiful SCARLET ADMIRAL, so well known by the broad scarlet stripes that are drawn over the wings ; the LARGE and SMALL TORTOISE-SHELL BUTTERFLIES ; the COMMA BUTTERFLY, so called from a comma-shaped white mark under the wings, and the rare and beautiful CAMBERWELL BEAUTY, are all British members of this genus.

DEATH'S-HEAD MOTH.—(*Acherontia Atropos.*)

THE second great division of the Lepidoptera is that of the MOTHS, distinguishable by means of the pointed tips of their antennæ, which are often furnished with a row of projections on either side, like the teeth of a comb ; and in the males are sometimes supplied with branching appendages.

The first family of the Moths is the Sphingidæ, a group which contains a great number of swift-winged insects, popularly and appropriately called Hawk-moths, from the strength and speed of their flight. In many instances the proboscis is of great length, sometimes equalling the length of the entire body, and in such instances it is found that the insect is able to feed while on the wing, balancing itself before a flower, hovering on tremulous wing, and extracting the sweets by suction.

One of the commonest species of this genus is the LIME HAWK-MOTH *Smerinthus Tiliæ*), so called because the larva feeds on the leaves of the lime-tree. It is a green caterpillar, thick bodied, covered with little protu-berances, and upon each side are some whitish streaks edged with red or yellow. Just at the end of the tail there is a short knobby protuberance, and the forepart of the body is rather narrow. When the larva has com-pleted its time of feeding, it descends to the ground, and buries itself about eight inches deep in the earth, whence the chrysalis may be extracted by the help of a trowel. Besides the lime, the elm and birch are favoured residences of this insect.

THE splendid insect appropriately named the DEATH'S-HEAD MOTH is tolerably common in our island, though, from its natural habits, and the in-stinct of concealment with which the caterpillar is endowed, it is not so frequently seen as many rarer insects. Owing to the remarkably faithful delineation of a skull and bones upon the back of the thorax, the insect is often an object of great terror to the illiterate, and has more than once thrown a whole province into consternation, the popular idea being that it was some infra-natural being that was sent upon the earth as a messenger of pestilence and woe, if not indeed the shape assumed by some witch residing in the neighbourhood.

The caterpillar of this moth is enormously large, sometimes measuring five inches in length, and being very stoutly made. It feeds on various plants, the jessamine and potato being its favourites, and may be best found by traversing potato-grounds in the night, and directing the light of a bull's-eye lantern among the leaves. It can be readily kept and bred, but requires some careful tending ; and it must be remembered that it will only eat the particular food to which it has been accustomed, and if bred among the potato will refuse the jessamine leaf, and *vice versâ.* When the caterpillar is about to change into its chrysa-lis state it should be placed in a vessel containing seven or eight inches of earth, which should be kept moderately damp by means of a moist sponge or wet piece of moss laid on the top. If this precaution be not taken, the shell of the chrysalis is apt to become so hard that the moth is unable to break its way out, and perishes in the shell. I have several specimens where the moth has thus perished.

HUMMING-BIRD MOTH.--(*Macroglossa stellatarum.*)

One of the most curious points in the history of the Death's-head Moth is its power of producing a sound, a faculty which is truly remarkable among the Lepidoptera. The noise is something like the grating, squeaking cry of the field-cricket, but not nearly so loud.

ALTHOUGH not gifted with the brilliant hues which decorate so many of the Hawk-moths, the HUMMING-BIRD MOTH is a more interesting creature than many an insect which can boast of treble its dimensions and dazzling richness of colour. This insect may be readily known by its very long proboscis, the tufts at the end of the abdomen, and the peculiar flight, which so exactly resembles that of the humming-bird, that persons accustomed to those feathered gems have often been deceived into the idea that England actually possesses a true humming-bird. Owing to its arrowy flight and the

piercing vision with which such flight is always accompanied, the capture of
the moth is a matter of no small difficulty, and when it settles, the quiet
sober hues of its plumage render it so similar in colour to the objects on
which it rests, that the eye can hardly distinguish its outline; and, being gifted
with an instinctive appreciation of the objects best suited for its concealment,
it is sure to alight on some surface which presents hues akin to those of its
body and wings.

IN the curious Moths of which the HYLAS is a good example, the wings
are as transparent as those of the bee tribe, and, indeed, the hymenopterous
idea seems to run through the whole of these creatures so thoroughly, that
the shape of their bodies, the mode of flight, even the manner in which they
move the abdomen, are so bee and wasp like, that an inexperienced observer
would certainly mistake them for some species of the Hymenoptera. Others
there are which bear an equal resemblance to the gnats, and are of corre-
spondingly small dimensions. These insects fly in a manner somewhat

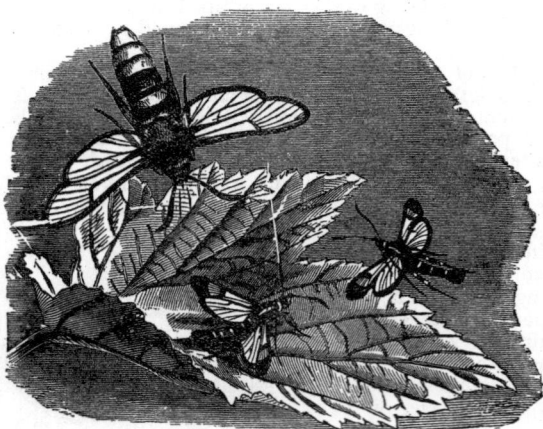

CLEARWING MOTHS.

resembling the movements of the Humming-bird Moth, and dart about
with considerable speed, though they are not so craftily wary as that insect,
and can be captured with comparative ease.

IN the Ægeriidæ, the wings are as transparently clear as in the Sesiadæ,
and the general aspect is equally like that of a moth. The species which is
shown in the illustration is very common in England, and is fond of haunting
currant-bushes, where it may be captured without much difficulty, being
rather dull and sluggish in taking to flight, though when once on the wing it
is quick and agile in its movements. On account of its resemblance to the
large gnats it is popularly called the GNAT CLEARWING (*Ægeria tipuli-
formis.*) The caterpillar of this insect feeds upon the pith of the currant-trees.

The large insect in the same illustration represents the LUNAR HORNET
CLEARWING (*Trochilium bembeciforme,*) an insect which is of tolerably, but
not very frequent occurrence. Its popular name is given to it in allusion to
its singular resemblance to a hornet, the similitude being so close as to
deceive a casual glance, especially when the insect is on the wing.

The larva of the present species feeds upon the willow, boring into the
young wood and sometimes damaging it to a serious extent. All these

insects inhabit, while in the larval state, the interior of branches or roots, and make a kind of cocoon from the nibbled fragments of the wood. Just before undergoing the transformation, the larva turns round so as to get its head towards the entrance of the burrow, and after it has changed into the pupal form, is able, by means of certain projections on the segments, to push itself along until the upper half of the body protrudes through the orifice, and permits the perfect moth to make its escape into the open air.

The wings of this insect are transparent, with orange-red nervures and dusky fringes. The head and thorax are shining brown black, with a yellow collar, and the abdomen is ringed with orange and dark brown.

THE well-known GOAT MOTH is, next to the Death's-head Moth, one of the largest of the British Lepidoptera, its body being thick, stout, and massive, and its wings wide and spreading.

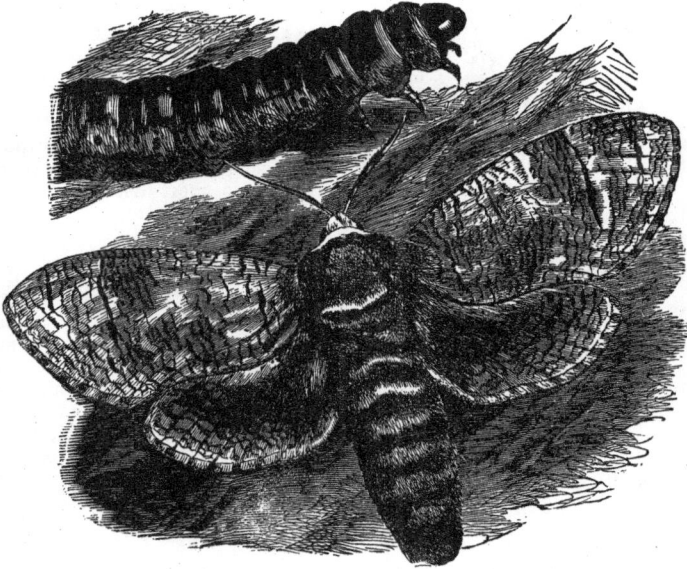

GOAT MOTH AND LARVA. —(*Cossus ligniperda.*)

The larva itself is but little smaller than that of the Death's-head Moth, and is by no means an attractive-looking creature. Its body is smooth and shining, mostly of dull mahogany-red tinged with ochreous yellow, and having a large oval patch of chestnut on the back of each segment. It is gifted with a curiously wedge-shaped head, and its muscular power is enormous, as may be proved by actual experiment during the life of the creature, or inferred from the marvellous arrangement of muscles which are made visible upon dissection.

It exudes a liquid of powerful and fetid odour, thought by some to resemble the unpleasant effluvium exhaled by the he-goat. Its influence extends to a considerable distance, and a practised entomologist will often detect the presence of a Goat-moth caterpillar simply by the aid of the nostrils.

The caterpillar passes three years in the larval stage of existence, and when the time approaches for its change, it ceases to burrow, and scoops out a convenient cell in the tree, lining it with a fabric of mixed wood-scrapings and silken threads. Before it emerges from this retreat, it pushes itself through its burrow like a sweep ascending a chimney, protrudes about half of the body, and then issues from the chrysalis shell, which it leaves within the burrow.

The WOOD LEOPARD MOTH, is a very prettily marked insect, though without the least brilliancy of colour. The caterpillar of this insect feeds upon the interior of many trees, seeming to prefer the wood of the apple, pear, and other fruit-trees. It is a naked, fleshy-looking larva, of a light yellow colour, and having a double row of black spots upon each segment. Like the Goat-moth, it prepares a cocoon-like cell when it is about to take the pupal form, but the lining is of stronger materials, cemented firmly together with a glutinous substance secreted by the insect. The moth is seldom seen until July, and is tolerably plentiful in some places, appearing to be decidedly local and rather intermittent in its visits.

THE family of the Bombycidæ includes several insects of inestimable value to mankind, the various silk-producing moths being included in its ranks. The common silk-worm is too familiar to need any notice, but as it is not generally known that upwards of forty silk-producing moths exist in different parts of the world, a short history will be given of some of them, together with a figure and a brief description of one of the finest species. All these insects secrete the silk in two large intestine-like vessels in the interior, which contain a gelatinous kind of substance, and become enormously large just before the caterpillar is about to change into a pupa. Both the silk organs unite in a common tube at the mouth, technically called the spinneret, and through this tube the semi-liquid is ejected. As soon as it comes into contact with air it hardens into that soft, shining fibre with which we are so familiar.

WOOD LEOPARD MOTH. —(*Zeuzera æsculi.*)

If a single fibre of silk be examined through a good microscope, it will be seen to consist of two smaller fibres laid parallel to each other, like the barrels of a double gun, this structure being due to the double secreting vessels. The goodness of silk chiefly consists in the manner in which these semi-fibres are placed together.

The caterpillar employs the silk for the purpose of constructing a cocoon in which it can lie until it has assumed the perfect form ; and proceeds with wonderful regularity and despatch in its work, its head passing from side to side, always carrying with it a thread, and the cocoon being gradually formed into the oval shape which it finally assumes. The few outermost layers are always rough and of poor quality ; these are stripped off, and the end of the thread being found, it is fastened to a wheel, and spun off into a hank of soft yellow fibre. The colouring matter is very variable, sometimes being hardly visible, and at others giving the silk a bright orange tint. It fades much on exposure to light.

THE family of the Arctiidæ, so called because some of the hair-covered larvæ have a bear-like look, is represented in England by many examples, some being really handsome insects, and others remarkable for some peculiarity in themselves or the larvæ,

SILK WORMS—MOTHS—COCOONS.   Page 554.
The Popular Natural History

THE TIGER-MOTH.—This common but beautiful moth is found in the beginning of autumn. It runs on the ground with such swiftness as to be often mistaken for a mouse. I have more than once seen a kitten chasing a tiger-moth among the flowers in a garden, evidently deceived by its resemblance to a mouse. The larva is popularly called "the woolly bear." It is rather large, and is surrounded with tufts of long elastic hairs of a reddish brown colour, which serve as a defence against many enemies. When disturbed, it rolls itself round, just as a hedgehog does, and if on a branch, suffers itself to fall to the ground, when the long hairy covering defends it from being injured by the fall. When the caterpillar is about to change to a pupa, it spins a kind of hammock, and lies there until it comes forth as a moth.

The colour and markings of this moth vary considerably. The usual tints are, the thorax brown, the body red, striped with black. The two anterior wings are cream colour, marked with bold patches of a deep brown : the posterior wings are bright red, spotted with bluish black.

Perhaps the most curious example of this family is a species which derives its popular name from its habits.

The HOUSE-BUILDER MOTH (*Oiketicus Sandersii*) is common in many parts of the West Indies, and is in some places so plentiful as to do considerable damage to the fruit-trees. As soon as the larva is hatched from the

TIGER MOTH.—(*Arctia caja.*)

egg, it sets to work in building its habitation : and even before it begins to feed, this industrious insect begins to work. The house is made of bits of wood and leaves, bound together with silken threads secreted in the interior. When the creature is small, and the house is of no great weight, it is carried nearly upright ; but when it attains size and consequent weight, it lies flat, and is dragged along in that attitude. The entrance of this curious habitation is so made that the sides can be drawn together, and whenever the creature feels alarmed, it pulls the cords and so secures itself from foes.

THE next family derives the name of Geometridæ from the mode of walking adopted by the larva. These creatures have no legs on the middle of the body, and are in consequence unable to crawl in the usual manner. Their mode of progression is popularly and appropriately termed "looping," and the caterpillars are called "loopers." When one of these larvæ desires to advance, it grasps the object firmly with its fore feet, and draws the hind feet close to them, forming the body into an arched shape, not unlike the attitude of a cat which meets a strange dog. The hinder feet then take a firm hold, the body is projected forward, until the fore feet can repeat the process. The whole action of the larva reminds the observer of the leech when crawling.

The power of grasp and general strength of muscle enjoyed by these larvæ are really surprising. Many of them can seize a branch with their hind feet, stretch out the body nearly horizontally, and remain in that position for hours. Some slight idea of the muscular force required to perform this action may be imagined by grasping an upright pole with the hands, and trying to hold the body out horizontally. Several of these cater-pillars are of dull brown hues, and being furnished with sundry projections in different parts of their bodies, they resemble dried sticks so closely that they can hardly be distinguished from the branches to which they cling.

The SWALLOW-TAILED MOTH is a well-known British species, very common in woods, and being mostly found among the underwood, whence it may be dislodged by beating the branches. The caterpillar feeds on many shrubs, but prefers th- willow the lime and elder trees, the elder being

SWALLOW-TAILED MOTH.—(*Ourapteryx sambucaria.*)

its chief favourite. Two specimens of the caterpillar are given for the purpose of showing its attitudes while at rest and while walking after the "looper" manner. The cocoon is also shown, made of withered leaves, and hanging from an ivy twig.

The colour of the larva is reddish brown, with some dark longitudinal lines, and before it changes into the pupal state it makes a slight silken cocoon among leaves. The colour of the wings is delicate sulphur yellow, fading to shining white at the base of the wings, and diversified with narrow streaks of brown.

THE family of the Hyponomeutidæ contains many beautiful species, none of which are of great size, but which, when magnified and a strong light thrown upon them, seem to be amongst the loveliest of the moth tribes.

Our last example of the Lepidoptera is the beautiful WHITE-PLUME MOTH, an insect which never fails to attract attention, on account of the singular elegance and beauty of its form.

This insect belongs to a small family which is remarkable for the fact that, except in one genus, the wings, instead of being broad membranous structures, are cleft into narrow rays, feathered in a most soft and delicate manner. The White-plume Moth is to be seen in the evenings, flying in a

WHITE PLUME.—(*Pterophorus pentadactylus.*)

curious uncertain manner, and looking not unlike a snow-flake blown casually by the wind. It seems never to fly to any great distance, settling quite openly on leaves or plants, without taking the precaution of clinging to the under side, as is the custom with so many of the smaller moths. When it rests, it folds the wings so that they only look like a single broad ray. The legs of this moth are very long and slender. The colour of this insect is pure white.

HOMOPTERA.

IN the next order are comprised some very grotesque insects, some of which have been thought to belong to other orders, and a few not being known to be insects at all until comparatively late years. They have rounded bodies, not more than three joints in the tarsi, and their wings are four in number, wholly membranous, the fore pair being larger than the hinder, but not overlapping in repose. The mouth forms a kind of tube, sometimes longer than the body, and often sufficiently hard and stiff to pierce the skin.

In this curious order are placed the Aphides, those little green insects that swarm upon roses and other plants, and are termed " blights " by gardeners, who employ that term in a strangely wide sense ; the Cicadæ, with their beautiful membranous wings, their large heads, and their loud voices ; the tribe of Hoppers, of which the Cuckoo-spit Insect, known in its perfect state under the name of Frog-hopper, and the beautiful Scarlet Hopper, are familiar British examples ; the wonderful Lantern-flies, also leapers, which are found only in hot climates ; the Wax Insects of China ; and lastly, the Scale Insects, or Coccidæ, from which the "lac" so important in commerce is obtained.

The Cicadæ have three joints to their feet, these members affording useful characteristics in settling the precise position of the various species. They are very large insects, sometimes measuring more than six inches between the

Pæciloptera circulata.   *Cicada flosfolia.*

tips of the expanded wings. Their mouth or beak is three-jointed, and very long, being tucked under the body when not required. The females are furnished with a curious apparatus, by which they are enabled to cut grooves in the branches of trees for the purpose of depositing their eggs therein.

The male Cicada has the power of producing a shrill and ear-piercing sound, so loud in many species that it can be heard at a considerable distance, and becomes a positive annoyance, like the same tune played for several hours without intermission. The organ by which the sound is produced is internal, but its position may be seen externally by looking at the under side of the body, just behind the last pair of legs, where a pair of horny plates may be seen. These plates are the protecting covers of the sound-producing apparatus, which consists of two drum-like membranes and a set of powerful muscles. The colour of the perfect insect is mostly of a yellowish cast, and the wings are firm, shining, and membranous, somewhat resembling those of the dragon-fly in texture, but having larger cells or spaces between the nervures.

One species of Cicada is a native of England (*Cicada Anglica*), and is to be found in the New Forest.

THE Cercopidæ, or Hoppers, are well known in this country, mostly from the habits of the larva, and the saltatorial powers of the perfect insect. The CUCKOO-SPIT, or FROG-HOPPER, is very plentiful in this country. The larva fixes itself upon various plants, and sucks their juices through its long beak, which it plunges into the soft substance. When the accumulation of froth is very great, which usually happens in the heat of the day, a drop of clear water begins to form at the lowest part, into which the froth drains itself, and is presently relieved by the falling of the drop. The scientific name of this insect is *Aphrophora spumaria.* The species which is represented in the engraving belongs to the same genus as the beautiful SCARLET HOPPER of England, so frequently found on ferns in the outskirts of woods.

Passing by the Psyllidæ, another family of this order, we come to the APHIDES, a family comprising a great number of species. The whole history of these insects is remarkable in the extreme, presenting many points which seem most incredible, which destroy several old-established opinions, and in all probability will serve, when fully investigated, in establishing a new basis on which to found a more perfect system. The Aphides are wonderfully prolific, crowding upon plants until they completely hide them from view, and all employed in sucking the juice by means of the peculiar beak. They haunt every part of the plant, the leaves and their stalks, the branches, and even the roots being infested by these persevering destroyers, which often do great damage, and even force the leaves and branches to twist themselves into extraordinary contortions. Some species raise certain excrescences which serve as habitations for the insects.

In many species there is a pair of tubercles towards the extremity of the insect, which exude a sweetish liquid in a manner analogous to the frothing of the Cuckoo-spit. This liquid falls upon the leaves of trees, and is then known by the name of honey-dew. Bees are very fond of this substance, and, wherever it is present in any quantities, may be seen licking up the sweet secretion. Ants are equally fond of honey-dew, but they go to the fountain-head at once, and lap it as it flows from the tubercles. Whole regiments of ants may be seen ascending trees in search of the Aphides ; and it is very amusing to see how they will search every atom of a tree on which the Aphides live, so as not to allow a single insect to escape them.

The COCHINEAL INSECT (*Coccus cacti*) belongs to the same order. This species is a native of Mexico, and lives upon a kind of cactus, called from its insect guest the *Cactus cochinelliferus.*

The LAC INSECT (*Coccus lacca*) resides in India and the hotter parts of Asia.

HETEROPTERA.

THESE insects are readily known by several conspicuous characteristics. The wings are four in number, and the front pair are very peculiar in their structure, the basal portion being horny, like the elytra of beetles, and the remaining portion membranous, like the hinder wings of the same insect. In some species, however, the wings are wanting, as in the common Bed-bug (*Cimex lectularius*). The body is always much flattened, the mouth is beak-like, and in the pupal stage the creature is active and resembles the perfect insect, except in its want of wings.

THE family of the Nepidæ is represented in England by the common WATER SCORPION, a very flat and leaf-like insect, which is found abundantly in slow-running streams, ditches, and ponds. It derives its popular name from its scorpion-like aspect, the two slender filaments appended to the abdomen representing the sting-tipped tail, and the raptorial fore-legs resembling the claws. It is with these legs that the Water Scorpion catches its prey, which, when once grasped in that hooked extremity, is never able to make its escape. The beak is short, but very strong and sharp, and is not bent under the thorax as is the case with that of the water boatman.

The next section of the Heteroptera includes insects which are mostly terrestrial, though some are fond of haunting the surface of water. The Hydrometridæ are well-known examples of the latter insects, and are popularly known by the name of Water-fleas. The common Gerris skims over the surface with wonderful rapidity, wheeling and turning as easily as a skater performing his manœuvres on the smooth ice. But the Hydrometra, a very slender creature, hardly thicker than a needle, and bearing a great resemblance to the well-known Walking-stick Insect, glides slowly over the surface, mostly keeping among the aquatic plants at the margin, and passing silently as a shadow over the water.

The family of the Cimicidæ is represented by the too common BED-BUG, a creature which is supposed to have been imported into England from America. This odoriferous, flat-bodied, rust-coloured insect has derived its very appropriate name from the old English word *bugge*, signifying a nocturnal spectre, and used in that sense by the old writers. These creatures are enabled, by means of their flat bodies, to creep into the smallest crevices; and when they have once taken possession of a room, can be with difficulty extirpated.

The Reduviidæ comprise a great number of terrestrial insects, mostly exotic, but a few being natives of our country. Some of them are very large, and one species, the WHEEL-BUG (*Arilus serratus*), is said to possess electric powers. Its popular name is derived from the curious shape of the prothorax, which is elevated and notched, so as to resemble a portion of a cog-wheel. One species, *Reduvius personatus*, inhabits houses, and is said to feed upon the bed-bug. The larva and pupa of this insect are difficult to discover, on account of their habit of enveloping themselves in a coating of dust. The HAMMATOCERUS belongs to this family. The insect is remarkable for the curious structure of the second joint of the antennæ, which consists of numerous small articulations. The generic title is derived from two Greek words, signifying "link-horned," and is given to the insect in allusion to this peculiarity.

APHANIPTERA.

WE now come to another order, deriving its name from the invariable absence of wings, the name being derived from two Greek words, the former signifying "invisible," and the latter "a wing." There are not many species belonging to this order, and they are all known by the popular name of Fleas.

THE FLEA. —(*Pulex irritans.*)

The strength and agility of the curious but annoying little insect the FLEA is perfectly wonderful. Many of my readers have doubtless seen the exhibition of the Industrious Fleas, who drew little carriages, and carried comparatively heavy weights with the greatest ease. The apparatus with which it extracts the blood of its victims is very curious, and forms a beautiful object under a microscope of low power. Its leap is tremendous in proportion to its size. This property it enjoys in common with many other insects, among which the Common Grasshopper, the Frog-hopper, and the Halticas, or Turnip-flies, are conspicuous. In all these insects the hinder pair of legs are very long and powerful.

DIPTERA.

WE now pass to the DIPTERA, or Two-winged Insects, which may be known not only by the single pair of wings, but by the little appendages at their base, called halteres, or balancers, and which are the only vestiges of the hinder pair of wings. Moreover, the wings are not capable of being folded. This order is of vast extent, and includes a whole host of species.

THE Tipulidæ are very familiar to us through the well-known insects called DADDY LONG-LEGS, or CRANE-FLIES, one of which is here represented. In their perfect state these insects are perfectly harmless, although ignorant people are afraid to touch them. But in their larval condition they are fearful pests, living just below the surface of the ground, and feeding on the roots of grasses. Whole acres of grass have been destroyed by these larvæ ; and

GREAT CRANE-FLY.—(*Tipula gigantea.*)

some years ago Blackheath Park was so infested with them, that in the beginning of autumn the ground was covered thickly with the empty pupa cases of the escaped insect.

GAD AND BOT FLIES.

The left-hand figure in the illustration represents the common BREEZE-FLY, a well-known British example of the Tabanidæ. It is also known by the popular names of GAD-FLY and CLEG. As in the gnats, the females are the only bloodsuckers, but they exert their sanguinary ability with terrible force.

On the right hand of the engraving may be seen a large and bold-looking fly. This belongs to the family of the Œstridæ, and is popularly known by the name of BOT-FLY. All these insects are parasitic in or upon animals. The larva of this Bot-fly resides in the interior of horses, and is conveyed there in a very curious manner. The parent fly deposits her eggs upon the hairs near the shoulders of the horse, where the animal is sure to lick them, in order to rid itself of the unpleasant feeling caused by agglutinated hairs.

The eggs are thus conveyed to the stomach, to the coats of which organ the larvæ cling, and there remain until they have attained their full growth. They then loosen their hold, and are carried, together with the food, through the interior of the animal, fall to the ground, and immediately begin to burrow. They remain underground until they have undergone their metamorphoses, and then emerge in the shape of the perfect insect.

The HUMBLE BEE-FLY.—This very curious insect

HUMBLE BEE-FLY.
(*Bombylius medius.*)

is found in the early days of spring, and may be seen hovering over the primroses and other spring flowers. It feeds in the same manner as the Humming-bird Moth, and much resembles that insect in many of its habits.

CRUSTACEA

HAVING now completed our brief survey of the Insects, we proceed to the CRUSTACEA, a very large class, in which are included the lobsters, crabs, shrimps, water-fleas, and a host of other familiar beings. Even the Cirrhipeds, popularly known under the name of Barnacles, are members of this large class, and a number of curious animals, which have until lately been classed with the spiders, are now ascertained to belong to the Crustacea.

These beings can be easily separated from the insects on account of their general structure, the head and throat being fused into one mass, called technically the cephalothorax ; the number of limbs exceeding the six legs of the insects ; and the mode of breathing, which is by gills, and not by air-tubes.

The name of Crustacea is sufficiently appropriate, and is given to these creatures on account of the hard shelly crust with which their bodies and limbs are covered.

The first section of these creatures are called the Podophthalmata, or Stalk-eyed Crustaceans, because their eyes are set upon footstalks. The first order is that of the Ten-legged Crustaceans, so called on account of the five pairs of legs that are set in each side. These are exclusive of the complicated apparatus of the mouth, and the jaw-feet which guard its entrance. The CRABS are placed first in the list of Crustaceans, and are technically called Brachyura, or Short-tailed Crustaceans, because their tails are of comparatively small size, and are tucked under the large shielded body. In the preliminary stages, however, the Crabs have tails as proportionately long as those of a lobster or a cray-fish.

As the shelly armour of the Crustaceans is, in most cases, so hard, strong, and unyielding, the mode of growth might be considered a problem not very easy of solution : for with the Crustaceans the growth continues during nearly the whole of life, or at all events for several years after they have passed through the various changes to which they are subjected in their imperfect stages of existence. Their increase of size and weight is marvellously rapid, and how it can be accomplished, without subjecting the Crustaceans to the lot of the starveling mouse, who crawled into a jar of corn, but could not crawl out again after feasting on its contents, seems to partake of the charater of an animated puzzle.

The answer to the problem is simply that the creature sheds its armour annually, expands rapidly while yet covered only by a soft skin, and is soon protected by a freshly-deposited coat of shelly substance. Even this answer contains a second problem little less difficult than that which it solves. How can a Crustacean, say a crab or a lobster, shed its skin? It is true that the cast shells are found, showing that the creature has escaped from its old and contracted tenement by a slit in some part of the body, such as the top of the carapace, and has left its shell in so perfect a state that it might easily be mistaken for the living animal.

COMMON THORNBACK CRAB.—(*Maia Squinado.*)

But how did it manage about the claws? We all know what large mus-

THORNBACK SPIDER-CRAB.

cular masses they are, how very small is the aperture in which the joint works, and how stiff and firm is the broad tendinous plate which is found in their interior. Examination shows that there is no opening on the claws

O O 2

through which the creature might have drawn the imprisoned limb, and it is also evident that the only method by which these members can be extricated is by pulling them fairly through the joints. As a preliminary step, the hard, firm, muscular fibres which fill the claw and give it the well-known pinching power, become soft, flaccid, and watery, and can thus be drawn through the comparatively small openings through which the tendons pass from one joint to another. The sharp and knife-like edges of the plates cut deeply through the muscle, which, however, is little injured, on account of its soft consistency, and heals with great rapidity as soon as the animal recovers its strength and is gifted with a new shell. In the common edible crab, the flesh is quite unfit for consumption during this process, as anyone can attest who has attempted to dress and eat a "watery" crab. Yet, in some of the exotic crustaceans, these conditions are exactly reversed, and the crabs are never so fit for the table as while they are soft and shell-less, after the old suit of armour has been thrown off, and before the new integument has received its hardening.

WE now come to the SPIDER-CRABS, scientifically termed Maiadæ.

A very common and a very useful British species is seen in the illustration, sitting on the rock. The common THORNBACK SPIDER-CRAB, or SQUINADO, is plentiful upon our coasts, and, as may be seen by reference to the figure, is not a very prepossessing creature in external appearance, its body being one mass of sharp and not very short spines, and its whole frame possessing a weird-like and uncomely aspect.

Ugly though it may be in an artistic point of view, it is one of the most useful inhabitants of the sea, acting as a scavenger for the removal of the decaying animal matter that is ever found in the seas. More especially along the shore, where the refuse of mankind, such as unsaleable fish and crustaceans, are continually being cast into the waves, the Squinado is found to perform the necessary office of removing all such substances. It is a voracious creature, and being gifted with an acute sense of smell, is sure to discover without delay any substance on which it can feed, and to make its way thereto without delay.

The Squinado, together with other crabs, sets to work boldly : with one claw he holds tightly to the banquet, and with the other tears off morsels and deftly feeds himself therewith, putting them into his comical jaws with the regularity of clockwork, and with a rapidity that reminds the observer of a Chinese flinging rice into his mouth with his chopsticks. The strength and sharpness of the claws are such that the toughest muscle cannot long withstand their power, and the flesh is torn from the bones as perfectly as if scraped away by a knife.

It is a curious fact that the back of this crab is generally a resting-place for sundry zoophytes, which often grow in such profusion as to hide the animal completely. A specimen half covered with zoophytes is

EDIBLE CRAB.—(*Cancer pagurus.*)

shown in one illustration, while in the other the Spider-crab is shown denuded of these appendages.

THE large family of the Canceridæ now comes before us, and is familiarly
known through the medium of the common EDIBLE CRAB.

This is a very common species, being plentiful around our rocky coasts,
and generally remaining in the zone just under low-water mark. The fishermen catch it in various ways ; but the most usual method, and that by which
the greatest number of these crustaceans are captured, is by means of certain
baskets, called crab-pots, cruives, or creels, according to the locality. These
baskets are round, and in shape something like a flattened apple, and have
an aperture at the top through which the crab gains access to the interior.
When once within the basket it cannot escape, because the opening is

GREEN CRAB.—(*Carcinus mænas.*)

guarded by an inverted cone of osiers, like the entrance to a common wire
mouse-trap, so that the elastic sticks yield to the expected prey while passing
downwards, but effectually prevent all upward movement.

Supplied with a number of these creels, a corresponding amount of rope,
floats, stones, and bait, the fisherman rows towards the best grounds, which
are always where the bed of the sea is rocky, and the depth from three to
twenty fathoms. The bait, consisting of haddock, skate, and other fish, is
placed in the basket, together with a few stones which serve to sink it, a line
is attached, and the creel lowered out of the boat. A buoy is attached to
the line and marked with the owner's name, so as to avoid mistakes as to the
proprietorship of the creel. The fisherman then rows to a little distance,
and sinks another baited creel, taking the precaution to place them so far
asunder that the lines cannot be entangled in each other.

Boys often employ their idle afternoons in crab-hunting, always going
among the rocks at low-water, and looking out for those rock masses that

are covered with heavy seaweeds. They are armed with a kind of lance, consisting of an iron hook fastened to a long stick, and with this they poke about in the crevices under the rocks, and twist out the crabs that have con-

NIPPER CRAB.—(*Polybius Henslowii.*)

cealed themselves. These crabs, however, seldom attain any great size, the larger specimens remaining in the deeper water. The boys call them "pungers," to distinguish them from the green crab.

FIGHTING CRAB.—(*Gelasimus bellator.*)

The shell of this crab is seldom found entirely clean, being generally encrusted with acorn-barnacles and various marine creatures. Sometimes, when the crab is a very old and large one, has ceased growing for several years, and consequently has needed no change of shell, it becomes absolutely loaded with all sorts of extraneous growths, and in many cases is almost invisible under its load. There is a very curious specimen in the large collection at the British Museum, where a number of oysters had affixed themselves to the shell, and consequently had been borne about with the crab in all its peregrinations.

WE now arrive at the family of the Portunidæ, or Swimming Crabs, which may be recognized by the construction of the last pair of feet, which are flattened sideways, and have the last joint dilated into a thin oblique plate, which answers as an oar or a fin, and enables the creature to propel itself through the water. The first example of this family is the GREEN or SHORE CRAB, so familiar to everyone who has passed even an hour on the coast between the time of high and low water. Although one of the commonest of our native crustaceans, it is at the same time one of the most interesting, and, owing to its diurnal habits, its fearless nature, and its love for the shallow waters, it is very easily observed. I have spent many a pleasant hour in watching the habits of this little creature, and could hardly have imagined the activity, the piercing sight, and the cleverness with which this crab is endowed.

The NIPPER CRAB is a really wonderful swimmer, being able, according to Mr. Couch's account, to ascend to the surface of the sea, and to pursue its prey through the waters. So well does this creature swim, and so voracious is its appetite, that it captures and eats even the swiftest sea-fish, having been known to pounce upon the mackerel and the pollack. Its method of proceeding seems to be to dart upon its prey, grasp it firmly with its sharply pointed and powerful claws, and retain its hold until the unfortunate victim is quite fatigued and falls an easy prey.

WE now arrive at another family, called the Ocypodidæ, or Swift-footed Crabs, from their extraordinary speed, which equals or even exceeds that of a man.

The figure in the engraving represents the FIGHTING CRAB, a creature whose name is well deserved. As the reader may observe, one of its claws is enormously large in proportion to the body, being indeed nearly equal in dimensions to the whole carapace, while the other claw is quite small and feeble. It is remarkable that sometimes the right and sometimes the left claw is thus developed. This animal is a most determined fighter, and has the art of disposing its limbs like the arms of a boxer, so as to be equally ready for attack or defence. The figure shows the crab in its posture of defence.

The Fighting Crab lives on the seashore or on the border of salt marshes, and burrows deeply in the earth, the holes being tolerably cylindrical and rather oblique in direction. In some places these holes are so close together

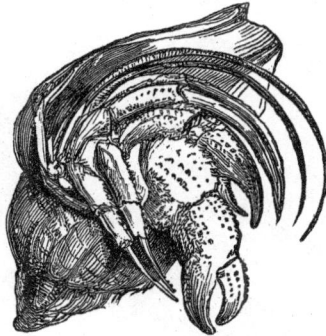

HERMIT CRAB.—(*Pagurus Bernhardus.*)

that the earth is quite honeycombed with them, and the place looks like a rabbit warren. Each burrow is tenanted by a pair of crabs, the male always remaining in the post of danger at the mouth of the tunnel, and keeping guard with his great claw at the entrance.

While running, it has a habit of holding the large claw aloft, and moving it as if beckoning to some one ; a habit which has caused one of the species to be named the Calling Crab. This action has in it something very ludicrous, and those who have watched the proceedings of a crab warren say that there are few scenes more ridiculous than that which is presented by the crustaceans when they are alarmed and go scuttling over the ground to their homes, holding up their claws and beckoning in all directions.

WE now come to a singular group of crabs, which are remarkable for their soft and shell-less tails and the mode employed to protect them. From their solitary habits they are called HERMIT-CRABS, and from their extreme combativeness they have earned the title of SOLDIER-CRABS.

The best known of these crustacea is the common HERMIT-CRAB of England (*Pagurus Bernhardus*), which is to be found plentifully on our shores. Like all its race, the Hermit-crab inhabits the shell of some mollusc, in which it can bury its unprotected tail, and into which it can retreat when threatened with danger. The Hermit-crab usurps the deserted home of various molluscs, according to its size, so that, when young and small, it is found in the shells of the tops, periwinkles, and other small molluscs ; and when it reaches full age, it takes possession of the whelk-shell and entirely fills its cavity.

To see a Hermit-crab fitting itself with a new shell is a very ludicrous sight. The creature takes the shell among its feet, twirls it about with wonderful rapidity, balances it as if to try its weight, probes it with the long antennæ, and perhaps throws it away. Sometimes, however, when the preliminary investigations have proved satisfactory, it twists the shell round until the tail falls into the opening, and then parades up and down for a little while. Perhaps it may be satisfied, and after twirling the shell about several times, whisks into it with such speed that the eye can scarcely follow its movements. Indeed it seems rather to be shot into the shell from some engine of propulsion than to move voluntarily into the new habitation. When the number of empty shells is great, the Hermit is very fastidious, and will spend many hours in settling into a new house.

In all these creatures the larger claw is very much developed ; so that when the Crab has withdrawn into the shell, the claw lies over the entrance and closes it like a living door, which has the further advantage of being used as an offensive weapon. The footstalks on which the eyes are set are moderately long, stout, and jointed, and enable their possessor to see in all directions.

LOBSTER.—(*Astacus gammarus.*)

WE next take the second great division of the Crustaceæ, namely those which have long and powerful tails. The LOBSTERS and SHRIMPS are examples of these creatures. In swimming rapidly through the water, the tail is the organ of propulsion which is employed, and a glance at its form will soon explain its use.

WE now come to the family of the Astacidæ, which includes two well-known and very similar creatures, the fresh-water cray-fish, and the salt-water Lobster. The general shape and appearance of the LOBSTER is too well known to need any description. The vast numbers of Lobsters which are annually brought to the London markets are largely supplied from Norway, although there are many parts of our own coasts where these creatures can be taken plentifully. The Lobster is not much of a rover, seldom straying far from the spot on which it was hatched.

The Lobsters are caught in creels or pots, like the crabs, but with greater ease and economy, as they are very fond of meat, be it fresh or tainted, and even if it should be putrefying will be attracted to it.

Like many other crustaceans, the Lobster is a most combative animal, quarrelling on the slightest pretext, and fighting most furiously. In these combats it mostly loses a claw or a leg, being obliged to discard entirely a wounded member. A fresh leg or claw sprouts from the scar, and it is to this circumstance that the frequently unequal size of lobster-claws is owing. Lobsters indeed part with these valuable members with strange indifference, and will sometimes shake them off on hearing a sudden noise.

If the fishermen find that they have wounded a Lobster, they have recourse to a very strange but perfectly efficacious remedy. Supposing one of the claws to be wounded, the creature would soon bleed to death unless

SAND-HOPPER..—(*T. litrus, saltator.*)

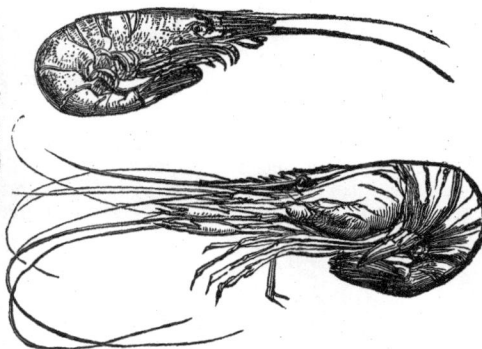

THE SHRIMP.—(*Crangon vulgaris.*)
THE PRAWN.--(*Palæmon serratus.*)

some means were taken whereby the flow of blood may be stopped. The method adapted by the fishermen consists in twisting off the entire claw. A membrane immediately forms over the wound, and the bleeding is stopped. The new limb that is to supply the place of that which was lost always sprouts from the centre of the scar.

THE next family includes the true SHRIMPS, and contains but one genus. The SHRIMP, which is so familiar on our tables, and which, until the marine aquaria became so common, was equally unknown in its living state, inhabits our shores, where it is produced in countless myriads. In every little pool that is left by the retiring tide the Shrimps may be seen in profusion, betraying their presence by their quick darting movements as they dash about in the water, and ever and anon settle upon some spot, flinging up a cloud of sand as they scuffle below its surface, their backs being just level with the surrounding sand. In consequence of this manœuvre, the fishermen call them " Sand-raisers." The small prawns are often confounded with the Shrimps and popularly called by the same title. They can, however, be easily distinguished from each other, the beak of the prawn being long, and deeply saw-edged, while that of the Shrimp is quite short and smooth.

OUR attention is now drawn to a very large group of crustaceans, called the Sessile-eyed Crustacea, because their eyes, instead of being placed on footstalks, are seated directly upon the shell. The body is divided with tolerable distinctness into three parts, for which the ordinary titles of head, thorax, and abdomen are retained as being more convenient and intelligible than the ingenious and more correct, though rather repulsive, titles that have lately been affixed to these divisions of the body.

They have no carapace, like the stalk-eyed crustaceans, nor do they breathe with gills, but by means of a curious adaptation of some of their limbs. None of the Sessile-eyed Crustacea attain any large size, an inch and a half being nearly their utmost limit in point of length. Most of these animals reside along the sea-shores, where they are of very great use in clearing away the mass of dead animal and vegetable matter which is constantly found in the sea.

In the illustration we have an example of the first family, called by the name of Orchestidæ, or Jumpers, because they possess the power of leaping upon dry ground. The figure represents the most familiar of these little crustaceans. This is the little SAND-HOPPER, or SAND-SKIPPER, which is seen in such myriads along all our sandy shores, leaping about vigorously just before the advancing or behind the retiring tide, and looking like a low mist edging the sea, so countless are their numbers. Paley has a well-known passage respecting this phenomenon, too familiar for quotation.

The leap of the Sand-hopper is produced by bending the body and then flinging it open with a sudden jerk—in fact, the exact converse of the mode of progression adopted by the lobster and shrimp. The Sand-hopper feeds on almost everything that is soft and capable of decay, and seems to care little whether the food be of an animal or vegetable nature. Decaying seaweed is a favourite article of food, and wherever a bunch of blackened and rotting seaweed lies on the sand, there may be found the Sand-hoppers congregated beneath it, and literally boiling out when the seaweed is plucked up.

The teeth of this creature are strong and sharp, as indeed is needful for the tasks imposed upon them. The Sand-hopper will eat anything; and on one occasion, when a lady had allowed a swarm of these little crustaceans to settle on her handkerchief, it was bitten to rags when she took it up. It is very fond of worms, will eat any kind of carrion, and sometimes, when pressed by hunger, has no scruple in eating its own kind.

THE two lower figures in the illustration represent the common WOODLOUSE, the latter figure being given in order to show the equal development of the legs. This creature is very plentiful in all damp places, and especially exults in getting under logs of wood or decaying timber. In cellars and outhouses

ARMADILLO WOODLOUSE.—(*Armadillo vulgaris.*
WOODLOUSE.—(*Porcellio scaber.*)

they are common, and are generally to be found in dark and damp localities. Fowls are very fond of them, and there is no surer way of extirpating these sharp-toothed creatures than by allowing some fowls to scrape and peck about in the places where they have taken up their residence. Under the bark of dead and decaying trees is a very favourite residence with the Woodlouse, and in such localities their dead skeletons may often be found, bleached to a porcelain-like whiteness. The colour of the Woodlouse is a darkish leaden hue, sometimes spotted with white.

The well-known PILL WOODLOUSE, or PILL ARMADILLO, are seen at the upper part of the illustration, one representing the creature as it appears while walking, and the other showing it when rolled up into a globular shape. In this attitude it bears a strong analogy to the common hedgehog, and a still stronger to the manis, as in the latter case the creature is defended by horny scales that protect it just as the external skeleton protects the armadillo. While rolled up, this creature has been often mistaken for a bead or a berry from some tree, and in one instance a girl, new to the country, actually threaded a number of these unfortunate crustaceans before she discovered that they were not beads.

WE now come to the last members of the Crustacea, creatures which were for a long time placed among the molluscs, and whose true position has only been discovered in comparatively later years. Popularly they are called Barnacles, but are known to naturalists under the general term Cirrhipedes, on account

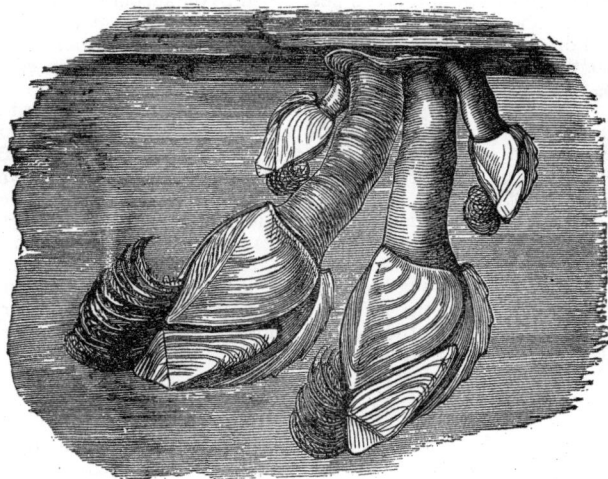

Lepas anatifera.

of the cirrhi, or bristles, with which their strangely transformed feet are fringed.

When adult, all the Cirrhipedes are affixed to some substance, being either set directly upon it, as the common acorn barnacle, so plentiful on our coasts; placed upon a footstalk of variable length, as in the ordinary goose-mussel; or even sunk into the supporting substance, as is the case with the whale barnacles. When young, the Cirrhipedes are free and able to swim about, and are of a shape so totally different to that which they afterwards assume, that they would not be recognized except by a practised eye.

Along the under surface are set six pairs of limbs not furnished with claws, but being developed at their extremities into two long filaments, joined and covered with hairs. By means of these modified limbs the cirrhipedes obtain their food.   The common acorn barnacle of our coasts affords a familiar and beautiful example of the mode by which this structure is made subservient to procuring a supply of food.   The closed valves at the upper part of the shell are seen to open slightly, a kind of fairy-like hand is thrust out, the fingers expanded, a grasp made at the water, and the closed member then withdrawn into the shell.

This hand-like object is in fact the aggregated mass of legs with their filaments.   As the limbs are thrust forward, they spread so as to form a kind of casting net ; and as they return to the shell, they bring with them all the minute organisms which were swimming in the water.   This movement continues without cessation, as long as the barnacles are covered with water, and appears to be as mechanically performed as the action of breathing is performed by the higher animals.

In the illustration is seen a group of the common GOOSE MUSSEL, or DUCK BARNACLE, so called on account of the absurd idea that was once so widely entertained, that this species of barnacle was the preliminary state of the barnacle-goose, the cirrhi representing the plumage, and the valves doing duty for the wings.

This barnacle is tolerably universal in its tastes.   It clings to anything, whether still or moving, and is the pest of ships on account of the pertinacity with which it adheres to their planks.   Its growth is marvellously rapid, and in a very short time a vessel will have the whole of the submerged surface coated so thickly with these cirrhipedes that her rate of speed is sadly diminished by the friction of their loose bodies against the water.

ACORN BARNACLE.
(*Balanus balanoides.*)

A good example of these creatures is afforded by the well-known ACORN BARNACLE, so plentiful on our coasts. They have no necks like those of the Goose Barnacle, but are sessile on the rocks.   Spots over which the tide only runs for a few hours are thickly studded with these Barnacles, and it is interesting to see how quickly they open their valves and fling out their arms as soon as the water covers them at each returning tide.   When the sea withdraws, they close their shells firmly, and retain within their interior a sufficiency of water wherewith to carry on the business of respiration until the next tide brings a fresh supply.   Total submersion seems to be hurtful to them.

ARACHNIDA.

ANOTHER class of animated beings now comes before us, which, under the general term of Arachnida, comprises the Spiders, Scorpions, and Mites.

These beings breathe atmospheric air, have no antennæ, and have four pairs of legs attached to the fore parts of the body.

In some of the higher Arachnida there is a bold division into thorax and abdomen, and the former portion of the body is clearly divided into separate segments.   By the earlier naturalists the Arachnida were placed among the insects, but may readily be distinguished by several peculiarities.   In the first place, they have more than six legs, which alone would be sufficient to separate them from insects.   They have no separate head ; the head and thorax being fused, as it were, into one mass, called the cephalothorax. In many of the lower species there is not even a division between the thorax and abdomen ; and the body, thorax, and abdomen are merged into one uniform mass, without even a mark to show their several boundaries.   They undergo no metamorphosis, like that of the insects, for, although the young Spiders change their skins several times, there is no change of form.

Beginning with the true Spiders, we find that their palpi (*i.e.*, the jointed antennæ-like organs that project from the cephalothorax) are more or less thread-like, and in the males are swollen at the extremity into a remarkable structure, as indicative of the sex as is the beard of man, the curled tail-feathers of the drake, and the gorgeous train of the peacock. In the different genera, these palpi are differently formed, and afford valuable indications for systematic zoologists.

In these strange creatures, the mandibles are furnished with a curved claw perforated at the extremity, sometimes like the poison-fang of a venomous snake, and used for a similar purpose. A gland furnishes a secretion which is forced through these organs, and is injected into any object that may be wounded by the sharp claw. The fluid which is secreted for the service of the fangs is nearly colourless, and is found to possess most of the properties that exist in the venom of the rattlesnake or viper.

They all spin those remarkable nets which we popularly call "webs," and which differ wonderfully in the various species. These webs are, in very many instances, employed as traps wherein may be caught the prey on which the Spider feeds, but in other cases are only used as houses wherein the

CRAB-SPIDER, OR MATOUTOU.—(*Mygale cancerides.*)

creature can reside. Some of the uses to which these wonderful productions are put, as well as some details of their structure, will presently be mentioned.

WE now pass to some typical species of these curious animals.

The Spiders belonging to the family of Mygalidæ may at once be known by the shape of their mandibles and the terrible claws which proceed from them. In the greater number of Spiders the claws are set horizontally, but in the Mygalidæ they are bent downwards, and strike the prey much as a lion clutches at his victim with his curved talons.

THE GREAT CRAB SPIDER belongs to the typical genus of this family, and is one of the formidable Arachnida that prey upon young birds and other small vertebrates, instead of limiting themselves to the insects, and similar beings, which constitute the food of the generality of the Spider race.

The talons of the spiders are **scientifically called by** the appropriate name of *falces*, the word being Latin, and signifying "a reaping-hook." By this name they will be called in the course of the following pages. The falces of the great Crab Spiders are of enormous size, and when removed from the creature and set in gold, they are used as tooth-picks, being thought to possess some occult virtue which drives away the toothache.

In the accompanying illustration is an example of the curious TRAP-DOOR SPIDER of Jamaica, erroneously called the Tarantula.

This spider digs a burrow in the earth and lines it with a silken web, but instead of merely protecting the entrance by a portion of the silken tube, it proves itself a more complete architect by making a trap-door with a hinge that permits it to be opened and closed with admirable accuracy. The door is beautifully circular, and is made of alternate layers of earth and web, and hinged to the lining of the tube by a band of the same silken secretion. It exactly fits the entrance of the burrow, and, when closed, so precisely corresponds with the surrounding earth that it can hardly be distinguished, even when its position is pointed out. It is a strange sight to see the earth open, a little lid raised, some hairy legs protrude, and gradually the whole form of the spider show itself.

THE curious and interesting WATER SPIDER is now far better known than was formerly the case, as the numerous aquaria that have been established over the kingdom have tended to familiarize us with this as well as with many other inhabitants of the water.

This creature leads a strange life. Though a really terrestrial being, and needing to respire atmospheric air, it passes nearly the whole of its life in the water, and, for the greater part of its time, is submerged below the surface. To a lesser degree, several other spiders lead a somewhat similar life, sustaining existence by means of the air which is entangled in the hairs which clothe the body. Their submerged existence is, however, only accidental, while in the Water Spider it forms the constant habit of its life.

TRAP-DOOR SPIDER—(*Cteniza nidulans*.)

The body of the Water Spider is profusely covered with hairs, which serve to entangle a large comparative amount of atmospheric air, but it has other means which are not possessed by the species already described. It has the power of diving below the surface, and carrying with it a very large bubble of air that is held in its place by the hind-legs ; and in spite of this obstacle to its progress, it can pass through the water with tolerable speed.

The strangest part in the economy of this creature is, that it is actually hatched under water, and lies submerged for a considerable time before it ever sees the land. At some little depth the mother spider spins a kind of egg or dome-shaped cell, with the opening downwards. Having made this chamber, she ascends to the surface, and there charges her whole body with air, arranging her hind-legs in such a manner that the bubble held between them cannot escape. She then dives into the water, proceeds to her nest, and discharges the bubble into it. A quantity of water is thus displaced, and the upper part of the cell is filled with air. She then returns for a second supply, and so proceeds until the nest is full of air.

In this curious domicile the spider lives, and is thus able to deposit and to hatch her eggs under the water without even wetting them. The reader will have noticed the exact analogy between this sub-aquatic residence and the diving-bell, now so generally employed. As to the spider itself, it is never wet; and though it may be seen swimming rapidly about in the water, yet the moment it emerges from the surface its hairy body will be found as dry as that of any land spider. The reason for this phenomenon is, that the minute bubbles of air which always cling to the furred body repel the water and prevent it from moistening the skin.

The eggs of this spider are inclosed in a kind of cup-shaped cocoon, not unlike the cover of a circular vegetable dish. This cocoon may be seen in the illustration in the upper part of the cell. It usually contains about a hundred little spherical eggs, which are not glued together.

The Water Spider is a truly active creature, and its rapid movements can be watched by means of placing one of these Arachnida in a vessel nearly filled with water. If possible, some water plant, such as the vallisneria or anacharis, should be also placed in the vessel. Here

WATER SPIDER.—(*Argyronetra aquatica.*)

the spider will soon construct its web, and exhibit its curious habits. It must be well supplied with flies and other insects thrown into the water. It will pounce on them, carry them to its house, and there eat them.

The limbs and cephalothorax of this species are brown, with a slight tinge of red; and the abdomen is brown, but washed with green. It is densely covered with hairs. On the middle of the upper surface of the abdomen are found round spots arranged in a square. The male is rather larger than the female, and his legs are larger in proportion. He may, however, be distinguished by the large mandibles and longer palpi.

WE now arrive at the Epeiridæ, a family containing some of the strangest members of the spider race. The best known of this family is the common GARDEN SPIDER, sometimes called the CROSS SPIDER, from the marks upon its abdomen. This is thought to be the best typical example of all the Arachnidæ. It is found in great numbers in our gardens, stretching its beautiful webs perpendicularly from branch to branch, and remaining in the centre with its head downwards, waiting for its prey. This attitude is tolerably universal among spiders ; and it is rather curious that the Arach-

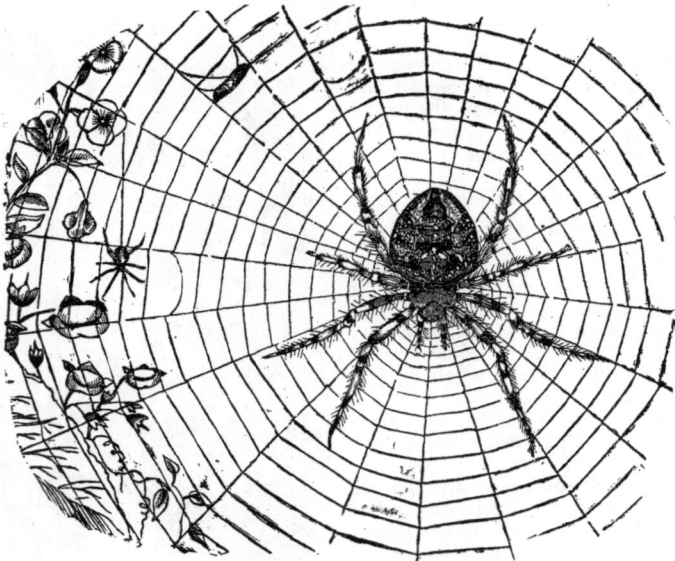

GARDEN SPIDER.—(*Epeira diadema.*)

nidæ should reverse the usual order of things, and assume an inverted position when they desire to repose.

The web of this spider is composed of two different kinds of threads, the radiating and supporting threads being strong and of simple texture. But the fine spiral thread which divides the web into a series of steps, decreasing in breadth towards the centre, is studded with a vast amount of little globules, which give to the web its peculiar adhesiveness. These globules are too small to be perceptible to the unassisted eye, but by the aid of a microscope they may be examined without difficulty. In an ordinary web, such as is usually seen in gardens, there will be about eighty-seven thousand of these globules, and yet the web can be completed in less than three-quarters of an hour. The globules are loosely strung upon the lines, and when they are rubbed off, the thread is no longer adhesive.

OF all the Spider race the SCORPIONS are most dreaded, and justly so. These strange beings are at once recognized by their large claws and the armed tail. This member is composed of six joints, the last being modified into an arched point, very sharp, and communicating with two poison glands in the base of the joint. With this weapon the Scorpion wounds its foes, striking smartly at them, and by the same movement driving some of the poison into the wound. The effect of the poison varies much, according to the constitution of the person who is stung, and the size and health of the Scorpion. Should the creature be a large one, the sting is productive

ROCK SCORPION.—(*Buthus afer.*)

of serious consequences, and in some cases has been known to destroy life. Generally, however, there is little danger to life, though the pain is most severe and the health much injured for the time, the whole limb throbbing with shooting pangs, and the stomach oppressed with over-powering nausea. The poison seems to be of an acrid nature, and the pain can be relieved by the application of alkaline remedies, such as liquid ammonia, tobacco ashes, &c. Melted fat is also thought to do good service, and the nausea is relieved by small doses of ipecacuanha. Some of the poison can mostly be brought to the surface by means of pressing a tube, such as a tolerably large key or the barrel of a small pistol, upon the spot, and the duration, if not the severity of the pain, is thereby mitigated.

In all these creatures the tail is composed of the last six joints of the abdomen, and the powerful limbs, with the lobster-like claws at the tips, are the modified palpi. The eyes of the Scorpions differ in number, some species having twelve, others eight, and others only six : these last constitute the genus Scorpio. On the lower surface of the Scorpions are seen two remarkable appendages, called the combs, the number of teeth differing in the various species. In the Rock Scorpion the teeth are thirteen in number, while in the red scorpion there are never less than twenty-eight. The Rock Scorpion is a large creature, measuring about six inches in length when fully grown.

MYRIAPODA.

IN accordance with the best systems of the present day, the MYRIAPODA are considered as a separate class.

The Myriapoda are without even the rudiments of wings, and possess a great number of feet, not less than twelve pairs ; and in some species there are more than forty pairs of legs. In allusion to their numerous feet the Myriapoda are popularly called Hundred-legs, and their scientific title is even bolder, signifying ten thousand feet. To this class belong the well-known centipedes, so plentiful in our gardens, and the equally well-known millepedes, found under decaying wood and in similar localities. In England none of the Myriapods attain to great dimensions, but in hot countries, and especially under the tropics, they become so large as to be positively formidable as well as repulsive.

WE now arrive at the true Scolopendræ, which, together with the allied genera, are popularly known by the name of CENTIPEDES. The genus Scolopendra is a very large one, containing about sixty species, most of them inhabitants of the tropics, and many attaining a large size. The great Scolopendræ are not only unpleasant and repulsive to the sight, but are really formidable creatures, being armed with fangs scarcely less terrible than the sting of the scorpion. These weapons are placed just below the mouth, and are formed from the second pair of feet, which are modified into a pair of strong claws, set horizontally in a manner resembling the falces of ordinary spiders, and terminated by a strong and sharp hook on each side. These hooks are perforated, and are traversed by a little channel leading from a poison gland, like that of the scorpion, so that the venomous secretion is forced into the wound by the very action of biting.

Arthronomalus longicornis.

The figure represents a centipede which is found in England, and in some localities is very common. It is represented of its natural size, and, as may be seen, is in no way conspicuous for its dimensions. It is, however, remarkable on another account. It has the power of giving out a tolerably strong phosphorescent light, which is only visible after dark, but is then very conspicuous, and has often caused the centipede to be mistaken for a glow-worm. It is not unfrequently found within peaches, apricots, plums, and similar fruits, when they are very ripe, and lies comfortably coiled up in the little space between the stone and the fruit, where the sweetest juices lie. The colour of this centipede is yellow ; its head is deep rust colour ; its antennæ are very hairy, and four times as long as the head segment. There are from fifty-one to fifty-five pairs of legs. Its length varies from two inches and a half to three inches.

ANNULATA.

A NEW class of animals now comes before us. These creatures are technically called ANNULATA, or sometimes ANNELIDA, on account of the rings, or annuli, of which their bodies are composed. They may be distinguished from the Julidæ by the absence of true feet, although in very many species the place of feet is supplied by bundles of bristles, set along the sides. The respiration is carried on either by means of external gills, internal sacs, or even through the skin itself. In most of the Annulata the body is long and cylindrical, but in some it is flattened and oval. The number of rings is very variable, even in the same species.

The group of worms which come first on our list is remarkable for the architectural powers of its members. In order to protect their soft-skinned body and delicate gills, they build for themselves a residence into which they exactly fit This residence is in the form of a tube, and in some cases, as in the Serpulæ, is of a very hard shelly substance, and in some, as the Terebella is soft and covered with grains of sand and fragments of shells.

The beautiful SERPULA is now very familiar to us, through the medium of marine aquaria, its white shell, exquisite fan-like branchiæ, and brilliant operculum, having lived and died in many an inland town where a living inhabitant of the ocean had never before been seen. The Serpula is able to travel up and down its tube by the bundles of bristles which project from the rings along the sides, and to retract itself with marvellous rapidity. It has no eyes, and yet is sensible of light. For example, if a Serpula be fully protruded, with its gill-fans extended to their utmost, and blazing in all its

Serpula contortuplicata.

scarlet and white splendour, a hand moved between it and the window will cause it to disappear into its tube with a movement so rapid that the eye cannot follow it. The gills, whose exquisitely graceful form and delicate colouring have always attracted admiration, are affixed to the neck, as, if they were set at the opposite extremity of the body or along the sides, they would not obtain sufficient air from the small amount of water that could be contained in the tube. The beautiful scarlet stopper ought also to be mentioned. Each set of gills is furnished with a tentacle-like appendage, one of which is small and thread-like, and the other expanded at its extremity into a conical operculum or stopper, marked with a number of ridges, which form a beautiful series of teeth around its circumference. The footstalk on which this stopper is mounted is a little longer than the gills, so that when the animal retreats into its tube the gills collapse and vanish, and the entrance of the tube is exactly closed by the conical stopper.

THE family of which the common EARTH-WORM is a very familiar example is distinguished by the ringed body without any gills or feet, but with bristles arranged upon the rings for the purpose of progression.

In the well-known Earth-worm, the bristles are short and very stiff, and are eight in number on each ring, two pairs being placed on each side; so that, in fact, there are eight longitudinal rows of bristles on the body, four on the sides and four below, which enable the creature to take a firm hold of the ground as it proceeds. Except that the worm makes use of bristles, and the snake of the edges of its scales, the mode of progress is much the same in both cases. The whole body of the creature is very elastic, and capable of

being extended or contracted to a wonderful degree. When it wishes to advance, it pushes forward its body, permits the bristles to hitch against the ground, and then, by contracting the rings together, brings itself forward, and is ready for another step. As in each full-grown Earth-worm there are at least one hundred and twenty rings, and each ring contains eight bristles, it may be imagined that the hold upon the ground is very strong.

The COMMON LEECH is almost as familiar as the earth-worm, and is one of a genus which furnishes the blood-sucking creatures which are so largely

COMMON LEECH.—(*Hirudo medicinalis.*)

used in surgery. It belongs to a large group of Annelida which have no projecting bristles to help them onward, and are therefore forced to proceed in a different manner. All these Leeches are wonderfully adapted for the purpose to which they are applied, their mouths being supplied with sharp teeth to cut the vessels, and with a sucker-like disc, so that the blood can be drawn from its natural channels ; while their digestive organs are little more than a series of sacs in which an enormous quantity of blood can be received and retained.

RADIATA.

ECHINODERMATA.

WE now arrive at a vast and comprehensive division of living beings, which have no joints whatever, and no limbs, and are called RADIATA, because all their parts radiate from a common centre. The structure is very evident in some of these beings, but in others the formation is so exceedingly obscure that it is only by anatomical investigation that their real position is discovered.

The highest forms in this division have been gathered together in the class Echinodermata. This word signifies " urchin-skinned," and is given to the animals comprising it because their skins are more or less furnished with spines resembling those of the hedgehog. In these animals the radiate form is very plainly shown, some of them assuming a perfectly star-like shape, of which the common star-fishes of our coasts are familiar examples. In some of the Radiates, such as the sea-urchin, the whole body is encrusted with a chalky coat, while in others it is as soft and easily torn as if it were composed of mere structureless gelatine.

The mode of walking, or rather creeping, which is practised by these beings, is very interesting, and may be easily seen by watching the proceedings of a common star-fish when placed in a vessel of sea-water. At first it will be quite still, and lie as if dead, but by degrees the tips of the arms will be seen to curve slightly, and then the creature slides forward without any perceptible means of locomotion. If, however, it be suddenly taken from the water and reversed, the mystery is at once solved, and the walking apparatus is seen to consist of a vast number of tiny tentacles, each with a little round transparent head, and all moving slowly but continually from side to side, sometimes being thrust out to a considerable distance, and sometimes being withdrawn almost wholly within the shell. These are the "ambulacræ," or walking apparatus, and are among the most extraordinary means of progression in the animal kingdom. Each of these innumerable organs acts as a sucker, its soft head being applied to any hard substance, and adhering thereto with tolerable firmness, until the pressure is relaxed and the sucker released. The suckers continually move forward, seize upon the ground, draw the body gently along, and then search for a new hold. As there are nearly two thousand suckers continually at work, some being protruded, others relaxed, and others still feeling for a holding-place, the progress of the creature is very regular and gliding, and hardly seems to be produced by voluntary motion.

We will now proceed to some examples of these curious beings.

WE first take a beautiful family of this order, called Echinidæ, because they are covered with spines like the quills of the hedgehog. Popularly they are known by the name of SEA-URCHINS, or SEA-EGGS.

In all these curious beings the upper parts are protected by a kind of shell, always more or less dome-shaped, but extremely variable in form. The shell is one of the most marvellous structures in the animal kingdom, and the mechanical difficulties which are overcome in its formation are of no ordinary kind. In the case of the common SEA-EGG, which is shown on the following illustration, the shell is nearly globular. Now, this shell increases in size with the age of the animal; and how a hollow spherical shell can increase regularly in size, not materially altering its shape, is a problem of extreme difficulty. It is, however, solved in the following manner. The shell is composed of a vast number of separate pieces, whose junction is evident when the interior of the shell is examined, but is almost entirely hidden by the projections upon the outer surface. These pieces are of a hexagonal or pen-

SEA-URCHIN.—(*Echinus sphæra.*)

tagonal shape, with a slight curve, and having mostly two opposite sides much longer than the others. As the animal grows, fresh deposits of chalky matter are made upon the edges of each plate, so that the plate increases regularly in size, still keeping its shape, and in consequence the dimensions of the whole shell increase, while the globular shape is preserved. If a fresh and perfect specimen be examined, the surface is seen to be covered with short sharp spines set so thickly that the substance of the shell can hardly be seen through them. The structure of these spines is very remarkable, and under the microscope they present some most interesting details, Moreover, each spine is movable at the will of the owner, and works upon a true ball-and-socket joint, the ball being a round globular projection on the surface of the shell, and the socket sunk into the base of the spine.

The Common Sea-urchin is edible, and in some places is extensively consumed, fully earning its title of Sea-egg, by being boiled and eaten in the same manner as the eggs of poultry.

LEAVING the Echini, we pass to the next large group of Echinodermata, called scientifically Asteriadæ, and popularly known as STAR-FISHES. These creatures exhibit in the strongest manner the radiate form of body, the various organs boldly radiating from a common centre.

Many of these creatures are exceedingly common upon our own coasts, so plentiful, indeed, as to be intensely hated by the fishermen. Of these, the common FIVE-FINGER, or CROSS-FISH, is perhaps found in the greatest numbers. All Star-fishes are very wonderful beings, and well repay a close and lengthened examination of their habits, their development, and their anatomy. There are sufficient materials in a single Star-fish to fill a whole book as large as the present volume, and it is therefore necessary that our descriptions shall be but brief and compressed. To begin with the ordinary habits of this creature. Everyone who has wandered by the sea-side has seen specimens of the common Five-fingers thrown on the beach, and perhaps may have passed it by as something too commonplace to deserve notice. If it be taken up, it dangles helplessly from the hand, and appears to be one of the most innocuous beings on the face of the earth. Yet, this very creature has in all probability killed and devoured great numbers of the edible molluscs, and has either entirely or partially excited the anger of many an industrious fisherman.

To begin with the former delinquency. It is found that the Star-fish is a terrible foe to molluscs, and although its body is so soft, and it is destitute of any jaws or levers, such as are employed by other mollusc-eating inhabitants of the sea, it can devour even the tightly shut bivalves, however firmly they may close their valves.

The second delinquency of the Star-fish is achieved as follows. By some wonderful power the Star-fish is enabled to detect prey at some distance, even though no organs of sight, hearing, or scent can be absolutely defined. When, therefore, the fishermen lower their baits into the sea, the Star-fishes and crabs often seize the hook, and so give the fisherman all the trouble of pulling up his line for nothing, baiting the hook afresh, and losing his time. The fishermen always kill the Star-fish in reprisal for its attack on their bait, and formerly were accustomed to tear it across and fling the pieces into the sea. This, however, is a very foolish plan of proceeding, for the Star-fish is wonderfully tenacious of life, and can bear the loss of one or all of its rays without seeming much inconvenienced. The two halves of the Asterias would simply heal the wound, put forth fresh rays, and, after a time, be transmuted into two perfect Star-fishes.

The movements of the Star-fish are extremely graceful, the creature gliding onward with a beautifully smooth and regular motion. It always manages

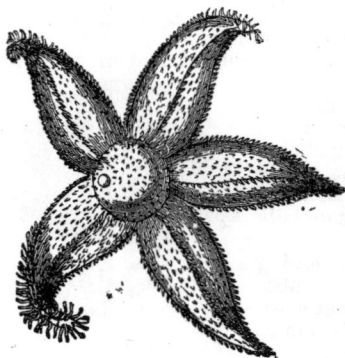

CROSS-FISH.—(*Uraster rubens.*)

to accommodate itself to the surface over which it is passing, never bridging over even a slight depression, but exactly following all the equalities of the ground. It can also pass through a very narrow opening, and does so by pushing one ray in front, and then folding the others back so that they may afford no obstacle to the passage. It also has an odd habit of pressing the points of its rays upon the bottom of the sea, and raising itself in the middle so as to resemble a five-legged stool.

ACALEPHA.

WE now arrive at a large and important class of animals. These beings are scientifically termed ACALEPHA, or Nettles, a word which may be freely rendered as Sea-nettles. The term is appropriate to many of the species which compose this large class, for a very great number of the Acalepha are possessed of certain poisoned weapons which pierce the skin and irritate the nerves as if they were veritable stinging-nettles floating about in the sea. Popularly, they are known by the familiar term Jelly-fishes, because their structure is so gelatinous, mostly clear and transparent, but sometimes semi-opaque or coloured with most beautiful tints.

On the right hand of the illustration on the following page may be seen a remarkable creature, called by the popular name of the SALLEE MAN, some-times corrupted, in nautical fashion, into SALLYMAN. In this curious animal the body is membranous, oval, and very flat, and may be at once recognized by the cartilaginous crest which rises obliquely from its upper surface.

The Velella is very widely distributed, and is found in every sea except those that are subject to the cold influences of the poles. It seldom approaches land, but may be met in vast numbers, sometimes being crowded together in large masses and of various sizes.

The upper figure represents the celebrated PORTUGUESE MAN-OF-WAR. This beautiful but most formidable acaleph is found in all the tropical seas, and never fails to attract the attention of those who see it for the first time. The general shape of this remarkable being is a bubble-like envelope filled with air, upon which is a membranous crest, and which has a number of long tentacles hanging from one end. These tentacles can be protruded or withdrawn at will, and sometimes reach a considerable length. They are of different shapes, some being short and only measuring a few inches in length, while the seven or eight central tentacles will extend to a distance of several feet. These long tentacles are formidably armed with stinging tentacles, too minute to be seen with the naked eye, but possessing venomous powers even more noxious than those of the common nettle. "It is in these appendages alone," writes Mr. D. Bennett, "that the stinging property of the Physalis resides. Every other part of the mollusc may be touched with impunity, but the slightest contact of the hand with the cable produces a sensation as painful and protracted as the stinging of nettles ; while, like the effect of that vegetable poison, the skin of the injured part often presents a white elevation or wheal."

The colours of the Physalis are always beautiful, and slightly variable, both in tint and intensity. The delicate pink crest can be elevated or depressed at will, and is beautifully transparent, grooved vertically through-out its length. The general hue of its body is blue, taking a very deep tint at the pointed end, and fading into softer hues towards the tentacles. A general iridescence, however, plays over the body, which seems in certain lights to be formed of topaz, sapphire, or aquamarine. The short fringes are beautifully coloured, the inner row being deep purple, and the outer row glowing crimson, as if formed of living carbuncle.

If the reader will now refer to the illustration he will see a long, flat, riband-like creature edged with a delicate fringe of cilia. This curious being is called VENUS' GIRDLE, and from its beauty fully deserves the name. This lovely creature is found in the Mediterranean, where it attains to the extraordinary length of five feet, the breadth being only two inches. Rightly,

PORTUGUESE MAN-OF-WAR.—(*Physalis pelagicus.*)
VENUS' GIRDLE.—(*Cestum Veneris.*)   SALLEE MAN.—(*Velella vulgaris.*)

the words breadth and length ought to be transposed, as the development is wholly lateral. The mouth of the Venus' Girdle may be seen in the centre of the body, occupying a very small space in proportion to the large dimensions of the creature to which it belongs.

WE now come to a very large order of acalephs, including all those beings which are so familiar under the title of JELLY-FISHES, SLOBBERS, and similar euphonious names. They are all united under the name of Discophora, or disc-bearers, because they are furnished with a large umbrella-like disc, by means of which they are enabled to proceed through the water.

In the accompanying illustration an example may be seen of the typical genus of this family, which is a native of our own seas. This is a sufficiently common species, and may be found plentifully on our shores, together with its kindred. There are few more beautiful sights than to stand on a pier-head or lie in a boat, and watch the Medusæ passing in shoals through the clear water, pulsating as if the whole being were but a translucent heart, trailing behind

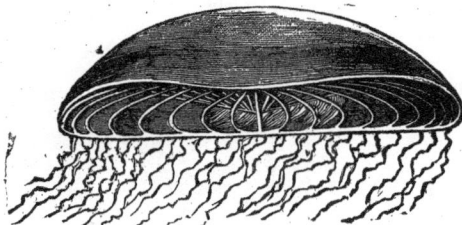

JELLY-FISH.—(*Medusa Æquorea.*)

them their delicate fringes of waving cilia, and rolling gently over as if in excess of happiness. At night, many of the Medusæ put on new beauties, glowing with phosphorescent light like marine fire-flies, and giving to the ocean an almost unearthly beauty that irresistibly recalls to the mind the " sea of glass mingled with fire."

ZOOPHYTES.

QUITTING the Acalephæ, we come to the vast class of ZOOPHYTES, or animal plants, so called because, though really belonging to the animal kingdom, many of them bear a singularly close resemblance to vegetable forms. Their substance is always gelatinous and fleshy, and round the entrance to the stomach are set certain tentacles, used in catching prey and conveying it to the stomach. These tentacles are armed with myriads of offensive weapons contained in little capsules, and capable of being discharged with great force. Organs of sight, smell, taste, and hearing seem to be totally absent, though it is possible that an extended sense of touch may compensate the creature for these deficiencies.

THE highest form of true Zoophyte is, undoubtedly, that which is so familiar under the name of SEA-ANEMONE—a name singularly inappropriate, inasmuch as the resemblance to an anemone is very far-fetched ; while that to the chrysanthemum, daisy, or dandelion is very close.

The widely-spread Anemone, with the circlet of pearl-like beads at the base of its tentacles, is the well-known BEADLET, the most common of all this order on the British coasts. It is a singularly hardy species, living mostly on the rocks that lie between high and low water mark. It is perhaps more variable in colour than any of the British Actiniæ, the body taking all imaginable hues, passing from bright scarlet to leaf-green, graduating from scarlet

Actinia mesembryanthemum.

to crimson, from crimson to orange, from orange to yellow, and from yellow to green. The spherical beads around its mouth are more persistent in colour than any other parts of the animal, being almost invariably a rich blue, just like a set of turquoises placed around the disc. These, however, are occasionally subject to change, and lose all colour, looking like pearls rather than turquoises.

LEAVING the sea-anemones, we now proceed to the next tribe, the Carophylliaceæ, in which there are many tentacles, in two or more series, and the cells many-rayed. Many of these beings deposit a corallum; but out of our British species, more than one-third are without this chalky support.

The ENDIVE CORAL is so called from the resemblance which its corallum bears to the crumpled leaves of that vegetable. The animal has no tentacles, and the cells are small, conical, and rather oblique. The corallum is fixed, sharply edged, and expanded from the base to the tip. All the living members of this pretty genus are to be found in the East and West Indian seas.

The three figures which occupy the left hand of the illustration represent

DEVONSHIRE CUP-CORAL.—(*Carophyllia Smithii.*)
(With side buds.)   TUFT CORAL.—(*Lophophelia prolifera.*)
DEVONSHIRE CUP-CORAL.—(and skeleton.)   ENDIVE-CORAL.—(*Euphyllia pavonia.*)

one of our few native Corals, shown under three aspects. The large, rounded figure in the lower corner exhibits the DEVONSHIRE CUP CORAL as it appears when the tentacles are fully expanded; that to the right shows the dead stony corallum of the same species, and the upper figure is given for the purpose of exhibiting the curious manner in which it multiplies itself by throwing off buds from its sides. It is not a very large, but it is a very pretty species, the colour of the corallum being generally pure translucent white, sometimes tinged with a delicate rosy hue, while that of the living animal is pearly white, variegated with rich chestnut and the palest imaginable fawn.

IN the family Oculinidæ, the corallum is branched and tree-like, and is here represented by our only known British form, the TUFT CORAL. It is very rare, and but seldom taken in our seas. As may be seen from the illustration, the corallum resembles a massive, thickly-branched tree. The individual corals are about half an inch in height, and the same in diameter.

WE now arrive at the Hydroida, which are known by the internal cavity being simple and the creature increasing by buds thrown out from the sides.

IN the Sertulariadæ the buds are inclosed in vesicles, and do not break away when adult. They are in cup-like cells, which have no footstalks.

The reproduction of these beings is very curious, for it is known that they

Sertularia filicula.   *Sertularia roseæ.*
Sertularia filicula (magnified.)

can be propagated by cuttings just like plants, as well as by cell vesicles, and that in the latter case the first stage of the young closely resembles that of the young Medusæ already mentioned. They also reproduce by offshoots ; and it is very likely that their capabilities in this respect are not limited even to these three methods.

Any of the common Sertulariæ affords a good example of this family ; and as they are easily procured, they are very valuable aids to those who wish to study the structure of these beautiful beings. Even the empty polypidon is not without its elegance, and is often made up into those flattened bouquets of so called sea-weeds, which are sold in such quantities at sea-side bathing towns. But when the whole being is full of life and health, its multitudinous cells filled with the delicate polypes, each furnished with more than twenty tentacles all moving in the water, its beauty defies description. These little polypes are wonderfully active and suspicious. At the least alarm, they retreat into their cells as if withdrawn by springs, and when they again push out their tentacles, it is in a very wary and careful manner.

ROTIFERA.

ALTHOUGH the Rotifera, or Wheel Animalcules, are generally placed among the Infusoria, on account of their minute dimensions and aquatic habits, it is evident, from many peculiarities of their formation, that they deserve a much higher place, and in all probability constitute a class by themselves.

They are called Wheel Animalcules on account of a curious structure which is found upon many of their members, and which looks very like a pair of revolving wheels set upon the head. These so-called wheels are two disc-like lobes, the edges of which are fringed with cilia, which when in movement

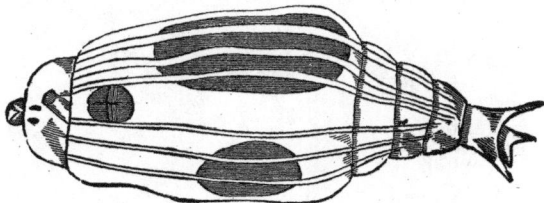

Rotifer citrinus.

give to the creature an appearance as if it wore wheels on its head, like those of the fairy knight of ballad poetry. These wheels can be drawn into the body at will, or protruded to some little extent, and their object is evidently to procure food by causing currents of water to flow across the mouth. All, however, do not possess these appendages, but have a row of cilia, mostly broken into lobes, extending all round the upper portion of the body.

These remarkable beings are mostly found in water that has become stagnant but is partially purified by the presence of the Infusorians, which always swarm in such localities.

The typical genus of this class is known by the name of ROTIFER, an example of which is seen in the illustration. In all the members of this genus the body is rather elongated, and furnished at the hinder end with a kind of telescopic tail, by means of which they can attach themselves at will to any object, and release themselves whenever they please. Sometimes they move their bodies gently about, while still grasping by the extremity of the tail; sometimes they are nearly motionless, while they frequently rock themselves backwards and forwards so violently that they seem almost to be testing the strength of their hold.

These creatures can both swim and crawl, the former act of locomotion being achieved by the movement of the cilia, and the latter by creeping along after the fashion of the leech, the head and tail taking alternate hold of the object on which they are crawling. The masticating apparatus is always conspicuous, whether the animal has the wheel protruded or withdrawn. It is situated behind the bases of the wheel-lobes, and looks, when the animal is at rest, something like a circular buckler with a cross composed of double lines drawn over its surface.

RHIZOPODA.

THE whole arrangement of the beings which we are now about to examine is still very obscure, and the best zoologists of the present time have declared that any system which has been hitherto adopted can only be considered as provisional.

These minute though beautiful beings exist in numbers that are only rivalled by the sands of the sea for multitude ; and the vast hosts of these creatures can be barely estimated even when we know that many large cities are built wholly of the dead skeletons of these microscopic beings, and that in a single ounce of sand from the Caribbean Sea nearly four millions of those shells have been discovered.

The first sub-class of these beings is the Foraminifera, so called on account of the tiny openings, or *foramina*, with which the pretty shells are pierced. Sometimes, however, this shell is wanting, and its place is supplied by a cover composed of matted sand-grains.

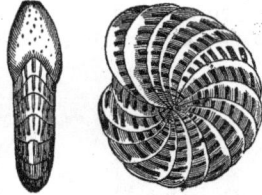

Polystomella Lessoniana.

PORIFERA.

WE now arrive at a large class of beings, which are by common consent allowed to form the very lowest link in the animal chain. The name PORIFERA is given to them because the whole of their surface is pierced with holes of various dimensions, the greater number being extremely minute,

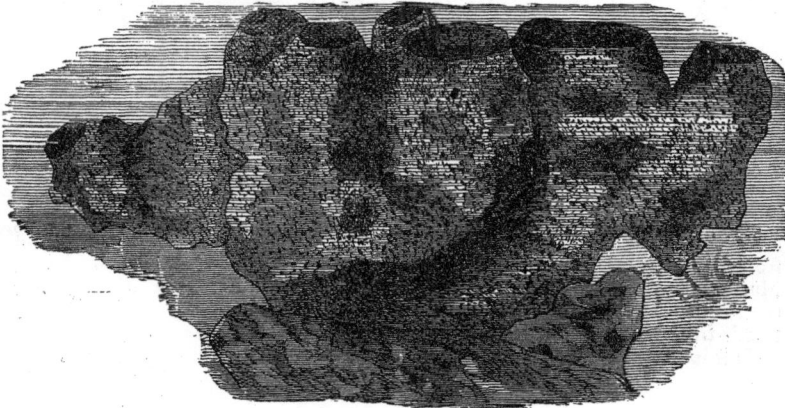

Grantia compressa.

while others are of considerable dimensions. The well-known Turkey SPONGE, so useiul for the toilet, will afford a good example of the porous structure.

The true living being which constitutes the Sponge is of a soft and almost gelatinous texture, to the unaided eye ; and with the aid of the microscope is found to consist of an aggregation of separate bodies like those of the Amœbæ, some of which are furnished with long cilia. By the constant action of the cilia a current of water is kept up, causing the liquid to enter by innumerable pores with which the surface is pierced, and to be expelled through the larger orifices. A Sponge in full action is a wonderful sight ; the cilia driving the water in ceaseless torrents, whirling along all kinds of solid particles, arresting those which are useful for digestion, and rejecting those with which it cannot assimilate.

The extraordinary object which is called by the appropriate name of NEPTUNE'S CUP is one of the most magnificent, as well as one of the most notable, of the Sponge tribe. It hardly looks like a Sponge; and when a

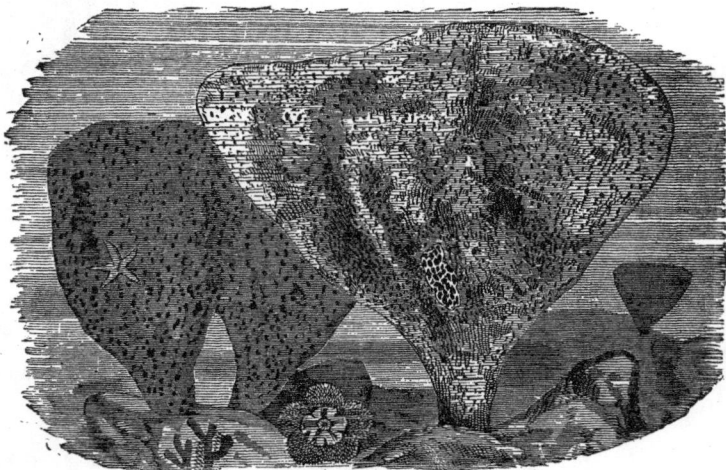

NEPTUNE'S CUP.—(*Thalassema Neptuni.*)

specimen is shown to persons who have no knowledge of the subject, they can hardly ever be made to believe that the exhibitor is not endeavouring to play a practical joke upon them.

The Neptune's Cup is of enormous dimensions, often measuring four feet in height, and having a corresponding width. Its exterior is rough, gnarled, and knotted like the bark of some old tree ; and if a portion were removed from the side, it might almost be mistaken for a piece of cork-tree bark. Many persons have imagined that the strangely-shaped object was made of the skin of an elephant's leg, and I have even heard a teacher telling her pupils that it was an old Roman wine-jar.

This is one of the exotic Sponges, being found only in the hotter seas.

INDEX.

Acalepha, 583.
Acanthopterygii, 462.
Accentor, 311.
——Hedge, 311.
Accentorinæ, 311.
Acorn Barnacle, 572.
Actinia Mesembryanthemum, 585.
Actiniæ, 585.
Adder, Puff, 433.
Adjutant, 390.
Admiral, Scarlet, 550.
Ægeriidæ, 552.
Agamas, 428.
Agouti, 145.
Ai, or Three-Toed Sloth, 225.
Albatros, Wandering, 408.
Alcinæ, 405.
Alligator, 422.
Alligatoridæ, 422.
Alpaca, 195.
American Monkeys, 19.
Ampelidæ, 324.
Ampelinæ, 324.
Anabas Scandens, 476.
Anaconda, 437.
Anchovy, 485.
Anemone, Sea, 585.
Animalcules, Wheel, 588.
Annelida, 578.
Annulata, 578.
Ant, White, 537.
Ant-Eater, Great, 220.
——————— Middle, 220.
——————— Little, 221.
——————— Porcupine, 223.
Antelopes, 168.
Ant-Lion, 540.
—— White, 421, 422.
Ants' Eggs, 545.
Ape, Baroary, 14.
Aphaniptera, 561.
Aphides, 557, 559.
Apteryx, 383.
Arachnida, 572.
Arctic Fox, 72.
Arctiidæ, 554.
Argonaut, 501.
Argus, 373.
—— Pheasant, 374.
Ariel, Petaurus, 110.
Armadillo, 219.
Arthronomalus Longicornis, 578.
Articulata, 524.
Aspalacidæ, 159.
Ass, 199, 200.
Asse, 74.
Auk, Great, 405.
Avicularidæ, 518.
Avocet, 393.
Aye-Aye, 33.

Baboons, 16.
Babyroussa, 209.
Bactrian Camel, 193.
Badger, 86.
—— Australian, 116.
Bajjerkeit, 218.
Balæna, 128.
Balænidæ, 128.
Bandicoot, Long-nosed, 116.
Barbel, 491.
Barnacle, Acorn, 572.
—— Duck, 571.
Barnacles, 570.
Bat, Long-eared, 37.
—— Vampire, 36.
Batrachians, 442.
Bay Bamboo Rat, 161.
Bear, 89.
—— Australian, 112.
—— Black, 89.
—— Brown, 89.
—— Grizzly, 90.
—— Malayan Sun, 91.
—— Polar, 91.
—— Syrian, 90.
Beaver, 141.
Bee, Honey, 546.
—— Eater, 284, 285.
—— Humble, 547.
Beetle, Burying, 527.
—— Bloody-nose, 532.
—— Eight-spot, 525.
—— Ground, 525.
—— Lamellicorn, 528.
—— Musk, 532.
—— Oil, 531.
—— Rove, 527.
—— Stag, 529.
—— Tiger, 524.
—— Violet Ground, 525.
—— Water, 526.
Bernicle Goose, 399.
Bird of Paradise, Emerald, 334.
Birds, 229.
Bison, 166.
Bittern, 388.
Bivalves, 506.
Blackbird, 320.
Blackcap Warbler, 306, 307.
Black Cock, 379.
—— Macaque, 15.
—— Yarke, 24.
Blattidæ, 534.
Bleak, 492.
Blennies, 475.
Blenny, Eyed, 476.
Blindworm, 424.
Blister, or Spanish Fly, 531.
Bloodhound, 63.
Bloody-nose Beetle, 533.
Boa Constrictor, 436.
Boar, 208.

Bohemian Waxwing, 325.
Boidæ, 436.
Bombycidæ, 554.
Boomslange, 439.
Bosch Vark, 209.
Bot Fly, 561.
Bower-bird, Satin, 335
Brachelytra, 527.
Brachyura, 562.
Bream, 491.
Breast-plate, 522.
Breeze Fly, 561.
Buffalo, 164.
—— Cape, 165.
Bug, Bed, 560.
—— Wheel, 560.
Bulldog, 66.
Bullfinch, 348.
Bull-head, 467.
Buprestidæ, 529.
Burying-beetle, 527.
Bush Hog, 209.
Bustard, Great, 384.
Butcher Birds. See Lanidæ.
Butterfly, Camberwell Beauty, 550.
—— Comma, 550.
—— Peacock, 549.
—— Swallow-tailed, 548.
—— Tortoiseshell, 550.
Buzzard, 245.

Caama. See Asse.
Cacajao, 24.
Cachalot, 129.
Cacomixle, 55.
Caddis-fly, 541.
Calamaries, or Squids, 504.
Calpidium ornatum, 522.
Camel, 192.
—— Bactrian, 193.
Camelopardalis, 183.
Campagnol, 139.
Campanulariæ, 587.
Canada Lynx, 49.
Canary, 344.
Canceridæ, 564.
Canis, 58—69.
Capercaillie, 379.
Capra, 177.
Capucin, 23.
Capybara, 146.
Carabidæ, 525.
Carabus, 525.
Cardiadæ, 520.
Carophylliaceæ, 585, 586.
Carp, 490.
Cashmir Goat, 179.
Cassowary, 382.
Cat, 48.
—— Civet, 53.

Cat, Wild, 48.
Catenicella hastata, 522.
———— cornuta, 522.
Cellularia Peachii, 522.
Centepedes, 578.
Cephalopoda, 501.
Cerastes, 434.
Cercopidæ, 559.
Certhidæ, 298.
Cetacea, 127.
Chacma, 16.
Chætodon, Beaked, 465.
Chætodontina, 465.
Chaffinch, 341.
Chambered Nautilus, 505
Chamelæon, 429.
Chameleonidæ, 429.
Chamois, 173.
Chati, 46.
Chatterer, Waxen, 225
Cheiroptera, 35.
Chetah, 50.
Chicken, Mother Cary's, 408.
Chimpansee, 4.
Chinchilla, 149.
Chiton, Marbled, 514.
Chitonidæ, 514.
Chœropus, 117.
Chondropterygii, 456.
Chough, 334.
Chrysochroa, 530.
Chrysomelidæ, 530.
Chub, 492.
Cicadæ, 558.
Cicada Anglica, 559.
———— Flosfolia, 558.
Cicindela, 524.
Cicindelidæ, 524.
Cimbe Femorata, 542.
Cimicidæ, 560.
Cirrhipedes, 571.
Civet Cat, 53.
Cleodora, 517.
Cleg, 561.
Clupeidæ, 484.
Coaita, Spider Monkey,
Coast Rat, 160.
Coati Mondi, 94.
Cobra di Capello, 440.
Coccidæ, 558.
Coccinellidæ, 533.
Cochineal Insect, 559.
Cockatoo, Sulphur-crested, 357.
——.—— Leadbeater's, 358,
359.
Cockchafer, 528.
Cockles, 520.
Cockroach, 534.
Cocktails. *See* Rove Beetles.
Cod, 480.
Coleoptera, 524.
Colobus, 11.
Colubrinæ, 437.
Colugo, 35.
Columba, 365.
Columbæ, 365.
Columbidæ, 365.
Columbinæ, 365.
Colymbidæ, 403.
Colymbinæ, 403.
Colymbus, 403.
Conchifera, 518.
Condor, 230.
Cone-shells, 510.
Cone, Admiral, 510.
———— Textile, 510.
Conger Eel, 483.

Conidæ, 510.
Conirostres, 327.
Coot, 397.
Coquette, Spangled, 296.
Coral, Devonshire Cup, 586.
———— Endive, 586.
———— Tuft, 586.
Cormorant, 411.
———————— Crested, 412.
Corncrake, 396.
Corvidæ, 327.
Corvinæ, 327.
Cowry, Money, 511.
———— Deep-toothed, 511.
Crab, Calling, 567.
———— Edible, 564.
———— Fighting, 567.
———— Green, 564.
———— Hermit, or Soldier, 567.
———— Nipper, 565.
———— Thornback-Spider, 563.
———— Swimming, 565.
———— Thornback, 563.
Crab-spider, or Matoutou, 573.
Crabro Cribrarius, 543.
Cramp-fish, 450
Crane, 386.
———— Demoiselle, 387.
Crane Fly, Great, 560.
Crested Currassow, 369.
Cricket, Field, 534.
———— House, 534.
———— Mole, 535.
Crocodile, 421.
Crocodilidæ, 420.
Crocodilus, 420.
Cross-bill, 349.
Cross-fish, 582.
Crotalidæ, 431.
Crow, 329.
———— Hooded, 332.
Crustacea, 562.
Cryptoprocta, 57.
Cuckoo, 364.
Cuckoo-spit, 559.
Curculionidæ, 532
Curlew, 393.
Cursores, 380.
Cushat, 367.
"Cuttle-bone," 505.
Cynipidæ, 544.
Cypræidæ, 510.

Dabchick, 404.
Dace, 492.
Daddy Long-legs, 561.
Dasypidæ, 217.
Dasyure, 118.
Death's Head Moth, 550.
Deer, Axis, 189.
———— Fallow, 188.
———— Kanchil, or Pigmy Musk, 191.
———— Musk, 190.
———— Red, 185, 188.
———— Rein, 187.
Demoiselle Dragon Fly, 540.
Dendrophidæ, 439.
Dendrosaura, 428.
Dermaptera, 414.
Devil's Coach-horse, 527.
Dicæum, Australian, 289.
Dipper, 317.
Diptera, 561.
Discophora, 584.
Diver, Great Northern, 403.

Dodo, 369.
Dog, Bull, 66.
———— King Charles's, 61.
———— Maltese, 61.
———— Mexican Lap, 63.
———— Newfoundland, 59.
———— Pomeranian, 60.
———— Prairie, 157.
———— Shepherd's, 65.
Dog-fish, Spotted, 457.
Dog-headed Monkeys. *See* Baboons.
Dolphin, 133.
Doris, 516.
Dormouse, 151.
Dory, John, 472.
Douroucouli, 25.
Dove, Ring, 367.
———— Stock, 366.
———— Turtle, 368.
Dragon, Flying, 428.
Dragon-fly, 539.
———————— Demoiselle, 540.
Duck Barnacle, 571.
Duck-bill, 222.
——. Eider, 402.
——. Wild, 401.
Dziggetai, 199.

Eagle, Golden, 238.
———— Bald, or Whiteheaded, 242, 245.
Earth-worm, 579.
Earwig, Giant, 533.
Echidna, 223.
Echinidæ, 581.
Echinodermata, 580.
Edible Crab, 564.
Eel, Conger, 483.
———— Electric, 483.
———— Sharp-nosed, 481.
Eggs, Sea, 581.
Eider Duck, 402.
Eland, 175.
Elephant, African, 205.
———————— Asiatic, 203.
Elephant, Sea, 126.
Elk, 185.
Emerald Bird of Paradise, 334.
Emeu, 381.
Endive Coral, 586.
Entellus, 10.
Entomophaga, 544.
Eolis, 516.
Erd Shrew. *See* Shrew Mouse.
Ermine, 81.
Eumenes Arcuatus, 545.

Falco, 248.
Falcon, Jer, 249.
———— Peregrine, 249.
———— Stone, 551.
Falconidæ, 238.
Falconinæ, 248.
Felidæ, 39.
Fennec, 73.
Ferret, 78.
Fieldfare, 320.
Finches, 338.
Fishes, 457.
———— Flat, 478.
———— Flying, 486.
———— Star, 581.
Fishing-frog, 473.
Fishing Hawk *See* Osprey
Fissurellidæ, 514.

Flamingo, 398.
Flea, 560.
—— Water, 560.
Flounder, 480.
Fly, Blister, or Spanish, 531.
—— Bot, 561.
—— Breeze, 561.
—— Caddis, 541.
—— Dragon, 539.
—— Great Crane, 560.
—— Humble Bee, 561.
—— May, 541.
Fly-catcher, Pied, 323.
——————— Spotted, 323.
Flying Dragon, 428.
—— Fish, 486.
—— Fox, 39.
—— Squids, 504.
—— Squirrel, 153.
Foraminifera, 589.
Fowl, domestic, 375.
Fox, 71.
—— Arctic, 72.
Foxhound, 64
Frigate, Pelican, 412.
Frog, 442.
—— Green, 444.
—— Pouched, 347.
—— Tree, 464.
Frog-hopper, 559.
Gadfly, 561.
Gallinæ, 369.
Gallinula, 396.
Gallinulinæ, 396.
Gannet, 410.
Gasteropoda, 514.
Gazelle, 169.
Gecko, 426.
Geissosaura, 424.
Gemsbok, 171.
Genett, Blotched, 54.
Geometridæ, 555.
Gerboa, 150.
Gibbon, Agile, 8.
Gibbons, 7.
Giraffe, 182.
Glow-worm, 530.
Gnoo, 173.
Goat, 177.
—— Cashmir, 179.
Goat-sucker, 267.
Goby, 473.
Golden Oriole, 322.
Goldfinch, 342.
Gold-fish, 490.
Goose, Bernicle, 399.
—— Mussel, 571.
—— Solan, 410.
Gorilla, 2.
Goshawk, 256.
Grantia Compressa, 589.
Grebe, Crested, 405.
—— Little, 404.
Greenfinch, 342.
Greyhound, 159.
Grivet, 12.
Grosbeak, 341.
Grouse, Black, 379.
Guanaco, 194.
Gudgeon, 491.
Guillemot, 407.
Guinea-fowl, 376.
Guinea-pig, 147.
Gull, Black-backed, 409.
Gurnard, Flying, 468.
———— Red, 467.

Hackee, or Ground Squirrel, 156.
Hag-fish, Glutinous, 496.
Hamster, 136.
Hammatocerus, 560.
Hare, 147.
Harp Shell, Imperial, 508, 509
Hawfinch, 341
Hedgehog, 104.
Hedge Sparrow, 311.
Helicidæ, 515.
Helix, 515.
Hen Harrier, 259.
Hermit Crab, 567.
Heron, 387.
Herring, 485.
Heterocera, 549.
Heteroptera, 559.
Hippopotamus, 216.
Hobby, 251.
Hog, Bush, 209.
—— Domestic, 208.
Homo, 1.
Homoptera, 557.
Honey-Eaters, 289.
Hoopoe, 286.
Hopper, Scarlet, 559.
Hoppers, 558, 559.
Hornbill, Rhinoceros, 350.
Hornwrack, 523.
Horse, 196.
—— River, 216.
—— Sea, 493.
Hound, Blood, 63.
—— Fox, 64.
—— Grey, 59.
Howler, Ursine, 22.
Humble Bee-fly, 561.
Humming-bird, Bar-tailed, 292.
——————— Copper-bellied Puff-leg, 294.
——————— Cora, 292.
——————— Double-crested, 292.
——————— Gould's, 292.
——————— Ruby-throated, 298.
——————— Ruby and Topaz, 297.
——————— Slender Shear-tail, 293.
——————— Spangled Coquette, 296.
——————— Vervain, 297.
——————— White-booted Racket-tail, 295.
Hunting Cat. *See* Chetah.
Hyalea, 517.
Hyæna, 53.
Hydroida, 587.
Hydrometra, 560.
Hydrometridæ, 560.
Hylas Moth, 552.
Hymenoptera, 543.
Hyponomeutidæ, 556.
Hyrax, African, 215.
—— Syrian, 216.

Ibex, 177.
Ibis, Sacred, 392.
Ichneumon, 55.
——————— Giant, 543.
——————— Indian, 56.
Ichneumon-fly, 543.
Ichneumonidæ, 544.
Iguana, 427.
Imperial Harp Shell, 508, 509
Indri. *See* Avahi.

Insecta, 524.
Insectivora, 96.
Invertebrata, 501.

Jacana, 395.
Jackal, 68.
Jackass, Laughing, 280.
Jackdaw, 331.
Jaguar, 44.
Jay, 327.
Jelly Fishes, 583.
Jer-Falcon, 249.
John Dory, 472.
Jungle Fowl, Australian, 371.
——————— Bankiva, 374, 375.

Kahau, 10.
Kanchil, or Pigmy Musk Deer, 191.
Kangaroo, 113.
——————— Rat, 115.
Kestrel, 253.
Kingfisher, 280, 282.
Kinkajou, 95.
Kite, 247.
—— Swallow-tailed, 246, 248.
Klip Das, 215.
Koala, 112
Koodoo, 174.
Kookaam. *See* Gemsbok.
Koulan. *See* Dziggetai.
Kuda-Ayer. *See* Tapir, Malayan.
Kukang, 30.

Lämmergeyer, 229.
Lampern, 495.
Lamprey, 494.
Lac Insect, 559.
Ladybirds, 533.
Lancelet, 497.
Landrail, 396.
Laninæ, 326.
Laniidæ, 326.
Lapwing, 384.
Lark, Sky, 347.
Laughing Jackass, 280.
Leaf Insect, 536.
Leadbeater's Cockatoo, 358.
Leech, 579, 580.
Lemming, 140.
Lemur, Flying, 35.
—— Ruffled, 29.
Lemurs, 28.
Leopard, 42.
——————— Hunting, 42.
Lepidoptera, 548.
Libellulidæ, 539.
Limacidæ, 515.
Limnæa, 516.
Limpets, 514.
Linnet, 342.
Lion, 39.
—— Ant, 540.
Litorinidæ, 512.
Little Chain, 522.
Lizard, Sand, 423.
—— Scaly, 423, 424.
—— Tree, 428.
Llama, 195.
Lobster, 568.
Locust, Migratory, 536.
Long-nosed Bandicoot, 116.
Loris, Slender, 29.
—— Slow-paced, 30.
Lynx, 49.

Q Q

Lynx, Canada, 49.
Lyre Bird, 301.

Macaques, 13.
Macaque, Black, 15.
Macaw, Blue and Yellow, 354.
Mackerel, 468.
Macropidæ, 108.
Magilus, 509.
Magot, 14.
Magpie, 332.
Maiadæ, 563.
Mallard, 401.
Maltese Dog, 61.
Mammalia, 2.
Man, 1.
Mandril, 17.
Manis, 217.
Mapach. *See* Racoon.
Marikina, 26.
Marimonda, 20.
Marmoset, 27.
Marmot, 159.
Marsupialia, 108.
Marten, Pine, 75.
Martin, Fairy, 274.
——— House, 275, 276.
——— Sand, 274, 275.
Mastiff, 67.
Mat, Sea, 522.
May-fly, 541.
Meadow Pippit, 317.
Meantia. 451.
Medusa, 585.
Menura. *See* Lyre Bird.
Menipea Fuguensis, 522.
Merlin, 251.
Mexican Lapdog. 63.
Miller's Thumb, 467.
Mink, 78.
Minnow, 492.
Mockigg Bird, 318.
Mole, 96.
——— Rat, 159.
Mollusca, 501.
Molluscs, Shore, 512.
Mongus, 56.
Monkey, Avahi. 31.
——— Aye-Aye, 33.
——— Black Macaque, 16
——— Black Yarke, 24.
——— Chacma, 17.
——— Capucin, 23.
——— Coaita, 20.
——— Colugo, 33.
——— Douroucouli, 25.
——— Entellus, 10.
——— Green, 12.
——— Grivet, 13.
——— Kukang, 30.
——— Magot, 14.
——— Marikina, 26.
——— Marimonda, 21.
——— Mandril, 18.
——— Papion, 19.
——— Proboscis, 10.
——— Ruffled Lemur, 29.
——— Slender Loris, 29.
——— Spider, 20.
——— Tarsier, 32.
——— Ursine Colobus, 11.
——— Ursine Howler, 22.
——— Wanderoo, 15.
Moose, or Elk, 185.
Moor Hen. *See* Water Hen.
Moschine Deer, 191.

Moth, Clear-wing, 552.
——— Death's-head, 550.
——— Goat, 553.
——— House-builder, 555.
——— Humming-bird, 551. |
——— Hylas, 552.
——— Lime Hawk, 550.
——— Lunar Hornet Clear-
         wing, 552.
——— Swallow-tailed, 556.
——— Tiger, 554.
——— White-plume, 557.
——— Wood Leopard, 554.
Mouse, 136.
——— Flying, 109.
——— Harvest, 136.
——— Short-tailed Field, 139.
——— Yellow-footed Pouched,
         119.
Mullingong, 222.
Murex, 507.
Muricidæ, 507.
Musk Ox, 168.
——— Rat. *See* Sondeli.
Musquash, or Musk Rat, 143.
Mussel, Edible, 519.
Mustela, 75.
Myriapoda, 577.
Myrmecobius, 119.
Myxine, 496.

Narwhal, 131.
Natterjack, 445, 446.
Nautilus, the Chambered, 505.
——— Paper, 501.
Necrophaga, 527.
Nennook. *See* Polar Bear.
Nepidæ, 559.
Neptune's Cup, 590.
Neritinæ, 511.
Neritina, Spined, 511.
Newfoundland Dog, 59.
——— Crested, 450.
——— Smooth, 450.
Neuroptera, 537.
Newt, 449.
Nightingale, 307.
Nudibranchidæ, 516.
Nuthatch, 299.
Nut Weevil, 532.
Nyctisaura, 425.
Nylghau, 176.
Nymphalidæ, 549.

Ocelot, 46.
Octopodiæ, 404.
Octopus, 504.
Oculinidæ, 586.
Ocypodidæ, 567.
Œstridæ, 562.
Oil Beetle, 531.
Ommastrephes, 504.
Ondatra. *See* Musquash.
Ophidia, 430.
Opossum, 121.
——— Merian's, 122.
——— Mouse, 109.
Orang-Outan, 5.
Orchestidæ, 569.
Oreosoma, 465.
Oriole, Golden, 322.
Orthoptera, 534.
Oryx, 171.
Osprey, 240.
Ostrich, 380.
Otter, 87.

Ouistiti. *See* Marmoset.
Ounce, 43.
Ousel, Water, 317.
Owl, Barn, 266.
——— Brown, or Tawny, 265.
——— Coquimbo, or Burrowing,
         262.
——— Scops-eared, 265.
——— Snowy, 261.
——— Virginian-eared, 263, 264.
Ox, 162.
——— Grunting, 167.
——— Musk, 168.
Oyster, 518.
——— Pearl, 519.

Pachydermata, 201.
Pachyglossæ, 426.
Paco. *See* Alpaca.
Palæornis, 352.
Paper Nautilus, 501.
Papilionidæ, 549.
Papion, 19.
Paradise, Emerald Bird of, 334.
Parinæ, 312.
Parrakeet, Ringed, 352.
——— Zebra, or Warbling
         Grass, 352.
Parrot, Grey, 354.
——— Green, 356, 357.
Pavonidæ, 372.
Peacock, 372.
——— Butterfly, 549.
Peccary, 211.
Pediculati, 474.
Peewit, 384.
Pelecanidæ, 410.
Pelecaninæ, 410.
Pelecanus, 412.
Pelican, White, 412.
Penguin, Cape, 407.
Pen-tail, 104.
Perch, 464.
——— Climbing, 476.
Perdicinæ, 376.
Perdix, 377.
Periwinkle, 513.
Pernis, 175.
Petaurus, Ariel, 110.
Petrel, Fulmer, 409.
——— Stormy, 408.
Phalangistines, 108.
Phatagin, 217.
Phasianidæ, 374.
Phasianinæ, 374.
Phasianus, 374.
Pheasant, 374.
——— Argus, 373.
Pheasant-shells, 513.
——— Shell, Australian, 513.
Philanthus Triangulum, 543.
Phocidæ, 123.
Pholades, 520.
Pholas, 520.
Physalis. *See* Portuguese Man-
         of-War, 584.
Picidæ, 359.
Picinæ, 359.
Pieridæ, 549.
Piddock, 520.
Pigeon, Blue Rock, 367.
——— Domestic, 367.
——— Passenger, 366.
——— Wood, 367.
Pike, 487.
Pilchard, 485.

Plaice, 480.
Pleuronectidæ, 478.
Plumularia, 587.
Podophthalmata, 562.
Poe Bird, 290.
Pœciloptera Circulata, 558.
Pointer, 64.
Polecat, 76.
Polystomella Lessoniana, 589.
Polyzoa, 521.
Pomeranian Fox Dog, 60.
Poodle, 62.
Porcupine, 144.
———— Canadian, 145.
Porifera, 589.
Porpoise, 142.
Portuguese Man-of-War, 583.
Potamobius, 401.
Potto. *See* Kinkajou.
———— ou. *See* Armadillo.
Prairie Dog, 157
Prawn, 569.
Proboscis Monkey, 10.
Proteus, 451.
Psyllidæ. 559.
Pteropoda, 517.
Puff Adder, 433.
Puffin, 406.
Puma, 45.
Pyrrhulinæ, 348.

Quadrumana, 2.
Quagga, 200.
Quail, 370.
Quata. *See* Coaita.

Rabbit, 148.
———— Rock, 217.
Racoon, 92.
Radiata, 580
Rat, 135.
———— Bay Bamboo, 161.
———— Coast, 160.
———— Kangaroo, 115.
———— Mole, 159.
———— Musk, 142.
———— Water, 137.
Ratel, Honey, 83.
Rattle-snake, 431.
Raven, 328.
Razor-shell, common, 520.
Redbreast, 311.
Redstart, 309.
Reduviidæ, 560.
Reduvius Personatus, 560.
Reindeer, 187.
Reptilia, 417.
Rhea, 382.
Rhinoceros Hornbill, 350.
Rhinoceros, Indian, 212.
———— Little Black, 213.
———— Two-horned, 214.
———— White, 214.
Rhinophryne, 448.
Rhizopoda, 588.
Rhopalocera, 549.
Rhyssa Persuasoria, 542.
Ring-dove, 367.
Roach, 492.
Rock-fish, 473.
Rock Scorpion, 577.
Rodents, 134.
Roebuck, 190.
———— Garrulous, 277.
Rollers, 277.
Rook, 331.

Rotifera, 588.
Roussette. *See* Flying Fox.
Rove-beetle, 527.
Ruff, 393.
Ruffled Lemur, 29.
Sable, 76.
Salamander, 448.
Salicornaria Farciminoides, 522.
Sallee Man, 583.
Salmon, 488.
Sandhopper, 569.
Sand Mole, 160.
Sand Wasps, 544.
Satin Bower Bird, . 5.
Saw-fish, 459.
Saw Flies, 542.
Scallop, 518.
Scansores, 350.
Scarlet Admiral, 550.
Scarlet-hopper, 559.
Scolopendræ, 578.
Scomberidæ, 468.
Scorpion, Rock, 577.
———— Water, 559.
Sea Mat, 522.
—— Horse, 493.
—— Nettles, 583.
—— Snails, 511.
—— Wolf, 475.
—— Hog. *See* Porpoise.
—— Anemone, 585.
—— Unicorn *See* Narwhal.
—— Urchins, 581.
Seal, 124.
—— Elephant, 126.
Secretary Bird 258.
Sepia, Webbed, 502.
———— Common, 503.
Sepiola, 504.
Serpula Contortuplicata, 579,
Sertulariadæ, 587.
Serpents, Tree, 439.
Sessile-eyed Crustacea, 569.
Shark, White, 459
———— Hammer-headed, 458.
Sheep, 180.
———— Spanish or Merino, 181.
Shetland Pony, 198.
Ship-worm, 521.
Shore Molluscs, 512.
Shrew Mouse, 99.
Shrew Mouse Elephant, 102.
———— Water, 101.
———— Oared, 102.
Shrike, Red-backed, 327.
Shrimp, 56.
Simpai, 9.
Siskin, 344.
Skate, Common, 460.
———— Thornback, 461.
Skinks, 424.
Skunk, 84.
Skylark, 3
Slender Loris, 29.
Slobbers. *See* Jelly Fishes, 584.
Sloth, 225.
Slowworm, 424.
Slug, Great Grey, 516.
Snail, Common, 515.
———— Edible, 515.
———— Water or Pond, 516.
Snails, Sea, 511.
Snake, Rattle, 431.
———— Ring 2d, 437.
Snipe. 394.
Solan Goose, 410.
Sole, 478.

Soldier Crab, 567.
Solenidæ, 520.
Solitary Wasps, 545.
Sondeli, 98.
Spangled Coquette, 296.
Spaniel, Water, 461.
Sparrow, 344.
———— hawk, 257.
———— hedge, 311.
Sphingidæ, 550.
Spined Neritina, 511.
Spider, Crab, 573.
Spider Crabs, 563.
———— Trap-door, 573.
———— Water, 574.
———— Garden, or Cross, 575.
Sponges, 589, 590.
Spoonbill, White, 389.
Springbok, 170.
Squinado, 564.
Squirrel, 152, 154.
———— Flying, 153.
———— Ground, 156.
———— Jelerang, or Javan, 153.
Stag Beetle, 529.
Staphylinidæ, 527.
Star Fishes, 581.
Starling, 337.
Starlings, 335.
Steinbok, 177.
Stickleback, 462.
Stoat, 81.
Stockdove, 366.
Stork, 390.
Strobilosaura, 427.
Sturgeon, 456.
Sucking-fish, 470.
Sun-Bears, 91.
———— Bird, Collared,
———— Birds, 287.
———— Javanese, 288.
———— Fish, 492.
Swallow, Common, 272.
———— Esculent, 271.
———— Sea, 409.
———— Tailed Butterfly, 548.
Swan, 399.
———— Black, 401.
———— Whistling, 400.
Swift, 269.
Swine, 208.
———— Fish, 478.
Sword-fish, 472.

Tabanidæ, 561.
Tadpoles, 443.
Tailor-bird, 304.
Tajacu, 211.
Talpidæ, 96.
Talpina, 96.
Tamanoir, or Ant Bear, 219, 220.
Tapir, 207.
———— American, 207.
———— Maylayan, 207.
Tapirus, 207.
Tarsier, 32.
Teal, 402.
Teledu, 84.
Tenthredinidæ, 542.
Tench, 491.
Tenthidæ, 504.
Termites, 539.
Tern, Common, 409.
Terrapins, 418.

Termes Bellicosus, 539.
Terebrantia, 544.
Terrier, English, 61.
——— Scotch, 68.
Tetrabranchiata, 505.
Thrush, Missel, 319.
——— Song, 321.
Thysanoptera, 537.
Tiger, 40.
——— Wolf, 53.
Tiger Moth, 554.
——— Beetle, 524.
——————— Eight-spot, 525.
Tipulidæ, 561.
Titmice, 312.
Titmouse, Blue, 313.
——————— Great, 313
——————— Long-tailed, 314.
Toad, 444.
Tody, Green, 278.
Top, Common, 513.
Torpedo, 460.
Tortoise, 417.
——————— Chicken, 419.
Tota. See Grivet.
Totaninæ. 297.
Toucan, Toco, 351.
Trap-door Spider, 573.
Tree Serpents, 439.
Trichoptera, 542.
Triglidæ, 467.
Trochilidæ, 291.
Trochilus, 291.
Trogon, Resplendent, 279.
Trogonidæ, 279.
Tropic Bird, 410.
Trout, 489.
Trumpeter, Golden - Breasted, 385.
Tunny, 469.
Turbot, 479.
Turkey, 375.
——————— Brush, 371.
——————— Buzzard. See Vulture Carrion.
Turritellidæ. 511.
Turtle, Hawksbill, 419.
——————— Green. 420.
Turtle-dove, 368.

Univalves, 506.
Upupa, 286.

Upupidæ, 286.
Urchin, Sea, 586.
Urchin. See Hedgehog.
Urocerus Gigas, 542.
Ursine Baboon. See Chacma.
——————— Colobus, 11.
——————— Howler, 22.
Urson. See Porcupine, Canadian.

Vampire Bat, 36.
Venus's Comb, 507.
——————— Girdle, 583.
Vertebrata, 1.
Vervet, 13.
Vespidæ, 546.
Vicugna, 194.
Viper, 434.
——————— Horned. 434.
Viperina, 431.
Viperidæ. 432.
Vulpes, 71.
Vulture, Alpine or Egyptian, 237.
Vulture, Arabian, 235, 236.
——————— Bearded, 229.
——————— Carrion, 235.
——————— Fulvous, or Griffin, 233.
——————— King, 231.
Vulpine, Phalangist, 110.

Wagtail, Pied, 315.
Walking Fish, 474.
Walrus, or Morse, 124.
Wanderoo. 15.
Warbler, Blackcap, 306, 307.
Warblers, 304.
Wasp, 546.
Wasps, Solitary, 545.
——————— Sand, 544.
——————— Wood, 544.
Water Hen, 396.
——————— Vole, or Rat, 137.
——————— Spaniel, 61.
——————— Spiders, 574.
——————— Ousel, 317.
Weevel, 531.
Weasels, 80.
Weaver Bird, Social, 339, 340.
Wentletrap, Staircase, or Precious, 511.

Wentletrap, Common or False, 512.
Whale, 127.
——————— Spermaceti, 129.
Whaup. See Curlew.
Wheel Animalcules, 588.
Wheatear, 309.
Whelk, 507.
White-headed Saki. See Black Yarke.
Widgeon, 401.
Wild Cat, 48.
Wish-ton-wish. See Prairie Dog.
Wolf, 70.
Wolverene, 84.
Wombat, 116.
Worm, Ship, 521.
Wood Wasps, 544.
Woodcock, 395.
——————— Thorny, 507.
Woodlouse, 570.
——————— Armadillo, 570.
——————— Pill, 570.
Woodpecker, Green, 362, 363.
——————— Great Black, 363.
——————— Northern Three-toed, 363.
——————— Seven-spotted, 363.
——————— Great Spotted, 360 361.
Wood Pigeon, 367.
Wren, 303.
Wrens, 301.
——————— Fire-crested, 307.
——————— Golden-crested, 305.
Wryneck, 363.

Yamma. See Llama.
Yak, 167.
Yellow-footed Pouched Mouse, 119.
Yellowhammer, 345, 346.

Zebra, 201.
——————— Parrakeet, 352.
Zebu, 163.
Zerda. See Fennec.
Zoophytes, 585.
——————— Bell, 587.